BEHAVIORAL SEX DIFFERENCES IN NONHUMAN PRIMATES

BEHAVIORAL SEX DIFFERENCES IN NONHUMAN PRIMATES

G. Mitchell

Department of Psychology
University of California, Davis
Davis, California

VNR VAN NOSTRAND REINHOLD COMPANY
NEW YORK   CINCINNATI   ATLANTA   DALLAS   SAN FRANCISCO
LONDON   TORONTO   MELBOURNE

599. 8055
M681

Van Nostrand Reinhold Company Regional Offices:
New York   Cincinnati   Atlanta   Dallas   San Francisco

Van Nostrand Reinhold Company International Offices:
London   Toronto   Melbourne

Library of Congress Catalog Card Number: 78-16628
ISBN: 0-442-24594-7

Manufactured in the United States of America

Published by Van Nostrand Reinhold Company
135 West 50th Street, New York, N.Y. 10020

Published simultaneously in Canada by Van Nostrand Reinhold Ltd.

15 14 13 12 11 10 9 8 7 6 5 4 3 2 1

Library of Congress Cataloging in Publication Data

Mitchell, G
   Behavioral sex differences in nonhuman primates.

   Includes bibliographies and index.
   1. Primates—Behavior.   2. Sexual behavior in
animals. I. Title.
QL737.P9M54   599'.8'055   78-16628
ISBN 0-442-24594-7

TO MY PARENTS

John D. Mitchell

and

Pauline L. Mitchell

for their

40th ANNIVERSARY

Preface

I wrote this book while on leave of absence from the Department of Psychology at the University of California at Davis in the Fall Quarter of 1977. On the whole, the book is not an unusual one. However, for a book on *primate behavior* it *is* unusual in that: (1) it is not a *general* introductory textbook on primate behavior, (2) it is not an edited volume, and (3) it is based upon a single narrow theme, sex differences.

Throughout this volume I have referred to many papers which were orally presented at conventions. I did this in order to make the material as up to date as possible. By the time this goes to press, many of these findings will probably have been published in journals or in books; however, I did not have access to such when I wrote the book.

For copies of those convention abstracts which have not been published since this writing, or for the addresses of the authors of those papers, I am supplying the following addresses for the four conventions most often cited: (1) the *American Society of Primatologists* meeting in Seattle, Washington in April of 1977, (2) the *American Association of Physical Anthropologists* meeting in Seattle, Washington in April of 1977, (3) the *International Primatological Society* meeting in Cambridge, England in August of 1976, and (4) the *Animal Behavior Society* meeting in University Park, Pennsylvania in June of 1977. Here are the four addresses:

(1) Dr. W. R. Dukelow, Treasurer-Membership
American Society of Primatology
Endocrine Research Unit
Michigan State University
East Lansing, Michigan 48823

(2) Dr. Frank E. Poirier, Secretary Treasurer
American Association of Physical Anthropologists
Department of Anthropology
Ohio State University
Columbus, Ohio 43210

(3) Allan M. Schrier, Secretary General
International Primatological Society
Psychology Department
Brown University
Providence, Rhode Island 02912

(4) Devra G. Kleiman, Secretary
Animal Behavior Society
National Zoological Park
Smithsonian Institution
Washington, D.C. 20009

Throughout the book I use the plural pronouns "we" and "us." "We" means "you" (the reader) and "I" (the writer). My intent here is to make you a more active participant in my sojourn through the data on "Behavioral sex differences in nonhuman primates." I do this, despite the fact that some of you may abhor the editorial "we," because I very quickly grow tired of writing such awkward phrases as, "It can now be said that . . ".

Finally, I thank all of those who supported me in this effort: Patricia Jones (who helped with the references and index), my children (Jody, Lisa and Gary), my editors (Terry Maple, Ashak Rawji and Alberta Gordon), Joe and Nancy Erwin, my typist (Marilee "Dee" Kindelt), and my graduate student (Nancy Caine). In addition Jaclyn Wolfheim read the chapter on vocalizations and made comments.

In my research as a psychologist-primatologist I have been strongly influenced by several people. George M. Haslerud and Harry F. Harlow were my academic advisors. William A. Mason and Donald Lindburg have influenced my attitudes toward laboratory and field research, respectively. My former graduate students are Joe Erwin, Terry Maple, Bill Redican, Jody Gomber, John Copp, and Barbara Sommer. My current graduate student is Nancy Caine. All of these individuals have affected me and my research. Undoubtedly their combined effect on me is seen to some extent in these pages.

I take the credit for the errors. I also take the blame for the style which may at times appear to be a reflection of my raw, undigested, aimless empiricism (having no clear line of purpose). If nothing else, the effort should be a prominent underachievement for its lack of theory. However, at least *most* of the facts regarding behavioral sex differences in nonhuman primates can be found between these covers (hopefully).

G. MITCHELL
Davis, California
December 25, 1977

Contents

Preface

SECTION 1. INTRODUCTION

1: **Introduction** / **1**

A. Definitions / 1
B. References for Chapter 1 / 4

2: **The Primate Order** / **5**

A. Sub-orders of Primates / 5
B. Infra-orders of Primates / 5
C. Superfamilies of Primates / 6
D. Families of Primates / 7
E. Sub-families of Primates / 8
F. Genera and Common Names / 9
G. References for Chapter 2 / 11

SECTION 2. PHYSICAL AND HORMONAL BACKGROUND

3: **Physical Sexual Dimorphism** / **12**

A. General Comments / 12
B. Sexual Dimorphism in Size / 14
C. Physical Dimorphism in Dentition / 19
D. Other Kinds of Physical Sexual Dimorphism / 21
E. References for Chapter 3 / 24

4: **Sexual Differentiation: Structure, Behavior and Nervous System** / **31**

A. Nonprimate Data / 31
B. Primate Prenatal Data / 33
C. Early Experiential Effects / 36
D. Adult Hormonal Effects / 37
E. Sex Hormones and the Nervous System / 37
F. Human Implications / 41
G. References for Chapter 4 / 44

5: **The Development of Play** / **49**

A. Prosimian Play / 49
B. New World Monkey Play / 50
C. Old World Monkey Play / 51
D. Macaque and Baboon Play / 53
E. The Play of Lesser and Great Apes / 57
F. Possible Neurological Correlates of Play / 60
G. The Function of Play / 60
H. Human Play / 61
I. References for Chapter 5 / 63

6: **Puberty and Adolescence** / **72**

A. General Comments on Sex Differences as a Process / 72
B. Prosimian Puberty / 73
C. Puberty in New World Monkeys / 74
D. Puberty in Old World Monkeys / 75
E. Puberty in Rhesus Monkeys (*Macaca mulatta*) / 76
F. Puberty in Other Species of Macaques / 78
G. Puberty in Baboons and Mangabeys / 80
H. Puberty in the Apes / 80
I. Comments Concerning Human Puberty / 82
J. Monogamy and the Delay of Puberty / 83
K. References for Chapter 6 / 83

7: **Sex Hormones in Adulthood: Prosimians, New World Monkeys, and Rhesus Macaques** / **90**

A. Review of Adult Sex Hormones / 91
B. Adult Sex Hormones in Prosimians / 91
C. Adult Sex Hormones in New World Monkeys / 92
D. Adult Sex Hormones in Rhesus Monkey Males / 94
E. Rhesus Monkey Adult Female Sex Hormones / 97
F. References for Chapter 7 / 102

8: Sex Hormones in Adulthood: Other Old World Monkeys,
Apes, Humans, Generalizations and Speculations / 109

A. Other Old World Monkeys — Adult Hormones / 109
B. Adult Sex Hormones in the Great Apes / 113
C. Adult Hormones in *Homo sapiens* (including
speculation) / 114
D. Neurology, Neurophysiology and Sex Hormones / 118
E. References for Chapter 8 / 119

9: Sexual Behavior: Prosimians, New World Monkeys,
and Rhesus Macaques / 125

A. General Comments / 125
B. Sexual Behavior in Prosimians / 126
C. Sexual Behavior in New World Monkeys / 128
D. Sexual Behavior in Rhesus Macaques / 132
E. References for Chapter 9 / 136

10: Sexual Behavior: Other Old World Monkeys, Apes,
Humans, Generalizations, and Speculations / 140

A. Sexual Behavior in Other Old World Monkeys / 140
B. Sexual Behavior in the Apes / 148
C. Comparisons of Ape to Human Sexuality / 153
D. General Comments on Natural Science and
Human Sexuality / 154
E. Incest Avoidance / 156
F. References for Chapter 10 / 157

SECTION 3. ONTOGENY

11: Male and Female Responses to Birth / 165

A. Prosimian Deliveries / 165
B. New World Monkey Deliveries / 166
C. Old World monkey Deliveries / 168
D. Lesser and Great Ape Deliveries / 173
E. *Homo sapiens* / 174
F. Overview / 174
G. References for Chapter 11 / 175

12: Infant Care: Prosimians, New World Monkeys,
and Macaques / 179

A. Prosimians / 181
B. Callithricidae / 183
C. Cebidae / 184
D. The Rhesus Monkey (*Macaca mulatta*) / 186
E. Japanese Macaques / 188
F. Crab-eating Macaques / 190
G. Stump-tailed Macaques / 190
H. Barbary Macaques / 191
I. Pigtail Macaques / 192
J. Bonnet Macaques / 192
K. References for Chapter 12 / 193

13: Infant Care: Other Old World Monkeys, Apes,
and Humans / 199

A. Other Old World Monkeys / 199
B. Lesser Apes / 205
C. Great Apes / 206
D. Thoughts About Human Infant Care / 210
E. References for Chapter 13 / 212

SECTION 4. INTRAGROUP ORGANIZATION

14: The Role of Gender in Social Spacing and Structure:
Prosimians, New World Monkeys, and Macaques / 219

A. Prosimians / 220
B. New World Monkeys / 221
C. Rhesus Macaques (*Macaca mulatta*) / 224
D. Other Macaques / 226
E. References for Chapter 14 / 229

15: The Role of Gender in Social Spacing and Structure:
Old World Monkeys, Apes, and Implications for Humans / 236

A. Other Old World Monkeys / 236
B. Lesser Apes / 240
C. Great Apes / 241
D. Thoughts About *Homo sapiens* / 242
E. References for Chapter 15 / 243

16: Dominance: Prosimians, New World Monkeys, Rhesus,
    and Japanese Macaques / 248

  A. Prosimians / 249
  B. New World Monkeys / 250
  C. Rhesus and Japanese Macaques (*Macaca spp.*) / 252
  D. References for Chapter 16 / 256

17: Dominance: Other Old World Monkeys, Apes,
    and Implications for Humans / 262

  A. Other Macaques / 262
  B. Other Old World Monkeys / 264
  C. Apes / 267
  D. Implications for Humans / 268
  E. References for Chapter 17 / 271

18: Leadership and Alliances / 278

  A. Leadership and Group Control / 278
  B. Alliances and Coalitions / 283
  C. References for Chapter 18 / 287

SECTION 5. EXTRATROOP BEHAVIOR

19: Extratroop Behavior: Vigilance, Protection,
    Territoriality, and Inter-troop Behavior / 293

  A. General Comments on Protection / 294
  B. Prosimians / 294
  C. New World Monkeys / 295
  D. Macaques (*Macaca spp.*) / 295
  E. Other Old World Monkeys / 297
  F. Apes / 299
  G. Human Considerations / 301
  H. References for Chapter 19 / 301

20: Feeding and Predation / 307
  A. General Comments on Feeding / 307
  B. Feeding in South American Monkeys / 308
  C. Feeding in Old World Monkeys / 309
  D. Ape Feeding / 310
  E. Sex Differences in Predation / 311
  F. References for Chapter 20 / 315

SECTION 6. COMMUNICATION

21: Sex Differences in Grooming / 319

A. Prosimians / 319
B. South American Species / 320
C. Old World Arboreal Monkeys / 321
D. Old World Terrestrial Monkeys / 322
E. Apes / 326
F. Implications for Humans / 327
G. References for Chapter 21 / 328

22: Sex Differences in Vocalizations / 335

A. General Comments / 335
B. "Male" Vocalizations / 336
C. "Female" Vocalizations / 343
D. Conclusion / 348
E. References for Chapter 22 / 349

23: Sex Differences in Visual Communication / 356

A. General Comments on Visual Communication / 356
B. Visual Orientation / 358
C. Prosimians / 359
D. New World Monkeys / 359
E. Old World Monkeys / 361
F. Apes / 367
G. Overview / 368
H. References for Chapter 23 / 370

**24: Aggression: Prosiminians, New World Monkeys
and Macaques / 374**

A. Prosimians / 375
B. New World Monkeys / 376
C. Macaques (*Macaca spp.*) / 377
D. References for Chapter 24 / 382

**25: Aggression: Other Old World Monkeys, Apes,
and Comments on *Homo sapiens* / 389**

A. Other Old World Monkeys / 389
B. Apes / 393
C. Comments on Human Aggression / 394
D. References for Chapter 25 / 396

SECTION 7. PATHOLOGICAL CONDITIONS

26: **Sex Differences in Response to Early Social Deprivation and Separation / 402**

A. Social Deprivation in Rhesus Macaques / 402
B. Social Deprivation in Other Monkeys / 407
C. Social Deprivation in the Apes / 408
D. Comments on People / 409
E. References for Chapter 26 / 410

27: **Fear, Stress, and Crowding / 415**

A. Stress, ACTH, and Sex Differences / 415
B. Crowding / 418
C. References for Chapter 27 / 420

28: **Vulnerability and Mortality / 423**

A. General Comments / 423
B. New World Monkeys / 424
C. Rhesus Macaques (*Macaca mulatta*) / 424
D. Other Old World Monkeys / 425
E. Apes / 427
F. Comments on *Homo sapiens* / 427
G. References for Chapter 28 / 429

SECTION 8. LEARNING AND THE BRAIN

29: **Learning and Performance / 433**

A. General Comments / 433
B. Monkeys, Apes and People / 434
C. References for Chapter 29 / 438

30: **Sex Differences in the Brain / 441**

A. General Comments / 441
B. Prosimians and New World Monkeys / 443
C. Old World Monkeys / 444
D. Thoughts on *Homo sapiens* / 446
E. References for Chapter 30 / 447

SECTION 9. VARIABILITY

31: Cyclical Variation / 451

A. General Comments / 451
B. Prosimians and New World Monkeys / 452
C. The Rhesus Monkey (*Macaca mulatta*) / 454
D. Other Old World Monkeys / 456
E. Apes / 458
F. Comments on Humans / 459
G. References for Chapter 31 / 460

32: Non-cyclical Variability / 467

A. General Comments / 467
B. Prosimian and New World Monkey Examples / 468
C. Rhesus Monkeys (*Macaca mulatta*) / 469
D. Other Old World Monkeys / 471
E. Apes / 473
F. Comments on Human Variability / 474
G. References for Chapter 32 / 474

33: Self-Awareness / 479

A. General Comments / 479
B. Individuality / 479
C. Species Identification or Species Identity / 480
D. Consequences of Early Isolation for Self-directed
   Behavior / 480
E. Self-Recognition / 481
F. Comments on Humans / 483
G. Self-Awareness and Gender Differences / 486
H. References for Chapter 33 / 486

34: Summary and Epilogue / 489

Appendix / 493

Author index / 497

Subject index / 509

BEHAVIORAL
SEX DIFFERENCES
IN NONHUMAN
PRIMATES

1.

Introduction

DEFINITIONS

Some definitions are needed to make the meaning of the title of this book ("Behavioral Sex Differences in Nonhuman Primates") as clear as possible. There are three important terms in the title—"behavior," "sex differences," and "nonhuman primates." Each of these terms needs to be defined specifically.

When we use the term "behavior" in this book we will usually be referring to specific movements, postures, facial expressions, and vocalizations exhibited by the animal subject or subjects in question. We are usually assuming that an animal did not *intend* to perform an "act." That is, we are assuming than an animal had no *conceptual* representation of its *actions*. We, like Reynolds (1976), recognize a difference between "behavior" and "action":

> . . .If we describe what people or animals do, without inquiring into their subjective reasons for doing it, we are talking about their *behaviour*. If we study the subjective aspects of what they do, the reasons and ideas underlying and guiding it, then we are concerned with the world of *meaning*. If we concern ourselves both with what people are, overtly and objectively, seen to do (or not do) and their reasons for so doing (or not doing) which relate to the world of meaning and understanding, we then describe *action*. (P.xv)

Reynolds (1977) thinks in terms of "behavior" when he speaks of nonhuman animals and in terms of "action" when he speaks of *Homo sapiens*. However, as Gallup (1971), Shafton (1976), and others have so ably noted, *some* nonhuman primates (most notably the great apes) are capable of self-recognition and of self-awareness. It may well be that chimpanzees, for example, *act* as well as *behave*. While there are no doubt quantitative and qualitative differences

1

between humans and the great apes in the degree to which they are self-aware, there is much to be gained from the knowledge that the great apes are capable of conceptualizing their own behavior.

Mason (1976) has also noted the similarities between great apes and people in representational processes:

> ... I am persuaded that the apes and man have entered a cognitive domain that sets them apart from all other primates. (P. 293)

The difference (in this capacity or in these capacities of awareness) between the great apes and the monkeys may or may not be an all-or-none matter. Old World monkeys, in particular, often display behaviors which suggest that they possess sophisticated levels of awareness. However, since most of the primate order does *not* seem to show the levels of awareness of the great apes, the term *behavior* instead of the term *action* has been used in the title of this book.

On the other hand, it is also clear that even people are not completely aware of everything they do. Very often we all behave at an unconscious level. There is, therefore, a repertoire of human *behavior* as well as of human *action*. What we discuss in this book on nonhuman primates will probably be of more relevance to the former than to the latter, although as we deal with the more advanced monkeys and particularly with the great apes, information from nonhuman primates also becomes relevant to human *action*.

The second important term in the title, "sex difference," is the major topic of this book. This term also has an alternative form: "gender difference." The term "sex difference" is used to refer to a difference which appears to have strong *biological* correlates or causes. It is assumed in this case that the bulk of the difference cannot be attributed to experiential, early environmental, cultural (or traditional) causes. The term "gender difference" refers to a difference between males and females that *can* be attributed to early experience, culture, or tradition, or to a difference for which no clear biological causes or correlates have been established.

It is recognized by many that boys and girls differ in levels of aggressiveness (cf. Maccoby and Jacklin, 1974). However, we cannot fully determine on the basis of extant data whether this difference in aggression level is a gender or sex difference. To do that, as Reynolds (1976) says, we need many cross-cultural studies, especially:

> ... in cultures where aggressiveness is discouraged and frowned upon, to indicate whether differences in overt aggressiveness between boys and girls are inevitably found. This would still not mean that they were true sex differences in aggressiveness, for it is hard to see how such differences in regard to

any human characteristic that involved a heavy component of intervention by adults could be clearly shown not to have gender elements. Human differences in action between boys and girls or between men and women are probably best regarded as gender differences in the present state of understanding. (P. 123)

In practice, of course, most human differences between males and females are *both* sex differences and gender differences. We have used the term "sex difference" in our title for reasons similar to our choice of "behavior." Since a greater part of the primate order is made up of prosimians and monkeys than of great apes and people, it is likely that *most* of the primates show "sex differences" more than "gender differences." Conversely, it is probable that more of the great ape differences, and probably most (though not all) of the differences in human beings are "gender differences" more than "sex differences." Of course, just as mankind has "behavior" as well as "action," there are also human "sex differences" as well as human "gender differences."

The third term in our title requiring further definition is the term "nonhuman primate." Deciding just which organisms should be included in the primate order has not been simple for taxonomists. Originally, the primate order was established on the basis of anatomical similarity. Later, however, taxonomists relied on the synthetic theory of evolution, on presumed lines of descent, on genetic relatedness, and on behavior in constructing different groups within the primate order.

The primates are ancient mammals. Fossil records have determined that primates have been around as long as most other mammals. The idea that they are a recent evolutionary development is therefore false (Napier and Napier, 1967).

Primates are generalists, not specialists. It is because of such generalized characteristics that the term "primate" is so difficult to define. There is, however, a pattern of characteristics to which one may refer in deciding whether or not a particular species is a primate. While any specific primate species does not fit all of the characteristics of the primate pattern, it should fit most of them. The primate characteristics are listed below.

1. Preservation of a generalized or primitive mammalian skeletal structure, particularly of the limbs.
2. A mobile set of grasping digits on hands and feet. The digits tend to have flattened nails (rather than claws) and sensitive tactile pads.
3. There is a reduction in the length of the muzzle (the baboon is an exception).
4. The eyes point forward and there is binocular vision in most forms. Vision is an important sensory mode for primates; smell is relatively less important.

5. There is a relatively primitive mammalian tooth pattern, but there is a loss of some of the most primitive mammalian teeth, particularly in the higher form of primates.
6. There is a remarkable elaboration of the brain, particularly of the cerebral cortex.
7. There is a longer period of gestation and a trend toward delivery of a single infant. Associated with these trends is a nutrition for the fetus prior to birth which is improved over that of other mammals.
8. There is prolongation of infant dependency in postnatal life.
9. There may have been a trend toward bipedalism within extinct members of the primate order, but only in human beings is bipedalism the primary mode of locomotion among extant primates.

For more explicit or detailed descriptions of these general characteristics of primates the reader may consult LeGros Clark (1962) or Napier and Napier (1967). It should be emphasized, however, that not all primates possess all of the characteristics listed above, and that this is particularly true of the more primitive primates.

REFERENCES

Clark, W. E. LeGros *The Antecedents of Man.* 2nd ed. Edinburgh: Edinburgh University Press, 1962.

Maccoby, E. E. and Jacklin, C. *The Psychology of Sex Differences.* Stanford, California: Stanford University Press, 1974.

Mason, W. A. Environmental models and mental modes: Representational processes in the great apes and man. *American Psychologist,* 1976, **31,** 284-294.

Napier, J. R. and Napier, P. H. *A Handbook of Living Primates.* New York: Academic Press, 1967.

Reynolds, V. *The Biology of Human Action.* San Francisco: Freeman, 1976.

Shafton, A. *Conditions of Awareness: Subjective Factors in the Social Adaptations of Man and Other Primates.* Portland, Oregon: Riverstone Press, 1976.

2.

The Primate Order

Now that we know the definition of the word primate, we can begin to list the primates we will be discussing in the remaining chapters of this book. There are over 200 species of primates. These different species are categorized into genera, subfamilies, families, superfamilies, infraorders, and finally into suborders. Let us start at the suborder level and work toward the species level, since there are only two suborders and over 200 species (cf. Bramblett, 1976).

A. SUBORDERS

1. Prosimii

   This suborder includes the most primitive of the primates. It contains no monkeys or apes. These animals have small brains, claws on many of their digits, an important sense of smell, and less well-developed binocular vision relative to other primates. They use their mouths more than their hands for grooming. In short, they have fewer of the primate characteristics than does the other suborder, Anthropoidea.

2. Anthropoidea

   This suborder includes true monkeys, apes, and people. As the name suggests, these animals are more humanlike than are the prosimians. The anthropoids have larger brains, eyes that are shifted forward, increased precision of the hands, and prolonged infant dependency, among other characteristics.

B. INFRAORDERS

The next level down in primate taxonomy is the *infraorder*. There are five of these; that is, the more than 200 species of primates can be divided into five different infraorders. Three of these infraorders belong to the suborder Prosimii, and two to the suborder Anthropoidea.

1. Prosimian infraorders
   a. Lemuriformes

   The Lemuriformes are found only in the Malagasy Republic, located on the island of Madagascar off the east coast of Africa. There are both diurnal (daytime) and nocturnal (nighttime) forms. They all have scent glands with which they scent-mark their home ranges, and are territorial. Most have elongated hind limbs for jumping.

   b. Lorisiformes

   The Lorisiformes infraorder is found in Asia and Africa. In Asia they are nocturnal and slow moving. They have the most opposable thumbs of any living primate. In Africa, they are all nocturnal except for one genus (Galago).

   c. Tarsiiformes

   The third and final prosimian infraorder, Tarsiiformes, is found in Southeast Asia and the Malay Archipelago. Tarsiers are small, nocturnal, have enlarged hind limbs, and large, relatively immobile eyes.

2. Anthropoidean infraorders
   a. Platyrrhini

   The Platyrrhini are found only in Central and South America. They have a broad nose with nostrils which are far apart and directed sideways, and a more primitive dentition with more teeth than Old World monkeys have. Some of the Platyrrhines have prehensile tails. All are arboreal.

   b. Catarrhini

   The last infraorder of primates, the Catarrhini, are also of the Anthropoidean suborder. The catarrhines are found only in the Old World. They have a more narrow nose with nostrils which are close together and directed downwards. They have thirty-two teeth in the same dental pattern as do people. They do not have prehensile tails (some have no tails at all) and many of them are terrestrial.

C. SUPERFAMILIES

There are a total of seven superfamilies. Two of these belong to the infraorder Lemuriformes, one to Lorisiformes, and one to Tarsiiformes, among the Prosimians. One superfamily belongs to the Platyrrhini, and two to the Catarrhini among the Anthropoidea.

1. Lemuriform superfamilies
   a. Lemuroidea

   This superfamily includes all lemurs as described under Lemuriformes,

except Daubentonidea.
  b. Daubentonidea
     This superfamily of lemurs includes only a single genus, *Daubentonia,* the aye-aye. It is almost extinct, resembles ancient fossil primates, has enlarged procumbent incisors, is nocturnal, and has a long wirelike compressed nail on one finger with which it feeds on insects and insect larvae.
  2. Lorisiformes, 3. Tarsiiformes, and 4. Platyrrhine superfamilies
     There is only one superfamily in each of these three infraorders. Their names are Lorisoidea, Tarsioidea, and Ceboidea, respectively.
  5. Catarrhine superfamilies
     a. Cercopithecoidea
        The first catarrhine superfamily, Cercopithecoidea, includes all of the Old World monkeys.
     b. Hominoidea
        The second catarrhine superfamily, Hominoidea, includes the apes and people.

D. FAMILIES

There are a total of twelve families in the primate order, six each among prosimians and anthropoidea. We will discuss them by superfamily.
  1. Lemuroidea families
     The superfamily lemuroidea is composed of three families.
     a. Lemuridae
        This includes both diurnal and nocturnal Madagascan common lemurs.
     b. Cheirogaleidae
        This family includes only nocturnal forms and some of the smallest lemurs belong in this group. (e.g., mouse lemur, dwarf lemur).
     c. Indriidae
        This family is composed of animals having large digestive tracts and stomachs which are specialized for a diet of large amounts of plants.
  2. Daubentonidea family
     a. Daubentonidae
        There is only one aye-aye, hence only one family.
  3. Loris family
     a. Lorisidae
        All lorises belong to one family, Lorisidae.
  4. Tarsioidea family
     a. Tarsiidae
        There is only one tarsier, hence only one family.

5. Ceboidea families
   a. Callithricidae
      This family of New World monkeys includes marmosets and tamarins which are smaller than the other Platyrrhines. They are unusual in that they often give birth to twins or triplets. They have sharp claws on all digits except the first two.
   b. Cebidae
      The second family of the Ceboidea superfamily is the Cebidae, monkeylike forms of the Americas. They are all arboreal, adept at climbing, and some have prehensile tails. There is one nocturnal form (*Aotus*). They have more teeth than do the Old World monkeys.
6. Cercopithecoidea families
   a. Cercopithecidae
      These are terrestrial and arboreal forms such as macaques, baboons, vervets, etc. Omnivores, they have digestive tracts which are structurally like those of people.
   b. Colobidae
      These are the leaf-eating monkeys which have large, sacculated stomachs much as do other cellulose-eating mammals, such as cows and some marsupials.
7. Hominoidea families
   There are two families in the superfamily Hominoidea.
   a. Pongidae
      This family is composed of all of the apes.
   b. Hominidae
      This family includes only humans (*Homo sapiens*).

E. SUBFAMILIES

The families Lemuridae, Cheirogaleidae, Indriidae, Daubentoniidae, and Tarsiidae are not subdivided into subfamilies. At the subfamily level the final three letters — dae, are changed to — nae.

1. Lorsidae subfamilies
   a. Lorisinae
      This subfamily includes all common lorises.
   b. Galaginae
      This subfamily includes only one genus, *Galago* for which there are different species and subspecies. *Galago* is a rapid-moving vertical clinger and leaper found in Africa in areas of both sparse and dense vegetation. Active both day and night, they are seen on the ground as well as in trees.

2. Callithricidae subfamilies
   The Callithricidae can be divided into two different subfamilies.
   a. Callithricinae
      This includes almost all of the common marmosets and tamarins.
   b. Callimiconinae
      This includes only Goeldi's marmoset.
3. Cebidae subfamilies
   a. Atelinae
      This subfamily includes spider monkeys, which have prehensile tails.
   b. Aotinae
      This subfamily includes the nocturnal and monogamous owl monkey
      as well as the diurnal and monogamous titi monkey.
   c. Cebinae
      This subfamily includes the relatively unspecialized or common cebids
      like the capuchins (organ-grinder monkeys) and the squirrel monkeys.
   d. Pitheciinae
      This is a little-studied subfamily including the bald-headed *Cacajao*.
   e. Alouattinae
      This subfamily includes all of the howling monkeys.
4. Cercopithecidae and Colobidae subfamilies
   These two families of Old World monkeys are typically not subdivided
   further except into genera, species, and subspecies.
5. Pongidae subfamilies
   a. Ponginae
      This subfamily is composed of all the great apes, two from Africa
      (the gorilla and the chimpanzee), and one from Asia (the orangutan).
   b. Hylobatinae
      This subfamily includes the two genera of lesser apes, *Hylobates* and
      Symphalangus. These are brachiating, monogamous, Asian apes.
6. Hominidae subfamilies
   There is only one family, one subfamily, one genus, one species of human
   being, Homininae, *Homo sapiens*.

F. GENERA AND COMMON NAMES

In Table 2-1 is a list of the genera and common names of the nonhuman primates
referred to most often in the remaining pages of this book. There is also an
Appendix of species and common names at the end of the book. The table, the
Appendix, and the taxonomic levels listed above will serve as guides to the reader
in placing each species at its appropriate level within the primate order. In addition, reference to Jolly (1972) and Napier and Napier (1967) will be helpful.

Pictures of each primate genus appear in Napier and Napier (1967). A guide for the pronunciation of primate names is presented in Riopelle and Fraga (1969).

Table 2-1. Subfamilies, Genera, and Common Names of Some Well-Known Nonhuman Primates.

Subfamily	Genus	Common Name
Lemurinae	Lemur	common lemur
	Hapalemur	gentle lemur
	Lepilemur	sportive lemur
Cheirogaleinae	Microcebus	mouse lemur
	Cheirogaleus	dwarf lemur
	Phaner	forked lemur
Indriinae	Indri	indri
	Avahi	wooly lemur
	Propithecus	sifakas
Daubentoniinae	Dabentonia	aye-aye
Lorisinae	Loris	slender loris
	Arctocebus	golden potto
	Perodicticus	potto
	Nycticebus	slow loris
Galaginae	Galago	bushbaby
Tarsiinae	Tarsius	tarsier
Callimiconinae	Callimico	Goeldi's marmoset
Callithricinae	Callithrix	common marmoset
	Cebuella	pygmy marmoset
	Saguinus	hairy faced tamarin
	Leontideus	lion tamarius
Atelinae	Ateles	spider monkey
	Brachyteles	wooly spider
	Lagothrix	wooly monkey
Aotinae	Aotus	night or owl monkey
	Callicebus	titi monkey
Cebinae	Cebus	capuchin
	Saimiri	squirrel monkey
Pitheciinae	Pithecia	saki
	Chiropotes	bearded saki
	Cacajao	uakari
Alouattinae	Alouatta	howler monkey
Cercopithecinae	Cercopithecus	Syke's monkey, guenons, vervets
	Erythrocebus	patas monkey
	Cercocebus	mangabey
	Cynopithecus	celebes black ape
	Macaca	macaques

Table 2-1 Continued

Subfamily	Genus	Common name
Cercopithecinae (con't.)	*Papio*	baboon
	Mandrillus	mandrills
	Theropithecus	gelada
	Miopithecus	talapoin
Colobinae	*Colobus*	colobus, guerezas
	Presbytis	langur
	Pygathrix	douc langur
	Nasalis	proboscis monkey
	Rhinopithecus	snubnosed langur
	Simias	Pagai island langur
Hylobatinae	*Hylobates*	gibbon
	Symphalangus	siamang
Ponginae	*Pongo*	orangutan
	Pan	chimpanzee
	Gorilla	gorilla
Homininae	*Homo*	people

REFERENCES

Bramblett, C. A. *Patterns of Primate Behavior.* Palo Alto, California: Mayfield, 1976.

Jolly, A. *The Evolution of Primate Behavior.* New York: Macmillan, 1972.

Napier, J. R. and Napier, P. H. *A Handbook of Living Primates.* New York: Academic Press, 1967.

Riopelle, A. J. and Fraga, A. M. The pronunciation of primate names. *Laboratory Primate Newsletter,* 1969, 8 (2), 1-5.

3.

Physical Sexual Dimorphism

As Phoenix, Goy, and Resko (1968) have pointed out, "the simple cataloguing of sex differences is not being pursued as an end in itself." (P. 46.) There are good reasons for cataloguing them.

Before any coverage of sex and/or gender differences in *behavior* can begin, differences in anatomy or physical characteristics, which, of course, affect behavior, must be noted and evaluated. In this chapter, over twenty different genera of nonhuman primates will be mentioned with regard to their respective degrees of sexual dimorphism in at least one of more than a dozen different physical characteristics.

Included in this chapter on physical characteristics (for which there are data on sex differences) are size (body weight and/or length), coloration, and dentition. Other physical characteristics such as those pertaining to the skeleton, hair, genitals, skin, physiology, brain, and shape are also discussed, but to a lesser extent than size, coloration, and dentition.

A. GENERAL COMMENTS

Many primatologists have attempted to explain why sexual dimorphism in size develops in most nonhuman primates. Baboon size has been related to an optimal distribution of biomass. It has been argued that, if large males are selected for, then small females must also be selected for to maintain an average biomass that is optimal. It has been assumed by such theorists that smaller females would require less food than larger females and hence would have higher reproductive success (Devore and Washburn, 1963). However, Coelho (1973) has shown that, in at least one species, the smaller female had a higher total energy expenditure than the larger male. We will reexamine the relationship between size, gender, and feeding in a later chapter.

Another frequently accepted theoretical position is that an adaptation to the greater hazards of terrestrial life (savannah or woodland-savannah) demanded larger more powerful males for troop protection (Russell and Russell, 1971). However, this theory has been questioned on the grounds that arboreal habitats may not be any less hazardous than terrestrial ones. In Bramblett (1976), for example, it is pointed out that "attacks by birds of prey on arboreal monkeys (such as martial eagle attacks on guenons) are just as deadly and possibly more frequent than leopard attacks on baboons." (P. 255.) In any event, there are some primarily arboreal species which show extreme sexual dimorphism (e.g., the orangutan).

It seems that many statements concerning the reasons for physical dimorphism between males and females have been made before the species in question has been adequately studied. The correct order of events should be to first completely specify those sex differences that do exist — both physical *and* behavioral — and then to try to explain them. It would make little sense, for example, to make the statement that males of a given species needed to be larger so that they could fight predators if males of that species had never been seen displaying such behavior. Patas males, for example, are larger than females yet they do not fight predators.

Blurton-Jones (1975) and others (Clutton-Brock and Harvey, 1976) believe that, in general, a marked size difference between the sexes is correlated with a high proportion of breeding females to breeding males. Clutton-Brock and Harvey (1976), summarizing data on 100 species of primates, also found that sexual dimorphism showed no obvious association with ecological variables but that, in general, the larger the species, the greater the dimorphism.

In a multivariate analysis done on twenty-one primate species, Jorde and Spuhler (1974) found that sexual dimorphism correlated positively with group size and negatively with arboreality. However, the orangutan (*Pongo pygmaeus*), a highly arboreal primate with small group size was not included in their analysis. The orangutan nevertheless, shows extreme sexual dimorphism.

Leutenegger (1977) has pointed out that, for the most part, there is a lower limit for female size below which it is adaptive to deliver and rear more than one offspring. The regular delivery of multiple infants does, in fact, occur only in the smallest sized species. The males are also small in these species. Sexual dimorphism, with larger size in males for defense, may not be needed in small species since enough young may be produced to replace those which die. There is also a relatively short period of dependency in such species. Examples of these small species are the marmosets and tamarins among which the females are slightly larger than the males (Ralls, 1976). For general information, the reader may consult Ralls (1976) who has published a collection of all mammals in which the females are larger than males. Few primates are included in her publication, however.

B. SEXUAL DIMORPHISM IN SIZE

1. Prosimians

Among the tree shrews (*Tupaia spp.*), weight gain stops at three months of age in females but continues to four months of age in males. Males are therefore larger than females at maturity. Among the lemurs, the sexes may differ very little in body size (as in *Lemur spp.*) or the females may be only slightly smaller than males (as in *Propithecus*) (Jolly, 1966).

In the lesser bushbaby (*Galago spp.*) females reach maturity earlier than do males (at 200 days of age) and there is only a little size dimorphism between the sexes in the expected direction (Doyle, 1974).

2. South American monkeys

In at least one species of marmoset, as already mentioned, females are, if anything, slightly larger than males (Ralls, 1976). Squirrel monkeys, titi monkeys, and night monkeys also show very little physical sexual dimorphism.

The male golden-haired Saki (*Pithecia pithecia*) is obviously different from the female but primarily in coloration (Stott, 1976). Among spider monkeys (*Ateles geoffroyi* and related species) there is less size dimorphism than in most ground-dwelling monkeys. The males are only slightly larger than the females (Eisenberg and Kuehn, 1966). In fact, Bramblett (1976) reports that females weigh 5 percent *more* than males.

It is among the howler monkeys that the most significant size difference between males and females is seen in the New World. In some populations of howler monkeys (e.g., northern Argentina) there is a strong enough sexual dimorphism in weight that the two distributions do not overlap (Pope, 1966). Males are *always* heavier than females. Males on Barro Colorado Island, in the Panama Canal Zone, are 20 percent larger than females.

Thus, among the New World monkeys, sexual dimorphism ranges from females slightly larger, through equivalent size for the two sexes, to males much larger. Since *all* New World monkeys are arboreal, this variability in dimorphism does not strongly support a negative relationship between dimorphism and arboreality.

3. Arboreal Old World monkeys

According to Dolhinow (1972), sexual dimorphism is not pronounced in the North Indian langur (*Presbytis entellus*) although the males are slightly larger and more robust. As early as eight monghs of age, however, male langurs are larger and stronger than females (Dolhinow and Bishop, 1970).

Among vervets, *Cercopithecus galeritus* and *Cercopithecus neglectus* show the greatest sexual dimorphism with *C. cephus, C. pogonias,* and *C. nictitans* showing intermediate dimorphism. Males of this genus also tend to have heavier birth weights (Valerio and Dalgard, 1974). The least dimorphism in this genus is shown by *C. talapoin* (also known as *Miopithecus talapoin*) (Wolfheim, in press). According to Wolfheim, among talapoins (*Miopithecus*), the males are not significantly larger than the females (Wolfheim, in press). Segre (1970) however, says the males are slightly larger than the females. According to Gautier-Hion (1975), who has made a study of this genus, the degree of sexual dimorphism has little to do with taxonomic relationships but *is* correlated with terrestrial life and "the more important and exclusive the social role of the adult male within the group, the greater are morphological differences between adult males and females. (P. 373.)

In the terrestrial patas monkey (*Erythrocebus patas*) of Africa, there is a pronounced sexual dimorphism with the male being much larger than the female (Gartlan, 1974). Males may weigh up to twenty-two pounds; females are much lighter (Gartlan, 1976). Similarly, in the arboreal proboscis monkey (*Nasalis larvatus*) of Borneo there is also pronounced sexual dimorphism. Older males can weigh over fifty pounds, more than *twice* the weight of the average female (Kern, 1964).

In summary, as we have seen with the arboreal New World monkeys, there is a great range of sexual dimorphism in size in the arboreal Old World monkeys. Clearly, tree-dwelling is not perfectly negatively related to dimorphism in size.

4. Terrestrial Old World monkeys

All of the adult monkeys in the genus *Macaca* show strong sexual dimorphism in size. This is true of rhesus (Bernstein, 1970), pigtails (Bernstein, 1970; Tarrant, 1975), crab-eaters (Bernstein, 1970), stumptails, in which males are often twice as heavy as females (Chevalier-Skolnikoff, 1975), toque macaques (Dittus, 1975), and Japanese macaques (Masui et al., 1973). The weight difference between males and females is usually evident at birth in rhesus (DiGiacomo and Shaughnessey, 1972; Valerio and Dalgard, 1975; Watts, 1977) and rhesus males outweigh females at every age (and significantly) between four and eight years (Kirk, 1972). Riopelle and Hale (1975) feel that the size difference at birth in rhesus males and females may be related to longer gestation length for males. According to these researchers (Riopelle, Hale and Watts, 1976), "the sole significant finding was the difference between the sexes. There was no significant effect of diet or of the interaction of diet and sex." (P. 207.) Rhesus males (470 grams) are simply heavier than females (416 grams) at birth. There is also a tendency for rhesus males to have greater length than females at birth (Riopelle, Hill, and Li, 1975).

The sex difference in birth weight is also apparent in pigtail macaques (*M. nemestrina*) (Tarrant, 1975), and crabeater macaques (*M. fascicularis*) (Valerio and Dalgard, 1975; Willes, Kressler, and Truelove, 1977). Male crabeaters also have been shown to consume more calories than females in the first sixty days of life (Willes, Kressler, and Truelove, 1977).

There is also substantial sexual dimorphism in *Cercocebus, Cynopithecus, Theropithecus,* and *Cercopithecus aethiops* (the most terrestrial of the vervets) (Bernstein, 1970).

Among baboons (*Papio spp.*) all species show pronounced sexual dimorphism in size. This is true for mandrills as well (Jouventin, 1975). In the yellow baboon (*Papio cynocephalus*), male neonates have greater initial weights than do females (McMahon, Wigodsky, and Moore, 1976). The large size difference between males and females was also found in forest-living baboons in Uganda (Rowell, 1966).

Despite the consistent and large difference in size between males and females among terrestrial Old World monkeys, it is not true that *every* male must necessarily be larger than every female. There is marked anatomic variability, for example, in *Macaca mulatta.* The anatomic variability in the rhesus equals or *exceeds* that of the human (Hromada, 1968). In addition, between two and four years of age in most macaques, female weights may not differ significantly from those of males since females reach maturity sooner than males (Kirk, 1972). In baboons also, the sexes mature at very different rates, females reaching maturity at three or four years and males at seven (Rowell, Din, and Omar, 1968).

Moreover, there appear to be differences in size within sex which are related to hormonal development. True sexual hermaphrodism has been reported for the rhesus monkey. There is probably a continuum of female-to-maleness (cf. Sullivan and Drobeck, 1966) rather than a discrete difference. What is more, van Wagenen (1966) has been able to produce smaller macaque males in the laboratory:

> With androgen treatment, using a greater dose than in the female, a small, readily handled adult male has been developed. His canines are not so large, and he is suggested for use in artificial insemination. One such male in our laboratory is 21 years old, but the body length is that of a 2-year-old. (P. 106)

5. Apes

There are, as we know from Chapter 2, both lesser apes and great apes. The two genera of lesser apes, *Hylobates* and *Symphalangus*, show little, if any, sexual dimorphism with respect to size (see van den Berghe, 1973). Among the great apes, on the other hand, there is some variability from genus to genus.

The common chimpanzee (*Pan troglodytes*) shows a dimorphism quite similar to that of man. Adult female chimpanzees are approximately 85-90 percent the weight of the males (van den Berghe, 1973). The adult male may weigh about 110 pounds (and often more than this) while the adult female weighs 90 pounds (Reynolds and Reynolds, 1965). The pygmy chimpanzee (*Pan paniscus*), however, shows the least sexual dimorphism of all the great apes. This is also true with respect to anatomical characteristics other than size. They are, therefore, more like people in this regard than are gorillas and common chimpanzees (Savage and Bakeman, 1976).

In the gorilla (*Gorilla gorilla*), the weight gain rate peaks at seven to eight years for the females, but not until eight to nine years for the males. As an adult, the male gorilla is about *twice* the size of the female gorilla (Scollay, Joines, Baldridge, and Cuzzone, 1975).

As in the gorilla, the average weight for the orangutan (*Pongo pygmaeus*) male is greater than is that of the female. At birth, the orangutan male neonate outweighs the female (Seitz, 1969) and adult males are much larger than adult females (de Silva, 1971; Rodman, personal communication; Rijksen, 1975). Since the orangutan is the most arboreal of the great apes, and the gorilla is the most terrestrial, there is no clear negative relationship between arboreality and dimorphism among the great apes.

6. Summary of Size Dimorphism Literature

Among invertebrates, females are usually larger than males. Females are also usually larger than males among many species of fish, in some species of amphibians and reptiles, and in some birds. Larger females are also seen in some mammals but probably not as frequently as in other animal classes. In the order Primates, it is a very rare occurrence to have larger females than males and when it does occur, the size difference between the male and female is slight. According to Ralls (1976), among mammals the phenomenon of females being larger than males is *not* correlated with an unusually large degree of male parental investment, nor with polyandry, nor with greater aggressiveness in females than in males, nor with female dominance or matriarchy. Ralls feels that selection pressures toward larger females probably occur in nonprimate animals because a bigger *mother* is often a better mother. It is rarely, if ever, the result of sexual selection (competition for males) acting upon the female sex. However, she believes that sexual selection *does* often operate on the males (even when the male is smaller than the female). Ralls (1976) asks the interesting question, "Why, for example, are females larger in only *some* species of marmosets if all have relatively large babies, a great deal of male parental care, and apparently similar social organization?" (P. 269.) At present there is no good answer to this question.

With regard to shape dimorphism, Wood (1976) examined five primate groups (including humans) and has concluded that:

> . . . the hypothesis that most of the considerable shape differences that exist between some male and female primates are due to underlying growth differences must be rejected. It is suggested that such differences are simply the result of disproportionate change in size. (P. 15)

Incidentally, with regard to "shape" and "size" components of sexual dimorphism, Wood (1976) found that, relative to other primates, *Homo sapiens* showed quite marked dimorphism but that, apart from a few exceptions, variables were consistently sexually dimorphic in all the primate groups examined. Interspecific differences in dimorphism were more a matter of degree than of pattern. Shape appeared to be a necessary factor to consider in size differences between males and females. For example, in a primate that has little sexual dimorphism in size such as *Pan,* shape represents 46 percent of the sex difference, whereas in *Papio,* which shows strong size dimorphism, shape represents only 31 percent (but of a larger *total* difference).

7. Humans

Aside from shape, the human is more unimorphic than dimorphic with respect to size. Adult males and females differ in size, proportionally at least, about as much as do male and female common chimpanzees (Birdwhistell, 1974; van den Berghe, 1973). In children from the ages of nine to twelve, girls have a faster growth rate than boys so that at these ages girls are often larger than boys. After these ages, however, boys grow faster than girls, and, in addition, boys grow at least two years after girls stop growing. As in at least some of the great apes, (chimp and gorilla) there is a puberty acceleration difference between boys and girls (Gavan and Swindler, 1966; Gizzen and Tijskens, 1971). The puberty growth spurt in humans lasts longer in the male.

Hamilton (1977), after a study of the magnitude of sexual dimorphism in size in five osteological American Indian populations showing different degrees of transition from hunting and gathering to agriculture, concluded that decreases in the magnitude of sexual dimorphism in size reflected the shift in patterns of subsistence in these archeological populations. That is, those tribes which hunted showed the most dimorphism and those that had shifted to agriculture showed the least.

Interestingly enough, the point has also been made (Harris, 1977) that very few osteological studies have evaluated biologic information separately by sex.[1] A commonly used "correction" for sex differences, it is felt, has not been good enough, since group relationships derived from each sex separately can produce markedly different theoretical orientations. For instance, Harris (1977) has stated that skeletal and dental traits are much less variable among females.

Moreover, there are many other measures of size besides weight and height. There are large sex differences in human physique on such dimensions as hand width, chest circumference, head circumference, thoracic height, sitting height, foot length, and upper limb length (Lin, 1977). In all of these measurements, males average larger than females. Body composition also differs. In children of ages six to fourteen, boys have a higher proportion of lean body mass to weight than do girls. This is because they have less fat, however, *not* more muscle. When girls and boys of similar size are compared, *absolute* muscle and lean body mass are similar (Bailey, 1977).

In summary, while there are many qualifications that can be made, in general, people are intermediate in weight dimorphism between such species as the baboons, where males may weigh twice as much as females, and gibbons, where males and females are nearly equal in size. There are very few species of primates in which the female is larger than the male. In at least one species of marmoset, where the female is larger than the male, the difference is very slight. Humans show strong dimorphism in shape (Wood, 1976).

C. PHYSICAL DIMORPHISM IN DENTITION

Only in the Old World monkeys and lesser and great apes do we find thirty-two teeth in the same dental pattern as in man. Butler (1975) has recently written an interesting article on the effects of age and sex on dental wearing which the interested reader might consult. Males and females do not use their thirty-two teeth in the same way. Nor are the teeth themselves the same in males and females.

[1] A similar point has been made by Siegel (1976).

Leutenegger and Kelly (1975) asked whether there was any relationship between dimorphism in canine size and dimorphism in body size. They found that among anthropoids (primates excluding prosimians) canine size dimorphism and body size dimorphism were positively correlated, but only to a moderate degree (r = 0.76). They believe that the degree of canine size dimorphism is closely related to the amount of male intrasexual selection (competition for females) operating in a given mating system. The degree of body size dimorphism on the other hand, may be determined by intrasexual selection but may also be modified by habitat, diet, foraging, antipredator behavior, and locomotory behavior. In any case, it is probably true that dimorphic males and females tend not to use their bodies and teeth in the same ways.

Among the North Indian langurs, for example, adult females have smaller canines than do adult males and the females do not grind their canines while males do (Dolhinow, 1972). Among African vervets, canine display is limited to males (the canines are also larger in males) (Durham, 1969). In general, among *Cercopithecus* species the greater the dimorphism in body size, the greater the dimorphism in canine size. Even in relatively unimorphic talapoins (*Miopithecus* or *Cercopithecus talapoin*) males display their canines more than do females (Segre, 1970) even though there is little if any dimorphism with respect to canine size.

Among Old World terrestrial monkeys, the earlier maturing rhesus female exhibits markedly less variability in dentition development than does the later maturing, less predictable male rhesus (Hurme and van Wagenen, 1956). The adult male rhesus has much larger canines than does the adult female. The crab-eating macaque male is slower than the rhesus in developing its adult canines (Hurme and van Wagenen, 1961). There are sex differences in dental development in the Japanese monkey which are similar to those in the rhesus (Nass, 1977). In fact, the general pattern of dental sexual dimorphism in the Cercopithecinae corresponds to the eruption sequence of their teeth. That is, in males, canines develop last and they get much larger. This relationship is not as clear in the Colobinae, the leaf-eating family of Old World monkeys (Sirianni, 1977). Among the Cercopithecinae, the mandibular first premolar, and the third molars also exhibit some degree of dimorphism (Sirianni, 1977). The pigtail macaques (*M. nemestrina*) show more sexual dimorphism in the molars than do the rhesus, and the rhesus show more than do the crabeaters (Swindler, Orlosky, and Robinson, 1974). Early androgen treatment leads to an earlier canine growth and to canines which are eventually not so large (van Wagenen, 1966).

There is also marked sexual dimorphism in the dentition of baboons, again especially in the canines. Post (1977) has suggested that the baboon sexual dimorphisms in dental proportions can be related to sexual subniche separation with respect to diet. The feeding behavior of *Papio cynocephalus* in Kenya revealed differences between the diets of males and females. (As we will see later, males probably eat more meat.)

Among the apes, pygmy chimpanzees show the least sexual dimorphism in dentition (Cramer and Zihlman, 1976). "The basic difference between dentitions of men and apes are in canine size and in the fact that both male and female hominids have small canines" (Pilbeam, 1968, P. 1335). Despite this lack of difference in canine size in humans, the canines of humans develop later in males than in females just as in macaques (Hurme and van Wagenen, 1961). Also, as in macaques, dental development and patterning is much less variable in female than in male humans (Harris, 1977).

Why have male hominids lost their large canines and hence the sexual dimorphism seen in many other primates? Pilbeam (1968) has written that decreased canine size was correlated with the use of tools and weapons and with changes in intergroup feuding. According to Zihlman (in press), in human evolution, males developed small canine teeth as they became *less aggressive* than male apes. They were closely integrated into their kin groups more than were apes – a first step toward "fatherhood." Gathering (as opposed to hunting) was important in defining motherhood and family; and, in order to protect their kin groups, females selected males who were less aggressive and more nurturing than their ape forebears. Small canine size and, incidentally, small body size may also have been selected by the females. Zihlman believes that it was possible that at this time in human evolution male–female pairs (monogamy) became institutionalized. Kinzey (1971), on the other hand, states that " . . . until such time as an Oligocene or Early Miocene ancestor of man is discovered having large canine teeth, it is much more conceivable that our ancestors never had them." (P. 689.)

D. OTHER KINDS OF SEXUAL DIMORPHISM

Many primatologists (e. g., Wickler, 1967) have emphasized the sociosexual signalling functions of sexual dimorphism. Differences in size, shape, and coloration may be of some advantage in sexual behavior. Certainly this is as conceivable a reason for dimorphism as group defense, diet, or hunting.

Wickler (1967), in his valuable treatise on sociosexual signals, assumes that in restricted groups of familiar primates there is a need for appeasement behavior and that sexual signals also serve this additional social function. They sometimes (like sexual presents) keep their original form but their motivation (in the sender) changes. (Wickler feels that a second main source for these appeasement signals is the brood-care context.) There is, therefore, a great social significance (other than sex) to both female and male sexual organs, and, marked dimorphism in these frequently develops.

On the other hand, many species of primates show sexual dimorphisms in coloration that are directly related to sexuality or reproduction (season, estrus, and/or attractiveness).

Among New World monkeys the male golden-headed Saki (*Pithecia pithecia chrysocephala*) is obviously different from the female in coloration. Even in infancy the male's face resembles that of the adult male and is different from the female. The male has a golden, buffy face and the female's face is graying black. The white-headed Saki is also dimorphic in face coloration, the male having a white face instead of gold and the female having a face much like the golden-headed Saki female. Not all Sakis are sexually dimorphic, however. In the hairy Saki, both sexes look like the females of other Saki Subspecies (Stott, 1976). (One might wonder whether this might cause some confusion for the different subspecies.)

Squirrel monkeys (also New World monkeys) change with the season. The male has an annual testis cycle wherein he gains weight (called the "fatted" condition) in association with increased spermatogenesis for reproduction (DuMond and Hutchinson, 1967).

In spider monkeys, the female has a pendulous clitoris larger than the male penis. (Eisenberg and Kuehn, 1966). In vervets, the male exhibits a bright red penis, bright blue scrotum, and white belly known as the "red, white, and blue display." The coloration occurs only during the breeding season (Struhsaker, 1967). Patas adult males also have a bright blue scrotum (Gartlan, 1974). The female patas is not only smaller than the male, she is also lighter in color (Gartlan, 1976) and the basic color of her nose patch (black) changes to pure white in midpregnancy (Loy, 1974). In some *Cercopithecus* species, only one adult male (the leader) acquires an adult male coat color, the other males retaining the juvenile color well into adulthood (Chevalier-Skolnikoff, 1972).

Also among the Cercopithecinae, the posterior coloration of the leader male mandrill (*Mandrillus sphynx*) may serve as a signal function to the troop as does his penile coloration (Jouventin, 1975). Jouventin (1975) has stated that the facial coloration of male mandrills has evolved through the rivalry of adult males for females and food. Rhesus males in India show a marked change of color of the sexual skin nearly two months before the onset of mating (Lindburg, 1971).

Among the Colobinae, the male proboscis monkey has a huge pendulous nose which is much larger than the female's (Kern, 1964). There are many other examples among monkeys.

As for the lesser apes, the gibbons are most famous for their color-phases. In the subspecies *Hylobates lar vestitus* which lives in Sumatra, there is no dimorphism in color. In *Hylobates lar lar* of Malaya there is a change in color with age but no sexual dimorphism. In *Hylobates lar pileatus* of Thailand there is sexual dimorphism with a dark adult male and a pale adult female. Both sexes are pale at birth and both darken with age, but the female reverts to the pale phase at sexual maturity (Fooden, 1969). Fooden believes that, in the gibbon, asexual dimorphism (age-related color changes) evolved from a more primitive monomorphism and that sexual dimorphism evolved from asexual dimorphism. He also points out that it may be significant that group size is smaller in dimorphic gibbons. For a general reference on the role of coloration in sexual dimorphism the reader may see Hamilton (1973).

In great apes, female chimpanzees show a reddening and pronounced swelling of the sexual skin with estrogen treatment (Birch and Clark, 1950). In the Budango forest, female sexual swelling first occurs at seven to eight years (Reynolds and Reynolds, 1965). There is also some (but much less) genital swelling in female gorillas which correlates with mating (Nadler, 1975). There is apparently little or no such swelling in the orangutan female (Rijksen, 1975).

Among male great apes, the gorilla group in the wild often contains only *one* old silver-backed male. The other adult males do not have silver backs (Marler, 1968). In orangutans there is marked sexual dimorphism in the shape of the face (the male has a facial mask including a beard, fatty cheeks, and a throat sac).

Physical sexual dimorphisms among nonhuman primates have also been found in: (1) maturation of hand-wrist skeletal complex (Tarrant, 1977); (2) nuclear chromatin in the prenatal amniotic fluid (Herbst, Taft, and Roffoy, 1976); (3) length of hair and diameter of hair in rhesus (Dolnick, 1969); (4) size of the bony pelvis of the squirrel monkey (Gingerich, 1972); (5) degree of baldness in stumptailed macaques (Orkin, 1967); and (6) brain size in rhesus macaques (Zuckerman and Fisher, 1938).

In *Homo sapiens,* there are also many other sexual dimorphisms in anatomy besides size and dentition. Females have larger breasts, more fat, a different larynx, and a different bodily shape. Males, on the other hand, exhibit baldness in adulthood, facial hair, broader chests and shoulders, and larger hands and feet.

Wickler (1967), in his review of sociosexual signals, states that some of the elements of sociosexual behavior in man are the same as in nonhuman primates. He believes that there is a continuous evolutionary line of dimorphism based on sexual signals from lower mammals to man, leading in the females from estrous swellings to the enlargement of the labia minora in sexually receptive female humans and to true as well as imitated human subcutaneous fat deposition in the female hindquarters. In the males, he believes that the line ascends from the laying of urine-marked tracks, through general urine marking, territory marking, threatening, and dominance demonstration to penile display in primates to various forms of phallus symbolism in humans.

The overview of five different groups (*Colobus, Papio, Gorilla, Pan, Homo*) of primates by Wood (1976), referred to earlier, puts at least skeletal dimorphism in *Homo sapiens* into proper perspective. As we have already said, Wood has shown that differences in shape are important to man. In his discussion, Wood states:

> The *Homo* group shows the least variation in strength of sexual differences and is exceptional among the groups studied in that in the case of some variables, such as canine size, it is not excessively dimorphic. Despite this, there were *more* cases of significant sexual differences in the *Homo* group (86% of comparisons) than in the *Gorilla* sample (83%). (P. 29)

Simply stated, humans have smaller sex differences but (apparently) more of them than do other primate groups.

In this chapter we have concentrated primarily on physical structures *other* than the genitals and other reproductive organs. We have also put aside any reference to specific sex differences in the central nervous system other than gross brain size. In our next chapter, however, we will deal with the prenatal development of physical dimorphism in precisely these two areas.

As we will see, behavioral differences between the sexes among nonhuman and human primates may very well have something to do with the development of physical differences between the sexes.

REFERENCES

Bailey, S. M. Sex differences in body composition for children of similar size. Paper presented at the *American Association of Physical Anthropologists* Meeting, Seattle, Washington, April, 1977.

Bernstein, I. S. Primate status hierarchies. In Rosenblum, L. A. (Ed.) *Primate Behavior: Developments in Field and Laboratory Research* (Vol. I). New York: Academic Press, 1970, pp. 71–109.

Birch, H. G. and Clark, G. Hormonal modification of social behavior IV. The mechanism of estrogen induced dominance in chimpanzees. *Journal of Comparative and Physiological Psychology,* 1950, **43**, 181–193.

Birdwhistell, R. L. Masculinity and femininity as display. In Weitz, S. (Ed.) *Nonverbal Communication.* New York: Oxford, 1974, pp. 144–149.

Blurton-Jones, N. Ethology, anthropology and childhood. In Fox, R. (Ed.) *Biosocial Anthropology.* New York: Wiley, 1975, pp. 69–72.

Bramblett, C. A. *Patterns of Primate Behavior.* Palo Alto: Mayfield, 1976.

Bramblett, C. A., Pejaver, L. D. and Drickman, D. J. Reproduction in captive vervet and Sykes monkeys. *Journal of Mammalogy,* 1975, **56** (4), 940–946.

Butler, R. J. Effect of age and sex on dental wearing. *American Journal of Physical Anthropology,* 1975, **40**, 132.

Chevalier-Skolnikoff, S. Sex-role differentiation in nonhuman primates, with implications for man. *Medical Anthropology Newsletter,* 1972, **3** (3), 9-13.

Chevalier-Skolnikoff, S. Heterosexual copulatory patterns in stumptail macaques (*Macaca arctoides*) and in other macaque species. *Archives of Sexual Behavior,* 1975, **4** (2), 199-220.

Clutton-Brock, T. H. and Harvey, P. A. A statistical analysis of some aspects of primate ecology and social organization. Paper presented at the *International Primatological Society* Meeting, Cambridge, England, August, 1976.

Coelho, A. M. Socio-bioenergetics and sexual dimorphism in Syke's monkey (*Cercopithecus albogularis kolbi*). Paper presented at the Annual Meeting of the *American Anthropological Association,* December, 1973.

Cramer, D. L. and Zihlman, A. L. Sexual dimorphism in pygmy chimpanzees (*Pan paniscus*). Cranial, dental and skeletal evidence. Paper presented at the *International Primatological Society* Meeting, Cambridge, England, August, 1976.

deSilva, G. S. Notes on the orangutan rehabilitation project in Sabah. *Maylayan Nature Journal,* 1971, **24**, 50-77.

DeVore, I. and Washburn, S. L. Baboon ecology and human evolution. In Howell, F. C. and Bourliére, F. (Eds.) *African ecology and human evolution.* New York: Viking Fund Publication #36, pp. 335-367.

DiGiacomo, R. F. and Shaughnessy, P. W. Estimation of gestational age and birth weight in the rhesus monkey (*Macaca mulatta*). *American Journal of Obstetrics and Gynecology,* 1972, **112**, 619-628.

Dittus, W. P. J. Population dynamics of the Toque monkey, *Macaca sinica.* In Tuttle, R. H. (Ed.) *Socioecology and Psychology of Primates.* The Hague: Moulton, 1975, pp. 125-151.

Dolhinow, P. J. The North Indian langur. In Dolhinow, P. J. (Ed.) *Primate Patterns.* New York: Holt, Rinehart and Winston, 1972, pp. 181-238.

Dolhinow, P. J. and Bishop, N. The development of motor skills and social relationships among primates through play. In Hill, J. P. (Ed.) *Minnesota Symposium on Child Psychology* (Vol. IV). Minneapolis: University of Minnesota Press, 1970, pp. 141-198.

Dolnick, E. H. Variability of hair growth in *Macaca mulatta.* In Montagna, W. and Dobson, R. L. (Eds.) *Advances in the Biology of Skin.* New York: Pergamon, 1969, pp. 121-128.

Doyle, G. A. The behavior of the lesser bushbaby. In Martin, R. D., Doyle, G. A. and Walker, H. C. (Eds.) *Prosimian Biology.* London: Duckworth, 1974, pp. 213-231.

DuMond, F. W. and Hutchinson, T. C. Squirrel monkey reproduction: The "fatted" male phenomenon and seasonal spermatogenesis. *Science,* 1967, **158**, 1467-1470.

Durham, N. M. Sex differences in visual threat displays of West African vervets. *Primates*, 1969, **10**, 91-95.

Eisenberg, J. F. and Kuehn, R. E. The behavior of *Ateles geoffroyi* and related species. *Smithsonian Miscellaneous Collection*, 1966, **151** (8), 1-63.

Fooden, J. Color-phase in gibbons. *Evolution*, 1969, **23** (4), 627-644.

Gartlan, J. S. Adaptive aspects of social structure in *Erythrocebus patas*. *Proceedings of the Fifth Congress of the International Primatological Society*. Tokyo, Japan, 1974.

Gartlan, J. S. Ecology and behaviour of the patas monkey. Film presented at the *International Primatological Society* Meeting, Cambridge, England, August, 1976.

Gautier-Hion, A. Dimorphisme sexual et organization sociale chez les Cercopithecines forestiers Africains. *Mammalia*, 1975, **39** (3), 365-374.

Gavan, J. A. and Swindler, D. R. Growth rates and phylogeny in primates. *American Journal of Physical Anthropology*, 1966, **24** (2), 181-190.

Gijzen, A. and Tijskens, J. Growth in weight of the lowland gorilla (*Gorilla gorilla gorilla*) and of the mountain gorilla (*Gorilla gorilla beringei*). *International Zoo Yearbook*, 1971, **11**, 183-193.

Gingerich, P. D. The development of sexual dimorphism in the bony pelvis of the squirrel monkey. *Anatomical Record*, 1972, **172**, 589-594.

Hamilton, M. E. Testing a model for the prediction of variation in human sexual dimorphism. Paper presented at the *American Association of Physical Anthropologists* Meeting, Seattle, Washington, April, 1977.

Hamilton, W. J. *Life's Color Code*. New York: McGraw-Hill, 1973.

Harris, E. F. Biologic implications of sexually dimorphic qualitative traits in the human dentition. Paper presented at the *American Association of Physical Anthropologists* Meeting, Seattle, Washington, April, 1977.

Herbst, A. L., Taft, P. D. and Robboy, S. J. Nuclear chromatin determination in amniotic fluid cells for prenatal sex prediction in the macaque. *American Journal of Obstetrics and Gynecology*, 1976, **124**, 761-768.

Hromada, J. Contribution to the question of variability in *Macaca mulatta* *Sbornik vedesky'ch* praci. *Lekarshk fakulty KU v Hradci Kralove*, 1968, **11** (1), 169-175.

Hurme, V. O. and van Wagenen, G. Emergence of permanent first molars in the monkey (*Macaca mulatta*): Association with other growth phenomena. *The Yale Journal of Biology and Medicine*, 1956, **28**, 538-566.

Hurme, V. O. and van Wagenen, G. Basic data on the emergence of permanent teeth in the rhesus monkey (*M. mulatta*). *Proceedings of the American Philosophical Society*, 1961, **105**, 105-140.

Jolly, A. *Lemur Behavior*. Chicago: University of Chicago Press, 1966.

Jorde, L. B. and Spuhler, J. N. A statistical analysis of selected aspects of primate demography, ecology, and social behavior. *Journal of Anthropological Research,* 1974, **30**, 197-224.

Jouventin, P. Les roles des colorations du Mandril (*Mandrillus sphinx*). *Zietschrift für Tierpsychologie,* 1975, **39**, 445-462.

Kern, J. A. Observations on the habits of the proboscis monkey, *Nasalis larvatus* (Wurmb) made in the Brunei Bay area, Borneo. *Zoologica,* 1964, **49**, 183-192.

Kinzey, W. G. Evolution of the human canine tooth. *American Anthropologist,* 1971, **73**, 680-694.

Kirk, J. H. Growth of maturing *Macaca mulatta. Laboratory Animal Science,* 1972, **22** (4), 573-575.

Leutenegger, W. Evolution of litter size in primates. Paper presented at the *American Society of Primatologists* Meeting, Seattle, Washington, April, 1977.

Leutenegger, W. and Kelly, J. T. Relationship of sexual dimorphism in canine size and body size to social, behavioral, and ecological correlates in anthropoid primates. *American Journal of Physical Anthropology,* 1975, **42**, 314.

Lin, P. M. Factors and sexual differences in human physique. Paper presented at the *American Association of Physical Anthropologists* Meeting, Seattle, Washington, April, 1977.

Lindburg, D. G. A field study of the reproductive behavior of the rhesus monkey (*Macaca mulatta*). Doctoral dissertation, Department of Anthropology, University of California, Berkeley, 1967.

Loy, J. Changes in facial color associated with pregnancy in patas monkeys. *Folia Primatologica,* 1974, **22**, 251-257.

Marler, P. Aggregation and dispersal: Two functions in primate communication. In Jay, P. C. (Ed.) *Primates: Studies in Adaptation and Variability.* New York: Holt, Rinehart and Winston, 1968, pp. 420-438.

Masui, K., Nishimura, A., Ohsawa, H. and Sugiyama, Y. Population study of Japanese monkeys at Takasakiyami I. *Journal of the Anthropological Society of Nippon,* 1973, **81** (4), 236-248.

McMahon, C. A., Wigodsky, H. S. and Moore, G. T. Weight of the infant baboon (*Papio cynocephalus*) from birth to fifteen weeks. *Laboratory Animal Science,* 1976, **26** (1), 928-931.

Nadler, R. D. Sexual cyclicity in captive lowland gorillas. *Science,* 1975, **189**, 813-814.

Nass, G. G. Intra-group variations in the dental eruption sequence of *Macaca fuscata fuscata.* Paper presented at the *American Association of Physical Anthropologists* Meeting, Seattle, Washington, April, 1977.

Orkin, M. Animal models (spontaneous) for human disease. *Archives of Dermatology,* 1967, **95**, 524-531.

Phoenix, C. H., Goy, R. W. and Resko, J. A. Psychosexual differentiation as a function of androgenic stimulation. In Diamond, M. (Ed.) *Reproduction and Sexual Behavior.* Bloomington: University of Indiana Press, 1968, pp. 33-49.

Pilbeam, D. The earliest hominids. *Nature,* 1968, **219**, 1335-1338.

Pope, B. L. The population characteristics of howler monkeys (*Alouatta caraya*) in Northern Argentina. *American Journal of Physical Anthropology,* 1966, **24**, 361-370.

Post, D. G. Baboon feeding behavior and the evolution of sexual dimorphism. Paper presented at the *American Association of Physical Anthropologists* Meeting, Seattle, Washington, April, 1977.

Ralls, K. Mammals in which females are larger than males. *Quarterly Review of Biology,* 1976, **51**, 245-276.

Reynolds, V. and Reynolds, F. Chimpanzees of the Budongo forest. In DeVore, I. (Ed.) *Primate Behavior.* New York: Holt, Rinehart and Winston, 1965, pp. 324-368.

Rijksen, H. D. Social structure in a wild orang-utan population in Sumatra. In Kondo, S., Kawai, M. and Ehara, A. (Eds.) *Contemporary Primatology.* Basel: Karger, 1975, pp. 373-379.

Riopelle, A. J. and Hale, P. A. Nutritional and environmental factors affecting gestation length in rhesus monkeys. *American Journal of Clinical Nutrition,* 1975, **28**, 1170-1176.

Riopelle, A. J., Hale, P. A. and Watts, E. S. Protein deprivation in primates: VII. Determinants of size and skeletal maturity at birth in rhesus monkeys. *Human Biology,* 1976, **48**, 203-222.

Riopelle, A. J., Hill, C. W. and Li, Su-Chen Protein deprivation in primates: V. Fetal mortality and neonatal status of infant monkeys born of deprived mothers. *American Journal of Clinical Nutrition,* 1975, **28**, 989-993.

Rowell, T. E. Forest-living baboons in Uganda. *Journal of Zoology, London,* 1966, **149**, 344-364.

Rowell, T. E., Din, N. A. and Omar, A. The social development of baboons in their first three months. *Journal of Zoology, London,* 1968, **155**, 461-483.

Russell, C. and Russell, W. M. S. Primate male behavior and its human analogue. *Impact of Science on Society,* 1971, **21**, 63-74.

Sauer, E. G. F. and Sauer, E. M. The South West African bushbaby of the *Galago senegalensis* group. *The Journal of the South West African Scientific Society,* 1962, **16**, 5-36.

Savage, E. S. and Bakeman, R. Comparative observations on sexual behavior in *Pan paniscus* and *Pan troglodytes.* Paper presented at the *International Primatological Society* Meeting. Cambridge, England, August, 1976.

Schwaier, Anita. The breeding stock of *Tupaias* at the Battelle Institute. *Laboratory Animal Handbook,* 1975, **6**, 141-149.

Scollay, P. A., Joines, S., Baldridge, C. and Cuzzone, A. Learning to be a mother. *Zoonooz*, 1975, **48** (4), 4-9.

Scott, J. P. Agonistic behavior of primates: A comparative perspective. In Holloway, R. L. (Ed.) *Primate Aggression, Territoriality and Xenophobia: A Comparative Perspective.* New York: Academic Press, 1974, pp. 417-434.

Segre, Amelia. Talapoins. *Animal Kingdom,* 1970, **73** (3), 20-25.

Seitz, A. Notes on the body weights of new-born and young orang-utans. *International Zoo Yearbook,* 1969, **9**, 81-84.

Siegel, M. E. Dento-facial relationships in sexually dimorphic populations, or why you should know the boys from the girls. Paper presented at the *International Primatological Society* Meeting, Cambridge, England, August, 1976.

Sirianni, J. E. Patterns of sexual dimorphism in primate tooth size. Paper presented at the *American Society of Primatologists* Meeting, Seattle, Washington, April, 1977.

Stott, K. Sakis: Imps of the rain forest. *Zoonooz,* 1976, **49** (4), 4-8.

Struhsaker, T. T. Behavior of vervet monkeys. (*Cercopithecus aethiops*). Berkeley: University of California Press, 1967.

Sullivan, D. J. and Drobeck, H. P. True hermaphrodism in a rhesus monkey. *Folia Primatologica,* 1966, **4**, 309-317.

Swindler, D. R., Orlosky, F. J. and Robinson, J. M. The significance of dental studies to the investigation of the taxonomy of living macaque species. *American Journal of Physical Anthropology,* 1974, **41**, 506.

Tarrant, L. H. Postnatal growth in the pig-tailed monkey (*Macaca nemestrina*). Paper read at the 44th Annual Meeting of the *American Association of Physical Anthropologists,* Denver, Colorado, April, 1975.

Tarrant, L. H. Sex differences in skeletal maturation at birth in the pig-tailed monkey. Paper presented at the *American Association of Physical Anthropologists* Meeting, Seattle, Washington, April, 1977.

Valerio, D. A. and Dalgard, D. W. Experiences in the laboratory breeding of non-human primates. *Laboratory Animal Handbook,* 1975, **6**, 49-62.

Van den Berghe, P. L. *Age and sex in human societies: A biosocial perspective.* Belmont, California: Wadsworth, 1973.

van Wagenen, G. Studies in reproduction (*Macaca mulatta*). In Miller, C. O. (Ed.) *Proceedings, Conference on Nonhuman Primate Toxicology.* Warrenton, Virginia: Department of Health, Education and Welfare, 1966, pp. 103-113.

Watts, E. S. Some guidelines for collection and reporting of nonhuman primate growth data. *Laboratory Animal Science,* 1977, **27**, 85-89.

Wickler, W. Socio-sexual signals and their intra-specific imitation among primates. In Morris, D. (Ed.) *Primate Ethology.* Chicago: Aldrine, 1967, pp. 69-147.

Willes, R. F., Kressler, P. L. and Truelove, J. F. Nursery rearing of infant monkeys (*Macaca fascicularis*) for toxicity studies. *Laboratory Animal Studies*, 1977, **27**, 90-98.

Wolfheim, J. H. Sex differences in behavior in a group of captive juvenile talapoin monkeys. (*Miopithecus talapoin*) *Behaviour*, in press.

Wood, B. A. Nature and basis of sexual dimorphism in primate skeleton. *Journal of Zoology*, 1976, **180**, 15-34.

Zihlman, A. Z. Motherhood in transition: From ape to human. In Miller, W. and Newman, L. (Eds.) *First Child and Family Formation*. Chapel Hill, N. C.: University of North Carolina (Not in press yet).

Zuckerman, S. and Fisher, R. B. Growth of the brain in the rhesus monkey. *Proceedings of the Zoological Society of London*, 1938, **107**, 529-538.

4.

Sexual Differentiation

In the next four or five chapters of this book, we will be discussing the role of sex hormones. There are four major groups of sex hormones to be considered: androgens, estrogens, progesterone, and gonadotrophins.

Androgens are primarily male sex hormones produced by the testes and adrenal glands in males and by the ovaries and adrenal glands in females. Males produce more androgens than do females. Some major androgens are testosterone and dihydrotestosterone.

Estrogens are primarily female sex hormones produced by the ovaries and adrenal glands in females and by the testes and adrenal glands in males. Females produce more estrogens than do males. Some major estrogens are estradiol-17β, estriol, and estrone.

Progesterone is also primarily a female hormone produced by the ovaries and adrenal glands in females and by the testes in males. Females produce more progesterone than do males.

Gonadotrophins are both male and female hormones produced by the pituitary gland via releasing factors in the basal hypothalamus of the brain. There are two major gonadotrophins — FSH and LH. FSH is an abbreviation for follicle stimulating hormone. LH is an abbreviation for leutenizing hormone. In the male, LH is often referred to as ICSH (Interstitial cell stimulating hormone). The source and major action of each of these hormones is given in Table 4-1.

A. NONPRIMATE PRENATAL HORMONES

Pfeiffer (1936) was among the first to show that structural sexual differentiation in mammals begins very early in development. He manipulated the sex hormones of rats by removing and then replacing their gonads. On the basis of such studies,

Table 4-1. Sex Hormones, Primary Sources and Major Actions.

(Jones, Shainberg, and Byer, 1977)

Primary Source	Hormone	Major Action
Male testes	Testosterone	Male secondary sex characteristics
	Androgens in general	Male and female sex drive
Female ovaries	Estrogen	Female secondary sex characteristics; menstrual cycle
	Progesterone	Breast development at puberty; maintenance of uterine lining during pregnancy
Anterior pituitary gland	FSH (follicle stimulating hormone; a gonadotrophin)	Stimulates growth of follicles containing ovum and produces estrogen in females; stimulates production of sperm in males
(Hypothalamic FSH and LH releasing factors control the release of the gonadotrophins; estrogen, progesterone, and testosterone levels, in turn, control the releasing factors in a feedback system.)	LH (leutenizing hormone) often termed ICSH (interstitial cell stimulating hormone; a gonadotrophin)	Ovulation, formation of corpus luteum in females (as LH); production of testosterone and other androgens in males (as ICSH)

he concluded, quite correctly, that the male hormone testosterone determined the direction in which sexual differentiation would proceed. However, he incorrectly assumed that testosterone worked by inducing changes in the pituitary gland.

Harris (cf. Levine, 1966), however, transplanted the male rat pituitary into a female and found that the female's reproductive structures, functions, and behavior remained normal. Moreover, a male rat with a female pituitary also remained malelike in reproductive structure and behavior.[1]

If testosterone did not affect sexual differentiation by changing the pituitary, how did it work? Harris and his associates (cf. Levine, 1966) eventually found that, regardless of the genetic sex of the animal, the presence of testosterone during a critical period of development not only changed reproductive structures to malelike structures but also had the effect of sensitizing the *brain* to the influence of male hormones. The absence of testosterone during this same critical

[1] I am indebted to Van Harrison for help on this topic.

period in early development sensitizes the brain to female hormones. The timing for this critical period varies with the mammalian species under study. In some species, it occurs during prenatal development; in others it occurs during early postnatal development. In rats and mice, it occurs between one and ten days of postnatal life. In guinea pigs the sensitive period is during pregnancy. In monkeys (rhesus), it is between the forty-sixth and ninetieth day of pregnancy (Neuman, Steinbeck, and Hahn, 1970).

B. PRIMATE PRENATAL DATA

The critical period for rhesus monkey sexual differentiation was discovered by Goy and Phoenix. Basing their first studies on the earlier work of van Wagenen, Hamilton, and Wells (cf. Phoenix, 1974), they injected pregnant female rhesus monkeys with testosterone propionate during various stages of gestation. If the androgenizing injections were given during a certain critical period of the female's pregnancy (forty-six to ninety days),[2] a genetically female fetus was born as a pseudohermaphrodite (having a mixture of male and female reproductive structure).

These pseudohermaphrodite rhesus monkeys had external genitalia that were those of a *male,* including a fully developed scrotum and a small but clearly developed penis, while internally, they had normal *female* reproductive structures including normally cycling ovaries. They even menstruated normally, with the exception that menarche arrived approximately seven months late, and menstrual flow occurred through the penile urethra. If the fetus was genetically male, it was in no way affected by the injections of testosterone.

So, the genitalia of female fetuses can be changed by the presence of prenatal androgen (in this case testosterone). What is the evidence for changes in the brain? Much of this evidence is behavioral and, hence, indirect. A minor digression is needed at this point to describe the nature of the behavioral data.

In the case of the rhesus monkey, prior to the Goy and Phoenix research, Harry Harlow and co-workers had already demonstrated that very early in the normal infant rhesus monkey's life there appeared certain sexually dimorphic *behavior* patterns which could be measured in terms of their frequency of occurrence (cf. Goy, 1966). Social threats, rough and tumble play, and initiation of play were displayed significantly more often by male than by female infant monkeys.

Goy and Phoenix compared their androgenized female (pseudohermaphrodite) infants with normal male and female infant rhesus monkeys with regard to

[2] It is possible to predict the fetal sex after the ninetieth day by amniocentesis, where the amniotic fluid is obtained and analyzed for nuclear chromatin (Herbst, Taft, and Robboy, 1976).

frequencies of threats, play, and play initiations. They found that their pseudo-hermaphrodites displayed these behaviors at a level that was intermediate to the normal male and female levels (cf. Goy, 1966).

Also in the early Harlow studies (cf. Goy, 1966) it was learned that prior to adolescence all infant rhesus monkeys display immature social and/or sexual mounts. The difference between the immature and mature mounts has to do with the placement of the feet. In the immature mount, as performed by *both* male and female normal infants and juveniles, the feet remain planted on the floor or ground, whereas, in the mature mount, as displayed by adult male rhesus, the feet are used to clasp the female's ankles during copulation. Goy and Phoenix found that their pseudohermaphrodites, unlike normal females, developed the mature mounting pattern of normal adult males.

The testes of the fetal rhesus monkey during late gestation (129–145 days) are capable of androgen biosynthesis and can bind and respond to gonatotrophin stimulation (Huhtaniemi et al., 1977).

In another project, Goy and Phoenix (1971) removed the testes of a male rhesus after the critical androgenization period (but still prenatal). Despite having no testes at birth, this male developed normal levels of male play behavior; and, although he mounted less than his control, he did use the mature foot clasp mount. Thus, testosterone (although usually present in late pregnancy) is not crucial for psychosexual masculinization after the second trimester of prenatal development.

Goy and Phoenix used quite high levels of testosterone. What levels normally occur when the fetus is genetically male? Resko (1974) compared the amounts of testosterone in normal male fetuses with normal female fetuses and found a significantly higher level of testosterone in the umbilical artery of the males than of the females. He also castrated three males on day 100 of gestation after which he measured their blood level of testosterone (on day 155). He found that the testosterone level in these castrated male fetuses did not differ significantly from that of normal female fetuses (Hagemenas and Kittinger, 1972). Thus, it appears to be the presence of testosterone (androgens) during early fetal development which determines many sexually dimorphic behavior patterns. Furthermore, the behaviors that are influenced by prenatal androgens are not limited to sexual behavior per se.

Do female hormones (estrogens and progesterone) play a role in sexual differentiation? Although androgens can be converted into estrogens by certain body tissues (e.g., neural), this conversion is not obligatory for organization or activation of masculine traits (Goy, Slimp, and Wallen, 1975). Progesterone, on the other hand, can in some circumstances act as an antagonist to androgens. Resko (1975) has suggested that there may be a reciprocal relationship between the levels of androgens and progesterone in the developing fetus. Progesterone levels in the umbilical vein and artery are higher in female fetuses. While it is definitely androgen which stimulates sexual differentiation, it is possible that progesterone may be indirectly involved by influencing the amount of testosterone present in fetal circulation. The levels of prenatal androgens can also be inhibited without removing the prenatal testes. "When cyproterone acetate (anti-androgen) is given to pregnant animals, the genetic males exhibit feminine changes in both anatomy and behaviour." (Brotherton, 1973, p. 175.)

In any case, prenatal not postnatal androgens are critical to sexual differentiation. Sex differences during infancy are *not* due to androgens or other hormones present after birth and before puberty. There is very little difference in the testosterone levels of males and females during the infant and juvenile periods (Kimble, 1973). Early male castrates still develop mounting and play patterns typical of normal males, although in adulthood they do not show erection, yawning or aggression as frequently as do normal males (Bielert, 1975-76).

Goy (1976) has experimented with various forms of androgen. For example, he found that although the two prenatally administered androgens, testosterone and dihydrotestosterone (DHTP) affected postnatal behavioral development similarly, dihydrotestosterone produced noticeably less genital virilization than did testosterone. There are apparently *some* differences in the hormonal requirements for genital and neural masculinization. Goy eventually hopes to demonstrate an even greater degree of independence for these two processes. Incidentally, it was by using DHTP as the androgen (which cannot be converted to estrogen) that Goy, Slimp, and Wallen (1975) showed that conversion of androgen to estrogen was not necessary for the differentiation of masculine behavioral traits. The reader may consult Herbert (1976) for an excellent general article on the hormonal basis of sex differences.

On the basis of evidence such as that presented above, Phoenix (1974) concludes that:

> Primates may have achieved some emancipation from hormonal control with an increase in cortical determination of behavior ... However, this evidence relates to the role of the gonadal hormones in regulating adult sexual behavior, and not to the importance of the steroidal hormones (especially testosterone) during prenatal life. (P. 30)

Phoenix also concludes that prenatal testosterone masculinizes the nervous system, probably in the hypothalamus but also elsewhere. However, in another article he cautions us that:

> ... androgens can be shown to act organizationally *only* upon those behaviors which exist as dimorphisms in the species. Those behavioral characteristics

which the sexes display with equal frequency are *not* influenced in any detectable manner by the early administration of androgens. (Phoenix, Goy, and Resko, 1968, P. 47)

There is also some evidence that sexual differentiation is not an all-or-none process but that different amounts of testosterone produce different levels of androgenization in the female. In fact, there is evidence that both the absolute level of testosterone and the ratio of testosterone to progesterone produce sexual differentiation (Resko, 1974).

Each of Goy and Phoenix's pseudohermaphrodites was masculinized to a different degree, both structurally and behaviorally. It is even possible for true hermaphrodism to occur naturally in the rhesus monkey (Sullivan and Drobeck, 1966) wherein gonads are characterized by having a central testicular portion surrounded by a rim of ovarian tissue.

Just as the degree of structural and behavioral sex differences attributable to prenatal androgenization may vary considerably with the timing of the hormonal administration and the species, behavioral dimorphism may also vary depending upon later hormonal experiences of the organisms involved. The ontogeny of certain hormonal processes may vary temporally from species to species. It may be that the pubertal process itself in some species extends from birth to initial ovulation (Foster, 1974). Jumping to hasty generalizations on the basis of research done on a species which shows marked physical and behavioral sexual dimorphism is clearly unwise. Even in normal rhesus monkeys, males sometimes present and females sometimes mount (Goy and Goldfoot, 1975).

C. EARLY EXPERIENTIAL EFFECTS

Aside from hormonal effects, there are other variables controlling gender differentiation. Social/experiential factors can strongly influence the behaviors which are expressed by either sex. Goldfoot (1976), by studying isosexual (all-male or all-female) as well as heterosexual groups of infant rhesus, has concluded that gender role is the resultant of behavioral predispositions induced by steroids prenatally combined with early social learning among *peers*. Social factors such as dominance rank among peers, for example, can modify prenatally determined tendencies in both sexes.

When normal female rhesus are reared only with other females they show increased mounting (although *not* more rough play). When normal male rhesus are reared with minimal peer contact they develop abnormal mating habits and less frequent mounting (Goy, 1976).

With regard to the effect of prenatal androgens on the central nervous system, Goy and Resko (1972) remark that considering the effects of early experience on sex differences it would probably be foolhardy to assume that there was a discrete neural locus of androgen action. They feel that androgens undoubtedly affect neural circuits that are involved in *experience* as well as those directly involved in sex differences.

In addition, although it is remarkable that one need only provide positive social experience as late as the second year of life in order to see the development of the foot clasp mount in a rhesus male, one needs to allow *long continuous* exposure to agemates to guard against the development of high levels of social aggression between animals which can adversely affect appropriate use of the foot clasp capability (Goy, Wallen, and Goldfoot, 1974). Obviously, behavioral sex differences do not depend on early hormones alone.

Joslyn (1973) has also presented what he has interpreted as evidence for an effect of experience on gender-related behavior. He injected androgen into *infant* female rhesus monkeys when they were between 6½ and 14½ months of age. The androgen increased their aggressiveness so that they replaced the top males in dominance. The social dominance lasted for one year after the last hormone injection. It had become so well learned during hormone treatment that it became independent of hormonal support. The sexual and play behaviors of the females remained "feminine."

A more complete treatment of sex differences in rhesus infancy will be given in later chapters of this book. At this point, we will just emphasize the fact that hormones alone do not explain all there is to know about sex differences in rhesus infants.

D. ADULT HORMONAL EFFECTS

In the adult male rhesus monkey, circulating testosterone is affected by circadian (diurnal) rhythms, by access to females, by seasonality, by alterations in social rank, and by agonistic (aggressive and fearful) encounters (Bernstein, 1974).

In all primate adults there are distinct behavioral patterns that serve to maintain the species. Most of these are closely correlated with reproductive cycles and differ in male and female individuals. Most evident are sex differences in sexual behavior. In adulthood, estrogens as well as androgens affect such behaviors (Neuman, Steinbeck, and Hahn, 1970). A summary of many of the effects of hormones on macaque and baboon adult behaviors may be found in Rowell (1972).

For the present we will go no further into the effects of sex hormones on adult primate behavior; however, in later chapters, we will discuss this topic in great detail.

E. SEX HORMONES AND THE NERVOUS SYSTEM

Throughout our discussion of prenatal sexual differentiation, we have mentioned the development of possible sex differences in the central nervous system resulting from the presence or absence of prenatal androgen between forty-six and ninety days of gestation in the rhesus monkey. The evidence which we have presented for the presence of gender-related neural differences has been indirect. If basic behaviors differed we have assumed that some underlying neurological difference mediated it. In the present section, we will consider the evidence for sex differences in the nervous system itself.

Much of the research that has been done on the primate brain has utilized the squirrel monkey (*Saimiri sciureus*) as a subject. In studies of this species, some sex differences have been found. Bubenik and Brown (1973) discovered "a marked sex difference in the nuclear diameter of neurons in the nucleus medialis amygdalae (NMA)." (P. 620.) The diameters were smaller in females than in males. The authors suggest that some constitutional differences exist that are further modified by circulating sex hormones. They found no sex differences in the size of the neurons of the suprachaismatic nucleus or in the size of neurons in the cerebral cortex.

Since gonadal hormones operate by way of a feedback mechanism involving gonadotrophic hormones and gonadotrophic hormone releasing factors in the hypothalamic and pituitary area, one might expect substantial sexual differentiation at that site if anywhere. A frontal cut in the medial preoptic area of the hypothalamus produces anovulation and amenorrhea in the baboon; however, the animal shows a temporary recovery of the regular release of LH, of ovulation, and of menstruation following the lesion. On the basis of these data, it has been suggested that, in the baboon at least, there are two levels in the hypothalamus which regulate gonadotrophin secretion and that the neurons in the preoptic area exert an influence on periodicity in the female (Hagino, 1976).

It is hypothesized that the interval between birth and puberty is the time required for the hypothalamus to mature and escape from the restraining influences of gonadal hormones. At puberty, there is a beginning to a new interaction between the sex hormones and certain cells of the hypothalamus. The most important activating sex hormones are testosterone and estrogen while progesterone appears to have an effect on the brain much like mild sedation (Hamburg and Lunde, 1966).

Estradiol (an estrogen) is concentrated and retained in certain neurons of the old periventricular brain (limbic and brainstem areas). The areas involved in the tree shrew, for example, are the preoptic-strial, basal hypothalamic, and amygdala areas (Keefer and Stumpf, 1975a). The distribution of these estrogen-concentrating neurons in the tree shrew and squirrel monkey brain is quite similar to the distribution for several other lower mammalian and nonmammalian species (e.g., rats, doves, etc.). The areas are interconnected to allow for interactions between both hormonal and sensory stimulation (Keefer and Stumpf, 1975b).

As for the amygdala, lesions there produce different effects in males and females. In male stumptailed macaques, there is decreased aggression and increased sexuality whereas in females there is sometimes heightened aggressivity. This sex difference has been linked to sex differences in gonadal hormones and to differential responses to the sex hormones in the preoptic neurons of the hypothalamus of males and females (Kling, 1974, 1975). Kling and Dunne (1976) report that in amygdala lesioned females:

> We have not yet observed any amygdala lesioned female that showed appropriate maternal behavior with survival of the infant. (P. 39)

Livrea (1972) has made the statement that it is highly probable that the hypothalamus is also the focal point of male hormone actions, although the action probably extends to extrahypothalamic levels, perhaps to the *entire* nervous system.

In the squirrel monkey, genital display (a display of penile erection indicating dominance as much as sex) is exhibited mainly by adult males. By stimulating the nervous systems of squirrel monkeys, MacLean (1975) has been able to elicit genital display from the globus pallidus. He concludes on the basis of his research that the globus pallidus mediates the display; that is, it organizes and programs the display but it is not responsible for the motor component.

Difficulties have arisen in the interpretation of the elicitation of gender-related dominance displays by brain stimulation. Maurus and Ploog (1971), for example, have not been able to explain why dominance gestures were elicited many more times from brain sites in *subordinate* females than from brain sites in dominant males, a finding opposite to what occurs in spontaneous behavior.

Perachio (1976) believes there also may be sites in the central nervous system for sexual *behavior* which are independent of *sexual* hormones and hormone release. Using electrical stimulation of the hypothalamus in *male* rhesus monkeys, Perachio elicited aggression and in other sites, ejaculation. He castrated some males and stimulated them again in the same sites. Castration did not eliminate evoked or spontaneous sexual and aggressive behavior, although the threshold for evoked behavior increased and testosterone treatment restored threshold levels. The behavioral effects, however, were not consistently related to given hormonal changes. In addition, stimulation of sites that produced aggression or sexual behavior did *not* induce hormone release. Perachio believes that hormones have a *supporting* but not a necessary effect on behavior in male rhesus monkeys and that this effect operates at specific preoptic and anterior levels of the hypothalamus.

Pfaff, Gerlach, McEwen, Ferin, Carmel, and Zimmerman (1976) have reported that the *rhesus* has sex hormone-concentrating neurons in preoptic hypothalamic and limbic structures. They point out how conservative these hormone-concentrating areas are by stating that fish, amphibians, birds, and various mammalian species also have them. Squirrel monkeys and tree shrews also have these hormone-concentrating areas (Stumpf et al., 1974).

The hypothalamic feedback system secreting leutenizing hormone (LH) in response to estradiol (estrogen) differs in male and female primates. Male primates are more sensitive to estradiol since it leads to a greater suppression of LH in males (Resko, 1975). In males, LH is called the interstitial cell stimulating hormone (ICSH) and it does just that − stimulates the interstitial cells of the testes to produce more hormone (testosterone, primarily). With increased titers of estradiol in the male, therefore, there is a diminution of androgen which results from the reduction of LH (ICSH) release. In the female, estradiol-17β acts directly on the pituitary to *inhibit* LH release and paradoxically on the brain to *facilitate* LH release (Spies and Norman, 1975).

Steiner et al. (1974) also point out that the hypothalamo-hypophyseal axis of males and females differs in sensitivity to the feedback action of estradiol. They believe that the difference is a function of the prenatal hormonal environment.

As we have seen, penile erection (as a display) has been evoked in the squirrel monkey by brain stimulation in the glubus pallidus. And as we have seen above, ejaculation has been evoked in rhesus monkeys. The highest probability of evoked penile erection from the rhesus was found in the preoptic area whereas the greatest proportion of all responses evoked was in the anterior cingulate region and the putamen (Robinson and Mishkin, 1968).

When the ventricular system of the brain of rhesus monkeys is infused with radioactively labelled testosterone, "uptake of labelled testosterone appeared to be greater in the pituitary and hypothalamus than in other areas of the brain." (P. 537.)

The hypothalamus, therefore, is strongly implicated in studies of sex differences in the brain. It seems to be organized differently depending upon the prenatal hormonal environment. It also appears to show sex differences in cyclicity (Valenstein, 1968). *Tonic* gonadotrophic release can be maintained by the medial basal hypothalamus alone, but connections with the anterior hypothalamus are necessary for cyclicity. According to Valenstein, one of the functions of early exposure to androgen is to disrupt this connection between the anterior hypothalamus and the medial basal hypothalamus.

In addition to testosterone and estrogen, progesterone is also concentrated by neurons of the hypothalamus. This suggests a direct action of progesterone on certain hypothalamic structures (Sar and Stumpf, 1973). Tabei and Heinrichs (1974) have made a study of the metabolism of progesterone by the brain and pituitary gland in baboon and rhesus males. The metabolic action of this hormone in the hypothalamus and pituitary did *not* change when males reached sexual maturity. Females were not studied.

In conclusion, as far as nonhuman primates are concerned, "the notion that the primates brain is emancipated from the influence of gonadal hormones has been overestimated." (Michael, 1965, P. 4.) However, as June Reinisch (1974) has so ably stated, "a fetal hormonal effect is influential in the subsequent development of sex differences, but in most cases its effects must be seen as limited in scope and of a diffuse quality rather than directly related to specific behaviors . . . the influence of the prenatal hormones is best understood as setting a bias on the neural substratum, which in turn predisposes the individual to the acquisition and expression of sexually dimorphic patterns of response and behavior." (P. 51.)

With neurological-hormonal sex differences being as complicated as they are, the finding of Zuckerman and Fisher (1938) that the adult male rhesus brain weighed an average of 89.45 grams and the adult female rhesus brain only 82.03 grams becomes relatively meaningless without some statement concerning the relative internal hormonal/neurophysiological workings of the respective organs.

While we have concentrated primarily upon the hypothalamus and the gonadal hormones in this section we did so primarily for the continuity this emphasis would provide from the earlier part of the chapter on the effects of prenatal testosterone. We do not wish to imply that these are the only neural factors of importance to sex differences. Spinal and vascular aspects, higher cerebral mechanisms, the limbic system, and neurotransmitters are also involved in sex differences in behavior (cf. Karczmar, 1975). Moreover, we have not even begun to cover, in all of its complexity, the topic of sex hormones in the brain. We only hope that the reader will get some idea of the differences that exist and the degree of complexity involved. For a more complete coverage of sex hormone receptors in the brain the reader may consult Pfaff (1976).

F. HUMAN IMPLICATIONS

As we have seen in the neurological data, the parts of the nervous system that are most implicated in matters of sex hormones and sex differences are portions of an old brain. Hormone-concentrating areas are essentially the same for fish, amphibians, birds, mammals, tree shrews, squirrel monkeys, macaques, and baboons. These areas in the human brain do not differ much from the same areas in the rhesus monkey brain. It is quite likely that much of the same neurohormonal research would apply to people if prenatal differentiation occurred in the same way in humans.

As far as human prenatal differentiation of the two sexes is concerned, we have, at best, only indirect evidence. For example, after the first two years of life, it is very difficult to change the psychological sexual identification of children, who, because they had atypical genitals were reared as the other sex would be reared. Prenatal and perhaps perinatal hormones apparently preset neural control mechanisms in the direction of sensitizing such persons to adopt cross-sex patterns (Eichorn, 1968). By age two, children know if they are male or female (Green, 1975).

Also in humans, there is a rare condition in which there is a malfunction of the adrenal glands causing an overproduction of androgens. The condition, known as the adrenogenital syndrome (AGS), is not recognized until after birth, although from birth on the chemical imbalance can be treated. Yet, due to the prenatal exposure to high levels of masculinizing androgens, females with AGS are born with masculinized genitalia. Ehrhardt (1974) studied seventeen female AGS patients and reported that:

> Fetally androgenized females were more often long-term "tomboys" showing intense energy expenditures in outdoor activities and low interest in parenting behavior. Yet, normal gender identification as a female seemed unaffected. The effects of prenatal androgen on human behavior in genetic females are modifications in *temperament* within a spectrum of acceptable feminine behavior. (P. 154)

A similar situation is also found among female fetuses whose mothers were given progestin in order to prevent miscarriage. Progestin was later found to have androgenizing effects. The "tomboy" results were found in ten young girls whose mothers were administered progestin during their pregnancies (Hamburg and Brodie, 1973). Such findings suggest that androgens may play a role in humans that is similar to the one they play in rhesus. An excellent book to read on case histories involving AGS and other sexual identity problems is Green (1974).

Harper and Sanders (1977) have pointed out that many students of human behavior have been reluctant to accept comparative evidence of an organismic contribution to the development of gender-typical activities in children. They emphasize that in people, just as in other animals, the development of behavioral sex differences depends upon *both* organismic and environmental factors.

Still, the number of species of primates that has been studied with regard to prenatal androgen is small. Almost all of the long-term research of relevance to this area has been done using the rhesus monkey, a species showing strong sexual dimorphism. In comparative studies of hormonal patterns, chimpanzees and humans, but *not* rhesus, have a "definitive feto placental unit" (i.e., there is a lack of serum estriol in the rhesus from the fetus and placenta). The levels of hormones are more alike in chimpanzees and humans because they have a prolonged LH peak not seen in the rhesus (Hobson et al., 1976). As a result, progesterone also increases remarkably in pregnant humans and chimpanzees during gestation but in macaques both estrogen and progesterone show only slight increases (Nomura and Ohsawa, 1976). At 129-145 days of gestation, the levels of testosterone in the rhesus male fetus are two to three times higher than for the human male fetus in the second trimester (Huhtaniemi et al., 1977). In addition, the concentration of testosterone in the adult rhesus male is 50 percent greater then in the adult male human. (The level in the infant rhesus male is about 1/100 the adult rhesus level.) Still *higher* than the rhesus in testosterone concentration are spider monkeys, squirrel monkeys, crab-eating macaques, capuchins, and baboons (Snipes, Forest, and Migeon, 1969).

Lanman (1977) in a paper on hormones of parturition in primates states that:

> . . . the macaque, at least as evidenced by the patterns of pregnancy hormones, does not appear to be the best animal. By these limited criteria, either the chimpanzee or marmoset is closer to the human. (P. 36)

The point is that facile generalization from the rhesus prenatal studies is *still* premature. If there is anything a primatologist knows for sure, it is that the 200 or more species of primates differ from one another *considerably*. Why should prenatal development be the one thing that does not vary?

On the other hand, there are the AGS data on humans; and, injections of human hormones (HMG) into rhesus females *do* result in perineal sexual skin development and in rhesus ovaries that are capable of increased steroid secretion and ovulation (Weiss, Rifkin, and Atkinson, 1976).

There are also other similarities between human and nonhuman primate hormonal response systems, although chimpanzees may be a better model for *Homo sapiens* than the rhesus. Serum FSH (follicle stimulating hormone) levels are higher in girls and immature female chimps than in boys and immature male chimps during infancy. The day-to-day variability for both FSH and LH is also higher in female infants (for both chimps and humans). In addition, the FSH levels for infant female (but not male) chimpanzees and humans show periodic variation with cycle periods ranging from 8.3 to 25.0 days (Faiman, Winter and Grotts, 1973).

But, despite such similarities, statements such as the following do *not* seem justified:

> Male and female roles, although overlapping, are biologically, hormonally, basically different from the start. No matter what the human primate may say or do females are born to be mothers. This is their biological evolutionary function. (Poirier, 1972, P. 12)

Nonetheless, the biological data that are available from primatology, endocrinology, neurology, and psychology do seem to support the overall conclusion which appeared at the end of a recent review of the area by Gadpaille (1972):

The research reviewed corroborates the commonsense folk knowledge that there are indeed innate psychological differences between normal men and women, that cultural and environmental influences do not and cannot operate as though such psychosexual dimorphism is nonexistent, and that psychosexual differentiation and the psychodynamics of development cannot usefully be comprehended except by taking into account the different bodies in which each sex, from conception forward, grows up. (P. 204)

Those readers wishing to read more in this area are referred to Hamburg (1974) and Hammer (1975).

REFERENCES

Bernstein, I. S. Behavioral and environmental events influencing primate testosterone levels. *Journal of Human Evolution,* 1974, **3**, 517-525.

Bielert, C. Social experience, hormones interact to shape sexual growth of male rhesus. *Primate Record,* 1975-76, **6** (2), 3-5.

Brotherton, J. Animal experiments with anti-androgens. *Bibliography of Reproduction,* 1973, **21**, 173-175.

Bubenick, G. A. and Brown, G. M. Morphologic sex differences in the primate brain areas involved in regulation of reproductive activity. *Experientia,* 1973, **15** (5), 619-621.

De Rousseau, C. J. Variability of accessory sex structures in *Macaca mulatta.* *American Journal of Physical Anthropology,* 1974, **41**, 475.

Ehrhardt, A. A. Androgens in prenatal development: Behavior changes in nonhuman primates and man. *Advances in the Biosciences,* 1974, **13**, 154-162.

Eichorn, D. H. Biology of gestation and infancy: Fatherland and frontier. *Merrill-Palmer Quarterly,* 1968, **14** (1), 47-81.

Faiman, C., Winter, J. S. D. and Grotts, D. Gonadotropins in the infant chimpanzee: A sex difference. *Proceedings of the Society for Experimental Biology and Medicine,* 1973, **144**, 952-955.

Foster, D. L. Endocrine development of the hypothalamohypophyseal-gonadal system of the nonprimate and subhuman primate female. *Third Ross Conference on Obstetric Research,* Columbus, Ohio: Ross Laboratories, 1974, pp. 23-33.

Gadpaille, W. J. Research into the physiology of maleness and femaleness: Its contributions to the etiology and psychodynamics of homosexuality. *Archives of General Psychiatry,* 1972, **26**, 193-206.

Goldfoot, D. A. Social and hormonal regulation of gender role development in rhesus monkey. Paper presented at the *International Primatological Society Meeting*, Cambridge, England, August, 1976.

Goy, R. W. Role of androgens in the establishment and regulation of behavioral sex differences in mammals. *Journal of Animal Science*, 1966, **25**, 21–35.

Goy, R. W. Hormonal and environmental influences on sexual behaviour in rhesus monkeys. Paper presented at the *International Primatological Society Meeting*, Cambridge, England, August, 1976.

Goy, R. W. and Goldfoot, D. A. Neuroendocrinology: Animal models and problems of human sexuality. *Archives of Sexual Behavior*, 1975, **4** (4), 405–420.

Goy, R. W. and Phoenix, C. The effects of testosterone administered before birth on the development of behavior in genetic female rhesus monkeys. In Sawyer, C. and Gorski, R. (Eds.) *Steroid hormones and brain function.* Berkeley, California: University of California Press, 1971, pp. 193–201.

Goy, R. W. and Resko, J. A. Gonadal hormones and behavior of normal and pseudohermaphroditic nonhuman female primates. *Recent Progress in Hormone Research,* 1972, **28**, 707–733.

Goy, R. W., Slimp, J. and Wallen, K. Endocrine aspects of sexually dimorphic behavior. *Acta Endocrinologica,* 1975, **199**, 432.

Goy, R. W., Wallen, K. and Goldfoot, D. A. Social factors affecting the development of mounting behavior in male rhesus monkeys. In Montagna, W. and Sadler, W. A. (Eds.) *Reproductive Behavior.* New York: Plenum, 1974, pp. 223–247.

Green, R. *Sexual Identity Conflict in Children and Adults.* Baltimore: Penguin Books, 1974.

Green, R. Sexual identity: Research strategies. *Archives of Sexual Behavior,* 1975, **4** (4), 337–352.

Hagemenas, F. C. and Kittinger, G. W. The influence of fetal sex on plasma progesterone levels. *Endocrinology,* 1972, **91**, 253–256.

Hagino, N. Hypothalamic regulation of ovulation and menstruation in the baboon (nonhuman primates). *Federation Proceedings,* 1976, **35**, 701.

Hamburg, D. A. The psychobiology of sex differences: An evolutionary perspective. In Friedman, R. C., Richart, R. M. and van de Wiele, R. L. (Eds.) *Sex Differences in Behavior.* New York: John Wiley, 1974, pp. 373–392.

Hamburg, D. A. and Brodie, H. K. H. Psychological research on human aggressiveness. *Impact of Science on Society,* 1973, **23**, 181–193.

Hamburg, D. A. and Lunde, D. T. Sex hormones in the development of sex differences in human behavior. In Maccoby, E. E. (Ed.) *The Development of Sex Differences.* Stanford, California: Stanford University Press, 1966, pp. 1–24.

Hammer, S. (Ed.) *Women, Body, and Culture.* New York: Harper & Row, 1975.

Harper, L. V. and Sanders, K. M. Preschool children's adult- and peer-contacts: Relationships between sex and setting. Paper presented at the *Animal Behavior Society* Meeting, University Park, Pennsylvania, May, 1977.

Herbert, J. Hormonal basis of sex differences in rats, monkeys, and humans. *New Scientist,* 1976, **70**, 284-286.

Herbst, A. L., Taft, P. D. and Robboy, S. J. Nuclear chromatin determination in amniotic fluid cells for prenatal sex prediction in the macaque. *American Journal of Obstetrics and Gynecology,* 1976, **124**, 761-768.

Hobson, W., Coulston, F., Faiman, C., Winter, J. and Reyes, F. Reproductive endocrinology of female chimpanzees: A suitable model of humans. *Journal of Toxicology and Environmental Health,* 1976, **1**, 657-668.

Huhtaniemi, I. T., Korenbrot, C. C., Serón-Ferré, M., Foster, D. B. Parer, J. T. and Jaffe, R. B. Stimulation of testosterone production *in vivo* and *in vitro* in the male rhesus monkey fetus in late gestation. *Endocrinology,* 1977, **100**, 839-844.

Jones, K. L., Shainberg, L. W. and Byer, C. O. *Sex and people,* New York: Harper and Row, 1977.

Joslyn, W. D. Androgen-induced social dominance in infant female rhesus monkeys. *Journal of Child Psychology and Psychiatry,* 1973, **14**, 137-145.

Karczmar, A. G. Neurotransmitters in the modulation of sexual behavior. *Psychopharmacology Bulletin,* 1975, **11**, 40-42.

Keefer, D. A. and Stumpf, W. E. Estrogen-concentrating neuron systems in the brain of the tree shrew. *General and Comparative Endocrinology,* 1975, **26** (4), 504-516. (a)

Keefer, D. A. and Stumpf, W. E. Atlas of estrogen-concentrating cells in the central nervous system of the squirrel monkey. *Journal of Comparative Neurology,* 1975, **160** (4), 419-441. (b)

Kimble, D. P. *Psychology as a Biological Science.* Pacific Palisades, California: Goodyear Publishing Company, 1973.

Kling, A. Differential effects of amygdalectomy in male and female nonhuman primates. *Archives of Sexual Behavior,* 1974, **3** (2), 129-134.

Kling, A. Brain lesions and aggressive behavior of monkeys in free living groups. In Fields, W. S. and Sweet, W. H. (Eds.) *Neural Basis of Violence and Aggression.* St. Louis, Missouri: Warren H. Green, Inc., 1975, pp. 146-160.

Kling, A. and Dunne, K. Social-environmental factors affecting behavior and plasma testosterone in normal and amygdala lesioned *Macaca speciosa. Primates,* 1976, **17** (1), 23-42.

Lanman, J. T. Parturition in nonhuman primates. *Biology of Reproduction,* 1977, **16**, 28-38.

Levine, S. Sex differences in the brain. *Scientific American,* 1966, **214** (4), 84-90.

Livrea, G. La base neuroendocrina del comportamento sessuale. *Archives of Scientific Biology,* 1972, **55**, 61–102.

MacLean, P. Role of pallidal projections in species-typical display behavior of squirrel monkey. *Transactions of the American Neurological Association,* 1975, **100**, 25–28.

Maurus, M. and Ploog, D. Social signals in squirrel monkeys: Analysis by cerebral radio stimulation. *Experimental Brain Research,* 1971, **12**, 171–183.

Michael, R. P. Some aspects of the endocrine control of sexual activity in primates. *Proceedings of the Royal Society of Medicine,* 1965, **58**, 595–598.

Neuman, F., Steinbeck, H. and Hahn, J. D. Hormones and brain differentiation. In Martini, L., Motta, M. and Fraschini, F. *The Hypothalamus.* New York: Academic Press, 1970, pp. 1–35.

Nomura, T. and Ohsawa, N. The use and problems associated with nonhuman primates in the study of reproduction. In Antikatzides, T., Erichsen, S. and Spiegel, A. (Eds.) *The Laboratory Animal in the Study of Reproduction.* New York: Gustav-Fischer, 1976, pp. 1–16.

Perachio, A. A. Hypothalamic regulation of behavioural and hormonal aspects of aggressive and sexual performance. Paper presented at the *International Primatological Society* Meeting, Cambridge, England, August, 1976.

Pfaff, D. W. The neuroanatomy of sex hormone receptors in the vertebrate brain. In Kumar, T. C. A. (Ed.) *Neuroendocrine Regulation of Fertility.* Basel: Karger, 1976, pp. 30–45.

Pfaff, D. W., Gerlach, J. L., McEwen, B. S., Ferin, M., Carmel, P. and Zimmerman, E. A. Autoradiographic localization of hormone-concentrating cells in the brain of the female rhesus monkey. *Journal of Comparative Neurology,* 1976, **170**, 279–293.

Pfeiffer, C. A. Sexual differences of the hypophyses and their determination of the gonads. *American Journal of Anatomy,* 1936, **58**, 195–226.

Phoenix, C. H. Prenatal testosterone in the nonhuman primate and its consequences for behavior. In Friedman, R. C., Richart, R. M. and van de Wiele, R. L. (Eds.) *Sex Differences in Behavior.* New York: Wiley, 1974, pp. 19–32.

Phoenix, C. H., Goy, R. W. and Resko, J. A. Psychosexual differentiation as a function of androgenic stimulation. In Diamond, M. (Ed.) *Reproduction and Sexual Behavior.* Bloomington, Indiana: University of Indiana Press, 1968, pp. 33–49.

Poirier, F. E. Primate socialization and learning. Paper presented at the Joint Meeting of the *American Ethnological Society, Society for Applied Anthropology,* and *Council on Anthropology and Education,* Montreal, April, 1972.

Reinisch, J. M. Fetal hormones, the brain, and human sex differences: A heuristic, integrative review of the recent literature. *Archives of Sexual Behavior,* 1974, **3**, 51–90.

Resko, J. A. Sex steroids in the circulation of the fetal and neonatal rhesus monkey: A comparison between male and female fetuses. *International Symposium on Sexual Endocrinology of the Prenatal Period*, 1974, **32**, 195-204. (a)

Resko, J. A. The relationship between fetal hormones and the differentiation of the central nervous system in primates. In Montagna, W. and Sadler, W. A. (Eds.) *Reproductive Behavior*. New York: Plenum, 1974, pp. 211-222. (b)

Resko, J. A. Fetal hormones and their effect on the differentiation of the central nervous system. *Federation Proceedings,* 1975, **34** (8), 1650-1655.

Robinson, B. W. and Mishkin, M. Penile erection evoked from forebrain structures in *Macaca mulatta*. *Archives of Neurology*, 1968, **19**, 184-198.

Rowell, T. E. *The Social Behaviour of Monkeys*. Baltimore: Penguin, 1972.

Sar, M. and Stumpf, W. E. Neurons of the hypothalamus concentrate [^{3}H^1] progesterone or its metabolites. *Science,* 1973, **183**, 1266-1268.

Sholiton, L. J., Taylor, B. B. and Lewis, H. P. The uptake and metabolism of labelled testosterone by the brain and pituitary of the male rhesus monkey (*Macaca mulatta*). *Steroids,* 1974, **24** (4), 537-547.

Snipes, C. A., Forest, M. G. and Migeon, C. J. Plasma androgen concentration in in several species of Old and New World monkeys. *Endocrinology,* 1969, **85** (5), 941-945.

Spies, H. G. and Norman, R. L. Interaction of estradiol and LHRH on LH release in rhesus females: Evidence for a neural site of action. *Endocrinology,* 1975, **97** (3), 685-692.

Steiner, R. A., Clifton, D. K., Spies, H. G. and Resko, J. A. Feedback control of LH by estradiol in female, male, and female pseudohermaphroditic rhesus monkeys. *Endocrinology,* 1974, **94** (Suppl.), 1-195.

Stumpf, W. E., Sar, M. and Keefer, D. A. Anatomical distribution of estrogen in the central nervous system of mouse, rat, tree shrew, and squirrel monkey. *Advances in Bioscience,* 1974, **15**, 77-88.

Sullivan, D. J. and Drobeck, H. P. True hermaphrodism in a rhesus monkey. *Folia Primatologica,* 1966, **4**, 309-317.

Tabei, T. and Heinrichs, W. L. Metabolism of progesterone by the brain and pituitary gland in subhuman primates. *Neuroendocrinology,* 1974, **15**, 281-289.

Valenstein, E. S. Steroid hormones and the neuropsychology of development. In Isaacson, R. L. (Ed.) *The Neuropsychology of Development*. New York: Wiley, 1968, pp. 1-39.

Weiss, G., Rifkin, I. and Atkinson, L. E. Induction of ovulation in premenarchial rhesus monkeys with human gonadotropins. *Biology of Reproduction,* 1976, **14**, 401-404.

Zuckerman, S. and Fisher, R. B. Growth of the brain in the rhesus monkey. *Proceedings of the Zoological Society of London,* 1938, **107**, 529-538.

5.
The Development of Play

We have seen in the previous two chapters that male and female primates often differ physically and that these physical differences are correlated with differences in behavior. The early work of Harry F. Harlow and collaborators at the University of Wisconsin (cf. Hansen, 1966; Harlow and Lauersdorf, 1974) had demonstrated that normal male rhesus monkey infants engaged in more rough-and-tumble play and initiated more play than did female infants. Goy's prenatally androgenized pseudohermaphroditic females displayed these behaviors at a level that was intermediate to the normal male and female levels (cf. Goy, 1966). Clearly, play behavior is one form of behavior in which sex differences can be seen (at least in the rhesus monkey). To what extent can these sex differences in infant and juvenile play frequencies and patterns be generalized to the rest of the primate order?

In order to answer this question we will review the literature on play behavior using a taxonomic format.

A. PROSIMIANS

Prosimian primates have not been studied as extensively as have Old World monkeys such as the rhesus. However, there have been a few published studies which have made reference to the play patterns of prosimians.

The slow loris (*Nycticebus coucang*), which occupies a nocturnal, arboreal niche and is indigenous to Southeast Asia and Indonesia, has been studied by Lockhard, Heestand, and Begert (1977). These researchers attempted to construct an ethogram of play behaviors in three juveniles (one male and two females). They observed play initiation, grappling, environmental play, and self play but reported no sex differences. In the same paper, reporting research done by

Charles-Dominique, Lockard et al. also described grappling in *Perodicticus potto* adults. The grappling is regarded as an *agonistic* wrestling match when it occurs between same sex adults but affiliative play when it occurs between two members of opposite-sexed adults prior to copulation. Thus, play is seen at least in adults in these prosimians; *and,* the gender of the animals involved in a play bout *does* make a difference. Surprisingly, Ehrlich and Musicant (1977) have reported that social play among slow lorises is *primarily* an adult activity. Social play among adults occurs often in *Nycticebus coucang, Perodicticus potto,* and *Propithecus verreauxi.*

Among bushbabies (*Galago senegalensis*) where there is a size difference between males and females, it was noted by Sauer and Sauer (1962) that the adult male played with the infant. Flinn and Nash (1975) noted that, as in the slow loris, interactions between same-sexed individuals were often agonistic, whereas interactions between opposite-sexed individuals were rarely agonistic and often included play. Very little else has been published on prosimian play.

B. NEW WORLD MONKEYS

The most primitive of the New World monkeys, the marmosets, show very little sexual dimorphism. As we have seen, in at least one species of marmoset, females are larger than males (Ralls, 1976). Juvenile marmosets have been seen to copulate before reaching maturity. However, the immature sexual behavior shown by these juveniles differs from the mature form. Unlike adult copulation, juvenile copulation is closely related to play. Juvenile males initiate play by encouraging contact while females initiate play by avoiding contact but encouraging pursuit by the male. Abbott (1976) believes that, even in these relatively monomorphic marmosets, an early sex bias in play may accompany development of the appropriate mature sexual behavior pattern. Unlike the slow lorises described by Ehrlich and Musicant (1977), however, marmoset play behavior occurs more frequently in juveniles and subadults than it does in adults (Ingram, 1976).

In an excellent review of the correlates of monogamy and monomorphism, Devra Kleiman (1977) points out that, as adults, marmoset and tamarin males do not get very involved in play with the young, although they carry and groom the young frequently. She asserts that with an increasing tendency toward monogamy and male parental investment there is a trend toward less and less male play.

As for more advanced New World monkeys, juvenile howler monkeys (*Alouatta palliata*) play actively and frequently while the play of older howlers is gentle. The mother and infant rarely play together, but when they do the style of their play is similar to that of the gentle play of older howlers (Baldwin and Baldwin, 1973).

As in some adult prosimians and in subadult marmosets, sexual play is seen in squirrel monkeys (*Saimiri sciureus*). In this sexual play, the adult or subadult male acts as pursuer and the adult or subadult female flees but provokes his further pursuit. The male usually initiates the interaction by approaching, while the female determines what course the interaction will follow (e.g., retreating play, or mating) (Latta, Hopf, and Ploog, 1967). Rosenblum and Lowe (1971) have shown that either age or familiarity affects partner preference and sex differences in play in squirrel monkeys. If there are familiar partners, female squirrel monkeys play less than do males.

As we can see above, play behavior in New World monkeys, as in prosimians, appears to be different in the two sexes; *and* play seems to be important in the development of sexual behavior.

C. OLD WORLD MONKEYS

In this section, we will omit macaques and baboons as they will be reviewed in the section which follows. Here, we will cover the following genera: *Cercopithecus, Miopithecus, Erythrocebus,* and *Presbytis.*

Among the species of the genus *Cercopithecus,* males tend to play more than do females (Bramblett, 1973). Bourliére et al. (1969) reported that while adult males of the species *C. campbelli* do not show as great an interest in infants as do females, they do permit the infants close to them when playing. In addition, he reports that members of the same species (Lowe's guenon) played with a pet mongoose.

Dolan (1976) has reported that subadult and juvenile males accounted for 65 percent of all play observed in a captive group of Syke's monkeys (*C. albogularis*). Dolan's juvenile females played less frequently than males that were both younger and older than themselves. Bramblett (1977) has reported similar sex differences in juveniles of this genus.

While females of the genus *Cercopithecus* may not play as often as males, it has been said that adult females serve to *reinforce* the play of immature animals. Adult males, on the other hand, (of the species *C. aethiops*) engage in rough play with older male juveniles. Males, in general, prefer to play within their own peer groups; however, older juvenile males as well as subadult females stimulate play in young infants (Fedigan, 1972). Subadults and juveniles are the most active, inquisitive, and playful members in Lowe's guenon (*Cercopithecus campbelli loweii*) (Hunkeler, Bourliére, and Bertrand, 1972).

There is some evidence that in reinforcing and stimulating play in young infants, females of *C. aethiops* are getting practice in the maternal role. Lancaster (1971) makes the point that there may be a variety of play behaviors and that "play mothering" may be but one variety. Males of *C. aethiops* show much less frequent and more variable play mothering than do the females. Male juveniles and subadults, in particular, are more solitary and initiate more chase and contact play. These sex differences in play mothering and contact play are seen in captivity as well as in the wild (Raleigh, 1977; Lancaster, 1971). Raleigh (1977), however, cautions us that cross-specific generalizations about sex differences in behavior are premature and that sex differences should always be viewed within the context of a group's social organization.

Studies on the talapoin (*Miopithecus talapoin*), a primate closely related to the genus *Cercopithecus*, show us that Raleigh's warning should be taken seriously. While Segre (1970) has reported that "each female talapoin has more in common behaviorally with her other female age-mates than with males or juveniles" and, while adult females *are* more infant oriented than are males (Segre, 1970), Wolfheim (1977) has reported that juvenile sex differences in this species are *not* always correlated with adult sex differences. *Juvenile* male talapoins in captivity are more active, assertive, and playful than are juvenile females, but *adult* male talapoins are *less* aggressive, more withdrawn, and *less* active than are adult females. However, talapoins also show less physical dimorphism than most of the vervets.

One other closely related genus should be discussed at this point, the patas monkey (*Erythrocebus patas*). The patas shows substantial physical sexual dimorphism and is predominantly terrestrial. Young males, as we might expect, play more often and longer than do females (Bramblett, 1973). Infant social play in the patas begins at two weeks of age and includes grappling, wrestling, play-biting, and jumping. By five months of age infants spend most of their time in play groups. As adults, however, and in intergroup squabbles at least, females, subadults, and juveniles show more aggression than do the males (Gartlan, 1976, film). Thus, sex differences in Old World monkeys are not always as predictable as some writers would lead us to believe.

The last genus we have selected to discuss in the present section is *Presbytis* (the langurs of Asia). In the lutong (*Presbytis cristatus*), as in most adult males of the genus *Cercopithecus*, the adult male seldom carries infants while adult females do. When a male is seen carrying an infant, it is usually the immature animal which initiates the interaction (the male remaining passive). In this species, juvenile males (more than juvenile females) are involved in the most vigorous play fighting. They sometimes play with both subadult and adult males (Bernstein, 1968). As Dolhinow and Bishop (1970) have reported, at eight months of age male langurs (*Presbytis entellus*) are already larger and stronger than are females. The more active running and wrestling of langur male infants brings them into contact with adult males more than does the quieter play of the female infants. Group play among large juveniles is almost completely in like-sex groups. According to Dolhinow and Bishop (1970), adult male langurs

(*Presbytis entellus*) sometimes, but seldom, play, whereas adult females *never* play.

Play is one of the first nonmother directed activities to develop, and the sex difference develops early in Nilgiri langurs (Poirier, 1972). In this species (*P. johnii*) the differences in play are both quantitative and qualitative. Males play more frequently and rougher than do females. Thus, among most Old World monkeys other than macaques and baboons, males are more playful than are females (see also Nagel and Kummer, 1974).

D. MACAQUE AND BABOON PLAY

1. *Macaca mulatta*[1]

As we have already mentioned several times, rhesus macaque males play more frequently and rougher than do rhesus macaque females (cf. Harlow and Lauersdorf, 1974). Social play in the rhesus monkey (*Macaca mulatta*) begins at about one month of age, is initially of the "rough-and-tumble" variety (Hinde and Spencer-Booth, 1967), and usually involves two or three individuals of similar age (Kaufman, 1966; Loy and Loy, 1974).

Infants (0-1 year), juveniles (1-2 years), and preadolescents/subadults (2-4 years) all engage in play. Juvenile play, the most frequent, consists of chasing, wrestling, and approach/withdrawal (Hinde and Spencer-Booth, 1967). Juvenile play occasionally occurs in groups of four to ten animals (Southwick, Beg, and Siddiqi, 1965). Rhesus infants and juveniles frequently play with their siblings (Lindburg, 1971; Loy and Loy, 1974).

Very early in life, male infants begin to initiate more play than do females and the play of males is more likely to be rough-and-tumble (Hansen, 1966; Seay, 1966). One-year-old males, between infancy and the juvenile period, play the most often, followed by two-year-old males, and one-year-old females (Hinde and Spencer-Booth, 1967; Loy and Loy, 1974). Two-year-old females are often busy play mothering (Chamove, Harlow, and Mitchell, 1967; Hinde, Rowell, and Spencer-Booth, 1964). The removal of adults can increase juvenile play as Bernstein and Draper (1964) found when all of their rhesus adults were removed from an enclosure. Apparently adults control the levels of play exhibited by immature monkeys..

In nature adult rhesus monkeys play far less frequently than do immature animals (Lindburg, 1971), but they *do* play (Breuggeman, 1977). Also in the wild, mothers appear to play with infants (more often with female infants) somewhat more than do adult males (Lindburg, 1971; Lichstein, 1973). However,

[1]I am indebted to Nancy Caine for help in the review of research on the macaque genus.

when mothers are removed, adult males increase their interactions with infants and much of this increase is in play (Spencer-Booth and Hinde, 1967; Redican and Mitchell, 1974). When given the opportunity, adult males play more with male infants than with female infants. Two adults (of either sex) have not been seen playing together (Lindburg, 1971; Southwick et al., 1965; Hinde and Spencer-Booth, 1967). If adult–adult play does occur it must happen infrequently.

These rhesus sex differences in play seem to be remarkably robust. They have been reported under natural conditions, in captivity, and under various degrees of social deprivation. Anderson and Mason (1974), for example, compared groups of socially deprived rhesus to socially experienced rhesus at eleven months and found that, while rough-and-tumble play was significantly higher in the experienced group, in *both* groups males more often engaged in rough-and-tumble play than did females and, in both groups, males responded to the other two participants in a triad in more of the triadic interactions than did females. Similarly, Brandt and Mitchell (1973) found that male rhesus infants, when paired with preadolescents, both elicited and emitted more play than did female infants. Males in a captive "nuclear-family" environment also played more than did females (Ruppenthal et al., 1974). In this same nuclear-family environment, adult males, but not adult females, reciprocated play with infants. Male infants played more with older infants and adult males, whereas female infants played more with younger infants (Suomi, 1972).

In studying the reactions of 1½ year old rhesus juveniles to both familiar and unfamiliar peers, Hansen, Harlow, and Dodsworth (1966) found that females exhibited less contact play and less noncontact play than did males under both conditions. Even when male rhesus are gonadectomized before puberty (or near puberty), they continue to play more than do females (Loy et al., 1977). And, in the original Wisconsin studies, where rhesus were reared in social isolation, even two-year-old social isolate males played more than did their female counterparts (Mason, 1960; see also Mitchell, 1968).

Obviously, in the rhesus at least, sex differences in play are to a large extent the result of behavioral predispositions induced by prenatal hormones, and these differences are remarkably resistant to, though not immune from, the effects of social learning (cf. Goldfoot, 1976). The sex difference in play even extends into adulthood (Mitchell, 1977).

2. Macaca radiata

As in the rhesus monkey, play in the bonnet monkey (*M. radiata*) develops very early in life. Four or more individuals are often involved in play groups involving chasing and grappling. Violent wrestling is usually confined to juvenile, subadult, and adult males. Less vigorous wrestling is seen when adult males play with immature males (Kaufman and Rosenblum, 1969; Nolte, 1955; Simonds, 1965; Sugiyama, 1971).

Sex differences are apparent as early as two months of age (Simonds, 1974), with same-sexed play predominating. If the only available playmates are of a different sex, sex differences do not appear to be as great. Adult females seldom play but, unlike rhesus males, adult male bonnet macaques play regularly and with all age groups, even with other adult males (Simonds, 1965; 1974). According to Rosenblum and Bromley (1976), however, the administration of gonadal hormones prior to adulthood decreases bonnet play behavior and activity levels drastically.

3. *Macaca sylvana*

The Barbary macaque (*Macaca sylvana*) is less like the rhesus than is the bonnet, yet evidence for a rhesuslike sex difference in play still appears in this species. Burton (1972) reports that Barbary macaques do not begin to engage in social play until three months of age, some two months later than rhesus. Barbary macaque infants prefer age-mates over other-aged playmates (MacRoberts, 1970).

Again, unlike rhesus and bonnets, sex differences in play do not appear until the juvenile stage (after one year of age) at which time males are more active and are more successful at initiating play bouts than are females (MacRoberts, 1970; Burton, 1972).

Contact between adult males and younger animals is frequent (Deag and Crook, 1971) but very little of it is play except in exceptional adult males. Adult females have never been seen playing (MacRoberts, 1970; Burton, 1972).

4. *Macaca nemestrina*

Pigtailed macaques (*M. nemestrina*) display a sex difference similar to that seen in the rhesus. Bernstein (1967, 1970, 1972) has reported that all age classes of pigtailed macaques play but that play is most common among juvenile males. Pigtails probably engage in less *social* play, but more *nonsocial* play than do bonnet macaques. The total percentage of time spent playing is greater in one-year-old bonnets than it is in one-year-old pigtails (Kaufman and Rosenblum, 1969). Kuyk, Dazey, and Erwin (1977), studying only female infants, found that overall play increased at least up to two months but that peer play increased between two and three months. Play with inanimate objects increased throughout this period.

5. Other Macaques

Crab-eating macaques (*M. fascicularis*) as well as stumptail macaques (*M. arctoides*) have been studied by Stevens (1967). The infant crab-eaters in her study preferred other infants as playmates although they also played with juvenile males. She observed no female juveniles playing with infants, however. Fady (1969) also found that crab-eater infants preferred peers to older animals, and that male play was rougher than female play. Adult males played with infants but adult females did not. Kurland (1973) saw no adult play in crab-eaters. Seay, Schlottman, and Gandolfo (1972) compared crab-eaters (*M. cynomologous*) to rhesus. Males of both species in their study engaged in more contact play than did the females but "behavioral sex differences are *more* evident among cynomologous babies" (p. 41, emphasis mine) than among rhesus.

In the study by Stevens (1967) alluded to above, stumptail macaque (*M. arctoides*) play was found to be similar to crab-eating macaque play in that infants preferred peers, and males played more than did females. However, there was a still higher incidence of male juvenile–infant play in the stumptails than in the crab-eaters. Again, juvenile female–infant play was nonexistent and there was no adult female reciprocation of infant-initiated play. In contrast to her crab-eater group, Stevens' stumptail group showed no adult male play. However, Bertrand (1969) has seen subadult and adult male play in stumptail macaques.

Among *Macaca fuscata* (the Japanese macaque), the infants initially show no preference for the sex of the playmate but, as the infants grow older, a preference for same-sexed peers develops (Imanishi, 1963). Japanese macaques play near their mothers and with peers until three years of age at which time they begin spending more time near adult males. By age four, 40 percent of their play is with *adult males*. Adult male play with infants and juveniles, however, becomes pronounced each year during the pregnancy and birth seasons. During these seasons adult–adult play is observed between males and between males and females, but not between females (Alexander, 1970).

In the genesis of object play in Japanese macaques, French and Candland (1977) observed that the primary innovators and users of a metal-rod swing in their group were a 3½ year old male and a yearling male.

The lion-tailed macaque (*M. silenus*) has not been studied very extensively; however, Bertrand (1969) has reported that nonadult lion-tailed macaques play more than do either *M. arctoides* or *M. mulatta* of nonadult status. The lion-tailed macaque is the most arboreal of the macaques. It would be interesting to see if the usual macaque sex differences in play extend up into the trees with these very playful primates.

6. Baboons

Unfortunately, there have been very few studies of sex differences in baboon play. Since baboons show extreme sexual dimorphism in size and dentition, and because they are closely related to the macaques, they are an important group to study.

G. H. Young (1977) of the University of Texas at Austin has studied infant baboons (*P. cynocephalus*) in captivity from birth to three months. At this age, sex differences in play are not found in a nursery environment, in a group of mothers with infants, or in a group containing adult males. Since the data do not go beyond the age of three months, it is not shown conclusively that sex differences in baboon play never develop. However, it is interesting that in some species of macaques, sex differences are already developed at the age of three months. If sex differences in play are shown not to exist in baboons at this age, this would present primatologists with an interesting problem (see also Young and Bramblett, 1977).

N. W. Owens (1975a, 1975b) has studied the play of olive baboons (*Papio anubis*) under more free-living conditions and in older animals than has Young. In olive baboons, males tend to play more, are rougher, and take a more active part in play than do females. Baboon infants of both sexes more often play the active part in play as they grow older. Adult male baboons play more than do adult females.

In the species of baboon studied by Owens (1975b) (*P. anubis*), play is different from aggression in that the mean length of *body contact* in play is greater than it is in aggression. Moreover, *chases* tend to be longer in aggression, particularly in adult male aggression. In male–male play and in adult male–adult male aggression, roles are frequently reversed; whereas, in female–female play and aggression, reversal of roles is uncommon. Owens (1975b) feels that play in males of this species may function as practice for aggression. The play allows for more prolonged contact to facilitate the learning process.

However, he believes that the selection pressures operating on females are different. Adult females do not come into estrus at the same time and do not need to compete for males. There is therefore a more stable hierarchy in females and a selection *against* severe fighting which protects the young. Even in pregnancy, fighting can, and does, occasionally cause stillbirth (Owens, 1975b).

E. PLAY IN LESSER AND GREAT APES

There are no data on sex differences in the development of lesser ape play. Berkson (1966) however, has reported that an adult male gibbon initiated play with an infant at the age of three months, that the infant gibbon at the age of five months initiated play with the adult male, and that the adult male did not initiate play between the ages of nine and twelve months but tolerated the infant's attempts to play. Mild adult male–infant play is quite common in the gibbon (Schaller, 1963). Maple (personal communication) is currently studying play in one male and two female adult gibbons (*Hylobates lar*) in captivity. Vigorous play between all members of the trio has been observed, but no sex differences have been detected.

In common chimpanzees (*Pan troglodytes*), mothers play with infants and so do males. One adult male chimpanzee observed by Goodall (1965) even wrestled with an adult male baboon. Reynolds and Reynolds (1965), however, have stated that adult females, especially the mothers, play with infants more than do adult males. Savage and Malick (1977) noted that mother chimpanzees in a captive group played more with *other* infants than with their own. Adult males often ignore or push infants away (van Lawick-Goodall, 1973).

Mason (1965) has shown in the laboratory that, for the two to four year old chimpanzee, play is more rewarding than is grooming or petting. The preference for play develops during the first year, continues to early puberty, *and* the preference is independent of conditions of rearing. Obviously, play is of great importance to the young chimpanzee.

The chimpanzee is one of the few mammals to remain relatively playful throughout its entire life. As we have already seen in the Mason (1965) study, play is one of the most highly preferred activities of young chimpanzees. Since play is still most common among infant and juvenile chimpanzees, it is not surprising that the adults appear to be increasingly playful around them. The mothers, however, begin play with infants before the infants play with peers. Infant-infant play seems to begin with the same responses an infant would normally direct toward its own mother. Long kisses (which remind one of extended mutual nursing) occur between infants. Mother chimpanzees, as noted above, play more frequently with other infants than with their own, yet they encourage the play of their own infant with others (adults or infants) by gently *shoving* them toward others. While encouraging such play, the mother rarely joins in. (Play bouts between three or more individuals are uncommon in chimpanzees.) In time, the infants also come to prefer playing with other adults rather than with their own mothers. In fact, in captivity, infants appear to temporarily *trade* mothers in play. The play, when it occurs, is also more active than it is with their own mothers. Tickling and embracing are especially frequent in mother-other infant play. In addition, sexual behavior occurs in these play sessions (Savage and Malick, 1977). The reader can refer to the excellent article by Savage and Malick for further information on this last topic.

Although female chimpanzees appear to play more than do males (especially with infants) when they are adults, sex differences in infancy seem to be the same as they are in most other primates (Savage and Malick, 1977). Nadler and Braggio (1974), for example, found more rough-and-tumble play in nursery-reared infant male chimpanzees than in females. Males were more assertive in social interactions whereas females tended to interact more with inanimate objects. Nadler and Braggio (1974) also found that early experience with adults was not necessary for the development of these sexually dimorphic patterns and that experience with peers appeared to *facilitate* the sex differences. Males initiated social interactions more frequently than females, and they participated in them for longer periods of time. There was definitely more play in males (Nadler and Braggio, 1974).

In another report (Braggio et al., 1976), the same authors found that males had significantly higher frequencies of aggressive contacts, rough-and-tumble play, *and* object manipulation than did females. All of the animals used in their study had not reached sexual maturity. The sex difference was found despite vastly different social experiences among their animals.

Nadler and Braggio (1974) and Braggio et al. (1976) reported the same infant sex differences for immature orangutans (*Pongo pymaeus*) as they had for the chimpanzees. Males play more often and they play rougher. Wilson et al. (1977) and Zucker et al. (1978) have described play between adult male orangutans and offspring in captivity. In one instance, an adult male orangutan rarely initiated play; however, extensive play was described between a nineteen-year-old adult male and his four-year-old male offspring (Zucker et al. 1978).

Okano (1971), on the other hand, described an adult female playing with a young adult male, although Okano, too, says that most of the play observed was seen among young males. Okano believes that at three to four years of age, orangutans in the wild form unisexual play groups and it is not until five to eight years of age that they begin to form heterosexual groups.

There have also been reports of orangutan *adult* heterosexual play. In these reports, the play is usually initiated by the *female* (Maple and Zucker, 1977). Self-motion play occurs more often in adult female than in adult male orangutans and this solitary play may be an indicator of female hormonal status (Maple and Zucker, in press).

In gorillas, on the other hand, there appears to be a preference for *heterosexual* juvenile play. In a study done by Freeman and Alcock (1973) heterosexual juvenile gorilla pairs played the most and homosexual female pairs played the least. In the same study, orangutan juveniles did not display this heterosexual preference. Male juveniles played more than female juveniles in both species (Freeman and Alcock, 1973).

In another paper comparing orangutans to lowland gorillas, it was reported that adult male gorillas take a more active interest in infants than do adult male orangutans (Wilson et al., 1977), and that the adult male gorilla in captivity may prefer to play with male infants. Schaller (1963) had earlier reported that infants in the wild are attracted to silver-backed male gorillas and often play on and around them. On the other hand, Hoff et al. (1977) saw reciprocal mother-infant play in gorillas as early as four months of age (usually initiated and terminated by the mother). The mother gorilla usually gently nudges the infant which results in the infant's wrestling with her. Mothers also play with infants of other females, but not as often as do chimpanzee mothers. Infant–infant play in the gorilla is first seen at six months of age (Hoff et al., 1977) with reciprocal peer play developing one month later. At eleven months of age, the play becomes rougher and further away from the mothers. Hoff et al. (1977) believe that male infants play more and earlier than do females (see also Maple and Zucker, in press).

In adulthood, gorillas sometimes play with each other. With a slight swelling of the sexual skin, an adult female gorilla will increase her play with an adult male. The play, however, consists of slow, gentle, caressing and wrestling (cf. Keiter, 1977).

In summary, the great apes differ somewhat from other primates in that adult females frequently play with infants (cf. Mitchell, 1977). In the chimpanzee, adult females probably play with infants more than do adult males. In the orangutan and gorilla there is undoubtedly a stage of social play between an infant and its mother that precedes solitary and peer play. Adult male great apes have the potential for play and sometimes show it, however, in infancy and in the juvenile period, males play more than do females in all of the great apes. In this sense, at least, they do not differ from macaques and many other primates.

F. POSSIBLE NEUROLOGICAL CORRELATES OF PLAY

Patricia S. Goldman (1976) has written an excellent review on the maturation of the mammalian nervous system. In this review she states that the orbital prefrontal cortex normally develops its functions earlier in male macaque monkeys than in females and that:

> . . . to the extent that prefrontal and anterior temporal cortical structures are critical for normal social interaction in the adult, they may also be an important component of the neural basis for socialization of the young, and in particular for the *playful* behavior that has been shown to be sexually dimorphic. (P. 62, emphasis mine)

G. THE FUNCTION OF PLAY

It is frequently stated or implicitly assumed by many primatologists that nonhuman primate play functions to establish a dominance order. Symons (1977) takes issue with this assumption, since function itself is difficult to demonstrate. As Symons says:

> However, play is defined, its functions are not immediately obvious (P. 14)

and that:

. . . social science concepts of function, and the attempt to find a moral order in nature, have impeded the study of function in primatology and may be responsible for much fruitless functional speculation. (P. 39)

Since social science concepts of function do not suffice, Symons (1977) espouses biological concepts of function. Since natural selection presumably eliminates most nonuseful (nonfunctional) behavior, the ultimate function of play would seem to be to produce variation for differential reproduction or to contribute to reproductive success. The first place we should look (based on Symons' reasoning) for functions of play would (therefore) be in the realm of sexual behavior.

As we have seen repeatedly in this chapter, sex differences in play often change at sexual maturity, and early mother–infant and infant–infant play is often sexual. The developing sex differences in play themselves almost "shape" the animals for future sexual interaction. At puberty, in particular, the nature of the play is often *primarily* sexual. It is a *sexual* hormone's prenatal appearance that prewires a sex difference in play.

With so many obvious elements in common between sex and play, it is surprising, or perhaps even ironic, that the fine paper on the function of play by Symons (1977) hardly mentions sexual play. We will return to this topic in our chapters on sexual behavior.

Baldwin and Baldwin (1977), in their excellent article on the role of learning and arousal in the development of play, acknowledge:

1. the influences of prenatal androgens in the development of sex differences;
2. the fact that males are often physically larger and stronger than females;
3. the role of "play-mothering" in the development of female play;
4. the possibility that early maturity in females may lead to an early diminution in active female play; and
5. a higher level of object manipulation play in females because objects are less arousing than are social activities.

Their explanations for sex differences in play include all of the above factors as well as the additional possibility than males can tolerate greater levels of arousal than can females.

H. HUMAN PLAY

Many studies of human children have suggested that there are sex or gender differences in human play very much like those seen in the macaque. Whiting and Pope (1974), for example, reported "a greater frequency of rough-and-tumble play among boys aged 3-6 in four out of six of the societies they studied." (Maccoby and Jacklin, 1974, P. 205.) Maccoby and Jacklin (1974), in their tome on sex differences in humans, state that "one of the best established sex differences is the much greater incidence of mock fighting (rough-and-tumble play) among boys." (P. 237.) (See also Omark and Edelman, 1973.) These differences are reflected in the toy preferences of young children (Benjamin, 1932). The gender differences are found as early as social play begins, at age 2 or 2½ according to Maccoby and Jacklin (1974). Goldberg and Lewis (1969) reported play differences in one-year-old infants, while Jacobsen (1977) reported that physical contact was more frequent in boys as early as 14½ months.

Braggio et al. (1976) made direct comparisons of young chimpanzees, orangutans, and children and found that, in all three species, males initiated more rough-and-tumble play than did females. Among preschool children, boys do more pushing, give more beatings, and do more chasing and laughing than do girls (McGrew and McGrew, 1970).

In a study of thirty-six *peer-reared* children, Milch and Missakian-Quinn (1977) reported that peer-rearing (minimal adult contact) tended to *reduce* contrast in gender differences in play roles. This is contrary to the Nadler and Braggio (1974) findings on chimpanzees and orangutans. In the latter study it was felt that experience with adult chimpanzees was not necessary for the development of sex differences in play and that experience with peers appeared to facilitate the sex or gender differences. On the other hand, studies of "together-together" (peer-only) reared rhesus macaques suggest that peer rearing depresses play (Harlow, 1969). Perhaps we are more like rhesus than like chimps in "needing" parents to maximize gender differences.

In three to five-year-old children, boys also display more frequent and more active *fantasy play* than do girls, and older children engage in more make-believe than do younger children. Moreover, boys (and older children) perform more make-believe play *outdoors* than do girls and younger children (Sanders and Harper, 1976).

There is evidence that human adults treat male and female children differently. Fathers, for example, report engaging in more rough-and-tumble play with their sons than with their daughters (Tasch, 1952). In the first year of life, infant humans respond more positively to father–infant play than to mother-infant play. Mothers apparently are more often engaged in caretaking functions, like holding, while fathers lean toward play behavior (Lamb, 1977).

In adult–adult interactions, too, men appear more likely than women to engage in physical activities and they are less likely to just talk (Caldwell and Peplau, 1977). The rougher style of male play is also revealed in adult humor and wit. Males appreciate hostile wit more than do females (O'Connell, 1958).

A number of authors, including Maccoby and Jacklin (1974), have assumed at least some role for prenatal androgens in shaping these human gender differences in play. Ehrhardt (1974) has noted, for example, that fetally androgenized human females (with adrenogenital syndrome) are more often long-term "tomboys," and Hamburg and Lunde (1966) have also emphasized prenatal determinants; *but* they also point out that there appears to be a critical period for environmentally determined human gender assignments between eighteen months and three years of age.

As we shall see later, the human's special ability of being able to conceptualize the *idea* of gender puts matters of play into a very different perspective than one sees in macaques, for example. Boys and girls as individuals, although perhaps primed in certain ways, also have a real *choice* about how they play and how often.

One other point deserves comment. Many researchers seem bent on proving the existence of human sex differences in play, or in other behaviors, in the very youngest of infants. It seems that the assumption in such cases is that the earlier one demonstrates sex differences, the more biological, prewired, inherent, or nonlearned they are (the less the environment or social learning can modify them), and the more animallike is *Homo sapiens.* However, a moment's reflection on this idea soon reveals how unsound it is. The older a human individual becomes, the *less* animallike he becomes. If we have learned anything from Piaget and other child psychologists, it is that the most "human" qualities of people start to develop *after* the second year. Language does not rapidly develop until age two, self-recognition is not apparent until age two, and recognition of one's own gender occurs at around age two. So, contrary to the assumption of such researchers, if we are to demonstrate *"animallike" sex* differences in people, we should at least some of the time expect a *better* (not a worse) chance of finding them before age two than after age two. An analogy might even be made here between the development of biological sex differences and the development of certain inborn reflexes like the clasping reflex and the Moro reflex. Certainly, *Homo sapiens* has a clasp reflex like that of other primates. *However,* people *lose* their clasp reflex with age as their nervous systems develop into the next hierarchical level of organization. We are not saying that biological determination of all human sex differences disappears in this fashion, only that it is possible that *some* of it does. There is already some evidence of sex differences *reversing* by adulthood even in nonhuman primates (Wolfheim, 1977; Savage and Malick, 1977). All of the evidence is not in regarding the biological determination of play differences beyond infancy.

REFERENCES

Abbott, D. H. Hormones and behaviour during puberty in the marmoset. Paper presented at the *International Primatological Society* Meeting, Cambridge, England, August, 1976.

Alexander, B. K. Parental behavior of adult male Japanese monkeys. *Behaviour*, 1970, **36** (4), 270-285.

Anderson, C. L. and Mason, W. A. Early experience and complexity of social organization in groups of young rhesus monkeys (*Macaca mulatta.*) *Journal of Comparative and Physiological Psychology*, 1974, **87** (4), 681-690.

Baldwin, J. D. and Baldwin, J. I. Interactions between adult female and infant howling monkeys (*Alouatta palliata*). *Folia Primatologica*, 1973, **20**, 27-71.

Baldwin, J. D. and Baldwin, J. I. The role of learning phenomena in the ontogeny of exploration and play. In Chevalier-Skolnikoff, S. and Poirier, F. E. (Eds.) *Primate Bio-Social Development*. New York: Garland, 1977, pp. 343-403.

Benjamin, H. Age and sex differences in the toy preferences of young children. *Journal of Genetic Psychology*, 1932, **41**, 417-429.

Berkson, G. Development of an infant in a captive gibbon group. *Journal of Genetic Psychology*, 1966, **108**, 311-325.

Bernstein, I. S. A field study of the pigtail monkey. *Primates*, 1967, **8**, 217-228.

Bernstein, I. S. The lutong of Kuola Selangor. *Behaviour*, 1968, **32**, 1-16.

Bernstein, I. S. Activity patterns in pigtail monkey groups. *Folia Primatologica*, 1970, **12**, 187-198.

Bernstein, I. S. Daily activity cycles and weather influences on a pigtail monkey group. *Folia Primatologica*, 1972, **18**, 390-415.

Bernstein, I. S. and Draper, W. A. The behavior of juvenile rhesus monkeys in groups. *Animal Behavior*, 1964, **12**, 84-91.

Bertrand, M. The behavioral repertoire of the stumptail macaque. *Bibliotheca Primatologica*, 1969, **11**.

Bourliére, F., Bertrand, M. and Hunkeler, C. L'ecologie de la mone de Lowe (*Cercopithecus campbelli Lowei*) en cote d'ivoire. *Extrait de la Terre et la Vie*, 1969, **2**, 135-163.

Braggio, J. T., Nadler, R. D., Lance, J. and Myseko, D. Sex differences in apes and children. Paper presented at the *International Primatological Society* Meeting, Cambridge, England, August, 1976.

Bramblett, C. A. Social organization as an expression of role behavior among old world monkeys. *Primates*, 1973, **14** (1), 101-112.

Bramblett, C. Sex differences in the acquisition of play among juvenile vervet monkeys. Paper presented at the *Animal Behavior Society* Meeting, University Park, Pennsylvania, June, 1977.

Brandt, E. M. and Mitchell, G. Pairing preadolescents with infants (*Macaca mulatta*). *Developmental Psychology*, 1973, **8**, 222-228.

Breuggeman, J. A. Toward an understanding of context and meaning in primate play. Paper presented at the *American Society of Primatologists* Meeting, Seattle, Washington, April, 1977.

Burton, F. D. The integration of biology and behavior in the socialization of *Macaca sylvana* of Gibraltar. In Poirier, F. (Ed.) *Primate Socialization.* New York: Random House, 1972.

Caldwell, M. A. and Peplau, L. A. Sex differences in friendship. Paper presented at the *Western Psychological Association* Meeting, Seattle, Washington, April, 1977.

Chamove, A., Harlow, H. F. and Mitchell, G. Sex differences in the infant-directed behavior of preadolescent rhesus monkeys, *Child Development,* 1967, **38**, 329-335.

Deag, J. M. and Crook, J. H. Social behavior and agonistic buffering in the wild Barbary macaque (*Macaca sylvana*). *Folia Primatologica,* 1971, **15**, 183-200.

Dolan, K. J. Meta communication in the play of a captive group of Syke's monkeys. Paper presented at the *American Association of Physical Anthropologists* Meeting, St. Louis, Missouri, April, 1976.

Dolhinow, P. J. and Bishop, N. The development of motor skills and social relationships among primates through play. In Hill, J. P. (Ed.) *Minnesota Symposia on Child Psychology* (Vol. IV). Minneapolis: University of Minnesota Press, 1970.

Ehrhardt, A. A. Androgens in prenatal development: Behavior changes in nonhuman primates and man. *Advances in the Biosciences,* 1974, **13**, 154-162.

Ehrlich, A. and Musicant, A. Social and individual behavior in captive slow lorises. *Behavior,* 1977, **60** (3-4), 195-220.

Fady, J. C. Les jeux sociaux: les compagnon de jeux chez les jeunes. Observations chex *Macaca irus. Folia Primatologica,* 1969, **11**, 134-143.

Fedigan, L. Social and solitary play in a colony of vervet monkeys, *Cercopithecus aethiops. Primates,* 1972, **13** (4), 347-364.

Flinn, L. and Nash, L. T. Group formation in recently captured lesser galagos. Paper presented at the Annual Meeting of the *American Association of Physical Anthropologists,* Denver, Colorado, 1975.

Freeman, H. E. and Alcock, J. Play behaviour of a mixed group of juvenile gorillas and orangutans. *International Zoo Yearbook,* 1973, **13**, 189-194.

French, J. A. and Candland, D. K. The genesis of "object play" in *Macaca fuscata.* Paper presented at the *Animal Behavior Society* Meeting, University Park, Pennsylvania, June, 1977.

Gartlan, J. S. Ecology and behaviour of the patas monkey. Film presented at the *International Primatological Society* Meeting, Cambridge, England, August, 1976.

Goldberg, S. and Lewis, M. Play behavior in the year-old infant: Early sex differences. *Child Development,* 1969, **40**, 21-31.

Goldfoot, D. A. Social and hormonal regulation of gender role development in rhesus monkey. Paper presented at the *International Primatological Society* Meeting, Cambridge, England, August, 1976.

Goldman, P. S. Maturation of the mammalian nervous system and the ontogeny of behavior. *Advances in the Study of Behavior,* 1976, **7**, 1-90.

Goodall, J. Chimpanzees of the Gombe Stream Reserve. In DeVore, I. (Ed.) *Primate Behavior.* New York: Holt, Rinehart and Winston, 1965, pp. 425-473.

Goy, R. W. Role of androgens in the establishment and regulation of behavioral sex differences in mammals. *Journal of Animal Science,* 1966, **25**, 21-35.

Hamburg, D. A. and Lunde, D. T. Sex hormones in the development of sex differences in human behavior. In Maccoby, E. E. (Ed.) *The Development of Sex Differences.* Stanford, California: Stanford University Press, 1966, pp. 1-24.

Hansen, E. W. The development of maternal and infant behavior in the rhesus monkey. *Behaviour,* 1966, **27**, 107-149.

Hansen, E. W., Harlow, H. F. and Dodsworth, R. O. Reactions of rhesus monkeys to familiar and unfamiliar peers. *Journal of Comparative and Physiological Psychology,* 1966, **61**, 274-279.

Harlow, H. F. Age-mate or peer affectional system. *Advances in the Study of Behavior,* 1969, **2**, 333-383.

Harlow, H. F. and Lauersdorf, H. E. Sex differences in passion and play. *Perspectives in Biology and Medicine,* 1974, **17**, 348-360.

Hinde, R. A., Rowell, T. E. and Spencer-Booth, Y. Behavior of socially living rhesus monkeys in their first six months. *Proceedings of the Zoological Society of London,* 1964, **143**, 609-649.

Hinde, R. A. and Spencer-Booth, Y. The behavior of socially living rhesus monkeys in their first two and a half years. *Animal Behaviour,* 1967, **15**, 169-196.

Hoff, M. P., Nadler, R. D. and Maple, T. The development of infant social play in a captive group of gorillas. Paper presented at the *American Society of Primatologists* Meeting, Seattle, Washington, April, 1977.

Hunkeler, C., Bourliére, F. and Bertrand, M. Le comportement social de la Mone de Lowe (*Cercopithecus campbelli Lowei*). *Folia Primatologica,* 1972, **17**, 218-236.

Imanishi, K. Social behavior in Japanese monkeys, *Macaca fuscata.* In Southwick, C. H. (Ed.) *Primate Social Behavior.* Princeton, N. J.: Van Nostrand, 1963, pp. 68-81.

Ingram, J. C. Social interactions within marmoset family groups (*C. jacchus*). Paper presented at the *International Primatological Society* Meeting, Cambridge, England, August, 1976.

Jacobsen, J. The development of peer interaction in human infants. Paper presented at the *Animal Behavior Society* Meeting, University Park, Pennsylvania, June, 1977.

Kaufman, I. C. and Rosenblum, L. A. The waning of the mother-infant bond in two species of macaques. In Foss, B. M. (Ed.) *Determinants of Infant Behavior* (Vol. IV). London: Methuen, 1969.

Kaufman, J. H. Behavior of infant rhesus monkeys and their mothers in a free-ranging band. *Zoologica,* 1966, **51**, 17-28.

Keiter, M. D. Reproductive behavior in sub-adult captive lowland gorillas (*Gorilla gorilla gorilla*). Paper presented at the *American Society of Primatologists* Meeting, Seattle, Washington, April, 1977.

Kleiman, D. G. Monogamy in mammals. *The Quarterly Review of Biology,* 1977, **52**, 39-69.

Kurland, J. A. A natural history of Kra macaques (*Macaca fascicularis*) at the Kutai Reserve, Kalimantan Timur, Indonesia. *Primates,* 1973, **14** (2-3), 245-262.

Kuyk, K., Dazey, J. and Erwin, J. Play patterns of pigtail monkey infants: Effects of age and peer presence. *Journal of Biological Psychology,* 1977, **18**, 20-23.

Lamb, M. E. Father-infant and mother-infant interaction in the first year of life. *Child Development,* 1977, **48**, 167-181.

Lancaster, J. B. Play-mothering: The relations between juvenile females and young infants among free-ranging vervet monkeys (*Cercopithecus aethiops*). *Folia Primatologica,* 1971, **15**, 161-182.

Latta, J., Hopf, S. and Ploog, D. Observation on mating behavior and sexual play in the squirrel monkey (*Saimiri sciureus*). *Primates,* 1967, **8**, 229-246.

Lichstein, L. Play in rhesus monkeys: I. Definition II: Diagnostic significance. *Dissertation Abstracts International,* 1973, **33** (8), 3985.

Lindburg, D. G. The rhesus monkey in North India. In Rosenblum, L. A. (Ed.) *Primate Behavior: Developments in Field and Laboratory Research* (Vol. 2). New York: Academic Press, 1971.

Lockard, J. S., Heestand, J. E. and Begert, S. P. Play behavior in slow loris (*Nycticebus coucang*). Paper presented at the *American Society of Primatologists* Meeting, Seattle, Washington, April, 1977.

Loy, J. and Loy, K. Behavior of an all-juvenile group of rhesus monkeys. *American Journal of Physical Anthropology,* 1974, **40**, 83-95.

Loy, J., Loy, K., Patterson, D., Keifer, G. and Conaway, C. H. Play behavior of gonadectomized rhesus monkeys. Paper presented at the *Animal Behavior Society* Meeting, University Park, Pennsylvania, June, 1977.

Maccoby, E. E. and Jacklin, C. N. *The Psychology of Sex Differences.* Stanford, California: Stanford University Press, 1974.

MacRoberts, M. H. The social organization of the Barbary apes on Gibraltar. *American Journal of Physical Anthropology,* 1970, **33**, 83-99.

Maple, T. and Zucker, E. L. Behavioral studies of captive Yerkes orangutans at the Atlanta Zoological Park. *Yerkes Newsletter,* 1977, **14** (1), 24-26.

Maple, T. and Zucker, E. L. Ethological studies of play behavior in captive great apes. In Smith, E. O. (Ed.) *The Structure and Function of Play in Primates.* New York: Academic Press, in press.

Mason, W. A. The effects of social restriction on the behavior of rhesus monkeys: I. Free social behavior. *Journal of Comparative and Physiological Psychology,* 1960, **53** (6), 582-589.

Mason, W. A. Determinants of social behavior in young chimpanzees. In Schrier, A., Harlow, H. F. and Stollnitz, F. (Eds.) *Behavior of Nonhuman Primates.* New York: Academic Press, 1965, p. 343.

McGrew, W. C. and McGrew, P. L. Group formation in preschool children. *Proceedings of the International Congress of Primatology,* 1970, **3**, 71-78.

Milch, K. H. and Missakian-Quinn, E. Longitudinal study of social play in peer-reared children. Paper presented at the *Animal Behavior Society* Meeting, University Park, Pennsylvania, June, 1977.

Mitchell, G. Persistent behavior pathology in rhesus monkeys following early social isolation. *Folia Primatologica,* 1968, **8**, 132-147.

Mitchell, G. Parental behavior in non-human primates. In Money, J. and Musaph, H. (Eds.) *Handbook of Sexology* (Chapter 54). Amsterdam: Elsevier, 1977, pp. 749-759.

Nadler, R. D. and Braggio, J. T. Sex and species differences in captive-reared juvenile chimpanzees and orangutans. *Journal of Human Evolution,* 1974, **3**, 541-550.

Nagel, U. and Kummer, H. Variation in cercopithecoid aggressive behavior. In Holloway, R. L. (Ed.) *Primate Aggression, Territoriality and Xenophobia.* New York: Academic Press, 1974, pp. 159-184.

Nolte, A. Field observations on the daily routine and social behavior of common Indian monkeys, with special reference to the bonnet monkey. *Journal of the Bombay Natural History Society,* 1955, **53**, 177-184.

O'Connell, W. E. A study of the adaptive functions of wit and humor. Unpublished doctoral dissertation, University of Texas, 1958.

Okano, J. A preliminary observation of orangutans in the rehabilitation station in Sepiloc, Sabah. *The Annual of Animal Physiology,* 1971, **21**, 55-67.

Omark, D. R. and Edelman, M. Peer group social interactions from an evolutionary perspective. Paper presented at the *Society for Research in Child Development* Conference, Philadelphia, 1973.

Owens, N. W. Social play behaviour in free-living baboons, *Papio anubis, Animal Behaviour,* 1975, **23**, 387-408. (a)

Owens, N. W. A comparison of aggressive play and aggression in free-living baboons, *Papio anubis, Animal Behaviour,* 1975, **23**, 757-765. (b)

Poirier, F. E. Nilgiri langur behavior and social organization. In Voget, F. W. and Stephenson, R. L. (Eds.) *For the Chief: Essays in Honor of Luther S. Cressman.* Eugene, Oregon: University of Oregon Anthropology Papers (No. 4), 1972, pp. 119-134.

Poirier, F. E. and Smith, E. O. Socializing functions of primate play. *American Zoologist,* 1974, **14**, 275-287.

Raleigh, M. J. Sex differences in the social behavior of juvenile vervet monkeys. Paper presented at the *American Society of Primatologists* Meeting, Seattle, Washington, April, 1977.

Ralls, K. Mammals in which females are larger than males. *Quarterly Review of Biology,* 1976, **51**, 245-276.

Redican, W. K. and Mitchell, G. Play between adult male and infant rhesus monkeys. *American Zoologist,* 1974, **14**, 295-302.

Reynolds, V. and Reynolds, F. Chimpanzees of the Budongo forest. In DeVore, I. (Ed.) *Primate Behavior.* New York: Holt, Rinehart and Winston, 1965, pp. 368-374.

Rosenblum, L. A. and Bromley, L. J. The effects of gonadal hormones on juvenile peer interactions in bonnet macaques. Paper presented at the *International Primatological Society* Meeting, Cambridge, England, August, 1976.

Rosenblum, L. A., Kaufman, I. C. and Stynes, A. J. Individual distance in two species of macaques. *Animal Behaviour,* 1964, **12** (2-3), 338-342.

Rosenblum, L. A. and Lowe, A. The influence of familiarity during rearing on subsequent partner preferences in squirrel monkeys. *Psychonomic Science,* 1971, **23**, 35-37.

Ruppenthal, G. C., Harlow, M. K., Eisele, C. D., Harlow, H. F. and Suomi, S. J. Development of peer interactions of monkeys reared in a nuclear family environment. *Child Development,* 1974, **45**, 670-682.

Sanders, K. M. and Harper, L. V. Free-play fantasy behavior in preschool children: Relations among gender, age, season, and location. *Child Development,* 1976, **47**, 1182-1185.

Sauer, E. G. F. and Sauer, E. M. The South West African bushbaby of the *Galago senegalensis* group. *The Journal of the South West African Scientific Society,* 1962, **16**, 5-36.

Savage, E. S. and Malick, C. Play and sociosexual behaviour in a captive chimpanzee (*Pan troglodytes*) group. *Behaviour,* 1977, **60**, 179-194.

Schaller, G. B. *The Mountain Gorilla.* Chicago: University of Chicago Press, 1963.

Schaller, G. B. Behavioral comparisons of the apes. In DeVore, I. (Ed.) *Primate Behavior.* New York: Holt, Rinehart and Winston, 1963, pp. 474-484.

Seay, B. M. Maternal behavior in primiparous and multiparous rhesus monkeys. *Folia Primatologica,* 1966, **4**, 146-168.

Seay, B. M., Schlottman, R. S. and Gandolfo, R. Early social interaction in two monkey species. *Journal of Genetic Psychology,* 1972, **87**, 37-43.

Segre, A. Talapoins. *Animal Kingdom,* 1970, **73** (3), 20-25.

Simonds, P. E. The bonnet macaque in South India. In DeVore, I. (Ed.) *Primate Behavior.* New York: Holt, Rinehart and Winston, 1965.

Simonds, P. E. Sex differences in bonnet macaque networks and social structure. *Archives of Sexual Behavior,* 1974, **3** (2), 151-165.

Southwick, C. H., Beg, M. A. and Siddiqi, M. R. Rhesus monkeys in North India, In DeVore, I. (Ed.) *Primate Behavior.* New York: Holt, Rinehart and Winston, 1965.

Spencer-Booth, Y. and Hinde, R. A. The effects of separating rhesus monkey infants from their mothers for 6 days. *Journal of Child Psychology and Psychiatry,* 1967, **7**, 179-197.

Stevens, C. W. Infant socialization in two species of macaques, *Macaca fascicularis* and *Macaca arctoides.* M. A. Thesis, University of California, Davis, 1967.

Sugiyama, Y. Characteristics of the social life of bonnet macaques. *Primates,* 1971, **12** (3-4), 247-266.

Suomi, S. J. Social development of rhesus monkeys reared in an enriched laboratory environment. *Abstract Guide of the 20th International Congress of Psychology.* Tokyo, Japan, 1972, pp. 173-174.

Suomi, S. J., Sackett, G. P. and Harlow, H. F. Development of sex preference in rhesus monkeys. *Developmental Psychology,* 1970, **3**, 326-336.

Symons, D. Effect and function in play. Paper presented at the *Animal Behaviour Society* Meeting, University Park, Pennsylvania, June, 1977.

Symons, D. Aggressive play and communication in rhesus monkeys. *American Zoologist,* 1974, **14**, 317-322.

Tasch, R. J. The role of the father in the family. *Journal of Experimental Education,* 1952, **20**, 319-361.

van Lawick-Goodall, Jane. Cultural elements in a chimpanzee community. In Menzel, E. W. (Ed.) *Symposia of the Fourth International Congress of Primatology (Vol. 1): Precultural Primate Behavior.* Basel: Karger, 1973, pp. 144-184.

Whiting, B. and Pope, C. A cross-cultural analysis of sex differences in the behavior of children aged three to eleven. *Journal of Social Psychology,* 1973, **91**, 171-188.

Wilson, M. E., Maple, T., Nadler, R. D., Hoff, M. and Zucker, E. L. Characteristics of paternal behavior in captive orangutans (*Pongo pygmaeus abelii*) and lowland gorillas (*Gorilla gorilla gorilla*). Paper presented at the *American Society of Primatologists* Meeting, Seattle, Washington, April, 1977.

Wolfheim, Jaclyn H. Sex differences in behavior in a group of captive juvenile talapoin monkeys (*Miopithecus talapoin*). *Behaviour,* 1977, **63**, 110-128.

Young, G. H. Paternal influences on the development of behavioral sex differences in infant baboons. Paper presented at the *American Association of Physical Anthropologists* Meeting, Seattle, Washington, April, 1977.

Young, G. H. and Bramblett, C. A. Gender and environment as determinants of behavior in infant common baboons (*Papio cynocephalus*). *Archives of Sexual Behavior,* 1977, **6**, 365–385.

Zucker, E. L., Mitchell, G. and Maple, T. Adult male-offspring play interactions within a captive group of orangutans (*Pongo pygmaeus*). *Primates,* 1978, **19** (2), 379–384.

6.

Puberty and Adolescence

A. GENERAL COMMENTS ON SEX DIFFERENCES AS PROCESSES

Up to this point, for continuity's sake, we have followed the development of sex differences in one behavior pattern (play) which appears to be largely determined by prenatal androgens. We have seen that there are, indeed, sex differences in play in nonhuman and human primates during infancy, and that there may be, at least in some cases, some carry-over of these differences into adulthood. However, as we all know, there are important changes in hormonal development that occur between infancy and adulthood. It is with these changes as well as with their behavioral correlates that we will be concerned in the current chapter. We will also follow the hormonal and behavioral changes into adulthood in our next four chapters on hormones and sexual behavior (Chapters 7, 8, 9, and 10). Starting with Chapter 11, we will then discuss behavioral sex differences in response to the birth process. We will complete our ontogenetic examination of sex differences with two chapters on factors which affect infant care.

Having set our course for the next few chapters, let us return to the topic of the moment which is puberty. First of all, some general comments concerning sex differences and puberty are in order. In physical and sexual maturation, the ontogeny of certain *processes* may vary temporally between species. It is very possible that the pubertal process in at least some species extends from birth to initial ovulation (Foster, 1974). Note that we have italicized the word processes. The development of sex differences *is* an ontogenetic process, not a static pattern. Sex differences develop and change with the situation and with age, sometimes appearing, sometimes disappearing, sometimes even *reversing* (cf. Mason, 1976; Wolfheim, 1977).

Prenatal hormones determine the path taken by the hypothalamus and pituitary at puberty. In male baboons, for example, metabolism of progesterone at these sites does not change at sexual maturity. Presumably, in females it does (Tabei and Henrichs, 1974).

Although the mechanism that controls the onset of puberty may well be present during fetal development and appears to be androgen dependent (Resko, 1975), the acquisition and development of sex differences in social behavior are not uniform, undifferentiated processes. Different behavioral sex difference processes undergo distinct but temporally consistent patterns. One team of investigators, employing cluster analysis, found eight distinct temporally consistent sex difference patterns in one species, *Macaca mulatta* (Smith, Fraser, and Begeman, 1977).

To give but one example of how sex differences may change between birth and puberty, we will use sex preference in rhesus monkeys. Suomi, Sackett, and Harlow (1970), using rhesus monkeys reared in partial social isolation, found that rhesus infants of both sexes, if under seven months of age, showed no sex preference in social interactions with age-mates. Initially, after seven months of age, they preferred their own sex, then gradually shifted toward a preference for the other sex, with females shifting earlier than males. For adults, instead of age-mates, however, all rhesus under three years of age preferred adult females to adult males, whereas after 3½ years of age, male rhesus preferred adult males while females preferred adult females. Clearly, the sex differences observed in partner preference depend upon both the age of the subject and the age of the partner. Since these monkeys were all reared in partial social isolation, there are apparently strong biological determinants of these different patterns of sex differences in partner preference.

The remainder of this chapter will concern itself with sex differences before, at, and after puberty as though the patterns were not as dynamic as they actually are. We should keep in mind, however, that ontogenetic changes in the process of sex differences have not been studied very extensively. We will, as usual, employ a taxonomic framework; but, for the sake of brevity, we will only include a few examples under each major grouping of primates.

B. PROSIMIAN PUBERTY

In general, prosimians reach sexual maturity much sooner than do monkeys and apes. In tree shrews (*Tupaia* spp.), for example, weight gain stops at three months of age in females and four months of age in males. Tree shrews only reach eight years of age in the laboratory before dying, so their entire life spans are accelerated relative to higher primates (Schwaier, 1975). At least one behavior displayed by tree shrews has been studied ontogenetically in an attempt to observe sex difference changes. A behavior known as "chinning," which accompanies scent-marking, does not occur prior to, but does occur after, puberty. Castration of adult males reduces chinning but, for some reason, testosterone administration does *not* change its frequency, while ACTH (adrenocortical trophic hormone from the pituitary gland) *does* increase chinning. This study provides us with one of many reminders that there are more hormones involved in sex differences than just gonadal hormones (von Holst and Buergel-Goodwin, 1975).

In the lesser bushbaby (*Galago* spp.), the difference in the full aggressive pattern seen in adult males (but not in females) does not mature until one year of age. (This pattern includes a special male call and urine-marking.) Females reach maturity earlier than do males. The onset of estrus, mating, and capability of conception occur at 200 days of age in the female (Doyle, 1974). Prosimian females differ from higher primates at puberty in that they do not show menstruation.

C. PUBERTY IN NEW WORLD MONKEYS

Rising levels of steroid sex hormones can be used to detect the onset of puberty in the common marmoset (*Callithrix jacchus*). Sex hormones appear in the blood plasma of both sexes at around 300 days of age and reach adult levels at 500 days of age at which time males successfully inseminate females. Testes also reach adult dimensions at 500 days of age. During adolescence or during the process of puberty (between 300 and 500 days of age), sub-adults are already seen forming marmoset-typical monogamous heterosexual pairs. However, a great deal of play is associated with copulation at this age. In adulthood there is little play between the members of the monogamous pair. In this pubescent play, there are sex differences, males initiating and approaching, females withdrawing but encouraging (Abbott, 1976). Glands in the skin of the circumgenital and sternal areas (in *both* sexes), which produce odor, also develop at puberty (Epple, 1975a) in *Cebuella pygmaea* and in *Saguinus fuscicollis* marmosets. Juvenile and pubescent marmosets show interest in infants and sometimes carry them. The *male* immatures carry the infants more than do the females in most groups (Epple, 1975b). The adult male also assumes the parental role more than does the female and castration has no dramatic effect on this parental performance (Epple, 1975b). Again, we see that something more than circulating gonadal hormones is involved in a sex difference.

Among the more advanced New World monkeys, in at least some species, the females begin to show what looks like primitive menstruation at puberty. There is an increase in the size of the pelvis in the female (but not the male) squirrel monkey (*Saimiri sciureus*) that does not occur until puberty (Gingerich, 1972). In squirrel monkeys, as in the rhesus, there is a sex difference in sex preference which changes at puberty. Subadult animals show no sex preferences and no sex segregation. At puberty, however, a preference for like-sex clumping develops. *Regardless* of hormonal status, adult squirrel monkeys show a like-sex preference (Coe and Rosenblum, 1974).

In the white-faced monkey (*Cebus capucinus*), young males but not females may leave the group for several years and join new groups. Solitary males have also been seen. Females reach sexual maturity by the end of the third year, later than do the more primitive marmosets. The females develop a very large clitoris at this time. Adult males, but not immatures, rub urine onto their arms, feet, and tail. Forehead patterns indicate the age and sex of each individual (Oppenheimer, 1969).

In the howler monkey (*Alouatta caraya*), females do not reach complete maturity until four to five years of age and males not until six to eight years of age (Benton, 1976). Clearly, the rates of maturation among New World monkeys are much slower than they are among prosimians and, consequently, their overall life spans are much longer. One twenty-five-year-old spider monkey (*Ateles* spp.) gave birth to a healthy infant and raised it (Sunderland, personal communication).

D. PUBERTY IN OLD WORLD MONKEYS

In this section, we will include discussions of *Colobus, Cercopithecus, Miopithecus, Erythrocebus,* and *Presbytis.* Macaques, mangabeys, and baboons will be discussed in the sections which follow.

Among the Colobinae, the transition from adolescence to adulthood is accompanied by a substantial amount of aggression, for males more than for females (Poirier, 1974). Territorial aggressiveness is most pronounced in younger subordinate males who frequently start the territorial clashes in *Colobus guereza caudatus* (Schenkel and Schenkel-Hulliger, 1967). As far as interest in infants is concerned, adult males of this genus will groom infants whereas juvenile males will not (Horwich and Manski, 1975).

In the genus *Cercopithecus,* the alpha adult male is the only member of a troop of Lowe's monkey (*C. cambelli lowei*) which emits alarm calls. Loud calls are exchanged between adult males of two groups when they encounter each other at the boundaries of their territories. Subadult males do not emit a loud call until they leave the troop at maturation (Hunkeler, Bourlière and Bertrand, 1972). Agonistic interactions occur between mature and maturing males in the same species. In another species (*C. aethiops*) young males entering puberty (as a group) receive more wounds than would be expected by chance (Struhsaker, 1966). Gautier-Hion and Gautier (1976) have also described changes in behavior at puberty in this genus.

In the talapoin (*Miopithecus talapoin*), as we have already seen, sex differences prior to puberty are not correlated with adult sex differences. Prepubescent males are more aggressive and active than are prepubescent females but adult males are less aggressive and active than adult females (Wolfheim, 1977).

In *Erythrocebus patas,* males do not stay in their natal groups. At puberty, they leave the group and may live a solitary existence. All-male groups are also seen. In fact, membership in an all-male group is a normal developmental phase for males of this species. Some males start this as early as eighteen months of age. Only young full adult males are solitary. At puberty, there are changes in the color of the male's scrotum and perianal skin. The scrotum becomes blue and the perianal skin scarlet. Immature males show a great interest in the blue scrotums of adult males (Gartlan, 1975). As in some species of the genus *Cercopithecus,* the onset of puberty in young patas males is followed by attacks on them by adult males (Hall and Mayer, 1967).

Finally, in langurs (*Presbytis* spp.), female age is not highly correlated with dominance and reproductive behavior. Female langur monkeys, unlike males, do not show a marked linear hierarchy so that age and reproductive value are not related. One does not see adult females attacking pubescent females. All Old World monkey females menstruate at puberty.

E. PUBERTY IN RHESUS MONKEYS (*Macaca mulatta*)

As in most other areas of primate behavior, puberty in rhesus monkeys has been studied more than has the puberty of any other primate. As we have seen in Chapter 4, many sex differences in rhesus are *not* due to androgens after birth but to prenatal androgens. Males suffering postnatal castration still develop mounting and show the normal pubertal decrease in play seen in noncastrates at puberty. However, they show decreases in penile erection, in yawning, and in aggression. Thus, the increases seen in normal male rhesus at puberty in these three behaviors *do* depend upon increased circulating levels of androgens at puberty. Adding androgens to the rhesus castrates before the normal age of puberty, however, does not bring out *all* of the sex-typical behaviors of normal postpubescent males. It does, on the other hand, produce precocious puberty in two-year-old males (Bielert, 1975).

Breuggeman (1973) described an adult male castrate in a free-ranging rhesus group who gave more parental care to young rhesus "than any other member of the group, male or female." (P. 199.) However, during the mating season, the male's behavior looked like that of any other male. Breuggeman believes that when parental care is given by *other* than the mother, the behavior may be less directly under the influence of hormonal control. We will discuss this topic in more detail in a subsequent chapter.

Brandt and Mitchell (1973) studying preadolescent–infant pairs in a laboratory setting, found that pubescent males were more aggressive toward infants than were pubescent females. Aggression (of all kinds) shows a rapid increase in rhesus at three to four years of age in both sexes; however, it increases more in males than in females (Cross and Harlow, 1965).

As we will see in a subsequent chapter, infant and juvenile female rhesus emit more calls for social contact than do males. However, when females have already reached puberty and the males have not, the sex difference disappears. It re-appears when the males also reach puberty (Erwin and Mitchell, 1973).

Wounding in rhesus peaks during the transition months between puberty and adulthood, particularly near the mating season. Females receive more wounds, but males' wounds are more likely to impair locomotion. Males show intergroup participation which increases up through four years of age, whereas the *inter*group participation of females starts decreasing after age two, just before female puberty (Hausfater, 1972).

At and after puberty, the canines of male rhesus (although larger than those of females) develop later than do the canines of female rhesus (Hurme and van Wagenen, 1961). In terms of weight, as we have seen, males significantly out-weigh females at every age except when females have reached puberty and males have not (Kirk, 1972).

Sexual maturity in male rhesus monkeys in captivity sometimes occurs before three years of age, females reaching puberty before males, sometimes as early as two years of age (Maple, Erwin, and Mitchell, 1973). There is an obvious change in the behavior of a male rhesus after his first ejaculation; however, in the first few months following the first ejaculation, the sexual performance of the still subadult male remains well below that of adults. In mounting, there is sometimes a lack of a foot clasp and there are fewer intromissions. Thus, in males at least, there is a prolonged transition period to adulthood. In four males observed by Michael and Wilson (1973), three had their first ejaculation between three and four years of age and one between four and five years of age. Two of six rhesus males observed by Maple et al. (1973), however, had impregnated females by three years of age.

In free-ranging rhesus monkeys on Cayo Santiago Island off Puerto Rico, both males and females begin to engage in consort activity at the age of three years. At three, consort activity is initially quite high but then *declines* in both males and females at ages four and five. This decrease in consort activity occurs in the males because at four and five years of age they join all-male peripheral groups, become solitary, or join nonnatal groups. Females of this age, however, remain in the natal group and, at four or five years of age, deliver their first offspring. This inhibits mating (Missakian-Quinn and Varley, 1977).

Also at approximately three years of age, the concentration of testosterone in the male rhesus monkey reaches the adult blood level of the hormone and shows no marked seasonal fluctuation over a period of 1½ years (Resko, 1967).

Dominance status also changes at puberty. As females become adult, they rank just below their own mothers. As males become adult, they initially rank near the mother but then either gain or lose rank or leave the group (Sade, 1967).

With increased estrogen levels at puberty in the females, possible sex differences in response to high levels of stress may develop. Estrogens can exert sensitizing or stimulating effects on the pituitary-adrenal axis, while testosterone appears to exert an inhibitory effect on this hormonal stress system. In one study of pubescent rhesus, "sex differences in ACTH-response became more pronounced as the initially subadult cagemates approached sexual maturity." (Sassenrath, 1970, p. 296.)

Normal rhesus monkeys, as we have seen, also show a marked reduction in play, the levels of which become very low by four to five years of age. If prepubescent rhesus are gonadectomized, they still show a slight decrease in play but continue to show moderate levels (instead of low levels) of play into their sixth and seventh years. Overall, gonadectomized prepubertal and peripubertal monkeys show more postoperative play than do controls; however, males *still* play more than females in both groups (Loy et al., 1977).

Following puberty and then adulthood, sex hormonal levels remain high in both male and female adult rhesus for years. If the female animal lives long enough, however, there is evidence that she will experience menopause. In the rhesus monkey, menopause occurs at around twenty-seven or twenty-eight years of age (Jones, 1975; van Wagenen, 1970).

F. PUBERTY IN OTHER SPECIES OF MACAQUES

In Japanese macaques (*M. fuscata*), as in rhesus macaques, males leave the natal troop at maturity (Kawamura, 1965). Males shift troops at four or five years of age, but some males, who are higher in rank, remain with the natal troop longer. Some males join all-male groups after puberty (Koyama, 1970) and some become solitary (Nishida, 1966).

Male Japanese macaques grow linearly in weight and length up to seven years of age. Males between three and seven years of age are often seen moving at a far distance from the troop. After seven years of age, growth slows down and stops at about eight years of age when the face and hips become bright scarlet.

Sugiyama and Ohsawa (1974) reported that nineteen out of twenty-three Japanese monkey males left the troop at four or five years of age, and three others left the troop before the age of twelve. They noted that no male spent his entire life in his natal troop. Most of the males do not leave separately but in groups of two or three. Some years of free movement follow until age seven when they approach a strange troop during the mating season, mate, and either leave or join the new troop. Even if they join, they leave the troop again within a few years. Males in the wild usually die at about fifteen to seventeen years of age (Sugiyama and Ohsawa, 1974).

Most females have had their first baby by the time they are five or six years of age but they are not fully adult until they reach the age of seven (Masui et al., 1973). Nigi (1976) reported that one female gave birth at the age of three years eleven months.

In the pigtail macaque (*M. nemestrina*), most females begin to swell by the age of three. Those who begin to swell later are usually of lower weight (Erwin and Erwin, 1976). At four years of age the female's growth curve shows a plateau but the male's growth curve continues to increase. Marked sexual dimorphism characteristic of adult pigtails does not begin until after four years of age (Tarrant, 1975). Other than these growth data, very little is known about puberty in the pigtail.

As in the rhesus, pubescent stumptail (*M. arctoides*) offspring rank just below the mother (Estrada, 1976). An interesting behavior is displayed by the stumptail macaque subadults when they witness adults copulating. They attempt to harass the copulating pair. The role of harasser is not unique to pubescent animals; it is also seen in juveniles, but subadults of both sexes are commonly involved in this behavior (Gouzoules, 1974). The age of sexual maturity in the stumptail has been reported to be as low as three years of age in the female and 3½ years of age in the male. The age of menarche in females born in some laboratories has steadily decreased through a period of years in captivity (Trollope and Blurton-Jones, 1975).

The bonnet macaque (*M. radiata*) differs from the rhesus and Japanese macaque in at least one important way. At puberty, bonnet males do *not* become solitary, perhaps because of stronger male–male bonds in this species (Parthasarathy, 1976). There are, however, sex differences in prepubertal play in bonnets similar to those seen in rhesus. Hormone treatment (administration of testosterone or estrogen) results in a dramatic decrease in play behavior as well as in activity levels, but in a dramatic increase in male–female interaction (Rosenblum and Bromley, 1976).

In the toque macaque (*M. sinica*), adult males and females differ in facial coloration. Females are adult at the age of 4½ to five years, males are adult at the age of seven to eight years. As in the bonnets, solitary living male toque macaques are never found, although males occasionally leave the troop for a few hours. Males do change troops, however. A male that lives to an old age may change troops five to six times in a lifetime. After maturity, females live an average of seventeen years, males only 10½ years. The mortality peak occurs in infancy in the females, and in adolescence, puberty, or subadulthood in the males (Dittus, 1975).

In the Barbary macaque (*M. sylvana*), the adult male leader helps reorient small infants away from their mothers by permitting subadult *males* to snatch the infant from them and to take it away to greater and greater distances. While under the care of pubescent or subadult males, the infant contacts juveniles and age-mates. Pubescent or subadult females, surprisingly, play a very small part in infant socialization. In the present-day Barbary macaques on Gibraltar, in contrast to all other macaques, it is the adolescent *females* that are largely isolated from other troop members. They are rebuffed by all adults and by subadult males. We use the qualification "present-day Barbary macaque" because in 1940 (cf. Burton, 1972) this population showed a greater role for subadult females, particularly in infant socialization. There has evidently been a "tradition drift" in the Gibraltar Barbary macaque. These sex difference patterns may *not* be as genetically fixed (Burton, 1972) as is often assumed.

G. PUBERTY IN BABOONS AND MANGABEYS

In the common or yellow baboon (*Papio cynocephalus*), sexual maturity is reached at about four years of age for both sexes, though males are not physically and socially mature until almost eight years of age (Hall and DeVore, 1965). Tail carriage in males changes with age; the proximal portion becomes more and more vertical with increasing age (Hausfater, 1977).

In *Cercocebus atys* (the sooty mangabey) young, subadult, and multiparous adult females frequently display "play mothering" or guardian behavior (Hunter, 1977). In *Cercocebus galeritus,* males emit very loud calls only after reaching sexual maturity. Quris (1975) has observed that young solitary males will travel alone away from the troop.

H. PUBERTY IN THE APES

Hylobates lar pileatus (a subspecies of gibbon) is sexually dimorphic in color. In adult males, the color is dark and in adult females, the color is pale. Both sexes are pale at birth and darken with age, but females revert to the pale color at puberty (Fooden, 1969). In both gibbons (*Hylobates* spp.) and siamangs (*Symphalangus* spp.), there is a peripheralization of maturing subadults of *both* sexes (Fox, 1972) at puberty.

In the great apes, there are also changes at puberty, but there is evidence that hormonal differences exist prior to puberty as well. Serum FSH (follicle stimulating hormone) levels are higher in female than in male infant chimpanzees (*Pan troglodytes*). The day-to-day variability (because of cyclicity in females) for both FSH and LH levels is also significantly higher in *infant* females.

In addition, the FSH levels for female chimpanzees in *infancy* show periodic variation. In male infants, there is no such cyclicity. The cycle periods range from eight to twenty-five days. Curiously, these sex differences occur only in infancy, *not* in childhood (Faiman, Winter, and Grotts, 1973). Here we have an excellent example of a sex difference disappearing and then returning again at puberty.

In the Gombe Stream Reserve, according to Goodall (1965), chimpanzee female estrus begins at six years of age. However, Reynolds and Reynolds (1965) state that female chimpanzees in the Budongo Forest start their sexual swelling at seven to eight years. The female chimpanzees stay with their mothers, perhaps for life. They continue to travel with their mothers until they become sexually receptive, but even between periods of estrus, the females return to their mothers. A female does not have her first infant until thirteen years of age (van Lawick-Goodall, 1973; Pusey, 1976).

According to Graham and McClure (1977), once menstruation begins at menarche in the chimpanzee, the female (except when pregnant) continues to have regular menstrual cycles to an advanced age. They reported a female who died at the age of thirty-nine years old who was still cycling at the age of thirty-eight years although breeding was unsuccessful at thirty-eight and ovarian senescence was advanced at this age. In one other female, regular menstrual cycles were still occurring at the age of forty-eight years, while all other evidence pointed to advanced age. Breeding attempts in this female failed. This evidence suggests that chimpanzee females apparently die before the age of fifty *without* cessation of menstrual cycles but they seem to become infertile in their forties.

The male chimpanzee reaches puberty at seven or eight years, both in Gombe and in the Budongo Forest. Males start to travel with other chimpanzees away from their mothers for short periods at around age seven, but even the male does not leave the mother for more than a few days until he is ten years old. *Social* maturity in the male is not achieved until fifteen years of age (van Lawick-Goodall, 1973; Reynolds and Reynolds, 1965; Goodall, 1965; Pusey, 1976).

In the lowland gorilla (*Gorilla gorilla gorilla*) and the mountain gorilla (*Gorilla gorilla beringei*), weight gain peaks at seven to eight years of age for females and eight to nine years of age for males. Sexual maturation occurs earlier in females than in males and the accompanying growth spurt does not last as long in females as in males (Gijzen and Tijskens, 1971). Among subadult lowland gorillas, young females with slight swelling increase their *play* behavior with males. The females initiate the play which is slow, gentle, caressing or wrestling. With additional swelling at puberty, the females display chest beating (typically a male pattern). In addition, they "mouth groom" young males by kissing them and sticking their tongues in the males' mouths. At this time, in sex play, there appears to be no competition among males. The pubescent or subadult female gorilla also rides the back of a male for anywhere from five seconds to ten minutes during which time the male has a visible erection. A male emits two short staccato grunts inviting a female to a back ride and she mounts him. During a back ride there is grunting by both male and female. There is also grunting and growling during copulation and there are facial expressions involving pursed lips and open mouths which change more in the male than in the female. Estrus begins at 5½ years in females, long before the weight gain reaches its peak. Intromission is seen for males beginning at five to seven years of age, again well before the male weight gain has peaked. During subadult estrus, the female initiates 80 percent of all courtship play preceding copulation (Keiter, 1977).

Faust (1977) has reported on the birth of an infant gorilla to an eight-year-old female raised in captivity at the Frankfurt Zoo. Other females in the zoo were over eleven years old at the time of their first baby. She seemed, according to Faust, to be an exceptionally young mother. Perhaps, as in macaques, earlier puberty in great apes can be related to rearing in captivity.

I. COMMENTS CONCERNING HUMAN PUBERTY

It has been suggested that the term menstrual cycle should be confined to the sex cycle of the human female (Butler, 1974), but it has also been used to describe the sex cycle of rhesus monkeys (Rowell, 1972). Butler's (1974) point is that nonhuman primates have *estrous* cycles not menstrual cycles because they are behaviorally different not only at different seasons but also within their monthly cycles. In contrast, Butler claims there is no evidence that receptivity is affected by ovulation in women.

One noticeable trend within the primate order is the increased delay to the onset of sexual maturity as we go from prosimians to monkeys, to apes, to people. However, in some species of monkeys in captivity, animals are reaching puberty at earlier and earlier ages. This finding is similar to what is referred to as the "secular trend" in humans. Throughout the world, humans are also reaching puberty sooner and sooner. This is probably a function of better diet (cf. McKinney et al., 1977).

According to Gavan and Swindler (1966), people are unique in showing a sex difference in growth rates. From ages nine to twelve, girls grow faster than do boys, after this boys grow faster than do girls. Human dimorphism is partly a result of a pubertal acceleration difference.

According to Poirier (1974), there is good evidence that humans show maximum aggression (like some Colobinae) in the transition from adolescence to adulthood and that human rates of aggression at this time are greater in males than in females.

In humans, as in other primates, it is likely that the onset of puberty involves an interaction between the sex hormones and certain cells of the hypothalamus. The interval between birth and puberty is the time required for the hypothalamus to mature in regard to the influences of gonadal hormones. As in chimpanzees, serum FSH levels are higher in infant girls than in infant boys, and infant girls also show cyclicity in FSH and LH (Faiman, Winter, and Grotts, 1973). Androgens increase dramatically at eight to ten years of age in *both* sexes but in boys they reach twice the level as in girls. At eight to nine years of age, estrogen increases in girls and this is especially true at around age eleven. However, the estrogens do not become cyclic until eighteen months prior to menarche. This occurs along with the growth spurt. Androgens, however, are apparently to some extent responsible for sexual desire in *both* sexes (Hamburg and Lunde, 1966).

The peripheral subadult male, seen in a great many nonhuman primates, has been said to have an analogous stage in humans. Lockard and Adams (1977) observed over 10,000 groups of people and analyzed their various age/sex groupings in public. It was found that there was a greater relative frequency of *exclusively* subadult male groups compared to subadult female groups.

Interesting discussions of puberty in humans appear in Katchadourian (1974), Hutt (1972), and Jones, Shainberg, and Byer (1977).

J. MONOGAMY AND THE DELAY OF PUBERTY

In her excellent review on monogamy, Devra Kleiman (1977) shows that monogamy sometimes exists because the female cannot rear a litter without aid from conspecifics and because the carrying capacity of the habitat allows for only one adult female in the same home range. In such cases:

> . . . (1) the young exhibit delayed sexual maturation in the presence of the parents, and thus only the adult pair breeds, and (2) the older juveniles aid in rearing younger siblings. (P. 39)

According to Kleiman (1977), marmosets, tamarins, gibbons, and siamangs provide examples of primates who show obligate monogamy. She also speculates on the possible beginnings of monogamy in *Homo sapiens,* noting that there is also delayed sociosexual maturation in humans, as well as the occurrence of juvenile baby-sitting.

REFERENCES

Abbott, D. H. Hormones and behaviour during puberty in the marmoset. Paper presented at the *International Primatological Society* Meeting, Cambridge, England, August, 1976.

Benton, L. The establishment and husbandry of a black howler (*Alouatta caraya*) colony at Columbia Zoo. *International Zoo Yearbook,* 1976, **16**, 149-152.

Bielert, C. Social experience, hormones interact to shape sexual growth of male rhesus. *Primate Record,* 1975, **76**, 6 (2), 3-5.

Brandt, E. M. and Mitchell, G. Pairing preadolescents with infants (*Macaca mulatta*). *Developmental Psychology*, 1973, **8**, 222-228.

Breuggeman, J. A. Parental care in a group of free-ranging rhesus monkeys (*Macaca mulatta*). *Folia Primatologica*, 1973, **20**, 178-210.

Burton, F. D. The integration of biology and behavior in the socialization of *Macaca sylvana* of Gibraltar. In Poirier, F. E. (Ed.) *Primate Socialization*. New York: Random House, 1972, pp. 29-62.

Butler, H. Evolutionary trends in primate sex cycles. *Contributions to Primatology*, 1974, **3**, 2-35.

Coe, C. L. and Rosenblum, L. A. Sexual segregation and its ontogeny in squirrel monkey social structure. *Journal of Human Evolution*, 1974, **3**, 551-561.

Cross, H. A. and Harlow, H. F. Prolonged and progressive effects of partial isolation on the behavior of macaque monkeys. *Journal of Experimental Research in Personality*, 1965, **1**, 39-49.

Dittus, W. P. J. Population dynamics of the Toque monkey, *Macaca sinica*. In Tuttle, R. H. (Ed.) *Socioecology and Psychology of Primates*. The Hague: Moulton, 1975, pp. 125-151.

Doyle, G. A. The behavior of the lesser bushbaby. In Martin, R. D., Doyle, G. A. and Walker, A. E. (Eds.) *Prosimian Biology*. London: Duckworth, 1974, pp. 213-231.

Epple, G. The behavior of marmoset monkeys (*Callithricidae*). In Rosenblum, L. A. (Ed.) *Primate Behavior* (Vol. 4). New York: Academic Press, 1975, pp. 195-239. (a)

Epple, G. Parental behavior in *Saguinus fuscicolliss* spp. (*Callithricidae*). *Folia Primatologica*, 1975, **24**, 221-238. (b)

Erwin, J. and Mitchell, G. Analysis of rhesus monkey vocalizations: Maturation-related changes in clear call frequency. *American Journal of Physical Anthropology*, 1973, **38** (2), 463-468.

Erwin, N. and Erwin, J. Age of menarche in pigtail monkeys (*Macaca nemestrina*): ·A cross-sectional survey of sex-skin tumescence. *Theriogenology*, 1976, **5**, 261-266.

Estrada, A. Social relations in a free-ranging troop of *Macaca arctoides*. Paper presented at the *International Primatological Society* Meeting, Cambridge, England, August, 1976.

Faiman, C., Winter, J. S. D. and Grotts, D. Gonadotropins in the infant chimpanzee: A sex difference. *Proceedings of the Society for Experimental Biology and Medicine*, 1973, **144**, 952-955.

Faust, R. Eighth gorilla born at Frankfurt. *AAZPA Newsletter*, 1977, **18**, 14.

Fooden, J. Color-phase in gibbons. *Evolution*, 1969, **23** (4), 627-644.

Foster, D. L. Endocrine development of the hypothalamohypophyseal-gonadal system of the nonprimate and subhuman primate female. *Third Ross Conference on Obstetrics Research.* Ross Laboratories, Columbus, Ohio, 1974, pp. 23-33.

Fox, G. J. Some comparisons between Siamang and gibbon behaviour. *Folia Primatologica,* 1972, **18**, 122-139.

Gartlan, J. S. Adaptive aspects of social structure in *Erythrocebus patas. Proceedings of the Symposium of the Fifth Congress of the International Primatological Society.* Tokyo: Japan Science Press, 1975, pp. 161-171.

Gautier-Hion, A. and Gautier, J. P. Croissance, maturite sexuelle et sociale, reproduction chez les cercopithecines forestiers Africains. *Folia Primatologica,* 1976, **26**, 165-184.

Gavan, J. A. and Swindler, D. R. Growth rates and phylogeny in primates. *American Journal of Physical Anthropology,* 1966, **24** (2), 181-190.

Gijzen, A. and Tijskens, J. Growth in weight of the lowland gorilla (*Gorilla gorilla gorilla*) and of the mountain gorilla (*Gorilla gorilla beringei*). *International Zoo Yearbook,* 1971, **11**, 183-193.

Gingerich, P. D. The development of sexual dimorphism in the bony pelvis of the squirrel monkey. *Anatomical Record,* 1972, **172**, 589-594.

Goodall, J. Chimpanzees of the Gombe Stream Reserve. In DeVore, I. (Ed.) *Primate Behavior: Field Studies of Monkeys and Apes.* New York: Holt, Rinehart and Winston, 1965, pp. 425-473.

Gouzoules, H. Harrassment of sexual behavior in the stumptail macaque, *Macaca arctoides. Folia Primatologica,* 1974, **22**, 208-217.

Graham, C. E. and McClure, H. M. Ovarian tumors and related lesions in aged chimpanzees. *Veterinary Pathology,* 1977, **14**, 380-386.

Hall, K. R. L. and DeVore, I. Baboon social behavior. In DeVore, I. (Ed.) *Primate Behavior.* New York: Holt, Rinehart and Winston, 1965, pp. 53-110.

Hall, K. R. L. and Mayer, Barbara. Social interactions in a group of captive patas monkeys (*Erythrocebus patas*). *Folia Primatologica,* 1967, **5**, 213-236.

Hamburg, D. A. and Lunde, D. T. Sex hormones in the development of sex differences in human behavior. In Maccoby, E. E. (Ed.) *The Development of Sex Differences.* Stanford, California: Stanford University Press, 1966, pp. 1-24.

Hausfater, G. Intergroup behavior of free-ranging rhesus monkeys (*Macaca mulatta*). *Folia Primatologica,* 1972, **18**, 78-107.

Hausfater, G. Tail carriage in baboons (*Papio cynocephalus*): Relationship to dominance rank and age. *Folia Primatologica,* 1977, **27**, 41-59.

Horwich, R. H. and Manski, D. Maternal care and infant transfer in two species of *Colobus* monkeys. *Primates,* 1975, **16** (1), 49-73.

Hunkeler, C., Bourliére, F. and Bertrand, M. Le comportement social de la Mone de Lowe (*Cercopithecus campbelli lowei*). *Folia Primatologica*, 1972, **17**, 218-236.

Hunter, J. L. Guardian behavior in the sooty mangabey (*Cercocebus atys*). Paper presented at the *Animal Behavior Society* Meeting, University Park, Pennsylvania, June, 1977.

Hurme, V. O. and van Wagenen, G. Basic data on the emergence of permanent teeth in the rhesus monkey (*M. mulatta*). *Proceedings of the American Philosophical Society*, 1961, **105**, 105-140.

Hutt, C. *Males and Females*. Baltimore: Penguin, 1972.

Jones, E. C. The post-reproductive phase in mammals. In van Keep, P. A. and Lauritzen, C. (Eds.) *Frontiers of Hormone Research* (Vol. 3). Basel: Karger, 1975, pp. 1-19.

Jones, K. L., Shainberg, L. W. and Byer, C. O. *Sex and People*. New York: Harper & Row, 1977.

Katchadourian, H. *Human Sexuality: Sense and Nonsense*. San Francisco: Freeman, 1974.

Kawamura, S. Matriarchal social ranks in the Minoo-B troop. In Imanishi, K. and Altmann, S. (Eds.) *The Japanese Monkey*. Atlanta: Altmann, 1965, pp. 105-112.

Keiter, M. D. Reproductive behavior in subadult captive lowland gorillas (*Gorilla gorilla gorilla*). Paper presented at the *American Society of Primatologists* Meeting, Seattle, Washington, April, 1977.

Kirk, J. H. Growth of maturing *Macaca mulatta*. *Laboratory Animal Science*, 1972, **22** (4), 573-575.

Kleiman, D. G. Monogamy in mammals. *The Quarterly Review of Biology*, 1977, **52**, 39-69.

Koyama, N. Changes in dominance rank and division of a wild Japanese monkey troop of Arashiyama. *Primates*, 1970, **11**, 335-390.

Laws, J. Vonder Haar, Dolhinow, P. J. and McKenna, J. J. Female hanuman langur monkey rank and reproduction. Paper presented at the *American Society of Primatologists* Meeting, Seattle, Washington, April, 1977.

Lockard, J. S. and Adams, R. M. Peripheral males: A primate model for a human subgroup. Paper presented at the *Animal Behavior Society* Meeting, University Park, Pennsylvania, June, 1977.

Loy, J., Loy, K., Patterson, D., Keifer, G. and Conaway, C. H. Play behavior of gonadectomized rhesus monkeys. Paper presented at the *Animal Behavior Society* Meeting, University Park, Pennsylvania, June, 1977.

Maple, T., Erwin, J. and Mitchell, G. Age of sexual maturity in laboratory-born pairs of rhesus monkeys (*Macaca mulatta*). *Primates*, 1973, **14** (4), 427-428.

Mason, W. A. Primate social behavior: Pattern and process. In Masterson, R. B. et al. (Eds.) *Evolution of Brain and Behavior in Vertebrates.* Hillsdale, N. J.: Laurence Erlbaum Associates, 1976, pp. 425-455.

Masui, K., Nishima, A., Ohsawa, H. and Sugiyama, Y. Population study of Japanese monkeys at Takasakiyami I. *Journal of the Anthropological Society of Nippon,* 1973, **81** (4), 236-248.

McKinney, J. P., Fitzgerald, H. E. and Strommen, E. A. *Developmental Psychology: The Adolescent and Young Adult.* Homewood, Ill.: Dorsey, 1977.

Michael, R. P. and Wilson, M. Changes in the sexual behaviour of male rhesus monkeys (*M. mulatta*) at puberty. *Folia Primatologica,* 1973, **19**, 384-403.

Missakian-Quinn, E. A. and Varley, M. A. Consort activity in groups of free-ranging rhesus monkeys on Cayo Santiago. Paper presented at the *American Society of Primatologists* Meeting, Seattle, Washington, April, 1977.

Nigi, H. Some aspects related to conception of the Japanese monkey (*Macaca fuscata*). *Primates,* 1976, **17** (1), 81-87.

Nishida, T. A sociological study of solitary male monkeys. *Primates,* 1966, **7**, 141-204.

Oppenheimer, J. R. Behavior and ecology of the white-faced monkey, *Cebus capucinus,* on Barro Colorado Island, C. Z. *Dissertation Abstracts,* 1969, **30**, Order Number 69-10, 811.

Parthasarathy, M. D. Social ostracism among bonnet monkeys. Paper presented at the *International Primatological Society* Meeting, Cambridge, England, August, 1976.

Poirier, F. E. Colobinae aggression: A review. In Holloway, R. (Ed.) *Primate Aggression, Territoriality, and Xenophobia: A Comparative Perspective.* New York: Academic Press, 1974, pp. 123-157.

Pusey, A. E. Sex differences in the persistence and quality of the mother-offspring relationship after weaning in a population of wild chimpanzees. Paper presented at the *International Primatological Society* Meeting, Cambridge, England, August, 1976.

Quris, Rene. Ecologie et organisation sociale de *Cercocebus galeritus agilis* dan le nord-est du Gabon. *Terre et la Vie, Revue d'Ecologie Appliquee,* 1975, **29**, 337-398.

Resko, J. A. Plasma androgen levels of the rhesus monkey: Effects of age and season. *Endocrinology,* 1967, **81**, 1203-1225.

Resko, J. A. Fetal hormones and their effect on the differentiation of the central nervous system. *Federation Proceedings,* 1975, **34** (8), 1650-1655.

Reynolds, V. and Reynolds, F. Chimpanzees of the Budongo forest. In DeVore, I. (Ed.) *Primate Behavior.* New York: Holt, Rinehart and Winston, 1965, pp. 368-424.

Rosenblum, L. A. and Bromley, L. J. The effects of gonadal hormones on juvenile peer interactions in bonnet macaques. Paper presented at the *International Primatological Society* Meeting, Cambridge, England, August, 1976.

Rowell, T. E. *The social behaviour of monkeys.* Baltimore: Penguin, 1972.

Sade, D. S. Determinants of dominance in a group of free-ranging rhesus monkeys. In Altmann, S. A. (Ed.) *Social Communication Among Primates.* Chicago: University of Chicago Press, 1967, pp. 99-114.

Sassenrath, E. N. Increased adrenal responsiveness related to social stress in rhesus monkeys. *Hormones and Behavior,* 1970, **1**, 283-298.

Schenkel, R. and Schenkel-Hulliger, L. On the sociology of free-ranging Colobus (*Colobus guereza caudatus* Thomas 1885). *First Congress of the International Primatological Society* (Frankfurt, 1966). Published in Starck, D., Schneider, R., and Kuhn, H. J. (Eds.) *Neue Ergebnisse der Primatologue.* Stuttgart: Gustav Fischer, 1967, pp. 183-194.

Schwaier, A. The breeding stock of *Tupaias* at the Battelle Institute. *Laboratory Animal Handbook,* 1975, **6**, 141-149.

Smith, E. O., Fraser, M. D. and Begeman, M. L. Allometric changes in behavioral development in captive rhesus macaques, *Macaca mulatta.* Paper presented at the *Animal Behavior Society* Meeting, University Park, Pennsylvania, June, 1977.

Struhsaker, T. T. Aggression and dominance among vervet monkeys (*Cercopithecus aethiops*). *Primates,* 1966, **7** (3), 402.

Sugiyama, Y. and Ohsawa, H. Life history of male Japanese macaques at Ryozenyama. Paper presented at the *Fifth International Congress of Primatology,* Nagoya, Japan, August, 1974.

Suomi, S. J., Sackett, G. P. and Harlow, H. F. Development of sex preference in rhesus monkeys. *Developmental Psychology,* 1970, **3** (3), 326-336.

Tabei, T. and Henrichs, W. L. Metabolism of progesterone by the brain and pituitary gland in subhuman primates. *Neuroendocrinology,* 1974, **15**, 281-289.

Tarrant, L. H. Postnatal growth in the pigtailed monkey (*Macaca nemestrina*). Paper presented at the 44th Annual Meeting of the *American Association of Physical Anthropologists,* Denver, Colorado, April, 1975.

Trollope, J. and Blurton-Jones, N. G. Aspects of reproduction and reproductive behavior in *Macaca arctoides. Primates,* 1975, **16** (2), 191-205.

vanLawick-Goodall, J. Cultural elements in a chimpanzee community. In Menzel, E. W. (Ed.) *Symposia of the Fourth International Congress of Primatology (Vol. 1): Precultural Primate Behavior.* Basel: Karger, 1973, pp. 144-184.

van Wagenen, G. Menopause in a subhuman primate. *Anatomical Record,* 1970, **166**, 392.

von Holst, D., and Buergel-Goodwin, U. The influence of sex hormones on chinning by male *Tupaia belangeri*. *Journal of Comparative Physiology*, 1975, **103**, 123-151.

Wolfheim, Jaclyn H. Sex differences in behavior in a group of captive juvenile talapoin monkeys (*Miopithecus talapoin*). *Behaviour*, 1977, **63**, 110-128.

7.
Prosimians, New World Monkeys, and Rhesus

As we have seen in Chapter 4, sex differences are at least partly determined by the presence or absence of prenatal male sex hormones (androgens). Prenatal administration of testosterone in the rhesus monkey, for example, results not only in more masculinized genitals but in more masculinized behavior in infancy as well. Some of the behaviors showing increases as a result of prenatal androgen administration are rough-and-tumble play, play initiation, mounting, threatening, and aggression. Evidently, high prenatal testosterone levels, normally coming from the fetal testes of a genetically male rhesus fetus, produce basic changes in the fetal brain so that play, male sexual posturing, and aggression are more likely to occur (see Chapter 4).

Thus, there is good developmental evidence for a *prenatal* role for sex hormones. However, sex hormones also play a role in behavioral development at later ages. In Chapter 6, we reviewed the changes in behavior correlated with the tremendous hormonal increases occurring at puberty. Puberty is a second critical period for hormonal effects. Earlier *prenatal* effects, however, are usually not affected by changes at puberty, although occasionally sex differences seen earlier in life may disappear or even reverse.

In adulthood, too, sex hormones affect the behaviors of male and female primates. These effects are often cyclical, the cycles often being daily, monthly, or annual. But we will reserve the major comments on the topic of cyclicity for later. During the present chapter, we will review, taxonomically, the evidence for a relationship between adult sex hormone levels and changes in behavior.

A. REVIEW OF ADULT SEX HORMONES

All primates have essentially the same basic sex hormones. In the male, as we have seen in Chapter 4, there are androgens (e.g., testosterone), in the female there are estrogens (e.g., estradiol) and progesterone. *All* of these hormones are secreted by the gonads of *both* sexes (by the testes in the male, and by the ovaries in the female) as well as by (to a lesser extent) the adrenal cortex in both sexes. The amounts of each hormone secreted depend upon the sex of the animal. In males, more testosterone is secreted; in females, more estrogen and progesterone are secreted. In pregnant adult females, the developing placenta also produces female sex hormones.

The sex hormone levels are regulated by a feedback system through the hypothalamus and pituitary gland and back to the gonads. FSH (follicle stimulating hormone) and LH (leutenizing hormone) are gonadotrophins released by the pituitary to, in turn, control release of the gonadal hormones. In males, LH is called ICSH (interstitial cell stimulating hormone). Table 4-1 in Chapter 4 should be consulted for the direct physiological role of each of these gonadotrophins in males and females.)

But how do these hormones differentially affect the brain? How are they related to sex differences in adult behavior? The first question has already been answered, in part, in our section on the nervous system in Chapter 4. However, it will be discussed again later in a different context. The second question is the one we will try to answer in the present chapter.

B. PROSIMIAN BEHAVIOR AND ADULT SEX HORMONES

Among prosimians, very little research has been done on the role of sex hormones in adult behavior. We will discuss adult hormonal correlates in only two genera: *Tupaia* and *Lemur.*

As in birds, mice, and rats, in the tree shrew (*Tupaia* spp.) estrogen (estradiol) is concentrated and retained in certain neurons of the old periventricular brain (limbic and brainstem areas). Major concentrating areas are in the preoptic-strial, basal hypothalamic, and amygdala areas (Keefer and Stumpf, 1975a). Since estrogen usually works in a negative feedback loop on the hypothalamus to inhibit gonadotrophin releasing factors which, in turn, inhibit the release of gonadotrophins from the anterior pituitary (see Table 4-1), this finding is not surprising. In contrast, for example, the hippocampus does not seem to retain estrogen (Stumpf, Sar, and Keefer, 1974).

In the tree shrew (*Tupaia belangeri*), "chinning," a movement required for scent-marking, occurs only in males that are not castrated. Testosterone injections into adult male tree shrews increase the "chinning" response in other males. Estrogen and progesterone injections, on the other hand, decrease the response. If adult females who scent-mark are injected with testosterone, chinning decreases in the male who is present even though testosterone injections in another male who scent-marks increase "chinning" in males who smell it. (von Holst and Buergel-Goodwin, 1975a). "Chinning" does not occur before puberty. It is surprising that while the testosterone injection does not appreciably increase the chinning response directly, castration reduces it. On the other hand, ACTH (adrenocorticotrophic hormone), a stress hormone, increases chinning (von Holst and Buergel-Goodwin, 1975b).

In prosimians, there is definitely a relationship between hormones and reproductive behavior. Mating in the adult ringtail lemur (*Lemur catta*), for example, is restricted to brief periods coinciding with vaginal estrus. Relative to "higher" primates, there is minimal emancipation of sexual behavior from hormonal influences (Evans and Goy, 1968).

C. NEW WORLD MONKEY BEHAVIOR AND ADULT SEX HORMONES

As we have already seen in the chapter on puberty (Chapter 6), the sex hormones testosterone and progesterone appear in the plasma of both sexes of the common marmoset (*Callithrix jacchus*) at around 300 days of age. They reach adult levels at 500 days of age and at this time females start to cycle and become pregnant. Males are also able to inseminate females at 500 days of age but not if they are in the presence of more adult males. With adult animals present there is a suppression of testosterone in juvenile males and a suppression of progesterone in juvenile females (Abbott, 1976).

Male marmosets (e.g., *Saguinus fuscicollus*) often carry newborn infants. Epple (1975) has reported that castration of male marmosets in adulthood has no dramatic effect on their parental performance. In this species, at least, the male hormones appear to be involved in behavioral sex differences other than infant care.

With regard to hormones of pregnancy, even though marmosets are relatively primitive primates, Lanman (1977) has argued that they have a pattern of pregnancy hormones closer to that of the human than do macaques. Unfortunately, very little work has been done on marmoset sex hormones and behavior.

In the squirrel monkey (*Saimiri sciureus*), administration of estradiol (estrogen) to adult females results in more affiliative behavior, more social contact, and more female grasping of males. Even though the males are not given any hormones, estrogen administration to females increases *male* approaches and male following. On the other hand, if females receive progesterone with the estradiol, grasping by the female and following by the male decline. There are also seasonal changes in behavior in squirrel monkeys that are hormonally mediated (Anderson and Mason, 1977), as we shall see later.

Despite these changes in social contact related to hormones, there is ordinarily a preference for like-sex clumping in squirrel monkeys that persists regardless of the hormonal status of the female. Adult females usually reject adult males. Yet, curiously, subadult squirrel monkeys do not show sex segregation (Coe and Rosenblum, 1974).

The male squirrel monkey undergoes an annual testis cycle wherein a spermatogenic phase is associated with the appearance of secondary sex characteristics termed the "fatted" condition. Apparently these physical changes are associated with a weather cycle such that increased precipitation is related to the "fatted" phenomenon. Active spermatogenesis (sperm production) is evident at this time in nearly all of the seminiferous tubules of the testes (DuMond and Hutchinson, 1967).

Nadler and Rosenblum (1972), however, have determined that androgens and castration affect the body weights of both female and male squirrel monkeys. Thus, the "fatted" seasonal condition in the field is regulated by androgen, and females share the responsiveness to androgen with the males.

As mentioned earlier in the text, there are neurological sex differences related to hormones in the squirrel monkey. For example, there is a marked sex difference in the size of certain neurons of the amygdala. (They are smaller in females than in males.) (Bubenik and Brown, 1973.)

In recording from single neurons in the midbrain and pons of female squirrel monkeys, Rose (1977) found that estradiol administration would increase the responsiveness of such cells. He applied genital and other somatosensory stimuli to the animals before and after administration of estradiol. The responses of those units responding to vaginal stimulation in the midbrain, pons, tectum, central gray, raphe nuclei, and reticular formation were facilitated if the female was estrogenized.

Vestergaard (1977) has stated that the important centers for hormonal-neural regulation of sexual behavior in the limbic system in the squirrel monkey involve biogenic amines (epinephrine, norepinephrine, serotonin, dopamine, etc.) as neurotransmitters and that the different sex hormones differentially affect this transmitter.

As in the tree shrew, the distribution of estrogen-concentrating cells in the squirrel monkey brain is similar to the pattern reported for many nonmammalian and lower mammalian species. The areas (including the amygdala referred to above) are interconnected so as to permit a functional network for hormonal and sensory stimuli (Keefer and Stumpf, 1975b; Stumpf, Sar, and Keefer, 1974).

Also in squirrel monkeys, there is a correlation between dominance and testosterone, although the evidence for this relationship is not as good as it is for primates such as macaques (Mazur, 1976).

Apparently, *absolute* levels of testosterone in the adult male of a species is not a good indicator of potential sex differences in structure or behavior. Squirrel monkey adult males, for example, have higher absolute levels of testosterone than do rhesus monkey adult males (Snipes, Forest, and Migeon, 1969).

Other than research of the squirrel monkey, very little has been published on the role of adult sex hormones in New World monkey behavior. It is known, however, that capuchin and spider monkey adult males (like the squirrel monkey) have higher absolute testosterone levels than do rhesus (Snipes, Forest, and Migeon, 1969). Since spider monkeys, in particular, show little sexual dimorphism, this finding is to some degree surprising.

D. RHESUS MONKEY ADULT MALE SEX HORMONES

Circulating testosterone levels definitely affect the behavior of adult male rhesus monkeys (*Macaca mulatta*). We have already seen that prenatal androgens in the rhesus determine, to a large extent, sexual differentiation of structure and behavior. As we will see, in addition to sexual effects, testosterone also affects circadian rhythms, access to females' annual rhythms (Rose, Gordon, and Bernstein, 1972), alterations in social rank, and successful and unsuccessful agonistic encounters (Bernstein, 1974).

When Bernstein, Gordon, and Rose (1973) put two groups of rhesus monkeys together, all four males in the group that was defeated showed depressed plasma testosterone levels while the alpha male of the victorious group showed an initial *threefold* increase in plasma testosterone which rapidly returned to baseline. The same results were obtained following defeat and victory in brief single male introductions.

If dihydrotestosterone proprionate (DHTP, an androgen) is given to a middle ranking male member of a rhesus monkey group for six weeks, it leads to an elevation of that monkey to the highest rank in the group. Aggressive behavior is increased, there is a gain in body weight, and an increase in yawning. With removal of the DHTP treatment, the dominance change persists but the weight and yawning changes do not (Cochran and Perachio, 1977).

Apparently, the structural changes resulting from the androgen are not necessary once the dominance rank has shifted. However, there is some evidence that physical secondary sex characteristics *do* affect dominance. On Cayo Santiago, the social status of adult male rhesus is associated with penile baculum and glans length (DeRousseau, 1974).

Direct sexual behavioral characteristics are also affected by testosterone. There is a rank order correlation of +0.87 between position in the dominance hierarchy and total observed ejaculatory mounting series in adult male rhesus (Gordon, Rose, and Bernstein, 1977). In addition, in a group of *males, mean* testosterone concentration increases sharply with the onset of mating and peaks when the highest levels of sexual activity are observed. However, there is no consistent relationship between an *individual* male's testosterone level and his *individual* level of sexual acitvity (Gordon, Rose, and Bernstein, 1977).

It is interesting that in the rhesus, one of the most sexually dimorphic of all primates, the potential for bisexuality persists even in those males with the highest testosterone levels. Males, for example, can still display femalelike presents after extensive in utero exposure to prenatal androgen (Goy and Goldfoot, 1975). In fact, in most higher primates, as we shall see, the potential for bisexual behavior is more conspicuous than it is in nonprimates. It is very likely that the sites of male hormone actions on the central nervous system of most higher primates extend beyond hypothalamic or even limbic levels, perhaps even to the entire central nervous system (Livrea, 1972).

Interestingly enough, as far as sexual behavior is concerned, the castration of a subadult (young adult) rhesus male decreases ejaculatory performance more than castration does in full adult males. In some adult male rhesus, ejaculation continues for a full two years following castration. Administration of testosterone restores lost sexual behavior in castrates, but not completely. The testes probably produce other important male hormones which are still missing following castration. *Still,* in sexual behavior, the male rhesus shows independence from hormones to some degree (Michael, 1971).

Another paradoxical effect related to male hormones is the change in play behavior seen at puberty and in adulthood. Remember that we have shown that prenatal androgens *increase* play initiation and frequency of play in infancy. At puberty, however, androgens evidently *decrease* play. Normal male rhesus show a marked reduction in play, reaching low levels at four or five years of age. Gonadectomized males, however, continue to show moderate levels of play into their sixth and seventh years. It is as if moderate levels of androgen *increase* play but large amounts decrease it (Loy et al., 1977).

The high levels of testosterone in adult male rhesus are more correlated with aggression than with play. The male hormone, however, does not so much cause as it does facilitate the aggressive behavior. In rhesus, testosterone is directly related to behavior aimed at *achieving* and/or *maintaining* dominance, but not to dominance status itself (Mazur, 1976).

In the adult rhesus male, there are *diurnal* changes in plasma testosterone levels. Testosterone is twice as high at 10:00 P.M. as it is at 8:00 A.M. Castration removes the diurnal rhythm (corticosteroid levels decline when testosterone levels rise) (Michael, Setchell, and Plant, 1974). In addition, rhesus males demonstrate a clear-cut *annual* behavioral rhythm that may depend upon an alteration in the threshold of relevant brain areas to gonadal hormones. In the nonbreeding season, there are minimal volumes of ejaculate with reduced numbers of sperm, many of which are immobile in wild rhesus. This is apparently not true of rhesus in captivity (Zamboni, Conaway, and van Pelt, 1974). However, castrated males *still* demonstrate the annual behavioral rhythm (Michael and Wilson, 1975). In other words, in India, there is a breeding season in rhesus. The timing of the annual increase (in the fall) has also been found to persist for two years in captivity (Michael and Zumpe, 1976; Plant et al., 1974). Some researchers, however, have found no seasonal fluctuation in captivity (Resko, 1967; Zamboni, Conaway, and van Pelt, 1974).

Another interesting cycle which takes place in the rhesus is the molting (loss of hair) cycle. This occurs annually and begins on the tail and crown, goes to the legs, hips, arms, and back and terminates on the flank. Adult male and non-pregnant female rhesus are the first to molt each year whereas prepubertal young and adult females with infants are the last. Molting begins at the end of mating season and extends to the end of the birth season. Hormones probably affect the molt, but there is no evidence of this as yet (Vessey and Morrison, 1970).

Thus, although not completely dependent upon hormones, male rhesus sexual behavior is sensitive to gonadal function. There *is* evidence that androgenic hormones maintain and nearly restore normal levels of sexual behavior in gonadectomized males. There is also a relationship between circulating blood levels of testosterone and aggressive as well as sexual behavior. Using electrical stimulation of fairly specific hypothalamic regions in males freely moving in a social group, Perachio (1976) was able to elicit directed acts of aggression and sequences of male ejaculatory behavior. Castration does not eliminate the evoked aggression or sex but postcastration thresholds for the evoked behaviors increase. The threshold effect is reversed by treatment with testosterone. The effects are not always consistent, however. Stimualtion of *some* brain sites that produce aggression or sexual behavior do not induce hormone release. Preoptic and anterior hypothalamic sites produce the most consistent results (Perachio, 1976). The uptake of radioactively tagged testosterone is greater in the pituitary and hypothalamus than in any other areas of the brain (Sholiton, Taylor, and Lewis, 1974).

Even more interesting is the finding that brain stimulation sometimes does not evoke sexual behavior if the male is in the presence of a more dominant male. Thus, dominance hierarchies, which are themselves related to testosterone levels, can control both spontaneous and electrically *evoked* male sexual behavior (Perachio, Alexander, and Marr, 1973).

These findings and others have forced Phoenix (1974) to conclude:

> However essential testosterone may be for *optimum* sexual performance, hormones alone are not enough .... (P. 256)

Female hormones in males also appear to vary with *male* sex behavior. After castrated male rhesus are treated with 19-hydroxytestosterone (an androgen), levels of estradiol-17β (an estrogen) are highest in those males that fail to recover intromission and ejaculation skills and in those that have the lowest mounting rates (ICSH levels, apparently "attempting" to drive the absent testes to produce more testosterone, increase) (Phoenix, 1976).

One final possible effect of testosterone on adult male rhesus should be mentioned. The evidence is not convincing, but it is possible that castration of adult male rhesus may increase male care of infants. Only two out of ten males showed this effect in one study (Wilson and Vessey, 1968). Breuggeman (1973) however, also reported increased infant care in a male castrate.

E. RHESUS MONKEY ADULT FEMALE SEX HORMONES

Before discussing the wealth of information on the effects of female rhesus hormones on female sexual behavior during the normal menstrual cycle, we will first present some information indicating the degree of freedom from hormones which the female rhesus may possess.

In *Macaca mulatta,* pairs in the laboratory will copulate when the female is pregnant and sometimes the male will copulate to ejaculation in these cases. Ejaculations in male matings with pregnant females peak between the sixth and tenth weeks of pregnancy. There is a relationship between the estradiol/progesterone ratio in the female and degree of sexual activity throughout the first eight weeks of pregnancy. However, sexual activity between pairs ceases during the second half of pregnancy despite the existence of a high estradiol/progesterone ratio in the female. While copulation during pregnancy is not independent of female hormones, there appear to be factors other than hormonal ones at work (Bielert et al., 1976).

Outside of rhesus pregnancy, it has been suggested by Richard Michael and collaborators that volatile constituents of rhesus female vaginal secretions affect rhesus sexual behavior. They believe that these vaginal secretions are responsible for *pheromonal* effects. That is, they believe that the secretions are smelled by males and that the smells increase sexual activity (Bonsall and Michael, 1971).

According to Michael and co-workers, the vaginal secretions are mixtures containing acetic, proprionic, isobutyric, butyric, and isovaleric acids. Estrogens presumably increase these secretions while progesterone suppresses them. The production of these pheromones depends upon the bacterial content of the rhesus vagina; and, the gonadal hormones supposedly control the secretions by working on the vaginal pH (Bonsall and Michael, 1971).

Not only does estrogen allegedly increase the pheromones, it also produces a preference for the female by rhesus males. Progesterone has the reverse effect according to Everitt and Herbert (1969). Thus, the sexual behavior of rhesus males is affected by ovarian hormones in the female via pheromones. Grooming of the female by the male is also increased by ovarian hormones (Herbert, 1967). Some females, according to Herbert, are much more sexually attractive to males than are others.

With regard to sexual attractiveness, rhesus females become more attractive to males when the females are given either estrogen *or* androgen. Estrogens enhance attractiveness in a simple fashion, but the androgens' effect depends on a complicated function of dose (Johnson and Phoenix, 1976).

In research on the sexual behavior of female rhesus, the terms *pro*ceptivity and *re*ceptivity are often used. In proceptivity, the female actively seeks the male. This behavior is increased by *both* estrogen and androgen. In receptivity, the female accepts the approaches of the male. This behavior is stimulated by estrogen but *not* by androgens. Johnson and Phoenix (1976) use this evidence to support their belief that testosterone is not the primary libidinal hormone of the female.

To get back to the vaginal secretion presumably increased by estrogens, application of extracts of these secretions to the sexual skin of adult female rhesus monkeys supposedly results in stimulation of male sexual activity as evidenced by increased male mounting, ejaculation, and grooming (Keverne and Michael, 1971). Because of data such as these, Michael (1965) believes that " . . . the notion that the primate's brain is emancipated from the influence of gonadal hormones has been overestimated." (P. 4.)

Michael (1969) has also presented evidence suggesting that anosmic (rendered unable to smell) males do not mate. He calls the vaginal secretions (which are presumably pheromones) "copulins."

The research of Michael and his collaborators regarding vaginal pheromones has been called into question by Goldfoot, Goy, and co-workers. They point out that ejaculate from a male rhesus can cause up to a fivefold elevation in vaginal aliphatic acid concentration. They also maintain that application of vaginal lavages from estrogenized donors does *not* increase copulation with spayed (gonadectomized) nonestrogenized females and that increases in concentrations of aliphatic acids are *not* always associated with increased copulation. They believe that increased sex depends more upon associative learning or upon extinction and disinhibition of sexual interest than upon pheromones:

> . . . the particular odor of a partner is *not* the determining characteristic which initiates sexual activity. Rather, either by innate mechanisms, or, much more likely, by associative learning, particular odors may be one additional cue, not always reliable, which tells the male something about his chances of success with a potential sexual partner. (Goldfoot et al., 1976)

Clearly, this presents a point of view quite divergent from that of Michael. Goldfoot (1977), however, has recently written that new data published by Michael support the Goldfoot findings. In Michael's new data, there are highly variable results; only three out of twelve males respond to the smell of aliphatic acids with more than one female. In addition, in 50 percent of Michael's new pairs, the male shows no arousal at all. In addition, one other difference between the results of Goldfoot and those of Michael has been clarified. Goldfoot had originally found a luteal not an ovulatory peak in vaginal acids while Michael had found an ovulatory peak. Now Michael *too* finds a luteal peak (cf. Goldfoot, 1977).

Aside from the pheromonal question, however, estrogens and progesterones *do,* in fact, affect adult female rhesus monkey sexual behavior. Adult rhesus female sexual solicitations (but not presents) peak at midcycle (Cochran, 1977). At ovulation (midcycle), time spent near a male increases, male ejaculation increases, and female sexual skin coloration (red) increases (Czaja and Bielert, 1975). It is felt by some that rhesus ovarian hormones have a greater influence on sexual behavior in a social context than in a standard pair test paradigm (Gordon, 1977).

In a captive field station situation, sexual receptivity (acceptance of male approaches) is confined to a discrete period of four to seventeen days midcycle and terminates abruptly as progesterone concentrations rise. Most rhesus females in such a situation are impregnated on the first ovulatory cycle of the breeding season (Gordon, Rose, and Bernstein, 1977). In this study, a breeding season *did* occur in captivity.

In other laboratories, sexual activity, while increasing in the fall, is low in the summer months. Relative amenorrhea in rhesus females where menstruation is infrequent and irregular is also common in the summer months (Keverne and Michael, 1970).

Female rhesus do not always elicit sexual behavior from males. On occasion, an adult male rhesus will spontaneously threaten or aggress a female, but it is more likely to occur in response to a female's threat or to a female's eating. Females sometimes, but less frequently, threaten males, usually in response to male threats or attempts to mount. This aggression between the sexes, however, usually becomes redirected aggression if *estrogen* levels are high and a sexual bond is being established. Attacks on females by females are more frequent at around menstruation (Sassenrath, Rowell, and Hendrickx, 1973). Ovariectomy increases male–female aggression, particularly in the male. Progesterone increases female aggression by decreasing her tolerance of males (Michael and Zumpe, 1970a). Threats made by the female toward the male are higher *away* from the ovulation period (midcycle) and less frequent during ovulation. Redirection of threats by the male are also more frequent at ovulation. When the female is treated with progesterone instead of estrogen, male threats are more likely to be directed toward than away from the female (Michael, 1970).

If females are trained to press a lever to admit a rhesus male, they will do so most readily at ovulation. Estrogen administration increases lever pressing for males, while progesterone treatment decreases it (Michael and Keverne, 1972). Both mounting and grooming by the male are related to estrogen in the female. High progesterone in female rhesus monkeys, on the other hand, decreases mounts and ejaculations. Even combined estrogen-progestin oral contraceptives given to females decrease male sexuality (Michael, 1969).

With regard to specific kinds of female sexual responsiveness, the sexual hand-reach, head-duck, and head-bob (sexual invitations by females) decrease after ovariectomy (Michael and Zumpe, 1970b).

When females are treated with estradiol, estrone (another estrogen), testosterone, and dihydrotestosterone at different times, the different hormones produce different changes. Ejaculations in males paired with these females increase when the females are treated with estradiol, estrone, *or* testosterone (but not dihydrotestosterone). The ejaculatory frequencies are not significantly greater after a female is treated with estradiol or estrone than when she is treated with the male hormone testosterone. Moreover, testosterone (but again not dihydrotestosterone) also increases female proceptivity (active sexual pursuit by the female) as much as do the two estrogens. However, the *pattern* of proceptivity is different. Under testosterone, rather than just increased proximity by the female, there is increased hand-slapping and shoulder flexing (female solicits). Female receptivity (accepting male advances) is little affected by all hormones except DHTP (dihydrotestosterone). The latter *decreases* receptivity. Interestingly enough, both androgens (testosterone and DHTP) reliably increase female yawning. Since yawning is a behavior typically seen in *males* during sexual behavior, the androgens apparently masculinize the female sexual response, one (testosterone) while increasing sexual interest, the other (DHTP) while decreasing sexual interest. Clearly the answer to the question of whether androgens increase rhesus female libido, depends upon the specific androgen in question (Wallen and Goy, 1977).

All of this is *not* to say that female hormones do not also affect female libido. They do. A clutching reaction often seen in adult female rhesus sexual behavior, assumed by some to be homologous to the human female orgasm, is decreased by bilateral ovariectomy. Estrogen treatment restores the response; progesterone decreases it (Zumpe and Michael, 1968). As already pointed out, female invitations for sex (presents, hand-reaches, and head-ducks) are affected in similar ways by ovariectomy, estrogen, and progesterone, respectively (Zumpe and Michael, 1970a). Moreover, redirected threats show the same pattern of differences (Zumpe and Michael, 1970b). Thus, the entire pattern of female sexual excitement and female sexual behavior can, indeed, be related to estrogen but not to progesterone.

But, getting back to the male hormones, some of these also affect female excitement. As we have seen, testosterone increases female libido while DHTP does not. Also we have seen that DHTP increases yawning (a male behavior) in female rhesus. It is interesting that in studies of prenatal effects of androgens, it has been found that DHTP produces less genital virilization but about as much behavioral masculinization in infancy as does testosterone (Goy, 1976).

If DHTP is given to a middle ranking adult female for six weeks, it increases her aggressive behavior and leads to her elevation to the highest dominance rank in the group. (The same would be true for middle ranking males, although they would not become more aggressive.) Also following DHTP treatment, adult females gain body weight. When treatment is terminated, the dominance change remains while yawning and weight return to baseline (Cochran and Perachio, 1977).

Assuming that we are no longer talking about DHTP alone, where would the androgen come from in a normal female if it ordinarily contributes to female sexual excitement? Both the ovaries and the adrenal glands produce testosterone in female organisms. It is known however, that adrenalectomy decreases female sexual receptivity in the rhesus. Evidently, testosterone increases female libido by affecting the female's central nervous system via the neurotransmitter 5-hydroxytryptamine (5-HT or serotonin), a biogenic amine. After adrenalectomy, 5-HT can reverse the effects on sexual receptivity (Gradwell, Everitt, and Herbert, 1975; Everitt and Herbert, 1969).

There is an interesting relationship between adult male sexual behavior, adult female sexual behavior, and the male and female sex hormones in rhesus. If adult female rhesus are allowed to press a bar for a male companion, both estrogen *and* testosterone treatment (given to her) will increase her pressing. However, while androgen increases both her bar pressing and her receptivity, it does not increase her attractiveness to the male. Only estrogen does this. When a female rhesus is given androgen and an opportunity to interact with a male, she will be interested in copulating, but the male will not (Michael and Keverne, 1972). It may be that, while androgen affects proceptivity through the female's central nervous system, estrogen affects attractiveness through her genitalia and external appearance (Trimble and Herbert, 1968).

Again, so as not to overstate our case for the role of testosterone on the neural mechanisms of female excitement and receptivity, let us examine the female nervous system with an eye out for the role of the female sex hormones. Bilateral lesions made in the ventral preoptic-anterior hypothalamic area in adult female rhesus monkeys block spontaneous ovulation and interfere with the hypothalamic-pituitary-gonadal hormone feedback system. Under these conditions, estrogen increases do not result in the release of LH (Norman, Resko, and Spies, 1976). Estradiol (an estrogen) has a dual action on the nervous system, however. It releases LHRH (leutenizing hormone releasing factor) from the hypothalamus which leads to LH release from the pituitary, *and* it has an inhibitory effect directly on the pituitary to inhibit *it* (Spies and Norman, 1975).

Depending upon early prenatal androgen, the hypothalamus is organized with a built-in cyclicity or with no cyclicity:

> . . . tonic gonadotrophic release can be maintained by the MBH (medial basal hypothalamus) alone, but connections with the anterior hypothalamus are necessary for cyclicity . . . one of the functions of early exposure to androgen is to disrupt this connection between the anterior hypothalamus and the medial basal hypothalamus. (Valenstein, 1968, P. 30)

The metabolism of female hormones by the brain (hypothalamus) and pituitary gland apparently changes at puberty in female primates; it does not change with sexual maturity in males (Tabei and Heinrichs, 1974). Thus, the male hypothalamus has neither cyclicity nor a pubertal change in responsiveness to estrogen and progesterone built in.

In addition, fish, amphibians, birds, mice, rats, tree shrews, and squirrel monkeys have sex-hormone-concentrating neurons in similar specific preoptic hypothalamic and limbic neural structures. It should not surprise any of us by now that rhesus monkeys also have them (Pfaff et al., 1976).

Evidence also exists which links sex differences in gonadal hormones to differential aspects of rhesus monkey male and female central nervous systems. Lesions in the amygdala affect adult male and female monkeys differently. There are increases in sexual behavior and arousal in both sexes but only in females is there a paradoxical increase in aggression (Kling, 1974). In males, there is a reduction in aggressiveness following lesions in the amygdala. Kling (1974) feels this sex difference is mediated via connections to preoptic neurons in the hypothalamus (see also Kling, 1975; Kling and Dunne, 1976). As we have seen earlier some primates other than rhesus show sex differences in the neurons of the amygdala. This may also be true of rhesus. In any case, the exact nature of the hormonally related central nervous system sex differences have not been demonstrated; it is only known that they do, in fact, exist.

Many of the same sex differences seen in rhesus monkeys with regard to responsiveness to different sex hormones are also apparent in other Old World monkeys and in the great apes. In our next chapter, we will discuss sex differences in adult sex hormones in these species as well as in men and women. Some general comments on the role of sex hormones in adult primates will also be presented.

REFERENCES

Abbott, D. H. Hormones and behaviour during puberty in the marmoset. Paper presented at the *International Primatological Society* Meeting, Cambridge, England, August, 1976.

Anderson, C. and Mason, W. A. Hormones and social behavior of squirrel monkeys (*Saimiri sciureus*): I. Effects of endocrine status of females on behavior within heterosexual pairs. *Hormones and Behavior*, 1977, **8**, 100-106.

Bernstein, I. S. Behavioral and environmental events influencing primate testosterone levels. *Journal of Human Evolution*, 1974, **3**, 517-525.

Bernstein, I. S., Gordon, T. P. and Rose, R. M. Plasma testosterone changes following merger of two rhesus monkey groups. *American Zoologist*, 1973, **13**, 1267.

Bielert, C., Czaja, J. A., Eisele, S., Scheffler, G., Robinson, J. A. and Goy, R. W. Mating in the rhesus monkey (*Macaca mulatta*) after conception and its relationship to oestradiol and progesterone levels throughout pregnancy. *Journal of Reproduction and Fertility*, 1976, **46**, 179-187.

Bonsall, R. W. and Michael, R. P. Volatile constituents of primate vaginal secretions. *Journal of Reproduction and Fertility*, 1971, **27**, 478-479.

Breuggeman, J. A. Parental care in a group of free-ranging rhesus monkeys (*Macaca mulatta*). *Folia Primatologica*, 1973, **20**, 178-210.

Bubenik, G. A. and Brown, G. M. Morphologic sex differences in the primate brain areas involved in regulation of reproductive activity. *Experientia*, 1973, **15** (5), 619-621.

Cochran, C. G. Reproductive strategies in female rhesus monkeys. Paper presented at the *Animal Behavior Society* Meeting, University Park, Pennsylvania, June, 1977.

Cochran, C. G. and Perachio, A. A. Dihydrotestosterone proprionate effects on dominance and sexual behaviors in gonadectomized male and female rhesus monkeys. *Hormones and Behavior*, 1977, **8**, 175-187.

Coe, C. L. and Rosenblum, L. A. Sexual segregation and its ontogeny in squirrel monkey social structure. *Journal of Human Evolution*, 1974, **3**, 551-556.

Czaja, J. A. and Bielert, C. Female rhesus sexual behavior and distance to a male partner: Relation to stage of menstrual cycle. *Archives of Sexual Behavior*, 1975, **4**, 583-597.

DeRousseau, C. J. Variabilty of accessory sex structures in *Macaca mulatta*. *American Journal of Physical Anthropology*, 1974, **41**, 475.

DuMond, F. V. and Hutchinson, T. C. Squirrel monkey reproduction: The "fatted" male phenomenon and seasonal spermatogenesis. *Science*, 1967, **158**, 1467-1470.

Epple, G. Parental behavior in *Saguinus fuscicollis ssp.* (*Callithricidae*). *Folia Primatologica*, 1975, **24**, 221-238.

Evans, C. S. and Goy, R. W. Social behavior and reproductive cycles in captive ringtailed lemurs (*Lemur catta*). *Journal of Zoology of London*, 1968, **156**, 181-197.

Everitt, B. J. and Herbert, J. Adrenal glands and sexual receptivity in female rhesus monkeys. *Nature*, 1969, **222**, 1065-1066.

Goldfoot, D. A. Olfactory influences on sexual behavior. *Primate News,* 1977, **15** (6), 3-7.

Goldfoot, D. A., Kravetz, M. A., Goy, R. W. and Freeman, S. K. Lack of effect of vaginal lavages and aliphatic acids in ejaculatory responses in rhesus monkeys: Behavioral and chemical analyses. *Hormones and Behavior,* 1976, **7**, 1-27.

Gordon, T. P. The influence of ovarian hormones in male sexual behavior in a social group of rhesus monkeys. Paper presented at the *Animal Behavior Society* Meeting, University Park, Pennsylvania, June, 1977.

Gordon, T. P., Rose, R. M. and Bernstein, I. S. Social and hormonal influences on sexual behavior in the rhesus monkey. Paper presented at the *American Society of Primatologists* Meeting, Seattle, Washington, April, 1977.

Goy, R. W. Hormonal and environmental influences on sexual behaviour in rhesus monkeys. Paper presented at the *International Primatological Society* Meeting, Cambridge, England, August, 1976.

Goy, R. W. and Goldfoot, D. A. Neuroendocrinology: Animal models and problems of human sexuality. *Archives of Sexual Behavior,* 1975, **4** (4), 405-420.

Gradwell, P. B., Everitt, B. J. and Herbert, J. 5-hydroxytryptamine in the central nervous system and sexual receptivity of female rhesus monkeys. *Brain Research,* 1975, **88**, 281-293.

Herbert, J. The social modification of sexual and other behaviour in the rhesus monkey. *Proceedings of the First International Congress of Primatology,* 1967, **1**, 232-246.

Johnson, D. F. and Phoenix, C. H. Hormonal control of female sexual attractiveness, proceptivity, and receptivity in rhesus monkeys. *Journal of Comparative and Physiological Psychology,* 1976, **90**, 473-483.

Jones, K. L., Shainberg, L. W. and Byer, C. O. *Sex and People.* New York: Harper & Row, 1977.

Keefer, D. A. and Stumpf, W. E. Estrogen-concentrating neuron systems in the brain of the tree shrew. *General and Comparative Endocrinology,* 1975, **26** (4), 504-516. (a)

Keefer, D. A. and Stumpf, W. E. Atlas of estrogen-concentrating cells in the central nervous system of the squirrel monkey. *Journal of Comparative Neurology,* 1975, **160** (4), 419-441. (b)

Keverne, E. B. and Michael, R. P. Annual changes in the menstruation of rhesus monkeys. *Journal of Endocrinology,* 1970, **48**, 669-670.

Keverne, E. B. and Michael, R. P. Sex-attractant properties of ether extracts of vaginal secretions from rhesus monkeys. *Journal of Endocrinology,* 1971, **51**, 313-322.

Kling, A. Differential effects of amygdalectomy in male and female nonhuman primates. *Archives of Sexual Behavior,* 1974, **3** (2), 129-134.

Kling, A. Brain lesions and aggressive behavior of monkeys in free-living groups. In Fields, W. S. and Sweet, W. H. (Eds.) *Neural Bases of Violence and Aggression.* St. Louis: W. H. Green, Inc., 1975, pp. 146-160.

Kling, A. and Dunne, K. Social-environmental factors affecting behavior and plasma testosterone in normal and amygdala lesioned *Macaca speciosa. Primates,* 1976, **17** (1), 23-42.

Lanman, J. T. Parturition in nonhuman primates. *Biology of Reproduction,* 1977, **16**, 28-38.

Livrea, G. La base neuroendocrina del comportamento sessuale. *Archives of Science and Biology,* 1972, **55**, 61-102.

Loy, J., Loy, K., Patterson, D., Keifer, G. and Conaway, C. H. Play behavior of gonadectomized rhesus monkeys. Paper presented at the *Animal Behavior Society* Meeting, University Park, Pennsylvania, June, 1977.

Mazur, A. Effects of testosterone on status in primate groups. *Folia Primatologica,* 1976, **26**, 214-226.

Michael, R. P. Some aspects of the endocrine control of sexual activity in primates. *Proceedings of Royal Society of Medicine,* 1965, **58**, 595-598.

Michael, R. P. Behavioral effects of gonadal hormones and contraceptive steroids in primates. In Solhanick, H. A. (Ed.) *Metabolic Effects of Gonadal Hormones and Contraceptive Steroids.* New York: Plenum, 1969, pp. 706-721.

Michael, R. P. The role of pheromones in the communication of primate behavior. *Recent Advances in Primatology,* 1969, **1**, 101-107.

Michael, R. P. Hormonal factors and aggressive behaviour in the rhesus monkey. *Proceedings of the International Society of Psychoendocrinology,* 1970 (Published in Ford, D. H. (Ed.) *Influence of Hormones on the Nervous System Symposium.* Basel: Karger, 1971, pp. 412-423.)

Michael, R. P. Determinants of primate reproductive behaviour. *Symposium on the Use of Nonhuman Primates for Research.* Sukhumi, USSR, December 13-17, 1971.

Michael, R. P. and Keverne, E. B. Differences in the effects of oestrogen and androgen on the sexual motivation of female rhesus monkeys. *Journal of Endocrinology,* 1972, **55**, 40.

Michael, R. P., Setchell, K. D. R. and Plant, T. M. Diurnal changes in plasma testosterone and studies on plasma corticosteroids in non-anesthetized male rhesus monkeys (*Macaca mulatta*). *Journal of Endocrinology,* 1974, **63**, 325-335.

Michael, R. P. and Wilson, M. I. Mating seasonality in castrated male rhesus monkeys. *Journal of Reproduction and Fertility,* 1975, **43**, 325-328.

Michael, R. P. and Zumpe, D. Aggression and gonadal hormones in captive rhesus monkeys (*Macaca mulatta*). *Animal Behaviour,* 1970, **18**, 1-10. (a)

Michael, R. P. and Zumpe, D. Sexual initiating behaviour by female rhesus monkeys (*Macaca mulatta*) under laboratory conditions. *Behaviour*, 1970, **36**, 168-185. (b)

Michael, R. P. and Zumpe, D. Environmental and endocrine factors influencing annual changes in sexual potency in primates. *Psychoneuroendocrinology*, 1976, **1**, 303-313.

Nadler, R. D. and Rosenblum, L. A. Hormonal regulation of the "fatted" phenomenon in squirrel monkeys. *Anatomical Record*, 1972, **173**, 181-187.

Norman, R. L., Resko, J. A. and Spies, H. G. The anterior hypothalamus: How it affects gonadotropin secretion in the rhesus monkey. *Endocrinology*, 1976, **99** (1), 59-71.

Perachio, A. A. Hypothalamic regulation of behavioural and hormonal aspects of aggressive and sexual performance. Paper presented at the *International Primatological Society* Meeting, Cambridge, England, August, 1976.

Perachio, A. A., Alexander, M. and Marr, L. D. Hormonal and social factors affecting evoked sexual behavior in rhesus monkeys. *American Journal of Physical Anthropology*, 1973, **38**, 227-232.

Pfaff, D. W., Gerlach, J. L., McEwen, B. S., Ferin, M., Carmel, P. and Zimmerman, E. A. Autoradiographic localization of hormone-concentrating cells in the brain of the female rhesus monkey. *Journal of Comparative Neurology*, 1976, **170**, 279-293.

Phoenix, C. H. The role of androgens in the sexual behavior of adult male rhesus monkeys. In Montagna, W. and Sadler, W. A. (Eds.) *Reproductive Behavior.* New York: Plenum, 1974, pp. 249-258.

Phoenix, C. H. Sexual behavior of castrated male rhesus monkeys treated with 19-hydroxytestosterone. *Physiology and Behavior*, 1976, **16**, 305-310.

Plant, T. M., Zumpe, D., Sauls, M. and Michael, R. P. An annual rhythm in the plasma testosterone of adult male rhesus monkeys maintained in the laboratory. *Journal of Endocrinology*, 1974, **62**, 403-404.

Resko, J. A. Plasma androgen levels of the rhesus monkey: Effects of age and season. *Endocrinology*, 1967, **81**, 1203-1225.

Rose, J. D. Facilitation by estradiol of single unit responses to genital and somatosensory stimuli in the midbrain and pons of the female squirrel monkey. *Anatomical Record*, 1977, **187**, 697-698.

Rose, R. M., Gordon, T. P. and Bernstein, I. S. Plasma testosterone levels in the male rhesus: Influences of sexual and social stimuli. *Science*, 1972, **178**, 643-645.

Sassenrath, E. N., Rowell, T. E. and Hendrickx, A. G. Perimenstrual aggression in groups of female rhesus monkeys. *Journal of Reproduction and Fertility*, 1973, **34**, 509-511.

Sholiton, L. J., Taylor, B. B. and Lewis, H. P. The uptake and metabolism of labelled testosterone by the brain and pituitary of the male rhesus monkey (*Macaca mulatta*). *Steroids*, 1974, **24** (4), 537-547.

Snipes, C. A., Forest, M. G. and Migeon, C. J. Plasma androgen concentration in several species of Old and New World monkeys. *Endocrinology*, 1969, **85** (5), 941-945.

Spies, H. G. and Norman, R. L. Interaction of estradiol and LHRH on LH release in rhesus females: Evidence for a neural site of action. *Endocrinology*, 1975, **97** (3), 685-692.

Stumpf, W. E., Sar, M. and Keefer, D. A. Anatomical distribution of estrogen in the central nervous system of mouse, rat, tree shrew, and squirrel monkey. *Advances in Bioscience*, 1974, **15**, 77-88.

Tabei, T. and Heinrichs, W. L. Metabolism of progesterone by the brain and pituitary gland in subhuman primates. *Neuroendocrinology*, 1974, **15**, 281-289.

Trimble, M. R. and Herbert, J. The effect of testosterone or oestradiol upon the sexual and associated behaviour of the adult female rhesus monkey. *Journal of Endocrinology*, 1968, **42**, 171-185.

Valenstein, E. S. Steroid hormones and the neuropsychology of development. In Isaacson, R. L. (Ed.) *The Neuropsychology of Development.* New York: Wiley, 1968, pp. 1-39.

Vessey, S. H. and Morrison, J. A. Molt in free-ranging rhesus monkeys, *Macaca mulatta. Journal of Mammalogy*, 1970, **51**, 89-93.

Vestergaard, P. Sexual mechanisms in the brain: Neurophysiological and neurochemical aspects. *Ugeskr. Laeg.*, 1977, **139**, 10-13.

von Holst, D. and Buergel-Goodwin, U. Chinning by male *Tapaia belangeri:* the effects of scent marks of conspecifics and of other species. *Journal of Comparative Physiology*, 1975, **103**, 153-171. (a)

von Holst, D. and Buergel-Goodwin, U. The influence of sex hormones on chinning by male *Tupaia belangeri. Journal of Comparative Physiology*, 1975, **103**, 123-151. (b)

Wallen, K. and Goy, R. W. Rhesus female sexual behavior: The effect of estradiol, estrone, testosterone, and dihydrotestosterone on female proceptivity, receptivity, yawning behavior, and male ejaculation. Paper presented at the *American Society of Primatologists* Meeting, Seattle, Washington, April, 1977.

Wilson, A. P. and Vessey, S. H. Behavior of free-ranging castrated rhesus monkeys. *Folia Primatologica*, 1968, **9**, 1-14.

Zamboni, L., Conaway, C. H. and van Pelt, L. Seasonal changes in production of semen in free-ranging rhesus monkeys. *Biology of Reproduction*, 1974, **11**, 251-267.

Zumpe, D. and Michael, R. P. The clutching reaction and orgasm in the female rhesus monkey (*Macaca mulatta*). *Journal of Endocrinology*, 1968, **40**, 117-123.

Zumpe, D. and Michael, R. P. Ovarian hormones and female sexual invitations in captive rhesus monkeys (*Macaca mulatta*). *Animal Behaviour*, 1970, **18**, 293-301. (a)

Zumpe, D. and Michael, R. P. Redirected aggression and gonadal hormones in captive rhesus monkeys (*Macaca mulatta*). *Animal Behaviour*, 1970, **18**, 11-19. (b)

8.

Generalizations, and Speculations

We have seen, in Chapter 7, that *testosterone* in adult rhesus males increases aggressiveness, dominance, and sexual behavior. In females, it increases sexual proceptivity and receptivity, but not sexual attractiveness. *Estrogen* in adult female rhesus increases sexual proceptivity, receptivity, and attractiveness, while *progesterone* decreases sexual attractiveness. In fact, in a number of species these generalizations appear to be true.

At ovulation, female sexual interaction increases. During the luteal phase of a female's cycle, when levels of progesterone are high, sexual interaction declines. The effect of progesterone has been attributed to a loss of female attractiveness more than to a loss in receptivity. Some say this is because progesterone acts directly on the vagina to reduce estrogen-induced vaginal olfactory cues for the male (Baum, 1976). We have also seen that the effect of an androgen depends upon the *kind* of androgen (e.g., DHTP or testosterone) we are talking about.

Now that we have reviewed what we have learned to this point, let us see to what extent we can generalize these findings on sex hormones in adults to other nonhuman primates.

A. OTHER OLD WORLD MONKEYS: ADULT HORMONES

The rhesus monkey (*Macaca mulatta*) has, as usual, been studied more than any other Old World monkey. However, there are data on other macaques, baboons, the talapoin, and one paper on the mangabey. Let us start with macaques.

In the pigtail macaque (*Macaca nemestrina*), among males living in captive heterosexual groups, testosterone levels are lower for sexually mature (but not fully mature) adult males than they are for fully mature adults. The testosterone levels of adult pigtail males are similar to those of rhesus males. Male adult pigtail monkeys, living without females show a rise in testosterone levels (roughly doubling) in two weeks following introduction of adult females to their group. Pigtail males in captivity have been reported to *not* show a seasonal testosterone response (Bernstein et al., 1977).

If pigtail adult males are immobilized or restrained, a spermatogenic arrest usually follows (Cockett et al., 1970) and testicular degeneration occurs (Zemjanis et al., 1970).

The preovulatory adult female pigtail monkey displays more sexual and social behavior toward males if she is socially dominant. Females low in social dominance rarely conceive and show low frequencies of sexual behavior *even* in the preovulatory condition (Goldfoot, 1971).

A female hormone not yet mentioned in our discussions of rhesus has been studied in pigtail females. For the first few days following delivery of an infant, increased levels of plasma *prolactin* are produced in female primates. Behavioral preference for her infant apparently coincides with the increased titers of prolactin in the pigtail female, particularly at around ten days postpartum (Gross, Schiller, and Bowden, 1977).

There is some evidence that pigtail macaques may provide better physiological models for human reproduction than the rhesus. Like the diurnal human sex hormone cycle, diurnal sexual behavior cycles in the pigtail are highest in the morning and lowest in the evening (Martenson et al., unpublished). Also:

> The similarity of plasma sex hormone changes during the menstrual cycle between women and the pigtail macaque suggested that this nonhuman primate should be a useful animal model for studying human reproduction. (Steiner et al., 1977, P. 217)

Among stumptail macaques (*Macaca arctoides*), plasma testosterone levels have been shown to be related to social rank in both males and females (Kling and Dunne, 1976). Testosterone has also been linked to baldness in adult males of the stumptail species. Baldness can be induced in adult males by long-term injections of testosterone (Takashima and Montagna, 1971). There are regional differences in the hair follicles of the scalp skin in the stumptail males with regard to testosterone metabolism (Takashima, 1974). For example, testosterone is catabolized much faster and is retained less by follicles above the forehead than by terminal follicles (Takashima, Adachi, and Montagna, 1970).

With regard to adult female hormones in the stumptail macaque, ovariectomy of adult females affects male–female interactions. While individual males differ greatly in the way they respond to spaying of their preferred sexual partners, in some cases there is a marked decline in male sexual activity. In this species, ovarian hormones apparently have variable activating effects on male–female interactions (Slob, 1975).

On the other hand, both male and female yawning peak during the female's luteal phase, male grooming of the female is lowest before menstruation, and female grooming of the male declines sharply just after ovulation. After ovariectomy, females usually show a decline in sexual presents and reaching back (the clutching reaction); however, they also often show a surprising *increase* in grooming and their consorting males, as already mentioned above, often continue to copulate and groom with them although they may approach and mount less often (Slob et al., 1975).

In the closely related Japanese monkey (*Macaca fuscata*), testosterone levels evidently do not correlate well with dominance or aggression within a troop. The adult male's level of androgen is apparently less important than are social stimuli (Eaton and Resko, 1974). During gestation, estrogen and progesterone do not markedly increase as they do in chimpanzees and humans (Nomura and Ohsawa, 1976).

The Japanese monkey has a breeding season. The hormonal factor determining the breeding or nonbreeding season is apparently not of pituitary origin but lies in higher parts of the central nervous system and involves leutenizing hormone releasing factors in the hypothalamus (Hayashi et al., 1975). There are either no menstrual cycles or irregular ones outside of the breeding season (Nigi, 1975).

In a study of sixty sexually mature female Japanese monkeys in a natural troop imported intact from Japan in 1972 to a ranch in south Texas, it was found that, even in this troop, the breeding season was retained (Wolfe, 1977).

Very little else has been published on other macaque species. Rosenblum and Bromley (1976), however, have noted dramatic changes in play, activity, and male–female interaction in the expected directions after treating male and female juvenile bonnet macaques (*M. radiata*) with estradiol or testosterone. In addition, Dang (1977) has noted that caesarian sections do not appear to deleteriously affect the menstrual cycles of crab-eater macaques. When caesarian sections were performed on thirty female crab-eaters (*Macaca fascicularis*) between the thirty-seventh and seventy-eighth days of pregnancy, ovulation occurred before menstruation and females were immediately fertile again. Menstruation reappeared within twenty-two to fifty-three days and cycles became regular immediately upon resumption of menstruation regardless of the age of the fetus at the time of the caesarian (Dang, 1977).

Among chacma baboons (*Papio ursinus*), females in estrus have a higher dominance rank than those out of estrus (Bolwig, 1959), while in gelada baboons (*Theropithecus gelada*) estrus has little effect on the behavior of either female or male adults. Neither sex spends any more time monitoring or interacting with the other during the estrous cycle. Moreover, social relationships in the troop as a whole are also unaffected except that estrous females are sometimes harassed by more dominant females (Dunbar, 1976).

As in the geladas, yellow baboon (*Papio cynocephalus*) *female* dominance also does not change during estrus although the number of agonistic bouts does change. Unlike geladas, however, male yellow baboon dominance changes *are* correlated with the presence of adult females in estrus. The number of agonistic bouts between males *decreases* during female estrus whereas the number of agonistic bouts between females *increases*. When bouts do occur during estrus, they are far more likely to lead to wounding than if they occur outside of estrus (Hausfater, 1975).

Among members of the species *Papio anubis,* estrus causes only slight changes in intrasexual social organization. It certainly cannot be regarded as a major disruptive factor, although there is some change in dominance during estrus in the low-ranking females (Rowell, 1969).

As in macaque immobilization, a period of quarantine or isolation affects hormonal levels in baboons. Lower conception rates and frank amenorrhea are common in the period immediately following quarantine (Hagino, 1976).

Ovulation and menstruation in baboons is regulated at the hypothalamic level. A frontal cut in the medial preoptic area produces anovulation and amenorrhea. There is, however, a temporary recovery of the cyclical release of LH, of ovulation, and of menstruation soon after the lesion. Apparently, there are two levels in the hypothalamus which regulate gonadotrophin release. One of them is outside of the preoptic area and it has an effect on hormonal periodicity (Hagino, 1976).

Male mangabeys (*Cercocebus albigena johnstoni*) show less frequent aggression and more grooming toward an adult female if she is sexually swollen (Chalmers, 1968). There are coloration changes (black to white or white to black) in the nose patch of the female patas monkey (*Erythrocebus patas*) which are correlated with hormonal changes (Loy, 1974). And, in talapoins (*Miopithecus talapoin*), testosterone stimulates mounting behavior in males but only in the most dominant males. The effects of gonadal hormones on female talapoins are apparently less affected by social status (Dixson, Herbert, and Rudd, 1972). Treating female talapoins with estradiol increases their aggressive behavior only slightly. Thus, in talapoins, gonadal hormones have effects that depend upon the preceding social structure (Dixson and Herbert, 1974). Partial support for this statement comes from a study by Scruton and Herbert (1972) in which treatment with estradiol of a new adult female talapoin introduced to an already stable group did not enhance the group's acceptance or rejection of her. (The resident female still attacks the strange female.) However, the estradiol did increase the stranger's sexual activity with the resident male.

B. ADULT SEX HORMONES IN THE GREAT APES

In reviewing the female sex hormones of the chimpanzee (*Pan troglodytes*) from a developmental perspective, we have seen that, as far as the pattern of pregnancy hormones is concerned, the chimpanzee female is closer to the human female than is the macaque (Lanman, 1977). Data comparing endocrinological patterns of sex hormones show that chimpanzees and humans, but not rhesus monkeys, have what has been referred to as a "definitive fetoplacental unit." (There is a lack of serum estriol from the fetus and placenta in the rhesus.) The levels of hormones are much more like human levels in the chimpanzee than in the rhesus. Midcycle LH peaks are longer in chimpanzees and humans (Hobson, 1976).

In infancy, as we have seen, serum FSH levels are higher in female than in male chimpanzees. The same is true for humans. The day-to-day variability for LH and FSH is greater in female than in male infant chimpanzees. Again, the same is true for humans. Finally, the FSH levels for infant chimpanzee females but not for males, show periodic variation in a cycle ranging from 8.3 to 25.0 days (Faiman, Winter, and Grotts, 1973).

During pregnancy, according to Nissen and Yerkes (1943), the female chimpanzee is more gentle, patient, friendly, and ingratiating with human attendants, especially if she is primiparous (carrying her first child). From one week to one or two months following delivery, however, she returns to her usual mean, irritable, or pugnacious self, presumably in concert with hormonal changes (from progesterone to estrogen?). Birch and Clark (1950) have noted that treatment with estrogen increases the dominance rank of female chimpanzees and that this change is correlated with reddening and swelling of the sex skin. Chimpanzee females also very clearly have an estrus associated with this swelling during which sexual behavior increases (Jensen, 1976). As we have seen, female menstrual cycles probably continue until death (at or before age fifty), there being no menopause as in humans or rhesus (Graham and McClure, 1977).

With regard to the male sex hormones, studies have indicated that changes in testosterone can be related to changes in dominance behavior but that dominance in the chimpanzee is less overt and violent than it is in macaques. In any case, the hormones do not cause, but rather facilitate or inhibit behavior (Mazur, 1976).

Adult hormones in the other great apes have been studied very little. Nadler (1977) has reported evidence of behavioral cyclicity in orangutans and Maple and Zucker (1977) have data which suggest that there may be cyclical proceptivity (female initiation) as well as receptivity in the adult female orangutan. There is also cyclicity in the tumescence of the perineal labia in female lowland gorillas (Nadler, 1975).

C. ADULT HORMONES IN *HOMO SAPIENS* (INCLUDING SPECULATION)

In humans, there are adult male sexual rhythms in sperm production and hormonal production (adrenal and testicular). Androgenic (male) hormones are at their highest in the morning, just after sunrise, for both men and women. This daily rhythm is different than for the rhesus monkey in which testosterone peaks at 10:00 P.M. (Southern and Gordon, 1975) but perhaps not different from the pigtail macaque (Martenson et al., 1977). This may explain the increased sexual feelings of many men and women early in the morning. The evidence for the morning peak in women, however, is not good unless the women are ovariectomized or on the pill (Curtis, 1972).

While male hormones are apparently stimulated by light, female hormones are at least sometimes stimulated by darkness (Jones, Shainberg, and Byer, 1977). Girls usually begin menstruating in the winter; women usually begin their monthly menstrual flow well into the darkness of night or early in the morning. Keeping a light on at night for two or three days thirteen days after menstrual flow will stabilize an irregular menstrual cycle at twenty-nine days (Jones, Shainberg, and Byer, 1977). However, estrogen and FSH, like androgens, are also higher in the early morning in both sexes (Curtis, 1972).

In addition to changes in sexual feelings and such obvious things as menstrual flow, other changes have been related to female hormones. Cycles of physical strength, energy, confidence, cheerfulness, and irritability have been reported to be altered with hormonal rhythms (Jones, Shainberg, and Byer, 1977).

Obviously, many physical changes can be related to human menstrual cycles and to the relative amounts of estrogen and progesterone in a woman's bloodstream. It has even been said that women who are larger breasted on the average have less body hair than women who are small breasted and that their hair is on the average finer in texture. These women presumably have more estrogen whereas others have more progesterone (Jones, Shainberg, and Byer, 1977). How correct these observations are, like all observations, can be debated. What is less obvious is that emotional changes are also related to female hormones.

In terms of sexual feelings and sexual behavior:

At the time of ovulation estrogen drops drastically. During this drop in estrogen some women feel a strong sexual drive. Since this is the time of ovulation, these feelings seem to be caused by a rise in androgen (malelike hormones) level in the blood, either from the ovary or from the adrenal gland. (Jones, Shainberg, and Byer, 1977, P. 83)

There are three times when there are sharp drops in both estrogen and progesterone: (1) at menstruation, (2) just after childbirth, and (3) during menopause.

The depression, irritability, and/or weepiness that often, though not invariably, occur at these times have been attributed to the correlated hormonal declines. However, all three of these events are also correlated with periods during which women often undergo strong social-emotional pressure.

Even if the hormones *are* more directly important to these emotions than are the social pressures, each woman responds differently to each hormone. Average responses in this area of research mean very little when we are dealing with as complicated, as intelligent, and as conceptually self-aware an *individual* as the adult human female. The most likely explanation for the mood changes with changes in hormonal levels will undoubtedly be that *both* social pressures and biological changes are involved, and that in some women the response of the husband or of society to her menstruation may be more important, whereas in others the hormone levels themselves may be more important.

Butler (1974), on the other hand, believes that the term menstrual cycle should be reserved for humans because humans, presumably, show no midcycle increase in sexual behavior and because there is no good evidence for an increase in human female receptivity at ovulation. There is considerable disagreement over this point. Rowell (1972), for example, feels that the similarities between some nonhuman primates and women warrant the use of the term menstrual cycle for those nonhuman primates (e.g., rhesus monkey). Others have also pointed out similarities between the menstrual cycles of nonhuman primates and humans (Steiner et al., 1977), including similarities in a midcycle increase in sexuality (cf. Jones, Shainberg, and Byer, 1977; also see Chapter 6).

In terms of sexual responsiveness, we should probably reserve judgment on whether or not human males and females behave independently of hormonal control. It is known, for example, that orgasmic response intensity decreases in post-menopausal women, and that there is an increased latency to orgasm. It is possible that human sexual behavior should be studied with regard to the concepts of attractivity, proceptivity, and receptivity since, even in nonhuman primates, each of these has a slightly different relationship with hormonal changes (Beach, 1976).

In addition, the evidence that human hormones function properly when injected into nonhuman primates should carry some weight. When five premenarchial rhesus females, for example, were injected with *human* gonadotrophins (HMG) their ovaries became capable of female sex hormone secretion, they ovulated, showed large elevations of serum estradiol, and developed swollen perineal sex skin (Weiss, Rifkin, and Atkinson, 1976). Also, it should be remembered that human birth control pills work for monkeys.

In human infants, as in chimpanzees, there is a sex difference in serum gonadotrophin levels. FSH is higher in girls and there is cyclicity in LH and FSH levels in girls but not in boys (Faiman, Winter, and Grotts, 1973). This difference disappears in childhood but reappears at puberty.

A few more points concerning menstrual cycles deserve review. Human females display menopause, usually in the late forties. During their menopause there is a cessation of menstrual cycles, menstrual flow, ovulation, and fertility but not of sexual behavior or of sexual arousal. Rhesus macaque females display a similar menopause, but in the twenties instead of in the forties (Jones, 1975). Rhesus have a true menopause, with cessation of bleeding. Curiously, chimpanzees show the absence of fertility in the forties but no cessation of the menstrual cycle (Graham and McClure, 1977). Hence, they do not have a true menopause. Thus, while the chimpanzee female is the best animal model for the human female in regards to many matters of reproductive biology, she will not suffice for all of them.

According to a review by Tiger (1975), there is a strong relationship between visual acuity and ability to perceive odors, and phases of the menstrual cycle in humans. Tiger (1975) also points out that birth control pills may interfere with some biological phenomena such as synchronization of menstrual cycles among roommates and close friends. In addition, he believes that the pill reduces sexuality (it apparently does in rhesus) and is related to depression in human females.

There is some good evidence (as mentioned above) that androgens are responsible for sexual desire in both sexes and in both nonhuman and human primates. With humans, removal of the ovaries produces no effect on sexual desire, whereas removal of the adrenals does. Apparently, the source of the androgen important for sexual desire in females is the adrenal cortex (Hamburg and Lunde, 1966).

In regards to other effects of androgens, the effects of testosterone are mediated by social-emotional cues. In macaques, increases in testosterone appear to be directly related to behavior aimed at achieving or maintaining dominance. In chimpanzees, changes in testosterone are also correlated with changes in dominance, although in a somewhat less overt and violent way. In humans, according to Mazur (1976), there is also sufficient evidence to consider testosterone as playing a role in the status processes of both males and females of human groups. However, Mazur (1976) cautions us that:

> Needless to say, the involvement of a "male hormone" in status processes provides no moral justification for the privileged position of males in most societies. (P. 224)

In heterosexual interactions among rhesus, redirection of aggression occurs frequently when consort bonds are being established. Threatening away by the male and female occurs most often if estrogen levels are high and decreases when progesterone levels are high. Michael (1975) believes similar things might occur in men and women.

Lionel Tiger (1975) views sex differences as "one of the clearest cases of biological structuring of bodily function" (P. 119) and that ". . . [testosterone] may be a significant predisposing or catalytic factor under appropriate social circumstances for the display of violent [human] behavior." (P. 124.) He believes that this phenomenon is particularly marked in adolescent males, pointing out that while adolescent females only double their testosterone secretions from childhood (from a lower level than males), boys increase plasma testosterone levels by at least twenty to thirty times (Tiger, 1975). There is, as we have seen, a substantial tradition of research implicating male sex hormones with aggression in primates (Hamburg and Brodie, 1973).

Since the controversial reports by Michael (1975) on the possibility of hormone-dependent olfactorily acting pheromones of vaginal origin in rhesus monkeys, several authors have speculated on the likelihood of human pheromones (Comfort, 1971). Mai (1972), for example, has said that humans may experience periods of heightened olfactory activity. Women are presumably more sensitive to musklike odors at menstruation and at ovulation, and the quality of these odors are apparently altered when progesterone levels are high. In addition, Michael (see Goldfoot, 1977) maintains that human adult females show a midcycle peak in vaginal aliphatic acids (similar to rhesus), but apparently two other laboratories disagree. Goldfoot (1977) points out that the presence of aliphatic acids does not constitute evidence for pheromones in any case.

But how then does one explain the reports that women living together and seeing males less than three times per week in a college dormitory, experience synchrony and suppression of the menstrual cycle, and that the introduction of a male to one of the females results in development of a normal menstrual periodicity and shorter cycles independent of the other females (cf. Mai, 1972). Mai posits that there are two pheromones, one released by the females to account for synchrony and suppression and one released by the male to account for release from suppression.

Apparently, the significant thing in synchrony is that the individual females of the group spend time together. Since those spending time with males have shorter cycles (suppression) — "there is *some* interpersonal physiological process which affects the menstrual cycle" (McClintock, 1971). Whether this change is pheromonal or not is open to question. Goldfoot (1977) has emphasized the role of learning for sexual activity:

> . . . it cannot be denied that in some persons body odors can strongly inhibit or facilitate sexual activity. I believe that this fact is the result of learned associations rather than pheromonal mechanisms. (P. 6)

However, it it not clear how learning could result in synchrony and suppression of menstrual cycles.

There are other cases in which social stimuli can produce changes in menstrual cycles. Quarantine (isolation) and a drastic change of environment can induce amenorrhea in nonhuman primates. A similar phenomenon occurs among airline stewardesses who travel long distances (Nieman, 1970; Greep, 1968).

D. NEUROLOGY, NEUROPHYSIOLOGY, AND SEX HORMONES

The important centers for sexual behavior in people are in the limbic system, the same as for squirrel monkeys and other primates (Vestergaard, 1977). Complete neural sexual dimorphism probably does not occur for any species of animals to say nothing of primates and people. Even in the rhesus monkey, a sexually dimorphic species, bisexual behavior is evident in males. In fact, male rhesus monkeys still display female presents after extensive *prenatal* exposure to testosterone. The female present is, of course, used for social bonds other than sex. Similarly, androgenized female rhesus still display feminine behaviors (e.g., presents). It is as if the female neural mechanisms are protected from the androgen effect (Goy and Goldfoot, 1975).

It is likely that there are target sites in certain hypothalamic structures of human brains for progesterone, estrogen, and testosterone (Sar and Stumpf, 1973) and that the basic neuroanatomy of sex hormone receptors and hormone-concentrating cells in the brain of *Homo sapiens* are quite similar to those of other primates (Pfaff, 1976). In addition, it is highly likely that the regulation of sexual behavior via the sex hormones involves the biogenic amines as neurotransmitters in man (Vestergaard, 1977; Karczmar, 1975):

> One of the major growing points in this general area concerns the problem of *how* hormones act within the CNS to modulate neural activity and hence induce alterations in behaviour. From a large number of experiments on sub-primate species . . . it has become clear that the biogenic amines noradrenaline, dopamine, and 5-hydroxytryptamine may play a fundamentally important role in mediating the effects of hormones on behaviour. (Everitt, 1976, abstract)

Psychopharmacology has shown us that drugs (other than sex hormones) which affect these neurotransmitters also induce sexual and aggressive behavior. Lesions of areas important to sex hormones made by neurotoxic drugs also implicate biogenic amine neurotransmission. In addition, sex hormones have a marked effect on cerebral biogenic amine metabolism. Psychological characteristics such as mood are also affected by these amines (Everitt, 1976), thus implicating the sex hormones in the realm of psychological health and emotion. In a sense, Freud was more correct than many modern psychologists believe; sex *is* of great importance to mental health and everyday life.

Outside of the realm of emotion and into the area of cognition (thinking, intelligence, etc), there are those who believe that sex hormones can be related to known gender differences in cognitive abilities. According to Broverman et al. (1968), males are superior to females in inhibitory perceptual restructuring tasks and this can be accounted for because estrogens (but not androgens) inhibit the activity of the enzymes choline acetylase and monoamine oxidase that synthesize the neurotransmitters acetylcholine and the biogenic amines.

Reports such as the one by Broverman et al. (1968) can cause controversy so stirring that it becomes as much political as scientific. Social learning theorists often see no role for neuroendocrine and central nervous system data on human gender differences. Neuroendocrinologists, meanwhile, emphasize differences in the brain. Both camps are guilty of bias and, because of "increased sophistication with which the adversaries wage intellectual battle" (Green, 1976, abstract) it is often difficult to separate the purely scientific-academic from the political. Two things are clear, however: (1) there are undoubtedly biologically based differences in the brains of human males and females, whatever that may entail, but (2) human beings are the most self-aware of all primates and their sex or gender differences cannot be explained in simple biolo*gistic* (or even sociobiolo*gistic*) terms.

For further reading on this topic, Gloor (1975) has written an article on the physiology of the limbic system; Karczmar (1975) has published a paper on neurotransmitters and sexual behavior; Gadpaille (1972) has written a review on possible physiological contributions to the understanding of homosexuality; and Newton (1973) in a magazine for the laity, has presented us with some interesting ideas on human female sexuality, implicating still another hormone (oxytocin) in sexual arousal.

REFERENCES

Baum, M. J. Inhibitory action of female progesterone on sexual interaction in primates. Paper presented at the *International Primatological Society* Meeting, Cambridge, England, August, 1976.

Beach, F. A. Sexual attractivity, proceptivity, and receptivity in female mammals. *Hormones and Behavior,* 1976, **7**, 105-138.

Bernstein, I. S., Gordon, T. P., Peterson, M. and Rose, R. M. Female influences on pigtail male testosterone levels. Paper presented at the *American Society of Primatologists* Meeting, Seattle, Washington, April, 1977.

Bielert, C. Social experience, hormones, interact to shape the sexual growth of male rhesus. *Primate Record,* 1975-76, **6** (2), 3-5.

Birch, H. G. and Clark, G. Hormonal modification of social behavior IV. The mechanism of estrogen-induced dominance in chimpanzees. *Journal of Comparative and Physiological Psychology,* 1950, **43**, 181-193.

Bolwig, N. A study of the behavior of the chacma baboon, *Papio urisinus. Behaviour,* 1959, **14**, 136-163.

Broverman, D. M., Klaiber, E. L., Kobayaski, Y. and Vogel, W. Roles of activation and inhibition in sex differences in cognitive abilities. *Psychological Review,* 1968, **75**, 23-50.

Butler, H. Evolutionary trends in primate sex cycles. *Contributions to Primatology,* 1974, **3**, 2-35.

Chalmers, N. R. The social behaviour of free living mangabeys in Uganda (*Cercocebus albigena johnstoni*). *Folia Primatologica,* 1968, **8**, 263-281.

Cockett, A. T. K., Elbadawi, A., Zemjanis, R. and Adey, W. R. The effects of immobilization on spermatogenesis in subhuman primates. *Fertility and Sterility,* 1970, **21** (8), 610-614.

Comfort, A. Likelihood of human pheromones. *Nature,* 1971, **230**, 432-433.

Curtis, G. C. Psychosomatics and chronobiology: Possible implications of neuroendocrine rhythms: A review of *Psychosomatic Medicine,* 1972, **34** (3), 235-256.

Dang, D. C. Resumption of menstruation and fertility after Caesarian in *Macaca fascicularis. Annals of Biology, Animal Biochemistry and Biophysics,* 1977, **17** (3A), 325-329.

Dixson, A. F. and Herbert, J. Gonadal hormones and aggressive behaviour in captive groups of talapoin monkeys (*Miopithecus talapoin*). *Journal of Endocrinology,* 1974, **61**, 46.

Dixson, A. F., Herbert, J. and Rudd, B. T. Gonadal hormones and behaviour in captive groups of talapoin monkeys (*Miopithecus talapoin*). *Journal of Endocrinology,* 1972, **57**, 41.

Dunbar, R. I. M. Oestrous behaviour and social relations among gelada baboons. Paper presented at the *International Primatological Society* Meeting, Cambridge, England, August, 1976.

Eaton, G. G. and Resko, J. A. Plasma testosterone and male dominance in a Japanese macaque (*Macaca fuscata*) troop compared with repeated measures of testosterone in laboratory males. *Hormones and Behavior,* 1974, **5**, 251-259.

Everitt, B. J. Growing points in research on sexual and aggressive behaviour. Paper presented at the *International Primatological Society* Meeting, Cambridge, England, August, 1976.

Faiman, C., Winter, J. S. D. and Grotts, D. Gonadotropins in the infant chimpanzee: A sex difference. *Proceedings of the Society for Experimental Biology and Medicine,* 1973, **144**, 952-955.

Gadpaille, W. J. Research into the physiology of maleness and femaleness: Its contributions to the etiology and psychodymanics of homosexuality. *Archives of General Psychiatry,* 1972, **26**, 193-206.

Gloor, P. Physiology of the limbic system. In Penry, J. K. and Daly, D. D. (Eds.) *Advances in Neurology* (Vol. 11). New York: Academic Press, 1975, pp. 27-55.

Goldfoot, D. A. Hormonal and social determinants of sexual behavior in the pigtail monkey (*M. nemestrina*). In Stoelinga, G. B. A. and van der Werff ten Bosch, J. J. (Eds.) *Normal and Abnormal Development in Brain and Behavior.* Leiden: University of Leiden Press, 1971, pp. 325-342.

Goldfoot, D. A. Social and hormonal regulation of gender role development in rhesus monkey. Paper presented at the *International Primatological Society* Meeting, Cambridge, England, August, 1976.

Goldfoot, D. A. Olfactory influences on sexual behavior, *Primate News,* 1977, **15** (6), 3-7.

Goy, R. W. and Goldfoot, D. A. Neuroendocrinology: Animal models and problems of human sexuality. *Archives of Sexual Behavior,* 1975, 4 (4), 405-420.

Graham, C. E. and McClure, H. M. Ovarian tumors and related lesions in aged chimpanzees. *Veterinary Pathology,* 1977, **14**, 380-386.

Green, R., The human primate: Development of sexuality and aggressivity. Paper presented at the *International Primatological Society* Meeting, Cambridge, England, August, 1976.

Greep, R. O. Hypothalamic-pituitary-ovarian relationships. In Hoffman and Kleinman (Eds.) *Advanced Concepts in Contraception.* Amsterdam: Excerpta Medica Foundation, 1968, pp. 108-110.

Gross, R. J., Schiller, H. S. and Bowden, D. M. Post partum serum cortisol and prolactin and their relationship with maternal and affective behaviors in the *Macaca nemestrina.* Paper presented at the *American Society of Primatologists* Meeting, Seattle, Washington, April, 1977.

Hagino, N. Hypothalamic regulation of ovulation and menstruation in the baboon (nonhuman primate). *Federation Proceedings,* 1976, **35**, 701.

Hamburg, D. A. and Brodie, H. K. H. Psychological research on human aggressiveness. *Impact of Science on Society,* 1973, **23**, 181-193.

Hamburg, D. A. and Lunde, D. T. Sex hormones in the development of sex differences in human behavior. In Maccoby, E. E. (Ed.) *The Development of Sex Differences.* Stanford, California: Stanford University Press, 1966, pp. 1-24.

Hausfater, G. Estrous females: Their effects on the social organization of the baboon group. *Proceedings of the Symposia of the Fifth International Congress of Primatology.* Tokyo: Japan Science Press, 1975, pp. 117-127.

Hayashi, M., Oshima, K., Yamaji, T. and Shimamoto, K. LH levels during various reproductive states in the Japanese monkey (*Macaca fuscata fuscata*). In Kondo, S., Kawaii, M. and Ehara, A. (Eds.) *Contemporary Primatology.* Basel: Karger, 1975, pp. 152-157.

Herbert, J. Hormonal basis of sex differences in rats, monkeys, and humans. *New Scientist*, 1976, **70**, 284-286.

Hobson, W. Reproductive endocrinology of female chimpanzees: A suitable model of humans. *Journal of Toxicology and Environmental Health*, 1976, **1**, 657-668.

Jensen, G. D. Comparisons of sexuality of chimpanzees and humans. Paper presented at the *International Primatological Society* Meeting, Cambridge, England, August, 1976.

Jones, E. C. The post-reproductive phase in mammals. In van Keep, P. A. and Lauritzen, C. (Eds.) *Frontiers of Hormone Research* (Vol. 3). Basel: Karger, 1975, 1-19.

Jones, K. L., Shainberg, L. W. and Byer, C. O. *Sex and People.* New York: Harper & Row, 1977.

Joslyn, W. D. Androgen-induced social dominance in infant female rhesus monkeys. *Journal of Child Psychology and Psychiatry*, 1973, **14**, 137-145.

Karczmar, A. G. Neurotransmitters in the modulation of sexual behavior. *Psychopharmacology Bulletin*, 1975, **11**, 40-42.

Kling, A. and Dunne, K. Social-environmental factors affecting behavior and plasma testosterones in normal and amygdala lesioned *M. speciosa*. *Primates*, 1976, **17** (1), 23-42.

Lanman, J. T. Parturition in nonhuman primates. *Biology of Reproduction*, 1977, **16**, 28-38.

Loy, J. Changes in facial color associated with pregnancy in patas monkeys. *Folia Primatologica*, 1974, **22**, 251-257.

Mai, L. Chemical communication in primates. *Anthropology*, UCLA, 1972, **4** (2), 57-83.

Maple, T. and Zucker, E. L. Behavioral studies of captive Yerkes orangutans at the Atlanta Zoological Park. *Yerkes Newsletter*, 1977, **14** (1), 24-26.

Martenson, J., Oswald, M., Sackett, D. and Erwin, J. Diurnal variation of common behaviors of pigtail monkeys (*Macaca nemestrina*). *Primates*, 1977, **18** (4), 875-882.

Mazur, A. Effects of testosterone on status in primate groups. *Folia Primatologica*, 1976, **26**, 214-226.

McClintock, M. K. Menstrual synchrony and suppression. *Nature*, 1971, **229**, 244-245.

Michael, R. P. Hormonal factors and aggressive behavior in the rhesus monkey. *Proceedings of the International Society of Psychoendocrinology* (Brooklyn, 1970). Basel: Karger, 1971, pp. 412-423.

Michael, R. P. Hormones and sexual behavior in the female. *Hospital Practice*, 1975, Dec., 69-76.

Michael, R. P., Bonsall, R. W. and Warner, P. Primate sexual pheromones. In Denton, D. H. and Coghlan, J. P. (Eds.) *Olfaction and Taste* (Vol. 5). New York: Academic Press, 1975, pp. 417-424.

Michael, R. P. and Wilson, M. Changes in the sexual behaviour of male rhesus monkeys (*M. mulatta*) at puberty. *Folia Primatologica,* 1973, **19**, 384-403.

Nadler, R. D. Cyclicity in tumescence of perineal labia of female lowland gorillas. *Anatomical Record,* 1975, **181**, 791-798.

Nadler, R. D. Sexual behavior of captive orangutans. *Archives of Sexual Behavior.* 1977, **6**, 457-475.

Newton, N. Trebly sensuous woman. In Tavris, C. (Ed.) *The Female Experience.* New York: Ziff-Davis, 1973.

Nieman, W. H. Comparative aspects of reproduction in primates. *Proceedings of the Third International Congress of Primatology,* 1970, **1**, 234-237.

Nigi, H. Menstrual cycle and some other related aspects of Japanese monkeys (*Macaca fuscata*). *Primates,* 1975, **16** (2), 207-216.

Nissen, H. W. and Yerkes, R. W. Reproduction in the chimpanzee: Report on forty-nine births. *Anatomical Record,* 1943, **86**, 567-578.

Nomura, T. and Ohsawa, N. The use and problems associated with nonhuman primates in the study of reproduction. In Antikatzides, T., Ericksen, S. and Spiegel, A. (Eds.) *The Laboratory Animal in the Study of Reproduction.* New York: Gustav Fischer, 1976, pp. 1-16.

Pfaff, D. W. The neuroanatomy of sex hormone receptors in the vertebrate brain. In Kumar, T. C. A. (Ed.) *Neuroendocrine Regulation of Fertility.* Basel: Karger, 1976, pp. 30-45.

Plant, T. M. and Michael, R. P. Diurnal variations in plasma testosterone levels of adult male rhesus monkeys. *Acta Endocrinologica Congress,* 1971, Abstract No. 69.

Rosenblum, L. A. and Bromley, L. J. The effects of gonadal hormones on juvenile peer interactions in bonnet macaques. Paper presented at the *International Primatological Society* Meeting, Cambridge, England, August, 1976.

Rowell, T. E. Intra-sexual behaviour and female reproductive cycles of baboons (*Papio anubis*). *Animal Behaviour,* 1969, **17**, 159-167.

Rowell, T. E. *The Social Behaviour of Monkeys.* Baltimore: Penguin, 1972.

Sar, M. and Stumpf, W. E. Neurons of the hypothalamus concentrate [^{3}H] progesterone or its metabolites. *Science,* 1973, **183**, 1266-1268.

Sassenrath, E. N. Increased adrenal responsiveness related to social stress in rhesus monkeys. *Hormones and Behavior,* 1970, **1**, 283-298.

Scruton, D. M. and Herbert, J. The reaction of groups of captive talapoin monkeys to the introduction of male and female strangers of the same species. *Animal Behaviour,* 1972, **20**, 463-473.

Slob, A. K.   Effects of ovariectomy on male-female interactions in the stump-tail macaque (*M. arctoides*). *Acta Endocrinologica*, 1975, **10**, 145.

Slob, A. K., Goy, R. W., Weigand, S. J. and Scheffler, G.   Gonadal hormones and behaviour in the stumptail macaque (*Macaca arctoides*) under laboratory conditions: A preliminary report. *Journal of Endocrinology*,   1975, **64**, 38.

Southern, A. L. and Gordon, G. G.   Rhythms and testosterone metabolism. *Journal of Steroid Metabolism*, 1975, **6**, 809-813.

Steiner, R. A., Schiller, H. E., Illner, P., Blander, R. and Gale, C. C.   Sex hormones correlated with sex skin swelling and rectal temperature during the menstrual cycle of the pigtail macaque (*Macaca nemestrina*). *Laboratory Animal Science*, 1977, **27**, 217-221.

Sullivan, D. J. and Drobeck, H. D.   True hermaphrodism in a rhesus monkey. *Folia Primatologica*, 1966, **4**, 309-317.

Takashima, I.   Studies of common baldness in the stump-tailed macaque V. Regional difference of testosterone metabolites in the hair follicles. *Journal of Dermatology*, 1974, **1** (1), 14-21.

Takashima, I., Adachi, K. and Montagna, W.   Studies of common baldness in the stump-tailed macaque IV. *In vitro* metabolism of testosterone in the hair follicles. *Journal of Investigative Dermatology*, 1970, **55** (5), 329-334.

Takashima, I. and Montagna, W.   Studies of common baldness of the stumptailed macaque (*Macaca speciosa*) VI.   The effect of testosterone on common baldness. *Archives of Dermatology*, 1971, **103**, 527-534.

Tiger, L.   Somatic factors and social behaviour. In Fox, R. (Ed.) *Biosocial Anthropology*. New York: Wiley, 1975, pp. 115-132.

Vestergaard, P.   Sexual mechanisms in the brain. Neurophysiological and neuro-chemical aspects. *Ugeskr. Laeg.,* 1977, **139**, 10-13.

Weiss, G., Rifkin, I. and Atkinson, L. E.   Induction of ovulating in premenarchial rhesus monkeys with human gonadotropins. *Biology of Reproduction*, 1976, **14**, 401-409.

Wolfe, Linda.   Behavior patterns of estrous females of the Arashiyama West troop of Japanese macaques (*M. fuscata*).   Paper presented at the *American Association of Physical Anthropologists* Meeting, Seattle, Washington, April, 1977.

Zemjanis, R., Gondos, B., Adey, W. R. and Cockett, A. T. K.   Testicular degeneration in *Macaca nemestrina* induced by immobilization. *Fertility and Sterility*, 1970, **21** (4), 335-340.

9.
Prosimians, New World Monkeys, and Rhesus Macaques

A. GENERAL COMMENTS

In our discussions of sexual behavior which follow, the importance of the concepts of attractivity, proceptivity, and receptivity in female primates will become evident. These concepts have been refined by Dr. Frank A. Beach (1976 a, b). Attractivity, we may remember, refers to the male's degree of interest in the female. A female may be attractive to a male and yet she may refuse his advances. Receptivity refers to the female's willingness to accept the male. Proceptivity refers to the female's active interest in the male. Included as a proceptive behavior, for example, is the initiation of sexual activity by the female. Thus, it is possible for a female to be attractive, receptive, and/or proceptive at the same time or at different times (Beach, 1976a). Sexual responsiveness in primates, as we have already seen, may be independent of, or dependent upon, one sex hormone or another depending upon whether we are discussing attractivity, receptivity, or proceptivity. It is known among primates which hormones underly sexual activity in males and females and, for the female at least, whether these hormones (estrogen, progesterone, testosterone) affect attractiveness, receptivity, or proceptivity (Everitt, 1976).

Another useful concept advanced by Beach (1976b) is the principle of S-R complementarity. In both sexes there are neural representations for both a male and a female sexual behavior pattern. Because of prenatal hormonal priming, however, there is also a sex-linked prepotency such that males are more likely to mount and females are more likely to present. The principle of S-R complementarity, however, states that, independent of the genetic sex of the animal, a masculine stimulus will elicit a feminine response and vice versa. So, if a male animal sees another male animal presenting, he is likely to mount (the complementary behavior).

As Beach says:

> The same individual cannot develop both a penis and a vagina, but the same brain can contain mechanisms for both male and female behavior. (1976b, P. 480)

Thus it is that the sexual present comes to have nonreproductive as well as reproductive meaning. Obviously, two males or two females cannot reproduce. The sexual present, then, comes to have a meaning of submission, appeasement, or friendship and we are left with what Beach (1976b) calls "the hypothesis of neural bisexuality" (the indented quote above). This hypothesis "encourages us to think in terms of varying degrees of masculinity and femininity within the *same individual.*" (P. 480.) It also suggests that male presenting is normal Beach (1968), moreover, has published an interesting chapter showing that for many mammals, female mounting behavior is widespread and not abnormal or atypical but a normal response in the repertoire of the species. Maple (1977) has presented a review suggesting that this is also true of primates.

In a hypothetical general species of primate, the adult female is said to be in *estrus* when she is physiologically and behaviorally ready for mating. This is usually during that part of her monthly cycle when she is fertile. Each species has its own typical set of reproductive postures and communicative acts related to copulation (cf. Bramblett, 1976). It is to these species differences in mating patterns that we will turn our attention in the next two chapters.

We presented information on adult male and female sex hormones in the last two chapters (Chapters 7 and 8). In Chapters 7 and 8, we discussed sexual behavior *only* in relation to those hormones. In Chapters 9 and 10, however, we will discuss the sexual acts themselves. As with the hormone chapters, we will first discuss prosimians, New World monkeys, and rhesus monkeys, then, in a subsequent chapter, the other Old World monkeys, apes, and humans.

B. PROSIMIAN SEXUAL BEHAVIOR

Among tree shrews (*Tupaia spp.*), breeding successes in the laboratory occur when animals are kept in adult heterosexual pairs. Females display a postpartum (following birth of her infants) estrus of from three to twenty-four hours. In the days directly before the female gives birth, the male shows increased sexual interest in her by following, nuzzling, and licking her. The female does not become receptive, however, until she has delivered her infant (Michael and Zumpe, 1971).

Receptive females of the genus *Tupaia* make short, darting runs to maintain about a meter's distance between her and her potential mate. In *T. tana*, the

the male performs a "courtship" dance, circling the female while stamping as the female vocalizes. The male's behavior includes licking the female's vulva, mounting, and gripping the fur on the neck of the female with his mouth. He places his forelimbs around the female's back, the female displays lordosis (arching of the back) for intromission, and he thrusts to ejaculation. After copulating, each animal grooms its own genitals. There may be as much as a dozen intromissions (Michael and Zumpe, 1971).

The female tree shrew remains attractive to the male for a longer period than she is receptive. This results in some aggression between the two toward the end of estrus. Michael and Zumpe (1971) have remarked that in tree shrew sex there is sometimes mounting of both males and females by estrous females (proceptivity). According to Autrum and von Holst (1968), however, the appearance of female mounting in tree shrews is also related to high stress levels.

Among lemurs, sexual behavior patterns, like social organization, vary with the genus and species. In *Cheirogaleus major* (the dwarf lemur), the female lies with her belly to a branch as the male mounts with his forelimbs around the female and his feet grasping her ankles. The male also waves his tail, licks the female's perineum and neck, and emits a warble sound. Each mount lasts a couple of minutes and there may be several mounts at ten minute intervals. *Microcebus* matings occur in a similar fashion (Michael and Zumpe, 1971), but only when the female's body weight is less than that of the male's (Perrit, 1977).

In the genus *Lemur,* there is great variation in frequency of male–female contact, and in *Lemur fulvus,* males and females scent-mark frequently with sexually different scents (Kress, 1975; Harrington, 1977).

Among the *Lorisidae,* there are descriptions available for the sexual behavior of the *Galago* genus. In *Galago senegalensis,* estrus includes swelling, flushing, vaginal opening, and white vaginal discharge. Estrus lasts for about three days, although the males show constant sexual interest in females all year long, attempting to follow them and to examine their genitals. The male scent-marks by urinating into his hand and rubbing the urine on his foot. Females are not receptive outside of estrus. When following the female around the male emits a cluck sound. He mounts with his arms around the female's middle, his chin on the back of her neck, and his feet on a branch or grasping the female's legs (Michael and Zumpe, 1971). Mounts can last for seven minutes, especially if they are ejaculatory mounts. While mounting the male emits a loud call and whistle, then remains motionless for a time just before dismounting. Genital self-grooming follows copulation for both the male and female (see also Doyle, Pelletier, and Bekker, 1967).

In *Galago alleni,* one male controls and mates with six to eight different females, all of which hold different territories within (and smaller than) his own. Courtship takes place independently of female estrus (Charles-Dominique, 1977).

In other lorisoids, sexual behavior can be quite different from the bushbaby pattern. The estrus female slow loris (*Nycticebus coucang*) suspends herself under a branch, upside down, as she presents. The male studies her genitals while she in in this posture after which he mounts her in this upside-down position (Michael and Zumpe, 1971).

In the potto (*Perodicticus*), as well as in the slow loris, play grappling between the sexes often precedes copulation (cf. Lockhard, Haestand, and Begert, 1977).

In the *Tarsier,* the female's genitals swell during estrus, which is very brief. Sexual activity includes vocalizations and self-grooming by both sexes and urine-marking by the male. The male chases the female, very rapidly approaches her from below, wraps his arms around her waist, and copulates. Self-grooming recurs following mating, at least in the male (Michael and Zumpe, 1971).

C. NEW WORLD MONKEYS

The monogamous marmosets and tamarins breed regularly in captivity and the female has from one to three young per litter. In the female marmoset, there are no visible external signs of swelling or physical estrus. A male of the species *Callithrix jacchus* follows the female three days after parturition and the female becomes receptive about three or four days after that (Michael and Zumpe, 1971).

Also in the common marmoset (*Callithrix spp.*) copulation is preceded by a display of courtship in which the members of a mating pair follow *each other, both* with arched backs, arms and legs extended, and piloerection (erect hair on back). Both sexes also scent-mark, after which the male approaches the female while lip-smacking, as the female crouches and lip-smacks at the male. Their tongues make contact ('kissing') and they lick each others' faces and genitals. The male then mounts the still crouching female by grasping the hair on her waist and making seven to ten rapid thrusts to ejaculation. Tongue movements are visible throughout copulation (Michael and Zumpe, 1971).

Even *juvenile* marmosets sometimes form pairs and copulate before puberty. However, as pointed out in Chapters 5 and 6, their copulation is closely related to play (Abbott, 1976). Dominant marmosets scent-mark more than do subordinate ones and dominant pairs will interfere with the sexual activity of subordinate pairs (Epple, 1970; Michael and Zumpe, 1971).

Tamarins (*Saguinus geoffroyi*) do not display the licking and arched backs, instead the male smells the female's scent-marked tail. However, rapid tongue movements are regarded as sexual invitations in many if not most of the marmosets and tamarins (Michael and Zumpe, 1971).

Dawson (1976) has noted that in tamarins, as in marmosets, there are no seasonal differences in the reproductive tracts of either males or females. In

Saguinus geoffroyi (the rufous-naped tamarin), the female at times appears to be sexually unsatisfied. This is indicated by her display of an upward tail curling (according to Moynihan, 1970). Males rarely perform upward tail curling.

As in the marmosets, if more than one male tamarin (*Leontopithecus rosalia*) is present in a group, only one male is sexually active and the female will affiliate more often with the sexually active male, unless the nonreproductive male is a relative like a brother or a son. Brothers or sons of a female are not socially inhibited in the presence of a sexual pair but they are reproductively inhibited. There seems to be an incest taboo and it seems to be a matter of female choice (Kleiman, 1977a).

In monogamous, monomorphic species like marmosets and tamarins, there is less dimorphism in sexual behavior than is seen in dimorphic species. Males and females perform many of the same courtship displays. Also, as reported by Kleiman (1977b), the adult male and female of a monogamous monomorphic species exhibit infrequent sociosexual interactions, relative to other species, except during the early stages of the pair bond formation (the honeymoon effect). While sexual behavior is less frequent, heterosexual grooming and sleeping together is *more* frequent. Also, according to Kleiman (1977b), juveniles exhibit delayed sexual maturation in the presence of parents relative to more dimorphic species. Referring to Kleiman:

> . . . a basic characteristic of monogamy may be the simple fact that a male and female sleep and groom together. (1977b, P. 59)

While sex is intense in the bond formation stage, it is infrequent thereafter.

Cebus monkeys (*Cebus spp.*) are not monogamous like marmosets and tamarins. Female cebus have a vaginal cycle of sixteen to twenty-three days, but there is only irregular external menstruation in some species. Copulation coincides with ovulation, and ovulatory cycles are apparent at least nine months of the year. Even though capuchins (*Cebus spp.*) are very common animals, there are no good descriptions of reproductive behavior in this genus (Michael and Zumpe. 1971).

Among wooly monkeys (*Lagothrix lagotrichia*) menstruation occurs regularly every twenty-three to twenty-six days in about 50 percent of the females. There are vaginal changes but no external swellings. Among night or owl monkeys (*Aotus trivirgatus*), copulation is not preceded by a courtship display although inspection and touching of the female's genitals by the male is common. Ejaculation occurs on the first mount after only three to four thrusts. Since the night monkey is monogamous it would be interesting to know if they fit the pattern displayed by marmosets and tamarins (cf. Michael and Zumpe, 1971). Most of the night monkey's activity occurs during the animal's dark cycle. Members of a pair stay close together, engage in extensive nose-to-nose touching, side-to-side touching, and much mutual heterosexual grooming (Leibrecht and Kelley, 1977).

Another monogamous South American genus, *Callicebus,* has been studied more extensively than has the night monkey. *Callicebus* (titi) monkeys do not precede copulation with a display. The male titi, like the male night monkey, inspects and manipulates the female's genitalia (Michael and Zumpe, 1971).

There is a strong and enduring emotional bond between individual males and females in titi monkeys. Titi monkeys show clear evidence of "jealousy" including marked enhancement of attraction by the male to his mate as a function of increasing proximity between that mate and a strange male. Female titi monkeys, on the other hand, are more attracted to their mates when other females are *not* around. The female titi shows a marked reduction in attraction for her male when a strange female is near him. Thus, there is a sex difference in jealousy behavior in this monogamous species with males showing more jealousy than females (Cubicciotti and Mason, 1977). Incidentally, in stable pairs, most of the following of the other-sexed mate is done primarily by the female titi monkey, not by the male. Thus, the female is primarily responsible for the close spatial coordination of the pair (Fragaszy, 1977). It is likely that sexual behavior in such a pair is much less frequent than in a pair just forming a bond.

In the squirrel monkey (*Saimiri sciureus*) there is cyclical sexual activity, sometimes accompanied by swelling of the female's external genitalia. The male squirrel monkey jumps back and forth between branches and his female. He opens his mouth and contracts his abdominal muscles (apparently emitting no sound). He also chases the female, bites her tail, and inspects her genitals prior to copulation. Females who are receptive direct genital displays toward the male partner by facing him, abducting one thigh, and displaying the erect clitoris, emitting a little urine as they display. The female initiates mounting by presenting the anogenital region toward the male (tail deflected). She looks back over her shoulder toward him (Michael and Zumpe, 1971).

Squirrel monkeys form temporary consort or courting pairs in which following, proximity, and mutual looking are common. When mounting, the male grasps the female's waist with his hands and her legs with his feet. After from ten to about twenty-five mounts, one to fifteen thrusts per mount, ejaculation occurs, after which both male and female inspect and groom their own genitals (Michael and Zumpe, 1971).

If the female squirrel monkey is not receptive she will move away from a courting male, but will not show agonistic behavior. Squirrel monkeys have a three-month mating season, the timing of which depends on geographical location. During mating season, males become "fatted," spermatogenic, aggressive, and excitable (Michael and Zumpe, 1971) and increase their marking behavior and sniffing of females (Hennessy et al., 1977). Females, as well as males, are physically affected by season (Baldwin, 1970; Nadler and Rosenblum, 1972).

Cubicciotti and Mason (1977) tested squirrel monkey pairs in the same jealousy tests that they ran on monogamous *Callicebus* (titi) pairs. Neither sex of *Saimiri* showed changes in attraction as a function of distance between their mate and a same-sexed competitor. Squirrel monkeys do not show "jealousy."

Also, as clearly as *female* titi monkeys maintained the stable, enduring bond with the males by following, so did *male* squirrel monkeys attempt to maintain their temporary consorts with a female by following her (Fragaszy, 1977). Thus, there is a sex difference in who follows whom which is different in these monogamous and nonmonogamous New World monkeys. In the squirrel monkey, the adult male acts as pursuer and the adult female flees but provokes his further pursuit. The male initiates sexual interaction by approaching while the female determines what course the interaction will take by retreating, or mating (Latta, Hopf, and Ploog, 1967). Evidently dominant squirrel monkey females are more active participants in sexual encounters than are subordinate females (Anschel and Talmage-Riggs, 1976).

There are clearly sexually dimorphic patterns in the sexual behavior of squirrel monkeys. Even when the brains are electrically stimulated to evoke behavior, touching the partner, advancing mouth toward partner's neck, and mounting are directed most frequently by males toward females. However, although straightening the body and thrusting the chin toward the partner are seen only in males when the brain is *not* stimulated to elicit them, these reactions are seen in *both* sexes that have had brain stimulation. In general, brain stimulated females can very often assume roles normally played by males (Maurus et al., 1975). This is good support for Beach's "hypothesis of neural bisexuality" (Beach, 1976b).

Two other species of New World monkeys deserve mention here: howler monkeys and spider monkeys. In the mantled howler monkey (*Alouatta villosa*), a sexually active consort pair spends a number of hours together, away from the group of up to forty-five individuals. As in some other New World monkeys, tongue movements and crouching are female invitations to the male for sex. Male howlers also make the tongue movements, and they mount as do other New World monkey males (dorsoventrally). During copulation, we see the phenomenon referred to as "reaching-back," wherein the howler female looks back at the mounting male and may reach back to touch him with her hand. Curiously, after copulation, female howler monkeys often roll on their backs. There is no evidence for a mating season in howler monkeys (Michael and Zumpe, 1971).

In black-handed spider monkeys (*Ateles geoffroyi*), there is menstruation every twenty-four to twenty-seven days. Adult males sometimes embrace adult females. The spider monkeys are of interest to us because they are not strongly dimorphic as are the howlers, yet they live in large groups, not in monogamous pairs. In addition, the adult female has a pendulous clitoris which is larger than the male's penis. Mounting and presenting for dominance and submission are lacking, and sexual behavior occurs relatively infrequently for a nonmonogamous primate (Eisenberg and Kuehn, 1966). The urine of female spider monkeys is attractive to males and males smell and lick their fingers after grabbing and/or fondling the female's pendulous clitoris. Klein and Klein (1971) believe that the large clitoris serves:

> . . . to retain some urine on her body in such a manner that it is almost constantly being deposited in places where males are likely to be searching for her. (P. 179)

D. RHESUS MACAQUES (MACACA MULATTA)

A review of the sexual behavior of free-ranging rhesus macaques was published by Agar and Mitchell (1975). Periods of estrus in the rhesus are stages of female receptivity which coincide with the ovulatory phase of the menstrual cycle. During female receptivity, males follow, females approach males, consort pairs form, heterosexual grooming between members of a consort pair increase, the female's skin on the face, nipples, and perineum turn red, and her sexual skin swells. During the consortship the female presents, there is increased sexual activity, copulation, sporadic arm reflexes in females during copulation, and ejaculation in the male after several mounts which leaves a coagulated plug in the female's vagina. All of this increases at or around the time of ovulation each month but occurs at other times as well; following delivery of an infant and at menstruation (Erickson, 1967), for example. One cannot predict estrous-type behavior in the rhesus from a knowledge of the hormonal state alone (Rowell, 1972).

At some time during her receptive stages, the female rhesus usually forms a consort. That is, she becomes involved with a male (or with several males in succession). Consorts vary in duration, in intensity, and in exclusivity. Some bonds are weak, others strong. Following by males or females or vice versa is seen only during consorts (Agar and Mitchell, 1975). Consort pairs (as in howler monkeys) are often found apart from the group. There is a mating season, but consort behavior occurs throughout the year.

Either sex can initiate a consort by following. Females, however, initiate about 70 percent of all close physical contact (Lindburg, 1971). Males lip-smack towards females, and females initially grimace and withdraw. The females are somewhat more apprehensive than the males.

Once established, a consortship can last for a single copulatory sequence or for each receptive period for *years* depending upon the individual or the dyad (Agar and Mitchell, 1975).

Aggression increases during the rhesus mating season and there is aggression directed by the larger males towards the females, especially in the early stages of a female's receptivity. As she becomes more receptive, however, the male and female pair *redirect* threats and mock attacks at virtually anything and anyone near them (Lindburg, 1971).

Grooming serves three roles in rhesus consorts: (1) as a means of displaying affection, (2) to strengthen the social bond, and (3) to sexually arouse the partner (Agar and Mitchell, 1975).

Presenting by the female involves turning the hindquarters toward the male and lifting the tail out of the way. Presenting may occur in sexual interactions or as an act of submission, appeasement, or friendliness. Mounting may also occur in sexual interactions, as a greeting response, as an indication of dominance, or as a way to redirect aggression (Agar and Mitchell, 1975).

Thus, not all presents and mounts lead to intromission, copulation, and ejaculation. Copulation includes a *series* of mounts, some of which include intromission and pelvic thrusts. Eventually on one of the mounts the male ejaculates. In mounting, the male holds the fur on the female's sides with his hands and clasps her legs with his feet. The female looks back and reaches back to touch his thigh or scrotum. During copulation the male often displays chewing and teeth gnashing movements (Hinde and Rowell, 1962). When the male ejaculates he bares his teeth and squeals while the female looks at him and lip-smacks. Sometimes the female also squeals and clutches the male. Michael and Zumpe (1971) believe this is homologous to the human female's orgasm.

Besides presenting, other rhesus female invitations include hand-reaches, quick jerky movements of one arm, head-bobs, head-ducks, and/or lateral movements of the head. The male yawns very frequently before and after mounts (Zumpe and Michael, 1970). Following an ejaculatory mount, the male eats the ejaculate remaining on his own genitals. He threatens the female if she tries to eat it (Agar and Mitchell, 1975; Michael and Zumpe, 1970).

Many of the basic findings presented above were made in field studies on a free-ranging colony of rhesus monkeys on Cayo Santiago, off Puerto Rico, by the late Clarence Ray Carpenter (1942). Carpenter published his findings on rhesus sexual behavior in 1942, over twenty-five years ago. They are still true today.

According to Lindburg (1975) female choice plays an important role in the formation of mating associations, and preferences of the females for certain males is related to factors other than male dominance.

Rhesus copulation is seasonal. Rhesus males in India show a change in the color of the sexual skin nearly two months before the onset of the primary mating season (Lindburg, 1971). Rhesus sex is also cyclical within the female's menstrual cycle, yet it sometimes occurs during pregnancy and at other times when we would not be expecting to see it based upon hormonal considerations. Pairs copulate to male ejaculation when the female is pregnant, but as we have seen, even during pregnancy there are *some* hormonal change contributions to sexuality (Bielert et al., 1976).

Even though the rhesus monkey shows some dependence on hormones and displays strong sexual dimorphism, *complete* sexual dimorphism in behavior does not occur in any species of primate. Beach's (1976b) hypothesis of bisexuality of the neural system seems to apply to rhesus. Bisexuality in male rhesus is extensive. According to Goy and Goldfoot (1975), for a given species, the greater the bisexuality in males the *less* the bisexuality in females and vice versa. (Among female rhesus monkeys, bisexuality is infrequent). According to Goy and Goldfoot (1975), the female neural mechanisms might be protected from androgens in both sexes. There are no normally reared rhesus females who show consistent rejection of males and/or consistent frigidity. While there is perhaps greater variability in males in regard to bisexuality, posturing, and so forth, males do not differ in maximal sexual expression as much as *pairs* do. In rhesus monkeys, impotence really means partner incompatibility or a bad bond. On the other hand, individual rhesus monkeys *do* differ in finickiness (Goy and Goldfoot, 1975).

Other evidence for greater variability in the sexual behavior of male rhesus than of female rhesus is seen in reports of more ambisexual or "homosexual" behavior in males than in females (Carpenter, 1942; Agar and Mitchell, 1975). While descriptions of "homosexual" behavior in this species usually make reference only to the actual physical act of a male mounting a male with anal intromission and ejaculation, with no evidence of a preference for male–male over male–female sex, Erwin and Maple (1976) describe such a preference in two laboratory-raised rhesus:

> . . . the relationship between the two males was primarily based on mutual affection, although a sexual component was unmistakably present. (P. 9)

Rhesus males also masturbate more often than do rhesus females (Agar and Mitchell, 1975). The fact that males, in the wild, will sometimes masturbate even when females in estrus are available, shows the importance of variables other than hormonal ones. Male–male sexual behavior and extended preferences for certain females (Everitt and Herbert, 1969), regardless of hormonal state, also indicate that adult hormones do not completely control rhesus sexual behavior.

Female "homosexual" behavior also occurs in rhesus monkeys, but less frequently than male homosexual behavior (Carpenter, 1942). Fairbanks and McGuire (1977) have noted that contact aggression directed by females toward females is correlated with female homosexual behavior. But, as already mentioned, malelike mounting is part of the normal repertoire of female rhesus and this behavior increases in females if males show no interest in them. It occurs in about one-third of rhesus females (Michael, Wilson, and Zumpe, 1974).

Adult female rhesus do not display as great a range of sexual behaviors as do adult males. Even in male–female dyads, different females with the same male produce a small range of behaviors whereas successive males with the same female result in a very large range of sexual performance (Michael and Saayman, 1967).

Sexual behavior and consort activity are also affected by social rank in both males and females. High-ranking males in free-ranging rhesus account for 49 percent of the consort activity; high-ranking females account for 45 percent (Missakian-Quinn and Varley, 1977). There is evidence, however, that in a troop undergoing a division, the hypothesis that dominance and reproductive success are linearly related needs revision (Sade, Chepko-Sade, and Schneider, 1977).

Even assuming that a particular male and female have a well-established consort of two days and their sexual behavior appears to be following the normal cyclical patterns related to the female's hormone levels, a brief social or nonsocial interruption of the consort can disturb the hormone-behavior synchrony. Following short-term (two day) separations of adult male–adult female rhesus pairs, patterns of grooming between consort pairs can be completely reversed (cf. Maple, Risse, and Mitchell, 1973; Maple, Erwin, and Mitchell, 1974).

One final point regarding rhesus sexual behavior deserves mention here. Again, it will point out the importance of factors other than hormonal ones. Among rhesus macaques there is an inhibition of mating between mothers and sons (Sade, 1968). In other words, mother–son mating is rare. There is *some* sexual activity with the mother when the son is sexually immature or is a young adult (Kortmulder, 1974); however, apparently the mother's dominance over her son is partly responsible for the inhibition.

The incidence of brother–sister mating is also low and, when it occurs, it also usually occurs in young animals. Of 406 consorts observed by Missakian-Quinn and Varley (1977), only 11 percent involved genealogically related pairs of animals. There were only nine instances of mother–son, fifteen instances of brother–sister, and nineteen instances of cousins, uncles–nieces, and other kin/related sex (also see Missakian, 1973).

The avoidance of mating among relatives is also maintained by the migration of maturing males from their natal groups, away from female relatives (Duggleby, 1977).

We will return to the topic of the inhibition of sexual behavior among kin/related individuals in the next chapter. However, the interested reader may also consult Demarest's (1977) article on incest avoidance among nonhuman primates.

REFERENCES

Abbott, D. H. Hormones and behaviour during puberty in the marmoset. Paper presented at the *International Primatological Society* Meeting, Cambridge, England, August, 1976.

Agar, M. E. and Mitchell, G. Behavior of free-ranging rhesus adults: A review. In Bourne, G. (Ed.) *The Rhesus Monkey* (Vol. I, Chapter 8). New York: Academic Press, 1975, pp. 323-342.

Anschel, S. and Talmage-Riggs, G. Social structure dynamics in small groups of captive squirrel monkeys. Paper presented at the *International Primatological Society* Meeting, Cambridge, England, August, 1976.

Autrum, H. and von Holst, D. Socialer "Stress" bei Tupajas (*Tupaia glis*) and seine Wirkung auf Washstum, Körpergewicht und Fortpflanzung. *Zeitschrift für vergleichende Physiologie,* 1968, **58**, 347-355.

Baldwin, J. D. Reproductive synchrony in squirrel monkeys (*Saimiri*). *Primates,* 1970, **11**, 317-326.

Beach, F. A. Factors involved in the control of mounting behavior by female mammals. In Diamond, M. (Ed.) *Perspectives in Reproduction and Sexual Behavior.* Bloomington, Indiana: Indiana University Press, 1968, pp. 83-131.

Beach, F. A. Sexual attractivity, proceptivity, and receptivity in female mammals. *Hormones and Behavior,* 1976, 7, 105-138. (a)

Beach, F. A. Cross-species comparisons and the human heritage. *Archives of Sexual Behavior,* 1976, **5**, 469-485. (b)

Bielert, C., Czaja, J. A., Eisele, S., Scheffler, G., Robinson, J. A. and Goy, R. W. Mating in the rhesus monkey (*Macaca mulatta*) after conception and its relationship to oestradiol and progesterone levels throughout pregnancy. *Journal of Reproductive Fertility,* 1976, **46**, 179-187.

Bramblett, C. A. *Patterns of Primate Behavior.* Palo Alto, California: Mayfield, 1976.

Carpenter, C. R. Sexual behavior of free-ranging rhesus monkeys (*Macaca mulatta*). II: Periodicity of estrus, homosexual, autoerotic and nonconformist behavior. *Journal of Comparative Psychology,* 1942, **33**, 143-162.

Charles-Dominique, P. Urine marking and territoriality in *Galago alleni* (Waterhouse, 1837 — Lorisoidea, Primates). A field study by radio-telemetry. *Zeitschrift für Tierpsychologie,* 1977, **43**, 113-138.

Cubicciotti, D. D., III and Mason, W. A. A comparison of heterosexual jealousy responses in *Callicebus* and *Saimiri.* Paper presented at the *American Society of Primatologists* Meeting, Seattle, Washington, April, 1977.

Dawson, G. A. Behavioral ecology of the Panamanian tamarin, *Saguinus oedipus* (Callithrichidae, Primates). *Dissertation Abstracts International,* 1976, **B37**, 645-646.

Demarest, W. J. Incest avoidance among human and nonhuman primates. In Chevalier-Skolnikoff, S. and Poirier, F. E. (Eds.) *Primate Biosocial Development.* New York: Garland, 1977, pp. 323-342.

Doyle, G. A., Pelletier, A. and Bekker, T. Courtship, mating, and parturition in the lesser bushbaby (*Galago senegalensis moholi*) under semi-natural conditions. *Folia Primatologica,* 1967, 7, 169-197.

Duggleby, C. R. Inbreeding in *Macaca mulatta* from Cayo Santiago. Paper presented at the *American Society of Primatologists* Meeting, Seattle, Washington, April, 1977.

Eisenberg, J. F. and Kuehn, R. E. The behavior of *Ateles geoffroyi* and related species. *Smithsonian Miscellaneous Collections,* 1966, 151 (8), 1-63.

Epple, G. Quantitative studies on scent marking in the marmoset (*Callithrix jacchus*). *Folia Primatologica,* 1970, 30, 48-62.

Erickson, L. B. Relationship of sexual receptivity to menstrual cycles in adult rhesus monkeys. *Nature,* 1967, 216, 299-301.

Erwin, J. and Maple, T. Ambisexual behavior with male-male anal penetration in male rhesus monkeys. *Archives of Sexual Behavior,* 1976, 5 (1), 9-14.

Everitt, B. J. Growing points in research on sexual and aggressive behaviour. Paper presented at the *International Primatological Society* Meeting, Cambridge, England, August, 1976.

Everitt, B. J. and Herbert, J. The role of ovarian hormones in the sexual preference of rhesus monkeys. *Animal Behaviour,* 1969, 17, 738-746.

Fairbanks, L. and McGuire, M. Homosexual behavior and female aggression in rhesus macaques. Paper presented at the *Western Psychological Association* Meeting, Seattle, Washington, April, 1977.

Fragaszy, D. M. Behavior in a novel environment in *Saimiri* and *Callicebus.* Paper presented at the *Western Psychological Association* Meeting, Seattle, Washington, April, 1977.

Goy, R. W. and Goldfoot, D. A. Neuroendocrinology: Animal models and problems of human sexuality. *Archives of Sexual Behavior,* 1975, 4 (4), 405-420.

Harrington, J. E. Discrimination between males and females by scent in *Lemur fulvus. Animal Behaviour,* 1977, 25, 147-151.

Hennessy, M. B., Coe, C. C., Mendoza, S. P., Lowe, E. L. and Levine, S. Seasonal fluctuations and sex differences in olfactory behavior of the squirrel monkey (*Saimiri sciureus*). Paper presented at the *American Society of Primatologists* Meeting, Seattle, Washington, April, 1977.

Hinde, R. A. and Rowell, T. E. Communication by postures and facial expressions in the rhesus monkey (*Macaca mulatta*). *Proceedings of the Zoological Society of London,* 1962, 138, 1-21.

Kleiman, D. G. The development of pair preferences in the lion tamarin (*Leontopithecus rosalia*): Male competition or female choice? Paper presented at the *Animal Behavior Society* Meeting, University Park, Pennsylvania, June, 1977. (a)

Kleiman, D. G. Monogamy in mammals. *The Quarterly Review of Biology,* 1977, **52**, 39-69. (b)

Klein, L. and Klein, D. Aspects of social behaviour in a colony of spider monkeys. *International Zoo Yearbook,* 1971, **11**, 175-181.

Kortmulder, K. On ethology and human behaviour. *Acta Biotheoretica,* 1974, **23** (3), 55-78.

Kress, J. H. Socialization patterns of *Lemur variegatus.* Paper presented at the *American Association of Physical Anthropologists* Meeting, Denver, Colorado, April, 1975.

Latta, J., Hopf, S. and Ploog, D. Observation on mating behavior and sexual play in the squirrel monkey (*Saimiri sciureus*). *Primates,* 1967, 8, 229-246.

Leibrecht, B. C. and Kelley, S. T. Some observations of behavior in breeding pairs of owl monkeys. Paper presented at the *American Society of Primatologists* Meeting, Seattle, Washington, April, 1977.

Lindburg, D. G. The rhesus monkey in North India: An ecological and behavioral study. In Rosenblum, L. A. (Ed.) *Primate Behavior: Developments in Field and Laboratory Research* (Vol. II). New York: Academic Press, 1971, pp. 1-106.

Lindburg, D. G. Mate selection in the rhesus monkey, *Macaca mulatta. American Journal of Physical Anthropology,* 1975, **42**, 315.

Lockard, J. S., Heestand, J. E., and Begert, S. P. Play behavior in slow loris (*Nycticebus coucang*). Paper presented at the *American Society of Primatologists* Meeting, Seattle, Washington, April, 1977.

Maple, T. Unusual sexual behavior of nonhuman primates. In Money, J. and Musaph, H. (Eds.) *Handbook of Sexology.* Amsterdam: Elsevier, 1977, pp. 1167-1186.

Maple, T., Erwin, J. and Mitchell, G. Separation of adult heterosexual pairs of rhesus monkeys: The effect of female cycle phase. *Journal of Behavioral Science,* 1974, **2**, 81-86.

Maple, T., Risse, G. and Mitchell, G. Separation of adult males from adult female rhesus monkeys (*Macaca mulatta*) after a short-term attachment. *Journal of Behavioral Science,* 1973, **1**, 327-336.

Maurus, M., Kühlmorgen, B., Hartmann-Wiesner, E. and Pruscha, H. An approach to the interpretation of the communicative meaning of visual signals in agonistic behavior ot squirrel monkeys. *Folia Primatologica,* 1975, **23**, 208-226.

Michael, R. P. and Saayman, G. S. Individual differences in the sexual behaviour of male rhesus monkeys (*Macaca mulatta*) under laboratory conditions. *Animal Behaviour*, 1967, **15**, 460–466.

Michael, R. P., Wilson, M. I. and Zumpe, D. The bisexual behavior of female rhesus monkeys. In Friedman, R. C., Richart, R. M. and van der Wiele, R. L. (Eds.) *Sex Differences in Behavior*. New York: Wiley, 1974, pp. 399–412.

Michael, R. P. and Zumpe, D. Sexual initiating behaviour by female rhesus monkeys (*Macaca mulatta*) under laboratory conditions. *Behaviour*, 1970, **36**, 168-185.

Michael, R. P. and Zumpe, D. Patterns of reproductive behavior. In Hafez, E. S. E. (Ed.) *Comparative Reproduction of Nonhuman Primates*. Springfield, Ill.: Charles C Thomas, 1971, pp. 205-242.

Missakian, E. A. Genealogical mating activity in free-ranging groups of rhesus monkeys (*Macaca mulatta*) on Cayo Santiago. *Behaviour*, 1973, **45**, 225-241.

Missakian-Quinn, E. A. and Varley, M. A. Consort activity in groups of free-ranging rhesus monkeys on Cayo Santiago. Paper presented at the *American Society of Primatologists* Meeting, Seattle, Washington, April, 1977.

Moynihan, M. Some behavior patterns of platyrrhine monkeys. II: *Saguinus geoffroyi* and some other tamarins. *Smithsonian Contributions to Zoology*, 1970, **28**, 1-77.

Nadler, R. D. and Rosenblum, L. A. Hormonal regulation of the "fatted" phenomenon in squirrel monkeys. *Anatomical Record*, 1972, **173**, 181-187.

Perrit, M. Influence du groupement social sur l'activation sexuelle saisonmiére chez le ♂ de *Microcebus murinus* (Miller, 1777). *Zeitschrift für Tierpsychologie*, 1977, **43**, 159-179.

Rowell, T. E. *The Social Behaviour of Monkeys*. Baltimore: Penguin, 1972.

Sade, D. S. Inhibition of son-mother mating among free-ranging rhesus monkeys. *Science and Psychoanalysis*, 1968, **12**, 18–38.

Sade, D. S., Chepko-Sade, B. D. and Schneider, J. Paternal exclusions among free-ranging rhesus monkeys on Cayo Santiago. Paper presented at the *Animal Behavior Society* Meeting, University Park, Pennsylvania, June, 1977.

Zumpe, D. and Michael, R. P. Ovarian hormones and female sexual invitations in captive rhesus monkeys (*Macaca mulatta*). *Animal Behaviour*, 1970, **18**, 293-301.

10.

Sexual Behavior: Other Old World Monkeys, Apes, Humans, Generalizations and Speculations

We will now continue our survey of sexual displays and sexual posturings in the primate order by describing the sexual behavior of Old World monkeys other than the rhesus and the sexual behavior of the lesser and great apes. We will have only a few words concerning the sexual behavior of *Homo sapiens*.

A. SEXUAL BEHAVIOR IN OTHER OLD WORLD MONKEYS

1. Other Macaques

Among the other macaques which have been studied most extensively with regard to sexual behavior are the Japanese macaque (*M. fuscata*), the pigtail macaque (*M. nemestrina*), and the stumptail macaque (*M. arctoides*).

Japanese macaques breed very well in captivity. Like rhesus, the females have a twenty-eight day cycle and show heightened receptivity near midcycle. During pregnancy, their estrogen and progesterone levels, like rhesus, but unlike chimpanzees and humans, show only slight increases (Nomura and Ohsawa, 1976). Females usually consort with several males and copulation involves eight to thirty mounts with one to eight thrusts per mount. As in the rhesus, the female reaches back with one hand to grab the male when he ejaculates. Mounts can be initiated by either the male or the female. Their sexual behavior is very much like that of rhesus monkeys (Michael and Zumpe, 1971).

In Japanese macaques, most males leave their natal troops by the time they reach the age of four or five years and no males spend their entire lives with their mother's troop. At about seven years of age they approach a strange troop during the mating season and mate (Sugiyama and Ohsawa, 1974). Agonistic behavior between males increases at the ages of six to nine; it also increases during the breeding season (Sugawara, 1968).

More dominant males tend to mate with more dominant females and lower ranked males with lower ranked females. If a female has not yet conceived she is more likely to mate with a dominant male. High-ranking Japanese macaque males will interfere with the courtship of lower ranking males (Stephenson, 1974).

As we know, Japanese monkeys have irregular cycles outside of their breeding season (Nigi, 1975). During the breeding season, some females occasionally mount males. All female Japanese monkeys seem to have partners they prefer, avoid sexual interactions with close kin, and sometimes display homosexual behavior (Wolfe, 1977). Mating behavior at the start of the breeding season is evidently stimulated by a heavy rainfall following a dry season. Day length does not seem to affect breeding, but a season of falling temperatures and the maturing of nut crops in the area of a troop seem to increase mating behavior (Azuma and Koford, 1966). There are increases in male displays such as shaking, kicking, leaping, tossing, and swinging during the mating season and these activities are correlated with ejaculation rank more than with dominance rank or attack frequencies. Thus, female choice of males appears to be related to male display (Modahl and Eaton, 1977).

Year-long affiliative bonds in Japanese monkeys are sometimes interrupted during the breeding season. Grooming partners outside of the breeding season are often not the same individuals as the consort partners during the mating season (Baxter and Fedigan, 1977).

The pigtail macaque (*M. nemestrina*) has a slightly longer menstrual cycle (thirty-two days) than does the rhesus. Also, unlike the rhesus, the pigtail female shows a very obvious sexual skin swelling for twenty-one of the thirty-two days of the cycle, and particularly so during the middle twelve days. Mounting occurs often during this period and females copulate with several males. For each pair, there is a series of mounts (eight to thirty-two) at three-minute intervals with around twelve thrusts per mount. Like rhesus females, pigtail females reach back and grasp the male's leg. At ejaculation, the male grimaces and emits a high-pitched squeal. In initiating courtship, the male retracts his ears and protrudes his muzzle and lips in a "len" or "flehmen" face (Michael and Zumpe, 1971).

As we can see from the minor differences between macaque species, no single species of macaque can adequately serve as an example for the entire genus as far as sexual behavior is concerned (Nadler and Rosenblum, 1973). For example, in contrast to rhesus monkeys, pigtail males do not show a seasonal response, nor do they need the presence of females to maintain testosterone levels (Bernstein et al., 1977).

In the pigtail, females low in social dominance rarely conceive and they evince low levels of sexual behavior even in a preovulatory condition (Goldfoot, 1971). Unlike rhesus, but like humans, the diurnal cycle of sexual behavior in pigtails is characterized by increased sex in the early morning (Martenson et al., in press). Female pigtails are more active in sexual behaviors than are males (Bernstein, 1972).

A third species of macaque for which there are many studies of sexual behavior is the stumptail macaque (*M. arctoides*). In the stumptail, the menstrual cycle is twenty-eight days long with barely visible bleeding at menses. There is no swelling of the sexual skin. Copulation involves only one mount, with many thrusts, ending in ejaculation. Thus, stumptails are referred to as single-mount ejaculators (SME). While mounting, the male lip-smacks while exposing the teeth, chatters, then barks. The female reaches back and, at ejaculation, the male mouths or bites the female's back. Prior to copulation, the male approaches, grabs the female's tail, pulls it aside, places a finger in the female's vagina, and sniffs his finger before mounting (Michael and Zumpe, 1971).

Among stumptails, males bite females during the later stages of copulation, but they also groom females more when the females are in estrus than when they are not in estrus (Blurton-Jones and Trollope, 1968). During copulation, subadult and juvenile monkeys (in particular) often harass the copulating couple. Animals of both sexes are commonly involved in the harassment (Gouzoules, 1974a).

As we shall see later, adult and juvenile males are often sexually aroused when a female is giving birth (Gouzoules, 1974b). This is a rather frequent occurrence in stumptails since in large groups in outdoor enclosures this species shows an unusual pattern of three birth peaks in a year (Estrada and Estrada, 1976).

Some of the most detailed descriptions of stumptail macaque sexual behavior have been published by Suzanne Chevalier-Skolnikoff (1975). She has noted, for instance, that adult male stumptail macaques display a "tieing" following ejaculation in a manner similar to, but not identical to, domestic dogs. Subadult males, however, do not manifest tieing. Also, Chevalier-Skolnikoff has noted that, while stumptail males are SME copulators, they are still capable of ejaculating three to five times per hour. During the male's ejaculation, there is evidence that the female experiences orgasm, usually at the time she reaches back to grasp the male. The male weighs approximately thirty pounds, the female only fifteen pounds. She frequently collapses under the male's weight during copulation (Chevalier-Skolnikoff, 1975).

Chevalier-Skolnikoff (1972) has also noted that stumptail macaques evidently learn adult sexual behavior in play and by observation of adults. The first occurrence of sexual behavior in her young stumptails was with an adult who actually aided the immature animal. An adult female actively inserts the young male's penis into her vagina. There is also frequent homosexual activity between adults and infants. For example, an adult male will sometimes mount a small male and achieve anal intromission. Oral sex between infants and adult males is also common. In closed colonies, in the laboratory at least, brother–sister, father–daughter, and even mother–son matings have been observed, and births have resulted from them (Trollope, 1976). The reader may consult Trollope and Blurton-Jones (1975) for other aspects of sexual behavior in *M. arctoides*.

In direct comparisons of the sexual response of males and females, Chevalier-Skolnikoff (1974) has noted that male orgasm is unusually obvious (being characterized by body rigidity, body spasms, facial expressions, and vocalizations). The female orgasm, on the other hand, is less obvious. However, orgasms in females are especially obvious during homosexual interactions. The female orgasms at this time are characterized by all of the features listed above for masculine orgasms. In heterosexual interaction, where female orgasm also occurs, it is much less masculine in form. The homosexual observations lead Chevalier-Skolnikoff (1974) to conclude that when stumptail females take an active role in sex, their potential for orgasm is much more similar to that of males than it is when they assume a passive role.

Behaviors such as those just described occur most often just before and during ovulation. For example, reaching back, one sign of impending female orgasm, is highest at ovulation. After ovariectomy, females show a decline in presenting and reaching back but an increase in grooming. Males, in general, continue to copulate but mount less often and groom more (Slob et al., 1975). However, there is marked individuality among stumptail males in the way they respond to the spaying of their preferred female partners. In some pairs, little or no change in sexual behavior can be detected (Slob, 1975).

Besides copulation, the stumptail macaque displays a great variety of sexual behavior, probably more so than any other species of macaque. Males, for example, sometimes engage in mutual masturbation, and females also engage in mutual vaginal stimulation to orgasm. There is also mutual fellatio (Kling and Dunne, 1976).

With regard to macaques other than rhesus, Japanese macaques, pigtails, and stumptails, less is known about sexual behavior. However, in the bonnet monkey (*M. radiata*) of southern India, it is known that single-mount ejaculation (as in the stumptail) occurs. Bonnet monkeys have a menstrual cycle of twenty-five to thirty-six days during which the menstrual flow may last ten days. There is no swelling of the sexual skin but there is a strong-smelling vaginal discharge. As in the other macaques, the female reaches back at ejaculation. There are no consort pairs formed in bonnet monkeys (Michael and Zumpe, 1971). In addition, there is less intergroup migration in bonnet males than in rhesus males, suggesting that there is more inbreeding in the bonnet (Wade, 1977). There is a mating season during which tensions are relatively high (Simonds, 1977).

The toque macaque (*M. sinica*) is closely related to the bonnet macaque. In this species, however, there is evidence that males attempt to mate in other troops by going on short sexual excursions. Solitary males, as are found in rhesus and Japanese macaques, are never found among toque macaques (Dittus, 1975a; 1975b).

The crab-eater macaque (*M. fascicularis*) also known as the Java macaque, common macaque, or kra, behaves much like the rhesus. Cycle length is twenty-eight days with a menstrual flow of two to seven days and a swelling and reddening of the sexual skin at puberty but not monthly. Copulation occurs primarily at mid-month (Michael and Zumpe, 1971). There is a high frequency of "inspecting" between males and females (Thompson, 1967). Crab-eater sexual behavior is disturbed by early rearing in social isolation (Chevalier-Skolnikoff and Poirier, 1977).

Almost nothing has been published about the arboreal, vocal and playful lion-tailed macaque (*Macaca silenus*). However, it is known that during sexual excitement the male calls and the female also has a "love call" (cf. Sugiyama, 1968).

All macaques have twenty-eight to thirty-two-day menstrual cycles with visible menses. In rhesus, Japanese, and pigtail macaques consort bonds are formed. These species are multiple-mount ejaculators (MME).

However, bonnet and stumptail monkeys, as noted above, are single-mount ejaculators (SME) who do not develop clear consort bonds. In MME species, both males and females initiate sexual activity. In SME species, the adult male usually initiates mounting. Only in the pigtail monkey, among macaques, is there a conspicuous monthly swelling (see Michael and Zumpe, 1971).

2. Baboons

In the yellow baboon (*Papio cynocephalus*), there is a thirty-one to thirty-five-day menstrual cycle with a menstrual flow of about three days. In contrast to most macaques there is a marked swelling of the sexual skin in all species of this genus. The yellow baboon is no exception (Michael and Zumpe, 1971). Among male yellow baboons, longevity and initial rank at the onset of a male's reproductive career are the most important factors determining the male's reproductive success (Saunders and Hausfater, 1977; also see Hausfater, 1975; Saunders and Hausfater, 1976).

In the chacma baboon (*P. ursinus*), copulation involves a series of mounts, each with one to twenty thrusts, only the last mount being an ejaculatory one. The mount occurs just as in the rhesus with foot clasp and facial grimace in the male. The female, however, emits a copulation call consisting of a number of sharp barks. After the mount, the female rapidly withdraws from the male (Michael and Zumpe, 1971).

Mounts in the chacma baboon are initiated by either sex. The female's present is like that of macaques. Females and males form consort pairs, but when the female shows maximal swelling, she consorts only with a dominant male. Males groom females more frequently at midcycle than at other times. This, too, is similar to the rhesus pattern (Michael and Zumpe, 1971). In one chacma troop observed by Saayman (1971), a very old male, ranking lowest in individual aggressive encounters, was the most sexually active.

Olive baboons (*P. anubis*) differ from chacma baboons in that ejaculation occurs in a single mount, males do not grimace during copulation, females do not give copulation calls, and the females do not immediately withdraw from the males following copulation (Michael and Zumpe, 1971). However, it is notable that two olive baboon troops from two different ecological conditions showed different sexual behavior patterns (Paterson, 1973).

The sacred baboon (*P. hamadryas*) lives in one-male groups. Each male has a harem of females. The male copulates only within his harem and only with females with sexual skin swelling. Hamadryas baboons are an MME species. In some harems an old male may be present but he never copulates (Kummer and Kurt, 1963). Old harem leaders lose their harem females mainly to their own young follower males with whom they are most intimate (Kummer et al., 1976).

In the other genus of baboons (*Theropithecus*), sexual behavior is much like that of the sacred baboon. The range of gelada harems is from a monogamous pair to a harem of eight females for one male. Although some gelada units include other adult males besides the leader, these additional males do not copulate with the females. Solitary or "free-lance" males emigrate from herd to herd. The leader male is a function of female choice (Mori and Kawai, 1975).

Interestingly, estrus has very little effect on either male or female gelada baboons. Sexual behavior is usually initiated by the female. The only change with estrus is in a tendency for dominant females to harass estrus females of lower rank (Dunbar, 1976).

3. Mangabeys

The arboreal mangabey (*Cercocebus spp.*) has a thirty-day menstrual cycle and there is a sexual swelling at midcycle. Unlike geladas, they live in multimale groups. Mangabeys are an MME species like rhesus. Mounts are initiated by either sex; and, at ejaculation, the female reaches back to grasp the male. There are no noticeable midcycle or seasonal increases in mangabey sexual behavior (Michael and Zumpe, 1971).

In the white-cheeked mangabeys, presenting is much more an activity of females than it is in baboons. In baboons, males often present, whereas in mangabeys they rarely do. In mangabeys, females also handle the genitals of other animals more than do males. Even though mounts may be initiated by either sex, sexual behavior in the mangabey is primarily a matter of male initiative in *Cercocebus albigena* (white-cheeked mangabeys) (Chalmers and Rowell, 1971).

4. Guenons, Vervets, and Talapoins

Most species of the genus *Cercopithecus* are arboreal and live in one-male groups. *Cercopithecus aethiops,* however, is primarily terrestrial and lives in multimale groups. Menstrual cycle length is thirty-three days, menses are not visible externally, and there is no sexual swelling. *C. aethiops* are single-mount ejaculators. At ejaculation the female looks back and grasps at the male (Michael and Zumpe, 1971). The alpha or control male of the vervet (*C. aethiops*) group initiates most of the sexual behavior (Fairbanks and McGuire, 1977); however, mounts may be initiated by the male or by the female who presents for a mount by crouching, deflecting her tail, and looking back over her shoulder (Gartlan, 1968).

Among the arboreal species of the genus *Cercopithecus* (guenons) there is only one adult male in each group (cf. Chevalier-Skolnikoff, 1972). The modal cycle length for *C. mitis* is thirty days, menses are not visible, and there is no external swelling. There is disagreement concerning presence or absence of midcycle peaks in sexual activity (Michael and Zumpe, 1971). There are probably seasonal changes as well (also see Gautier-Hion and Gautier, 1976).

In the closely related talapoin monkey (*Miopithecus talapoin*) there is a definite midcycle swelling of the sexual skin, a thirty-three day cycle, and increased copulation at midcycle. Furthermore, male talapoins become "fatted" during a May through August mating season (Michael and Zumpe, 1971). Among talapoins, adult females usually dominate males (Dixson, Herbert, and Rudd, 1972):

> Talapoin courtship seems to involve a high level of fear on the part of the male. Female talapoins are aggressive. (Wolfheim and Rowell, 1972, P. 254)

Some male talapoins were *killed* by females in a captive group during the mating period (Wolfheim, 1977). Despite this dominance and aggression, high-ranking females are the most preferred sexual partners of the talapoin males (Keverne, 1976). High-ranking females sometimes mount males and, less frequently, other females. However, dominant females sometimes prefer to *be* mounted by others (Dixson, Scruton, and Herbert, 1975).

Both male and female talapoins masturbate (Dixson, Scruton, and Herbert, 1975). The hormone estradiol, given to a female talapoin, will increase her sexual activity with males (Scruton and Herbert, 1972).

5. Patas Monkeys

The patas monkey (*Erythrocebus patas*) lives in one-male groups. They have thirty day menstrual cycles with no midcycle swelling, but there is increased sexual activity at midcycle. Multiple mounts are required before ejaculation and breeding is probably seasonal (Michael and Zumpe, 1971).

The female solicits mounting by making a cringing half-run toward the male with her tail down, eyebrows lowered, lips closed, and with vocalizations from inflated cheek-pouches (Michael and Zumpe, 1971).

6. Langurs

Langurs (*Presbytis spp.*) are seen in both one-male and multimale groups. They have a twenty-seven day cycle with a two to four-day menstrual flow, and there is no external swelling at midcycle. The langurs are an MME species but the males do not use a foot clasp when they mount. All mounts are initiated by the female who drops her tail to the ground and shakes her head from side to side. Sexually excited males emit repeated low-pitched sounds. They have a breeding season (Michael and Zumpe, 1971).

Males who take over new groups among Hanuman langurs (*P. entellus*) often systematically kill off all the young infants of the group, which advances the estrus of the females of the group with whom he then copulates (Hrdy, 1977). Infanticidal males apparently have a reproductive advantage over noninfanticidal males (Chapman and Hausfater, 1977). However, the taking over of groups by a male, followed by infanticide, does not occur in all areas (Curtin, 1977; Jay, 1963):

> . . . vulnerability of a troop to male takeovers, not human disturbance of the habitat, appears to determine the differences in infant survivorship between troops. (Hrdy, 1977, letter)

During copulation, subadult male langurs sometimes harass the copulating pair (Jay, 1963).

Since many langur groups contain only one male, it may not be surprising that many langur males (*P. entellus*) live in all-male groups. In these groups, males learn female sexual behaviors as well as male sexual behaviors. As Weber and Vogel (1970) note:

> . . . our data seem to indicate that certain sexual functions of females in bisexual troops are transferred to subordinate males in only-male groups. (P. 79)

Males who take over bisexual troops often come from such all-male groups. However, it should not be assumed that, in taking over a troop, langur males are aggressive toward all infants. They do not kill those infants that are in their second year. Indeed, Horwich (1974) has described some very gentle infant-directed behaviors in one species of langur (*P. obscurus*).

B. SEXUAL BEHAVIOR IN THE APES

1. Lesser Apes

Gibbons (*Hylobates spp.*) and siamangs (*Symphalangus spp.*) are the mono-gamous arboreal Asian lesser apes. Among these lesser apes, sexual behavior is rich and varied when it occurs, but it occurs infrequently relative to nonmono-gamous species (cf. Kleiman, 1977). Gibbons and siamangs probably engage in less sexual activity than do the nonmonogamous great apes.

In *Hylobates lar*, the sex partners of a pair seem to recognize each other by the "duet" song which also reinforces the pair bond (Apfelbach, 1972). Mating postures can be dorsoventral or ventroventral.

In siamangs observed by Koyama (1971), the mating posture was first ventro-ventral (face-to-face) then dorsoventral (like rhesus). Copulation apparently oc-curs in the morning, after morning calls, with juveniles and infants observing (Koyama, 1971).

About sixty instances of free-ranging siamang sexual behavior were seen by Chivers and Aldrich-Blake (1976). These authors believe that the ventroventral position is relatively rare for siamangs. In the siamang, the males may mount with or without ejaculating, hence they are probably an MME species. The evi-dence presented by Chivers and Aldrich-Blake (1976) also suggests that there are monthly cycles in sexual activity and a mating season.

2. The Common Chimpanzee

As we have already seen, common chimpanzees (*Pan troglodytes*) have regular menstrual cycles and a pronounced midcycle swelling. The menstrual cycle is about thirty-five days long with external menstrual flow occurring on three of these days for up to forty-eight years (Graham and McClure, 1977). Marked genital swelling occurs for eighteen of the thirty-five days in the middle of the cycle and sexual activity is maximal at this time. The common chimpanzee is an SME species. The female crouches and the male thrusts from five to twenty times to ejaculation in a single mount. The female often looks back at the male and makes squealing or screaming sounds. Breeding does not appear to be sea-sonal (Michael and Zumpe, 1971).

As we have seen earlier, the male common chimpanzee does not leave his mother for more than a few days until he is ten years old. Females stay even longer with their mothers (van Lawick-Goodall, 1973). Savage and Malick (1977) have shown that mothers sometimes sexually present to their own male infants but more often present to male infants *not* their own. The male infants, in turn, thrust at adult fe-males. Mothers also masturbate themselves while handling a male infant (not their own). Juvenile males sometimes direct sexual behavior towards infants and often copulate with adult females. Mother–son copulation *is* seen, but it is infrequent, does not occur at midcycle, the thrusting is slower and longer, and it occurs when the son seems distressed. In addition, mother–son copulation in the common chimpanzee occurs without the mother's presenting and often with the mother trying to prevent it. The little male, in captivity, will sometimes display a temper tantrum until his mother permits it. Furthermore, the entire precopulatory display is only seen in males with adult females other than their mothers.

In play behavior, infants appear to kiss each other, manipulate each other's genitals with fingers and/or lips, *and* thrust at each other (Savage and Malick, 1977). Infant males sometimes have erections through an entire play session. The kisses between infants are often very long in duration (Savage and Malick, 1977). Group play appears to center around the oldest male infant.

When a male chimpanzee reaches adolescence, he starts to sleep alone unless a female nearby is in estrus, in which case he sleeps with her (Riss and Goodall, 1976).

Chimpanzee sexual behavior is individualistic (van Lawick-Goodall, 1973). Both males and females, young and old, may mount another animal. The usual posture for common chimpanzees is dorsoventral, but van Hooff (1973), for example, observed one young male mounting another male's head and thrusting into its face. Maple (1976) has published a review of unusual sexual behaviors of nonhuman primates. Common chimpanzees provide some of the most unusual examples in his review.

The common chimpanzee is, for the most part, a promiscuous animal. However, even with regard to their promiscuity, they have been unpredictable. Temporary monogamous consort pairs have been seen by Tutin (1974) in which the members of the pair exhibited possessive behavior. Apparently parous (those who have had a baby) females are more possessive than are nulliparous females. Possessiveness in males correlates with the amount of time the male is seen grooming the female or sharing food with her. In such monogamous pairs:

> If female choice is involved, it is of interest to note that the selection criteria appear to be social and caretaking abilities of the males and not their dominance status. (Tutin, 1974, P. 448)

While male common chimpanzees are generally dominant over female adult chimpanzees there is said to be a reversal in dominance when the female is at the height of her sexual swelling. In addition, the administration of estrogen to a female increases her dominance (Mason, 1970):

> In chimpanzee sexual relations, the roles are not as sharply differentiated as in the case with monkeys . . . . (Mason, 1968, P. 26)

3. The Pygmy Chimpanzee

There is more than one kind of chimpanzee. The pygmy chimpanzee (*Pan paniscus*) is smaller, and differs from *Pan troglodytes* in regards to location of female genitalia, preferred copulatory postures, and copulatory communication (Savage and Bakeman, 1976). Unlike common chimpanzees, pygmy chimpanzees engage in copulation during all phases of the sexual cycle. The female genitals remain turgid even following menses.

Female homosexual *ventroventral* thrusting bouts are common among pygmy chimpanzee females and they occur with quite visible clitoral erection and with gestures and vocalizations commonly observed in male-female copulation. Copulation in *Pan paniscus* occurs frequently during, or prior to, feeding. It is often associated with food sharing (Savage and Bakeman, 1976).

Both males and females exhibit thrusting during copulation. Copulation is also accompanied by direct eye-to-eye contact and mutual face peering (Savage and Bakeman, 1976; see also Savage, 1976).

For other references to chimpanzee sexual behavior see Bauer (1976) and Martin (1976).

4. The Mountain Gorilla

Autoerotic behavior (masturbation) has been seen in both males and females in captive mountain gorillas (*Gorilla gorilla beringei*) (Schaller, 1963). In the wild, sexual behavior between immature mountain gorillas is rare and between adults and immatures it almost never occurs. Swelling is only slight and can usually be seen only in very young adult females. The estrus cycle is twenty-eight days and the period of receptivity only two to three days. Females will copulate when they are pregnant and the females themselves initiate most of the sexual contact that occurs between them and the male (Harcourt and Stewart, 1976).

During mating both males and females thrust and make vocalizations and facial expressions. Orgasm in females is assumed to occur and homosexual behavior, particularly in females, is said to occur regularly. Most gorilla females leave their natal group; however, Harcourt and Stewart (1976) reported a case where at least one wild female gorilla was known to have been impregnated by either her father or her brother. Schaller (1963) reported that captive mountain gorillas sometimes engage in ventroventral copulation.

5. The Lowland Gorilla

In the lowland gorilla (*Gorilla gorilla gorilla*), copulation has been reported to occur after conception and sporadically throughout pregnancy (Nadler, 1975).

Outside of pregnancy, captive female lowland gorillas show only slight swelling when receptive. At swelling, captive subadult female lowland gorillas increase their play with males. The females initiate the play with slow, gentle, caressing, and wrestling. As swelling increases, subadult female lowland gorillas even display chest beating, usually a male activity. They also "mouth groom" the males by kissing them and they stick their tongues into the male's mouths (Keiter, 1977).

Subadult male lowland gorillas respond to these advances by their captive female partners by "back riding" the females. The females get on the backs of the males and ride them for five seconds to ten minutes during which time the males have a visible erection. The males even invite the females to ride on their backs by giving two short staccato grunts. During such riding there is grunting by both males and females. It is interesting that during copulation there is also grunting and growling (Keiter, 1977).

While copulating, the captive male and female subadult lowland gorillas show "pursed lips" and "mouth open" facial expressions. The facial expressions change more in the male than in the female (Keiter, 1977).

In the lowland gorilla, estrus begins at 5 1/2 years in the female. Intromission begins between the ages of five and seven years in subadult males. During estrus, the female initiates 80 percent of the courtship play which precedes copulation (Keiter, 1977). Nadler (1975) has found that lowland gorillas mate in a cyclic manner which is closely related to the degree of female genital swelling.

6. The Orangutan

As in the gorilla and pygmy chimpanzee, ventroventral as well as dorsoventral copulation has been observed in captive orangutans (*Pongo pygmaeus*) (Michael and Zumpe, 1971). Sexual behavior is seen early in the life of young orangutans in captivity. Mothers of male infants establish genital contact and engage in mounting with pelvic thrusting toward their infants, sometimes as early as the second week of life. The mothers also provide oral-genital contact. (Maple et al., 1977).

Research on captive orangutans has raised questions concerning the assumptions that: (1) the adult male is primarily responsible for initiating sex, and (2) the adult male may even have to force himself on the female. In captivity the female may show a *cyclic* sexual assertiveness resembling that of the female gorilla, if conditions allow it. Thus, the orangutan female not only shows receptive behavior, she also may show proceptive sexual behavior. The male's interest in the female can also be cyclic (Maple and Zucker, 1977), but there is some doubt about this issue. The female, however, shows no obvious sexual swelling (Rijksen, 1975).

Female orangutan proceptive sexual behaviors include play, fast brachiation, hand and head slapping, contact grappling, oral-genital contact, hand-genital contact, body positioning, genital-genital contact, and pelvic thrusting. Apparently, appropriate sexual behavior in adulthood is related to early mother-infant play and sexual behaviors (Maple et al., 1977). During much sex play the male infant also thrusts against his mother.

In attempting to put captive orangutans back into their native habitat in an orangutan rehabilitation project in Sabah, de Silva (1971) noted that females were capable of proceptive sexual behavior. The female in one mating episode was the more active partner in the mating itself and appeared to be the most upset when the mating was interrupted. However, in this particular case the male was not a full adult (de Silva, 1971).

As is commonly known, the only long-term orangutan social unit is the adult female with her dependent offspring. Adult males generally forage alone. Juvenile females stay with their mothers for several years whereas juvenile males tend to range further away at an earlier age than do juvenile females. Sometimes solitary juvenile males are observed in the wild (Horr, 1972). Apparently, at five to eight years of age, young males and females sometimes gather to make sexual groups (Okano, 1971). Although there is on occasion a great deal of adult male aggression toward the female during sexual interactions, orangutans appear to have a diverse sexual repertoire involving female proceptivity and sexual relations with immature animals (Maple and Zucker, 1977).

According to Nadler (1976), it is possible that the orangutan may resemble people in sexual behavior more than do the other great apes. It is Nadler's view that the male is the primary initiator of sexual activity and that, while cyclicity does occur along with female proceptivity, the male appears to initiate sexual intercourse irrespective of the phase of the female's cycle and despite female resistance. Also, in orangutans, the duration of each copulation is longer than for other primates and a variety of positions and postural adjustments occur during each copulation (Nadler, 1976).

Up to this point, we are left with the impression that either the male forces sex upon the female or the female persistently attempts to get sex from the male. Actually, there are a great many occasions when each sex appears to be attractive, proceptive, and receptive to the other. Orangutans are unpredictable. The sexual behavior of the orangutan male *does* tend toward extremes, however—from near lethargy to daily copulation (Zucker et al., 1977).

Aside from direct sexual contacts with the mother, infant orangutans also have opportunities to learn sexual behavior by observing adults. Juveniles show a very strong interest in adult copulation. However, it is the opinion of at least one group of researchers that the developing male infant acquires sexual behaviors primarily from contact with his mother. There are few sexual interactions between males and infants (Zucker et al., 1977; Wilson et al., 1977).

Aside from sociosexual behavior, orangutans have also been observed using inanimate objects in sex and they also masturbate occasionally. According to Nadler and Braggio (1974), females interact more with inanimate objects than do males, at least in the juvenile period.

7. Final Comments on Great Apes

Only in the orangutan, among the great apes, has the male been known to force himself onto the female (Pitcairn, 1974). The chimpanzee female is the only ape female to show an obvious sexual swelling at midcycle. There may also be a preference for heterosexual play in the great apes, particularly in the gorillas (Freeman and Alcock, 1973), but data are not yet sufficient on this point. Sex appears in play between infants and mothers as well as between peers. There appear to be far more unpredictable and unusual sexual behaviors among great apes than among other primates (Maple, 1976). There is not only more variety from episode to episode, there is also more variety within a sexual episode in apes than in monkeys.

C. COMPARISONS OF APE TO HUMAN SEXUALITY

Jensen (1976; in press) has noted a number of similarities between chimpanzee and human sexual behavior. These similarities include: (1) the eye contact courtship gesture of male toward female; (2) copulation thrusting patterns; (3) genital arousal in play; (4) harassment by infants during copulation; (5) inhibition of sexual behavior in a male when a more dominant male is present; (6) privacy-seeking in consort relationships; (7) sleeping together during consort relationships; and (8) repeated selection of the same partner for sex. Jensen (1976; in press) feels that people differ from the apes in not having an estrus. In humans there is continued rather than periodic sexual activity and there is the use of language and fantasy.

With regard to the loss of estrus and other hormonal effects in humans, it is possible that this difference may not be as great as Jensen suggests. For example, while apes show cyclicity they also copulate during pregnancy and outside of estrus. Also, the likelihood of such phenomena as human pheramones is still in question (Comfort, 1971). In people, as in apes and monkeys, there is evidence for a role of sexual arousal at the time of birth (Newton, 1973; Tanzer, 1973; Tavris, 1973); and, as in the rhesus, there is the potential for sexual arousal at menstruation (Erickson, 1967; Rowell, 1972).

On the other hand, in an attempt to relate male testosterone levels to sexual behavior in twenty adult men it was concluded that:

> . . . rising levels of circulating testosterone probably are not involved in precipitating the desire for initiating sexual activity. (Doering et al., 1974, pp. 427–428).

In addition, among human females there is the evidence that excessive amounts of estrogens at times may even depress female sexual behavior (Eichorn, 1968).

With regard to sexual behavior in infancy we are more like the apes than many of us realize. In a study of three human female infants, Bakwin (1973) found that, at less than one year of age, these infants were masturbating by using friction of the thighs, by rubbing the body against an object, or by direct manipulation. These autoerotic episodes (with erections) were accompanied by grunting and flushing which progressed to what appeared to be orgasm followed by general relaxation, pallor, sweating, and deep sleep.

Bakwin (1973) also observed masturbation culminating in orgasm (but with no ejaculation) in male infants. There was a remarkable capability for repeated orgasm in a short time. One boy, aged eleven months, had ten orgasms in one hour, another had fourteen in thirty-eight minutes. Orgasm was inferred from grunting, flushing, etc.

D. GENERAL COMMENTS ON NATURAL SCIENCE AND HUMAN SEXUALITY

As we have seen in previous chapters, sex differences in bodily size, coloration, and shape are of some importance to sexual behavior (Wickler, 1967). No one would doubt that this is also true for people. However, aside from obvious physical differences, there are also differences in behavior that may be important for sexual arousal. Jensen (1976) has noted, for example, that apes and humans use eye contact in sexual courtship. It is known that both sexes of people prefer silent eye-to-eye contact with people of the other sex to such contact with their own sex. Thus, eye contact is important in heterosexual love. It is also known that male and female people use eye contact differently (Exline, 1971). Females tolerate more eye contact than do males.

Throughout our coverage of sex differences in play, hormones, and sexual behavior, we have repeatedly been exposed to studies which show sex differences in aggression. We will turn to the subject of aggression in more detail later; however, it is important at this time to briefly raise the possibility of a role for aggression in human sex differences in sexual behavior. In a study of eighty-five undergraduate males and forty-five undergraduate females, Baron and Bell

(1977) and Baron (1977) have examined the relationship between sexual arousal and aggression. In these studies the subjects were first angered or not angered by a confederate of the same sex and then exposed to erotic or nonerotic stimuli. There *was* a gender-related difference but it was a surprising one. Males responded to mildly arousing erotic stimuli by inhibiting aggression. Very arousing erotic stimuli had no effect on the aggressiveness of the men. Females also responded to the mildly erotic stimuli by inhibiting their aggressiveness. However, very erotic stimuli *increased* the subsequent aggression of the women. Thus, women responded with increased aggression to types of erotic stimuli which were found to inhibit aggression in males. Baron and Bell (1977) conclude:

> There is indeed an important link between sexual arousal and aggressive behavior. The nature of this relationship, however, seems to be substantially more complex than was once suspected. (P. 87)

Results such as those above should remind us that facile generalization of nonhuman primate sex differences to people are unwarranted. As we have seen repeatedly in this book on the diverse species of nonhuman primates, the best way to learn about a particular species is to study *that species*. Generalizing, for example, from one ape to another is no substitute for studying each ape species separately. People are no exceptions.

Whitsett (1977) has recently presented a critique of the natural science approach to human sexuality. He argues that the conceptual rules of natural science are inappropriate for the study of human sexuality. The "sex difference" approach, he says, divides human sexuality into female and male sexuality based on anatomical distinctions. He criticizes such a deterministic model and argues for a phenomenological "human science" approach to generate existential, experiential descriptions of sexuality. It is our opinion that he is responding to oversimplified biologistic paradigms and that a responsible, properly *scientific*, natural science approach is indeed appropriate. The information supplied in this chapter should convince us all of that. We only hope that someone will synthesize the available human data on sexuality soon, and relate it responsibly to the primatological data. In fact, we would like very much to do that in this volume. Unfortunately, to do a good job of this would mean that we would have to double an already long section on sex differences in sexuality. It would also lead us away from our main subjects, *nonhuman* primates.

After having attempted to examine the published material on sex differences on human sexuality, we can now understand and respect the decision made by Maccoby and Jacklin (1974) in their esteemed tome on human sex differences:

> Sexual behavior per se may of course be the sphere of behavior most affected by the biology of sex. The reader should be forewarned, however,

that this book does not deal with sexual behavior in the usual sense of the term.   Sexual behavior is a topic widely written about, and we leave it to the sexologists. (P. 3)

If the experts on sex differences in people could not possibly include human sexual behavior in their volume, it is understandable that a humble book on the nonhuman primates could do likewise.   However, we can supply the reader with several beginning references on the topic (Jones, Shainberg, and Byer, 1977; Katchadourian, 1974; McCary, 1967) in the hopes that something more substantial on this topic will also be read.

E. INCEST AVOIDANCE

As we have seen throughout the last two chapters, many species of nonhuman primates (e.g., rhesus, Japanese monkeys, chimpanzees, etc.) have behavioral mechanisms by which they avoid inbreeding.   There is, for example, active avoidance of mother-son incest, primarily by the mother, in chimpanzees (cf. Savage and Malick, 1977).   In rhesus, "the chances of mating occurring *within* a genealogy are considerably lower than the chances of it occurring with animals not related by uterine ties." (Fox, 1975, p. 22.)

Many authors (cf. Kortmulder, 1974; Demarest, 1977) have pointed out that in people it is deemed boring or dull to marry a brother or a sister.   In the Israeli kibbutzim, where unrelated children are raised together, people almost never marry those with whom they have grown up.   Thus, there is some evidence for what might be called a stimulus saturation or habituation theory of incest avoidance.   This may be the reason that apes captured at infancy and raised together fail to breed as adults (Maple, in press).

There is also the possibility that role conflict may be of importance in human incest avoidance.   If the individual is already a sibling or a parent or a daughter, how can they also be a lover or a spouse?   Or, perhaps aggression is needed for sex and marriage and aggression is socialized out of the sibling and parent-child relationships.   According to this theory, courtship must have conflict in it (Kortmulder, 1974).

Of course, in people there is not only an avoidance of incest, there is also a proscription or taboo against it.   Yet, simply because people have rules and nonhuman primates do not, does not mean that nonhuman primate outbreeding is of no relevance to *Homo sapiens*.   As Robin Fox (1975) has said:

I take for granted that we have rules since we are an animal that makes rules; the real question is do rules represent more than a "labelling" procedure for behavior that would occur anyway. (P. 10)

Demarest believes that there are indeed some basic biological incest avoidance responses which nonhuman primates and people have in common. Among these are lack of interest because of early habituation, and the tendency for the young, especially males, to leave the natal group as they near or pass puberty.

REFERENCES

Apfelbach, R. Electrically elicited vocalizations in the gibbon *Hylobates lar* (*Hylobatidae*) and their behavioral significance. *Zeitschrift fur Tierpsychologie,* 1972, **30**, 420-430.

Azuma, S. and Koford, C. B. The influence of weather on mating time in rhesus and Japanese macaques. *Primates,* 1966, **7** (3), 404.

Bakwin, H. Erotic feelings in infants and young children. *American Journal of Diseases of Childhood,* 1973, **126**, 52-54.

Baron, R. A. Heightened sexual arousal and physical aggression: An extension to females. Unpublished manuscript, 1977.

Baron, R. A. and Bell, P. A. Sexual arousal and aggression by males: Effects of type of erotic stimuli and prior provocation. *Journal of Personality and Social Psychology,* 1977, **35**, 79-87.

Bauer, H. R. Sex Differences in aggregation and sexual selection in Gombe chimpanzees. *American Zoologist,* 1976, **16**, 209.

Baxter, M. J. and Fedigan, L. M. A comparative analysis of grooming and consort partner selection in a troop of Japanese monkeys. Paper presented at the *American Society of Primatologists* Meeting, Seattle, Washington, April, 1977.

Bernstein, I. S. Daily activity cycles and weather influences on a pigtail monkey group. *Folia Primatologica,* 1972, **18**, 390-415.

Bernstein, I. S., Gordon, T. P., Peterson, M. and Rose, R. M. Female influences on pigtail male testosterone levels. Paper presented at the *American Society of Primatologists* Meeting, Seattle, Washington, April, 1977.

Blurton-Jones, N. G. and Trollope, J. Social behaviour of stump-tailed macaques in captivity. *Primates,* 1968, **9**, 365-394.

Chalmers, N. R. and Rowell, T. E. Behaviour and female reproductive cycles in a captive group of mangabeys. *Folia Primatologica,* 1971, **14**, 1-14.

Chapman, M. and Hausfater, G. Infanticide in langurs: A mathematical analysis. Paper presented at the *American Society of Primatologists* Meeting, Seattle, Washington, April, 1977.

Chevalier-Skolnikoff, S. Sex role differentiation in nonhuman primates, with implications for man. *Medical Anthropology Newsletter,* 1972, **3** (3), 9-13.

Chevalier-Skolnikoff, S. Male-female, female-female, and male-male sexual behavior in the stumptail monkey, with special attention to the female orgasm. *Archives of Sexual Behavior,* 1974, **3** (2), 95-116.

Chevalier-Skolnikoff, S. Heterosexual copulatory patterns in stumptail macaques (*Macaca arctoides*) and in other macaque species. *Archives of Sexual Behavior,* 1975, **4**, (2), 199-220.

Chevalier-Skolnikoff, S. and Poirier, F. S. (Eds.) *Primate Bio-Social Development: Biological, Social and Ecological Determinants.* New York: Garland, 1977.

Chivers, D. J. and Aldrich-Blake, F. P. Sexual behaviour in free-ranging siamang. Paper presented at the *International Primatological Society* Meeting, Cambridge, England, August, 1976.

Comfort, A. Likelihood of human pheromones. *Nature,* 1971, **230**, 432–433.

Curtin, R. A. Socioecology of langurs in the Nepal Himalaya. Paper presented at the *American Association of Physical Anthropologists* Meeting, Seattle, Washington, April, 1977.

Demarest, W. J. Incest avoidance among human and nonhuman primates. In Chevalier-Skolnikoff, S. and Poirier, F. E. (Eds.) *Primate Biosocial Development.* New York: Garland, 1977, pp. 323-342.

de Silva, G. S. Notes on the orangutan rehabilitation project in Sabah. *Maylayan Nature Journal,* 1971, **24**, 50-77.

Dittus, W. P. J. Dynamics of migration in *Macaca sinica* (Linnaeus, 1771) and its relation to male dominance, reproduction and mortality. *American Journal of Physical Anthropology,* 1975, **42**, 298. (a)

Dittus, W. P. J. Population dynamics of the toque monkey, *Macaca sinica.* In Tuttle, R. H. (Eds.) *Socioecology and Psychology of Primates.* The Hague: Mouton, 1975, pp. 125-151. (b)

Dixson, A. F., Herbert, J. and Rudd, B. T. Gonadal hormones and behaviour in captive groups of talapoin monkeys (*Miopithecus talapoin*) *Journal of Endocrinology,* 1972, **57**, 41.

Dixson, A. F., Scruton, D. M. and Herbert, J. Behaviour of the talapoin monkey (*Miopithecus talapoin*) studied in groups in the laboratory. *Journal of Zoology.* (London), 1975, **176**, 177-210.

Doering, C. H. Brodie, H. K. H., Kraemer, H., Becker, H. and Hamburg, D. A. Plasma testosterone levels and psychologic measures in men over a 2-month period. In Friedman, R. C., Richart, R. M. and van de Wiele, R. L. (Eds.) *Sex Differences in Behavior.* New York: Wiley, 1974, pp. 413–431.

Dunbar, R. I. M. Oestrous behaviour and social relations among gelada baboons. Paper presented at the *International Primatological Society* Meeting, Cambridge, England, August, 1976.

Eichorn, D. H. Biology of gestation and infancy: Fatherland and frontier. *Merrill-Palmer Quarterly,* 1968, **14** (1), 47-81.

Erickson, L. B. Relationship of sexual receptivity to menstrual cycles in adult rhesus monkeys. *Nature,* 1967, **216**, 299-301.

Estrada, A. and Estrada, R. Birth and breeding cyclicity in an outdoor living stumptail macaque (*Macaca arctoides*) group. *Primates,* 1976, **17** (2), 225–231.

Exline, R. V. Visual interaction: The glances of power and preference. *Current Theoretical Research on Motivation,* 1971, **19**, 163–206.

Fairbanks, L. A. and McGuire, M. T. Inhibition of control male behavior in vervet monkeys. Paper presented at the *American Society of Primatologists* Meeting, Seattle, Washington, April, 1977.

Fox, R. Primate kin and human kinship. In Fox, R. (Ed.) *Biosocial Anthropology.* New York: Wiley, 1975.

Freeman, H. E. and Alcock, J. Play behaviour of a mixed group of juvenile gorillas and orangutans. *International Zoo Yearbook,* 1973, **13**, 189–194.

Gartlan, J. S. Sexual and maternal behavior of the vervet monkey, *Cercopithecus aethiops. Journal of Reproduction and Fertility,* 1969, **6**, 137–150.

Gautier-Hion, A. and Gautier, J. P. Croissance, maturité sexuelle et sociale, reproduction chez les cercopithécinés forestiers Africains. *Folia Primatologica,* 1976, **26**, 165–184.

Goldfoot, D. A. Hormonal and social determinants of sexual behavior in the pigtail monkey (*Macaca nemestrina*). In Stoelinga, G. B. A. and van der Werff ten Bosch, J. J. (Eds.) *Normal and Abnormal Development in Brain and Behavior.* Leiden: University of Leiden Press, 1971, pp. 325–342.

Gouzoules, H. Harassment of sexual behavior in the stumptail macaque, *Macaca arctoides. Folia Primatologica,* 1974, **22**, 208–217. (a)

Gouzoules, H. Group responses to parturition in *Macaca arctoides. Primates,* 1974, **15**, 287–292. (b)

Graham, C. E. and McClure, H. M. Ovarian tumors and related lesions in aged chimpanzees. *Veterinary Pathology,* 1977, **14**, 380–386.

Harcourt, A. H. and Stewart, K. J. Sexual behaviour of wild mountain gorilla. Paper presented at the *International Primatological Society* Meeting, Cambridge, England, August, 1976.

Hausfater, G. Dominance and reproduction in baboons (*Papio cynocephalus*): A quantitative analysis. In *Contributions to Primatology* (Vol. 7). Basel: Karger, 1975, pp. 1–150.

Horr, D. A. The Borneo orangutan. *Borneo Research Bulletin,* 1972, **4**, 46–50.

Horwich, R. H. Development of behaviors in a male spectacled langur (*Presbytis obscurus*). *Primates,* 1974, **15**, 151–178.

Hrdy, S. B. Infanticide as a primate reproductive strategy. *American Scientist,* 1977, **65**, 40–49.

Hrdy, S. B. Reply to Dolhinow. *American Scientist,* 1977, **65**, letter.

Jay, P. The Indian langur monkey (*Presbytis entellus*). In Southwick, C. H. (Ed.) *Primate Social Behavior.* Princeton, N.J.: Van Nostrand, 1963, pp. 114–123.

Jensen, G. D. Comparisons of sexuality of chimpanzees and humans. Paper presented at the *International Primatological Society* Meeting, Cambridge, England, August, 1976.

Jensen, G. D. Sexual patterns of chimpanzees: Human comparisons. *Journal of Sex Research,* in press.

Jones, K. L., Shainberg, L. W. and Byer, C. O. *Sex and People.* New York: Harper & Row, 1977.

Katchadourian, H. *Human sexuality: Sense and Nonsense.* San Francisco: W. H. Freeman, 1974.

Keiter, M. D. Reproductive behavior in subadult captive lowland gorillas (*Gorilla gorilla gorilla*). Paper presented at the *American Society of Primatologists* Meeting, Seattle, Washington, April, 1977.

Keverne, E. B. Dominance, aggression and sexual behaviour in social groups of talapoin monkeys. Paper presented at the *International Primatological Society* Meeting, Cambridge, England, August, 1976.

Kleiman, D. G. Monogamy in mammals. *The Quarterly Review of Biology,* 1977, **52**, 39-69.

Kling, A. and Dunne, K. Social-environmental factors affecting behavior and plasma testosterone in normal and amygdala lesioned *M. speciosa. Primates,* 1976, **17** (1), 23-42.

Kortmulder, K. On ethology and human behaviour. *Acta Biotheoretica,* 1974, **23** (2), 55-78.

Koyama, N. Observations on mating behavior of wild Siamang gibbons at Fraser's hill Malaysia. *Primates,* 1971, **12**, 183-189.

Kummer, H., Abegglen, J. J., Abegglen, H., Bachman, H., Falett, J. and Sigg, H. Intimacy and fear of aggression in baboon social relationships. Paper presented at the *International Primatological Society* Meeting, Cambridge, England, August, 1976.

Kummer, H. and Kurt, F. Social units of a free-living population of hamadryas baboons. *Folia Primatologica,* 1963, **1-2**, 4-18.

Letson, G. W. and Ellis, J. E. Analysis of sexual behavior in captive orangutans. Paper presented at the *Animal Behavior Society* Meeting, University Park, Pennsylvania, June, 1977.

Maccoby, E. E. and Jacklin, C. N. *The Psychology of Sex Differences.* Stanford, California: Stanford University Press, 1974.

Maple, T. Unusual sexual behaviors of nonhuman primates. In Money, J. and Musaph, H. (Eds.) *Handbook of Sexology.* Amsterdam: Elsevier, 1976, pp. 1167-1186.

Maple, T. in press.

Maple, T. and Zucker, E. L. Behavioral studies of captive Yerkes orangutans at the Atlanta Zoological Park. *Yerkes Newsletter,* 1977, **14** (1), 24-26.

Maple, T., Zucker, E. L., Hoff, M. P. and Wilson, M. E. Behavioral aspects of reproduction in the great apes. Paper presented at the *American Association of Zoological Parks and Aquariums* Meeting, San Diego, California, September, 1977.

Martenson, J., Jr., Oswald, M., Sackett, D. and Erwin, J. Diurnal variation of common behaviors of pigtail monkeys (*Macaca nemestrina*). *Primates*, in press.

Martin, D. Apes, aphrodisiacs and artificial insemination. *Yerkes Newsletter*, 1976, **13**, 20-24.

Mason, W. A. Chimpanzee social behavior. In Bourne, G. (Ed.) *The Chimpanzee*. New York: Karger, 1970, pp. 265-288.

Mason, W. A., Davenport, R. K., Jr. and Menzel, E. W., Jr. Early experience and the social development of rhesus monkeys and chimpanzees. In Newton, G. and Levine, S. (Eds.) *Early Experience and Behavior*. Springfield, Ill.: Charles C Thomas, 1968, pp. 1-41.

McCary, J. L. *Human Sexuality*. New York: Van Nostrand-Reinhold Co., 1967.

Michael, R. P. and Zumpe, D. Patterns of reproductive behavior. In Hafez, E. S. E. (Ed.) *Comparative Reproduction of Nonhuman Primates*. Springfield, Ill.: Charles C Thomas, 1971, pp. 205-242.

Modahl, K. B. and Eaton, G. G. Display behavior in a confined troop of Japanese macaques (*Macaca fuscata*). Paper presented at the *American Society of Primatologists* Meeting, Seattle, Washington, April, 1977.

Mori, U. and Kawai, M. Social relations and behavior of gelada baboons. In Kondo, S. et al. (Eds.) *Contemporary Primatology*. Basel: Karger, 1975, pp. 470-474.

Nadler, R. D. Sexual cyclicity in captive lowland gorillas. *Science*, 1975, **189**, 813-814. (a)

Nadler, R. D. Second gorilla birth at the Yerkes Regional Primate Research Center. *International Zoo Yearbook*, 1975, **15**, 134-137. (b)

Nadler, R. D. Sexual behavior of captive orangutans. Paper presented at the *International Primatological Society* Meeting, Cambridge, England, August, 1976.

Nadler, R. D. and Braggio, J. T. Sex and species differences in captive-reared juvenile chimpanzees and orangutans. *Journal of Human Evolution*, 1974, **3**, 541-550.

Nadler, R. D. and Rosenblum, L. A. Sexual behavior of male pigtail macaques in the laboratory. *Brain, Behavior and Evolution*, 1973, **7**, 18-33.

Newton, N. Trebly sensuous woman. In Tavris, C. (Ed.) *The Female Experience*. New York: Ziff-Davis, 1973.

Nigi, H. Menstrual cycle and some other related aspects of Japanese monkeys (*Macaca fuscata*). *Primates*, 1975, **16** (2), 207-216.

Nomura, T. and Ohsawa, N. The use and problems associated with nonhuman primates in the study of reproduction. In Antikatzides, T., Ericksen, S. and Spiegel, A. (Eds.) *The Laboratory Animal in the Study of Reproduction.* New York: Gustav-Fischer, 1976, pp. 1-16.

Okano, J. A preliminary observation of orangutans in the rehabilitation station in Sepilok, Sabah. *The Annual of Animal Physiology,* 1971, **21**, 55-67.

Paterson, J. D. Ecologically differentiated patterns of aggressive and sexual behavior in two troops of Ugandan baboons, *Papio anubis. American Journal of Physical Anthropology,* 1973, **38** (2), 641-647.

Pitcairn, T. K. Aggression in natural groups of pongids. In Holloway, R. L. (Ed.) *Primate Aggression, Territoriality, and Xenophobia.* New York: Academic Press, 1974, pp. 241-272.

Rijksen, H. D. Social structure in a wild orangutan population in Sumatra. In Kondo, S., Kawai, M. and Ehara, A. (Eds.) *Contemporary Primatology.* Basel: Karger, 1975, pp. 373-379.

Riss, D. and Goodall, J. Sleeping behavior and association in a group of captive chimpanzees. *Folia Primatologica,* 1976, **25**, 1-11.

Rowell, T. E. *The Social Behavior of Monkeys.* Baltimore: Penguin, 1972.

Saayman, G. S. Behavior of the adult males in a troop of free-ranging chacma baboons. *Folia Primatologica,* 1971, **15**, 36-57.

Saunders, C. D. and Hausfater, G. Demographic influences on sexual selection in baboons: A computer simulation. Paper presented at the *American Society of Primatologists* Meeting, Seattle, Washington, April, 1977.

Saunders, D. and Hausfater, G. Sexual selection in baboons (*Papio cynocephalus*): A computer simulation of differential reproduction with respect to dominance rank in males. Paper presented at the *International Primatological Society* Meeting, Cambridge, England, August, 1976.

Savage, E. S. Pygmy chimpanzees: Update. *Yerkes Newsletter,* 1976, **13**, 25-27.

Savage, E. S. and Bakeman, R. Comparative observations on sexual behaviour in *Pan paniscus* and *Pan troglodytes.* Paper presented at the *International Primatological Society* Meeting, Cambridge, England, August, 1976.

Savage, E. S. and Malick, C. Play and socio-sexual behaviour in a captive chimpanzee (*Pan troglodytes*) group. *Behaviour,* 1977, **60**, 179-194.

Schaller, G. *The Mountain Gorilla.* Chicago: University of Chicago Press, 1963.

Scruton, D. M. and Herbert, J. The reaction of groups of captive talapoin monkeys to the introduction of male and female strangers of the same species. *Animal Behaviour,* 1972, **20**, 463-473.

Simonds, P. E. Aggression and social bonds in an urban troop of bonnet macaques (*Macaca radiata*). Paper presented at the *American Association of Physcial Anthropologists* Meeting, Seattle, Washington, April, 1977.

Slob, A. K. Effects of ovariectomy on male-female interactions in the stumptail macaque (*Macaca arctoides*). *Acta Endocrinologica,* 1975, **10**, 145.

Slob, A. K., Goy, R. W., Wiegand, S. J. and Scheffler, G. Gonadal hormones and behaviour in the stumptail macaque (*Macaca arctoides*) under laboratory conditions: A preliminary report. *Journal of Endocrinology,* 1975, **64**, 38.

Stephenson, G. R. Social structure of mating activity in Japanese macaques. *Symposium of the Fifth Congress of the International Primatological Society.* Nagoya, Japan: Japan Science Press, 1974, pp. 63-115.

Sugawara, K. Analysis of the social relations among adolescent males of Japanese monkeys (*Macaca fuscata fuscata*) at Koshima Islet. *Journal of the Anthropology Society of Nippon,* 1968, **83** (4), 330-354.

Sugiyama, Y. The ecology of the lion-tailed macaque (*Macaca silenus*): A pilot study. *Journal of the Bombay Natural History Society,* 1968, **65** (2), 283-292.

Sugiyama, Y. and Ohsawa, H. Life history of male Japanese macaques at Ryozenyama. *Symposia of the Fifth Congress of the International Primatological Society.* Nagoya, Japan: Japan Science Press, 1974, pp. 407-410.

Tanzer, D. Natural childbirth: Pain or peak experience. In Tavris, C. (Ed.) *The Female Experience.* New York: Ziff-Davis, 1973.

Tavris, C. (Ed.) *The Female Experience.* New York: Ziff-Davis, 1973.

Thompson, N. S. Some variables affecting the behaviour of irus macaques in dyadic encounters. *Animal Behaviour,* 1967, **15**, 307-311.

Trollope, J. Reproduction in a closed colony of *Macaca arctoides.* Paper presented at the *International Primatological Society* Meeting, Cambridge, England, August, 1976.

Trollope, J. and Blurton-Jones, N. G. Aspects of reproduction and reproductive behavior in *Macaca arctoides. Primates,* 1975, **16** (2), 191-205.

Tutin, C. E. G. Exceptions to promiscuity in a feral chimpanzee community. *Symposia of the Fifth Congress of the International Primatological Society.* Nagoya, Japan: Japan Science Press, 1974, pp. 445-449.

van Hooff, J. A. R. A. M. A structural analysis of the social behaviour of a semi-captive group of chimpanzees. In von Cronach, M. and Vine, I. (Eds.) *Social Communication and Movement.* London: Academic Press, 1973, pp. 75-162.

van Lawick-Goodall, J. Cultural elements in a chimpanzee community. In Menzel, E. W. (Ed.) *Symposia of the Fourth Congress of the International Primatological Society* (Vol. 1). *Precultural Primate Behavior.* Basel: Karger, 1973, pp. 144-184.

Wade, T. Inbreeding and kin selection in the evolution of macaque social behavior. Paper presented at the *American Society of Primatologists* Meeting, Seattle, Washington, April, 1977.

Weber, I. and Vogel, C. Sozialverhalten in ein-und zweigeschlechtigen Langurengruppen. *Homo,* 1970, **21**, 73-80.

Whitsett, G. A critique of the natural science approach to human sexuality. Paper presented at the *Animal Behavior Society* Meeting, University Park, Pennsylvania, June, 1977.

Wickler, W. Socio-sexual signals and their intra-specific imitation among primates. In Morris, D. (Ed.) *Primate Ethology.* Chicago: Aldine, 1967, pp. 69-147.

Wilson, M. E., Maple, T., Nadler, R. N., Hoff, M. and Zucker, E. Characteristics of paternal behavior in captive orangutans (*Pongo pygmaeus abelii*) and lowland gorillas (*Gorilla gorilla gorilla*). Paper presented at the *American Society of Primatologists* Meeting, Seattle, Washington, April, 1977.

Wolfe, L. Behavior patterns of estrous females of the Arashiyama West troop of Japanese macaques (*Macaca fuscata*). Paper presented at the *American Association of Physical Anthropologists* Meeting Seattle, Washington, April, 1977.

Wolfheim, J. H. and Rowell, T. E. Communication among captive talapoin monkeys (*Miopithecus talapoin*). *Folia Primatologica,* 1972, **18**, 224-255.

Wolfheim, J. G. A quantitative analysis of the organization of a group of captive talapoin monkeys (*Miopithecus talapoin*). *Folia Primatologica,* 1977, **27**, 1-27.

Zucker, E. L., Wilson, M. E., Wilson, S. F. and Maple, T. The development of sexual behavior in infant and juvenile male orangutans (*Pongo pygmaeus*). Paper presented at the *American Society of Primatologists* Meeting, Seattle, Washington, April, 1977.

11.

Male and Female Responses to Birth

Several reviews of behavior related to birth in primates have been published in the last eight years (Brandt and Mitchell, 1971; Brandt and Mitchell, 1973; Mitchell and Brandt, 1975; Atwood, 1976; Lanman, 1977; Caine and Mitchell, in press). We will refer to these reviews frequently within the present chapter. While these are, admittedly, primarily secondary sources, the number of primary sources cited in each one of these reviews is large. For example, there are over 100 primary references in the Brandt and Mitchell (1971) review alone. Clearly, this chapter cannot refer to all of the primary literature, even when we are covering only the topic of sex differences.

The response of others to the adult female's delivery of her infant has drawn increasing interest in primatology over the last five years. Much of this interest has been generated by the belief that initial responsiveness to an infant could tell the researcher something basic about the socialization tendencies of a given species, or for that matter, of a particular gender.

Much has been written in the 1970s about the mother's behavior and/or physiology during delivery, but it has been only recently that the behavior of other animals present at parturition has been carefully monitored. The birth process itself is difficult to study, particularly in the wild. Births usually occur at night. It is understandable, therefore, that most of the data on reactions to birth come from laboratory observations. Let us now examine what we know about the response to birth in male and female nonhuman primates from a taxonomic framework.

[1]This chapter is based on work done with Edna Brandt and Nancy Caine.

A. PROSIMIAN DELIVERIES

1. Tree shrews (cf. Brandt and Mitchell, 1971)

Females of *Tupaia spp.* deliver a litter of from one to three young, usually during the summer months and in the morning as soon as the sun comes up. Gestation is nearly fifty days, the female's abdomen obviously swelling at fifteen days prior to parturition. During these fifteen days, in some species (e.g., *T. belangeri*), she becomes more passive. The male attempts to copulate with her at this time but is usually rejected. In other species of *Tupaia* (e.g., *T. longipes, T. glis*) pregnant females are active and aggressive.

Birth itself usually occurs in a nest (or nest box in the laboratory) sometimes made of inedible food parts, leaves, etc. There are separate nests for the parents and the young in most species (but not in *T. glis*).

In some species, postpartum heat occurs and the male does not copulate with the female again until after the next birth (e.g., *T. belangeri*). Cannibalism of the newborn young, especially by adult males, is a problem in animals in captivity. However, even the mothers themselves sometimes eat their own young in captivity.

The tree shrew mother, if she does not eat her infants, behaves like most primates during the first thirty minutes following parturition. However, eventually she leaves the nest and does not return for forty-eight hours. Subsequently she returns at dawn once every forty-eight hours to nurse for four to ten minutes during which the infants' stomachs become distended with milk. Avoidance of the infants' nest by both parents, except for suckling, is a normal tree shrew behavior.

Occasionally, an adult female will protect another female while the latter is in labor, keeping the adult male out of the parturition nest box. (See Brandt and Mitchell, 1971, for numerous primary references.)

2. Lorises (cf. Brandt and Mitchell, 1971)

Depending upon the species, gestation in the Lorisidae may range from 120 to 175 days. Births of multiple offspring occur often. Postpartum estrus may also occur for two or three days. The mother is generally quite protective of her offspring, and cleans and cares for them.

The pregnant female begins to gain weight about thirty days prior to parturition and, at this time, she may become more aggressive toward the male. At birth, the female may initially become more active and carry nest materials in her mouth. She, at first, decreases her interactions with others, then joins the others in the nest to groom. She is very restless and examines her genitals frequently. The actual delivery of the first infant takes about thirty seconds once it begins. Other animals usually do not interfere or assist in the birth. After delivery, she eats the afterbirth, lessens her activity, and sleeps with her infant or infants.

In some cases, loris females in captivity may neglect their newborns and males may cannibalize them. In *Galago senegalensis,* the female often actively avoids the male while protecting her offspring and emitting a strange, protesting cry at the male.

Twins are common in *Galago. Galago* is a nocturnal genus in which births occur during the day. The births occur in a sleeping area with the entire group present. The others always seem very interested in the process. Sometimes an older female, perhaps the mother of the delivering female, watches the nest and carries the babies away from danger. Even younger siblings take care of newborn young on occasion. Males may be interested in the birth process but are probably more interested in having sex with the female just prior to parturition. Females show the most consistent interest in parturition and in the neonate. Females sometimes groom a mother in labor and may even help her eat the afterbirth. As already mentioned there is a danger, especially in captivity, of the adult male eating the offspring (see Brandt and Mitchell, 1971; Caine and Mitchell, in press).

3.   Lemurs (cf. Brandt and Mitchell, 1971)

Depending on the species, lemurs may typically have one or up to four neonates at a time and gestation may last from as short as 59 days to as long as 150 days. In general, the smaller species of lemurs have more young and shorter pregnancies (e.g., *Microcebus spp.*).

Leutenegger (1977) has pointed out that:

. . . there is strong selection for low reproductive turnover in most, if not all, primates species. But due to scaling effects there exist maternal size limits below which the production and delivery of single offspring is intolerable. (Abstract)

Multiple births are associated with environmental instability, retention of primitive mammalian characteristics, and with a derived character correlated with phyletic dwarfism (Leutenegger, 1977).

Microcebus females make nests out of grass and dry leaves. Following delivery there is an occasional postpartum estrus and the female eats the placenta.

Curiously, females of the species *Lemur variegatus* pull out their flank and thigh hair one day before birth, leaving one of their sides hairless. They use the hair to build a nest. *All* members of captive lemur groups take great interest in the birth process.

B. NEW WORLD MONKEY DELIVERIES

1. Callitrichidae (cf. Brandt and Mitchell, 1971; Caine and Mitchell, in press)

Among marmosets and tamarins, monogamous and monomorphic species, twins and triplets are common and births always occur at night. By the next morning, the father usually has possession of the infant or infants. In fact, the father assists at the birth by receiving and washing the babies. Gestation in marmosets is about 150 days. Births usually, but not always, occur in the spring.

Occasionally, cannibalism occurs even in marmosets and tamarins. This may be the reason that females sometimes withdraw themselves from the group prior to parturition.

If present, all group members, including adult and juvenile males, show great interest in the neonates. The adult male attempts to have sexual relations with the female; and, juveniles, regardless of sex, may wrestle the afterbirth from the female and eat it. Actually, the interest displayed by adult males in the infants ranges from complete indifference to intense concern. Adult males have never been seen to eat the placenta.

While other animals show interest in the newborn, the female typically does not relinquish it until at least one hour after delivery. At this time, she lets the male take it. At postpartum, the male makes more attempts to groom the female than vice versa (Rothe, 1974). In these species, the male is often in a more highly agitated state than is the delivering female. Only one female in a group reproduces (Epple, 1975). The young exhibit delayed sexual maturation in the presence of the parents, and thus only the adult pair breeds (Kleiman, 1977).

In the marmosets, which have larger infants than do tamarins, the male starts carrying the infant at birth; in tamarins, the male sometimes waits until three weeks postpartum until he starts carrying the infants. The lion tamarin male, however, will present food to the offspring (Kleiman, 1977) during these three weeks.

As we have already seen, the marmosets and tamarins provide interesting exceptions to normal behavioral rules for nonhuman primates. They are also interesting because their pattern of pregnancy hormones is closer to that of *Homo sapiens* than is that of, for example, the macaque (Lanman, 1977).

2. Cebidae (cf. Caine and Mitchell, in press)

In the howler monkey (*Alouatta pallita*), a circle of adult females and/or juveniles surround the mother when she is delivering. They seem to be more interested in the infant than in the mother. As these observing animals try to touch the infant, the mother avoids them by turning away. Female dominance increases at the birth of her infant (Glander, 1975). Males seem to show little or no interest in delivery.

In the woolly monkey (*Lagothrix lagothrica*), however, all group members, and *especially* the adult male, gather around and watch the laboring female intently. They do not physically interfere or aid in the birth process. The male shows a protective demeanor toward the mother and infant shortly after the delivery.

In the Saki monkey (*Pithecia pithecia*), the adult male may be so interested in the newborn that he may attempt to steal it repeatedly. Sakis may be monogamous, however (Kleiman, 1977).

In contrast to males of the New World species discussed above, squirrel monkey males actively *retreat* from a female in labor. Young females, however, stay near the mother throughout delivery and may attempt to smell and grab the neonate. What is surprising is that juvenile males may also attempt to grab the newborn even though adult males run from the birth.

C. OLD WORLD MONKEY DELIVERIES

1. Baboons (cf. Caine and Mitchell, in press)

Among gelada baboons (*Theropithecus gelada*), females do not isolate themselves from the group when in labor. Instead, other animals withdraw from her at the moment of delivery. Following parturition, first the infants, then adult and juvenile females, approach the mother. A related juvenile may be allowed to touch and groom the mother but the mother is unlikely to allow another animal to touch the neonate unless the animal is her juvenile daughter.

The hamadryas baboon female (*Papio hamadryas*) sits approximately two meters from the rest of the group when in labor. Adult females are nearby and juveniles watch intently.

Similarly, birth in the chacma baboon (*P. ursinus*) occurs under a bush while others in the group watch, and in olive and yellow baboons (*P. anubis* and *P. cynocephalus*) the mother is surrounded by other animals. Older juveniles and subadult females seem to be most interested while young adult and juvenile males evince only passing interest. Older adult male baboons, however, may approach and even touch the infant. Male infants usually have greater initial weights than do females (McMahan, Wigodsky, and Moore, 1976).

2. Vervets

Adult female vervets (*Cercopithecus aethiops*) do not isolate themselves from the rest of the group; in fact, the delivering female is seen in close proximity to other animals, primarily females and juveniles. The adult males show little interest in newborns.

3. Langurs (see Caine and Mitchell, in press)

Adult and subadult female langurs (*Presbytis entellus*) show a great deal of interest in the birth process. The mother does not withdraw from the others as long as they do no more than touch, smell, or lick the neonate. At times, another adult female is permitted to hold the neonate a few hours following its delivery. Langur males are indifferent to neonates even when the newborn shows distress.

In *P. johnii* births, juveniles and subadults of both sexes show interest but primarily in the mother rather than in the neonate. This is particularly true of close kin. Again, adult males show little interest, although they sometimes assume a protective stance.

The neonate is passed around from female to female within hours after birth in many langurs (*P. obscurus, Pygathrix nemaeus*). In fact, Dolhinow (1977) reported such behavior in *P. entellus*. Older nulliparae exceed all other classes of females in the frequency with which they attempt to obtain infants. Nevertheless, even females who had already given birth to one or more offspring attempt to take infants, especially if such females are pregnant or lactating. Oppenheimer (1976) also describes high female interest in the birth process in *P. entellus* (also see Caine and Mitchell, in press). Contrary to what we might expect, however, these older, more experienced, and sometimes pregnant females are far more likely to mistreat the neonate than are the younger and older inexperienced females (Hrdy, 1976). The dominance status of a female langur increases following delivery of her infant (vonder Haar Laws, Dolhinow, and McKenna, 1977).

A langur neonate may be taken and remain apart from its mother for as much as 48 percent of its first day of life. Immature males also occasionally handle the neonates (Hrdy and Hrdy, 1977). Adult male langurs ignore neonates. However, it would be interesting to see how males of the species *Presbytis potentziani* (the Mentawai leaf monkey) respond, since it is felt by some (Kleiman, 1977) that this species is monogamous.

4. Macaques (see Mitchell and Brandt, 1975)

Captive groups of stumptail macaques (*M. arctoides*) show great interest in the birth of an infant. Juvenile males, for example, may show sexual arousal and the alpha male may attempt to copulate with the laboring female even when the neonate's head is crowning. The female withdraws and screams in response to such advances.

Subadult and young adult female stumptails show an interest in the placenta, but only the mother herself eats it. Approaches by other females are avoided by the new mother. The alpha male, however, sometimes sits near the new mother and grooms her and/or she grooms him. A juvenile son may show the most interest in the delivery itself. Surprisingly, adult females show the *least* interest. Stumptails, incidentally, show an unusual pattern of three birth peaks throughout the year (at least they do when they are in an outdoor enclosure) (Estrada and Estrada, 1976). Also, in a closed colony, births have resulted from incest between father and daughter, brother and sister, and mother and son (Trollope, 1976). After giving birth, females rise in dominance status, unless they deliver a dead fetus (Weisband and Goy, 1976).

Pigtail macaque (*M. nemestrina*) mothers appear uneasy with the increased attention given to them at delivery. Following delivery in the pigtail macaque, there is a striking positive relationship between maternal-infant preference and a rise in serum prolactin and cortisol levels in the mother (Gross, Schiller, and Bowden, 1977). Both of these factors peak at ten days following delivery.

At birth, the female infant pigtailed monkey is already significantly ahead of the male (as in humans) in the maturation of the hand-wrist skeletal complex (Tarrant, 1977). Premature and high risk male neonates are more vulnerable and more retarded in food intake than are their female counterparts (Sackett et al., 1974; Sackett, Holm, and Landesman-Dwyer, 1975). Male neonates are more vulnerable, in general, than are female neonates (Dazey and Erwin, 1976).

Bonnet monkey (*M. radiata*) group members also show great interest in a birth, but the mother is not as uneasy as is the pigtail mother. She permits a great deal of freedom with the neonate. Females with new infants sit closer to a central adult male than do pregnant females (Judge and Rodman, 1976). Social tensions are reduced during the bonnet birth season (Simonds, 1977).

In *M. fascicularis* (crab-eating macaque), caesarean deliveries do not affect subsequent ovulation and menstruation (Dang, 1977), although there is a tendency in rhesus for caesarean females to reject or not accept their neonates as readily (see Mitchell and Brandt, 1975). Pregnancy can be diagnosed as early as twenty-one to twenty-eight days following conception in the crab-eating macaque (Mahoney, 1975).

Parturition is more prevalent in females who have previously lost an infant in Japanese macaques (*M. fuscata*) (Tanaka, Tokuda, and Kotera, 1970). At birth, the wild Japanese macaque group usually maintains its distance from a mother-neonate dyad.

More is known about rhesus monkey (*M. mulatta*) births than about parturition in any other nonhuman primate. During a delivery, other *pregnant* females seem to show little interest in the mother and neonate. Sometimes they sleep throughout a delivery, although they may occasionally wake up and groom the new mother. Male yearling rhesus have been reported to aid their mothers and lick the afterbirth from their neonate siblings.

At times, delivering rhesus females are closely attended by other females, particularly by "childless" females. Such females sometimes arch their backs and finger their vaginas as they witness an infant being delivered. Males show sexual excitement at the event, including attempted mounts during and following delivery. Although other animals seem interested in the afterbirth, only the mother eats it. Nulliparous females have been known to taste the placenta, however. Favorite individuals of the delivering mother are sometimes permitted to look closely at her infant but not to touch it. Other adult females may display grooming, lip-smacking, intense curiosity, and protective responses toward the mother and neonate. In some cases, adult females attempt to and even succeed in permanently kidnapping the infant from the new mother.

Adult male rhesus often seem nervous when a female is delivering. They may attempt to groom and inspect the mother and show some manual interest in the placenta. They do not taste the afterbirth, however. The delivering female tries to avoid the male. For more complete descriptions of these phenomena in rhesus, the reader may see Brandt and Mitchell, 1971; Brandt and Mitchell, 1973; Mitchell and Brandt, 1975; Caine and Mitchell, in press.

It is interesting that, in the rhesus, the birth weight of the neonate tends to increase with increasing parity (DiGiacomo and Shaughnessy, 1972). The birth of a first offspring, typically at four or five years of age in free-ranging groups, apparently inhibits mating during the following breeding season (Missakian-Quinn and Varley, 1977). Primiparous females are much more clumsy with their infants during delivery and immediately thereafter than are multiparous females (Mitchell and Schroers, 1973); however, they are also more restraining and protective (Mitchell and Stevens, 1969).

The onset of parturition in the rhesus monkey is a function of rising levels of maternal oxytocin, a decrease in maternal progesterone (which, when present, blocks the uterine distending effects of estrogen), increasing volume of amniotic fluid and size of fetus, and maturation of fetal hormonal systems including possibly increasing fetal estrogens (cf. Mitchell and Brandt, 1975; Resko, 1974).

5. Colobus Monkeys and the Patas

King colobus monkeys (*Colobus polykomos*) "share" the neonate shortly after its birth. Siblings of the neonate sometimes hold the newborn, and all animals except adult males participate in sharing. In the black and white colobus (*C. guereza*), only adult females and juveniles show interest, the females handling and caring for the neonate within hours following delivery. Mothers do not object to this. The birth of an infant has a socially cohesive effect in the colobus troop (Horwich and Wurman, 1977).

Adult female patas monkeys (*Erythrocebus patas*) are greatly attentive toward mother and neonate.   The mother, however, does not permit other females to touch her infant. Adult males show no interest whatsoever (see Caine and Mitchell, in press).

D. LESSER AND GREAT APE DELIVERIES

1.   Gibbons and Siamangs

Free-ranging female siamangs (*Symphalangus syndactylus*) evidently isolate themselves from their monogamous groups before birth and rejoin their groups twenty-four hours after parturition.   The oldest offspring of the group typically departs on the day of the birth (cf. Caine and Mitchell, in press).   In the gibbon (*Hylobates spp.*), during birth, the adult male pays no attention to the delivery itself but makes attempts to mount the coy female.   The male does not touch the infant for two or three days after birth, at which time he takes an interest in the infant's anogenital area (Poglayen-Neuwall, 1977).

2.   Orangutans (see Caine and Mitchell, in press)

Male orangutans (*Pongo pygmaeus*), if present during a delivery in captivity, usually do not interfere in the process.   However, on occasion, a male may become sexually aroused and attempt to copulate with a laboring female.   When such a male sees the infant's head emerging from the vagina, however, his behavior changes.   A male orangutan has been seen to assume a calm, nonsexual demeanor, position himself behind the birthing female, put his mouth over the emerging neonate's head, and slowly pull until the head was completely free. At this point, the male used his hands to deliver the rest of the infant, waited until the female turned around, and then handed the newborn to its mother (Ullrich, 1970).   Terry Maple, at Emory University, has seen an orangutan male eat and repeatedly regurgitate the placenta.   According to Horr (1977), an orangutan female has been seen blowing into the nostrils of its newborn infant, apparently administering artificial respiration.

3.   Chimpanzees (cf. Brandt and Mitchell, 1971)

Captive chimpanzees have also been seen using their mouths following delivery. New female chimpanzee mothers (*Pan troglodytes*) have been seen placing their mouths over the mouths of their newly emerged infants and blowing conspicuously as if using artificial respiration.   However, according to Maple (personal communication), putting the mouth over the face is a species-typical greeting in chimpanzees. Other females appear to be interested in the delivery of an infant. Occasionally, they kidnap and keep the new infant for themselves, particularly if they are young and childless.

4. Gorillas

Faust (1977) reported the birth of an infant to a very young (eight years old) adult female gorilla (*Gorilla gorilla*). It was obviously her first baby and the birth occurred in the evening. The mother initially defended the infant from her companion but eventually became restless and careless with it so that the baby had to be hand-raised. The mother herself was hand-raised. Maple et al. (1977) warn us that young apes must be raised with their own mothers if at all possible. If and when rejection does occur, it is essential that the young apes be given a foster mother of the same species. Otherwise, captive-rearing may not produce normal mothering or, for that matter, even normal mating. There has been successful breeding for gorillas in captivity, however. At Yerkes Primate Center, for example, three gorillas were born in one month (Nadler, 1976) in a large outdoor enclosure.

E. *HOMO SAPIENS*

Recent obstetrical advances have found that the human father's role in the birth process can be an important one. The husband's presence is necessary for a woman to experience a "peak" experience (even orgasm) when she delivers. Women feel safer with their husbands present; in fact, the most helpful technique in the Lamaze method of child birth is reported to be the presence of the husband at delivery. Furthermore, following the delivery of an infant, the father can play a far more active role than is traditionally expected, especially if he has participated in the birth (see Caine and Mitchell, in press, for relevant studies).

F. OVERVIEW

Marmoset and tamarin males show more interest in neonatal delivery than any other genera of nonhuman primates. They are monogamous; and, as we know, monogamous males have a greater interest in their offspring and show more parental care in general. However, the monogamous lesser apes show this interest at a much later infant age than do the marmoset and tamarin males. They certainly do not seem to show a great deal of interest at parturition. There is no explanation for this difference.

On the other hand, the epitome of the polygynous male, the squirrel monkey, literally runs away when a female delivers an infant. The male squirrel monkey also shows little interest in infants later in their lives. The patas male shows a similar pattern.

Strong female interest in neonates and in the birth process is seen in langurs. Kin selection, "practice" at mothering, or strong female alliances in social organization of langurs may be related to this effect.

Juveniles of both sexes seem to be interested in birth regardless of the species, especially juveniles related to the delivering mother. Apparently sexual maturity is necessary to produce the pattern of sex differences typically seen across taxa.

Sexual excitement on the part of males—the so-called postpartum estrus—is seen at almost all taxonomic levels, in the orangutan as well as the tree shrew. Perhaps this sexual attraction brings males into the vicinity of females when these females are especially vulnerable and in need of protection. Perhaps human males also have this potential for sexual attraction to the birthing female.

REFERENCES

Atwood, R. J. Parturitional posture and related birth behavior. *Acta obstetricia et gynecologica Scandinavica,* 1976, Supplement 57, 1-25.

Brandt, E. M. and Mitchell, G. Parturition in primates: Behavior related to birth. In Rosenblum, L. A. (Ed.) *Primate Behavior: Developments in Field and Laboratory Research* (Vol. II). New York: Academic Press, 1971, pp. 177-223.

Brandt, E. M. and Mitchell, G. Labor and delivery behavior in rhesus monkeys (*Macaca mulatta*). *American Journal of Physical Anthropology,* 1973, **38** (2), 519-522.

Caine, N. G. and Mitchell, G. Behavior of primates present at parturition. In Erwin, J., Maple, T. and Mitchell, G. (Eds.) *Captivity and Behavior* (Vol. 1). New York: Van Nostrand-Reinhold, in press.

Dang, D. C. Resumption of menstruation and fertility after caesarean in *Macaca fascicularis. Annals of Biology, Animal Biochemistry and Biophysics,* 1977, **17**, 325-329.

Dazey, J. and Erwin, J. Infant mortality in *Macaca nemestrina:* Neonatal and postnatal mortality at the Regional Primate Center, University of Washington, 1967-1974. *Theriogenology,* 1976, **5**, 267-279.

Di Giacomo, R. F. and Shaughnessy, P. W. Estimation of gestational age and birth weight in the rhesus monkey (*Macaca mulatta*). *American Journal of Obstetrics and Gynecology,* 1972, **112**, 619-628.

Dolhinow, P. J. Caretaking patterns of the Indian langur monkey. Paper presented at the *American Association of Physical Anthropologists* Meeting, Seattle, Washington, April, 1977.

Epple, G. The behavior of marmoset monkeys (*Callithricidae*). In Rosenblum, L. A. (Ed.) *Primate Behavior* (Vol. 4). New York: Academic Press, 1975, pp. 195-239.

Estrada, A. and Estrada, R. Birth and breeding cyclicity in an outdoor living stumptail macaque (*Macaca arctoides*) group. *Primates*, 1976, **17** (2), 225-231.

Faust, R. Eighth gorilla born at Frankfurt. *American Association of Zoos, Parks, and Aquaria Newsletter*, 1977, **18**, 14.

Glander, K. E. Dominance in mantled howling monkeys. *American Journal of Physical Anthropology*, 1975, **42**, 303.

Gross, R. J., Schiller, H. S. and Bowden, D. M. Post partum serum cortisol and prolactin and their relationship with maternal and affective behaviors in the *Macaca nemestrina*. Paper presented at the *American Society of Primatologists* Meeting, Seattle, Washington, April, 1977.

Horr, D. A. Orang-utan maturation: Growing up in a female world. In Chevalier-Skolnikoff, S. and Poirier, F. E. (Eds.) *Primate bio-social development*. New York: Garland, 1977, pp. 289-322.

Horwich, R. H. and Wurman, C. The ontogenic implications of infant birth on troop sociality in colobus monkeys. Paper presented at the *Animal Behavior Society* Meeting, University Park, Pennsylvania, June, 1977.

Hrdy, S. B. Allomaternal care and abuse of infants among Hanuman langurs. Paper presented at the *International Primatological Society* Meeting, Cambridge, England, August, 1976.

Hrdy, S. B. and Hrdy, D. B. Allomaternal choice of charges among Hanuman langurs. Paper presented at the *American Association of Physical Anthropologists* Meeting, Seattle, Washington, April, 1977.

Judge, D. S. and Rodman, P. S. *Macaca radiata:* Intragroup relations and reproductive status of females. *Primates*, 1976, **17**, 535-539.

Kleiman, D. G. Monogamy in mammals. *The Quarterly Review of Biology*, 1977, **52**, 39-69.

Lanman, J. T. Parturition in nonhuman primates. *Biology of Reproduction*, 1977, **16**, 28-38.

Leutenegger, W. Evolution of litter size in primates. Paper presented at the *American Society of Primatologists* Meeting, Seattle, Washington, April, 1977.

Mahoney, C. J. Practical aspects of determining early pregnancy, stage of fetal development, and imminent parturition in the monkey (*Macaca fascicularis*). *Laboratory Animal Handbook*, 1975, **6**, 261-274.

Maple, T., Zucker, E. L., Hoff, M. P. and Wilson, M. E. Behavioral aspects of reproduction in the great apes. Paper presented at the *American Association of Zoological Parks and Aquariums* Meeting, San Diego, California, September, 1977.

McMahon, C. A., Wigodsky, H. S. and Moore, G. T. Weight of the infant baboons (*Papio cynocephalus*) from birth to fifteen weeks. *Laboratory Animal Science,* 1976, **26**, 928-931.

Missakian-Quinn, E. A. and Varley, M. A. Consort activity in groups of free-ranging rhesus monkeys on Cayo Santiago. Paper presented at the *American Society of Primatologists* Meeting, Seattle, Washington, April, 1977.

Mitchell, G. and Brandt, E. M. Behavior of the female rhesus monkey during birth. In Bourne, G. H. (Ed.) *The Rhesus Monkey* (Vol. 2). New York: Academic Press, 1975, pp. 231-244.

Mitchell, G. and Schroers, L. Birth order and parental experience in monkeys and man. In Reese, H. W. (Ed.) *Advances in Child Development and Behavior* (Vol. 8). New York: Academic Press, 1973, pp. 159-184.

Mitchell, G. and Stevens, C. W. Primiparous and multiparous monkey mothers in a mildly stressful social situation: First three months. *Developmental Psychobiology,* 1969, **1**, 280-286.

Nadler, R. D. Three gorillas born at Yerkes in one month. *Yerkes Newsletter,* 1976, **13**, 15-19.

Oppenheimer, J. R. *Presbytis entellus:* Birth in a free-ranging primate troop. *Primates,* 1976, **17**, 541-542.

Poglayen-Neuwall, I. Parturition in a hand-reared, primiparous gibbon. *Zoological Garten,* 1977, **47**, 57-58.

Resko, J. A. Sex steroids in the circulation of the fetal and neonatal rhesus monkey: A comparison between male and female fetuses. *International Symposium on Sexual Endocrinology of the Perinatal Period,* 1974, **3**, 195-204.

Rothe, H. Allogrooming by adult *Callithrix jacchus* in relation to post partum oestrus. *Journal of Human Evolution,* 1974, **3**, 535-540.

Sackett, G. P., Holm, R. A., Davis, A. E. and Fahrenbruch, C. E. Prematurity and low birth weight in pigtail macaques: Incidence, prediction, and effects on infant development. *Symposium of the Fifth Congress of the International Primatological Society.* Nagoya, Japan: Japan Science Press, 1974, pp. 189-205.

Sackett, G. P., Holm, R. A. and Landesman-Dwyer, S. Vulnerability for abnormal development: Pregnancy outcomes and sex differences in macaque monkeys. In Ellis, N. R. (Ed.) *Aberrant Development in Infancy: Human and Animal Studies.* Hillsdale, N.J.: Laurence Erlbaum Associates, 1975, pp. 59-76.

Simonds, P. E. Aggression and social bonds in an urban troop of bonnet macaques (*Macaca radiata*). Paper presented at the *American Association of Physical Anthropologists* Meeting, Seattle, Washington, April, 1977.

Tanaka, T., Tokuda, K. and Kotera, S. Effects of infant loss on the interbirth interval of Japanese monkeys. *Primates,* 1970, **11**, 113-117.

Tarrant, L. H. Sex differences in skeletal maturation at birth in the pig-tailed monkey. Paper presented at the *American Association of Physical Anthropologists* Meeting, Seattle, Washington, April, 1977.

Trollope, J. Reproduction in a closed colony of *Macaca arctoides*. Paper presented at the *International Primatological Society* Meeting, Cambridge, England, August, 1976.

Ullrich, W. Geburt und naturliche Geburtshilfe beim Orang-utan. *Der Zoologische Garten,* 1970, **39**, 284-289.

vonder Haar Laws, J., Dolhinow, P. J. and McKenna, J. J. Female Hanuman langur monkey rank and reproduction. Paper presented at the *American Society of Primatologists* Meeting, Seattle, Washington, April, 1977.

Weisband, C. and Goy, R. W. Effect of parturition and group composition on competitive drinking order in stumptail macaques (*Macaca arctoides*). *Folia Primatologica,* 1976, **25**, 95-121.

12.

Infant Care: Prosimians, New World Monkeys, and Macaques

In many ways, the next two chapters should be of great interest to most of us. In these two chapters we will be discussing the possible role of the parents in determining gender-related behavior. Up to this point, we have been mostly concerned with the more direct effects of biology upon behavior (e.g., hormonal effects). However, it is, by now, beyond question that the early experiences of nonhuman primates affect the development of gender-related activities including the most basic ones, those related to sexual behavior. In the next two chapters we will be concerned with the role of the *parents or other significant animals* in the development of behavioral sex differences but *not* with the effects of abnormal early isolation or deprivation. The latter topics will be discussed in chapters subsequent to these.

When we refer to significant animals other than the parents (in discussing infant care) we are recognizing the fact that infants receive care from adoptive parents, from close kin, and even from immature animals. Spencer-Booth (1970) has written a review of the extant relationships between mammalian young and conspecifics other than mothers and peers. In general, among mammals, there are sex differences in infant care with females providing more of the care for infants than males. There are also differences in the responses of mothers and other adults to male and female young (Bekoff, 1972).

Among primates, instead of direct infant care, the more typical role for males is defense of the females and young. In some species, when an infant is threatened by an outsider, or by a predator, an adult male may attack while all other animals in the group retreat (Rowell, 1974). In fact:

> The data suggest that consideration of defensive rather than hunting behavior of human males might be a more profitable approach to understanding

behavioral changes in prehuman and early human societies. (Rowell, 1974, P. 143)

Redican (1976), on the other hand, has emphasized the great variability in care of infants by males in nonhuman primate societies. He believes that, to the extent that males interact directly with infants, they respond to sex differences in the infants more than do mothers. Also, according to Redican, the degree of male-infant interaction appears to be inversely related to the degree of maternal restrictiveness toward infants. That is, if the mother permits a female to contact her infant, the male is likely to contact it also. Others have also discussed the role of nonmaternal care of infants in nonhuman primates (Raphael, 1969).

Mitchell (1969) and Mitchell and Brandt (1972) have reviewed the role of the primate male in infant care. In general, male care of infants depends upon kinship, familiarity with the infant's mother, conditions of crowding or captivity, the female's stage of estrus, social change, sex of infant, dominance of mother, age of infant, his consort relations with the infant's mother, cultural tradition, time of the year, an infant being orphaned, the number of males in a group, hormonal factors, the male's own early experience, the male's own interest in the center of the troop, and the species of the male in question (Mitchell, 1969; Mitchell and Brandt, 1972). We will see how each of these factors can operate as we review the sex differences in infant care which appear in the primate order.

There are a few, but not many, studies of neurological correlates of infant care. Goldman (1976) has found that there is a decrease in infant care after lesions of the anterior association cortex, while others have reported deficiencies in infant care following amygdalectomies (Kling, 1974; Kling and Dunne, 1976). Incidentally, there have been sex differences reported for some species in certain neurons of the amygdala (Keefer and Stumpf, 1975).

Despite the increasing numbers of articles published on nonmaternal infant care, there is *typically* a major sex difference in parental interest in the young. In most species of primates, males may show some interest in the young, but their interest is more variable than is the interest of females. The variability appears from species to species as well as from one time to another for an individual. Female interest, on the other hand, is both strong and persistent (Hamburg, 1968; also see Redican, 1976).

Thus, in most nonhuman primates, but not in all of them, the mother is the more significant parent. There is also, however, a great range in the quantity and quality of maternal care. This variability is particularly large among the prosimians. At one extreme are the tree shrews (*Tupaia spp.*), in which the litter of two or three infants live in a separate nest from the parents; at the other extreme is the brown lemur, in which the infant develops very slowly but is carried, wrapped like a belt, around the mother's waist.

A. *PROSIMIANS*

As noted above, tree shrew mothers give birth to their litter of two or three infants in a nest separate from their own. The tree shrew mother then abandons the neonates until forty-eight hours later, at which time she goes to their nest, squirts milk into their mouths for ten minutes, then leaves again for forty-eight hours. The mother's persistence at infant care does not seem to be that great. If she does not find her infants in the nest within two minutes of her visit, she abandons them forever (cf. Jolly, 1972).

The male tree shrew is perhaps even less dedicated to the young than is the female. While he may construct the nest in which the young are delivered and in which they will subsequently live (Martin, 1966), he may (in captivity) cannibalize the young if the mother does not defend their nest or nest box (see Mitchell and Brandt, 1972; Sorenson, 1970). These findings regarding "paternal" aggression do not apply to all tree shrew species, however. There are differences in social organization in tree shrews which undoubtedly affect sex differences in infant care. Some species of tree shrews live in polygamous family groups with paternal dominance while others appear to be asocial. There is also territoriality in some species but apparently not in others (Sorenson, 1974). In *Tupaia longipes,* one adult female will protect a nest box against an adult male, while a second female gives birth (Sorenson and Conaway, 1966).

In bushbabies *(Galago spp.),* two infants may be placed together and left temporarily by the mother; or, one of the infants may be left as many as forty yards away while the mother visits the other infant. This behavior is known as "parking the infant" (Doyle, 1974). Bushbabies are found individually, in pairs, or in large groups. The natural group is probably a "family group" but it is unlike the more typical monogamous families (Doyle, 1974). The male has many separate females who maintain separate territories. He "visits" each of them from time to time. In zoos, bushbabies *(G. senegalensis)* live peaceably in large groups of five males and five females. The very first young delivered in such a situation (usually twins) may be attended by their father who sits in the nest and covers the young with his body when they are threatened (Gucwinska and Gucwinski, 1968). Rosenson (1972) reported that infants received some interest and even grooming from adult males of another bushbaby species *(G. crassicaudatus).* Rosenson, however, noted that the infants (aged seven days) elicited such maternallike acts from males only rarely. Buettner-Janusch (1964) has also published information on bushbaby behavior, noting that males in captivity sometimes kill and eat the offspring.

There is, as noted above, substantial variability in prosimian infant care, partly because of the great variability in social organization. For example, *Daubentonia* and *Lepilemur* are solitary animals, *Microcebus* form sleeping groups, while *Propithecus* and *Indrii* form family groups. Even within the one genus *Lemur* there is great variability. Lemurs, in general, form very large groups. In *Lemur catta,* animals other than the mother contact the infant even before the infant is one week old: In *Lemur fulvus* this outside contact does not occur until the infant is four weeks old, whereas in *Lemur variegatus* there is little attention from the mother and *no* attention from others until the infant is seven to eight weeks old. In *Lemur fulvus,* however, an infant becomes disturbed by a week-long separation from others. Also in *Lemur fulvus,* an infant becomes independent earlier if there is an adult male present. Thus, even within one genus of the prosimians one must be cautious in generalizing from one species to another with regard to patterns of infant care (Klopfer, 1972).

The infants of the species *Lemur variegatus* do not cling to their mothers but are deposited in nests. When they are carried, they are carried in the mother's mouth not on her ventrum. Multiple births are the rule (Klopfer and Dugard, 1976):

> . . . in the mother's absence young turn more to other adults than to their siblings, who are, apparently, not adequate social substitutes for adults . . . the infants tend to become explorative and independent sooner if reared with other adults than only with their mother. (Klopfer and Dugard, 1976, P. 210, 220)

However, there are probably friendlier relations between males and females, as well as between males and infants, in *L. fulvus* than in *L. variegatus* (Kress, 1975).

Males of the species *Lemur catta* apparently ignore newborn infants, whereas monkey lemur males (*Propithecus verreauxi*) crowd around the delivering female and try to groom the infant shortly after birth. Mature males of this species also sometimes hold and cuddle an infant while the mother feeds. There is a family group in *Propithecus* and the adult male has an active role in infant care (cf. Mitchell, 1969).

Among other prosimians, both male and female care of the neonate has been reported for *Tarsius syrichta.* The female tarsier clutches the newborn to her breast most of the time. Occasionally, the infant is seen on the back of the adult male (Schreiber, 1968).

Thus, male care in prosimians ranges from infanticide with cannibalism of infants, through indifference and nest building, to playing with, holding, and cuddling of infants. Female care ranges from nest building and nursing only once every forty-eight hours, through infant "parking" (leaving infants alone), to extensive defense, holding, and cuddling of extremely helpless infants in the monkey lemur. Overall, female prosimians provide more care for infants than do male prosimians. There are probably no species of prosimians in which males care for infants more than do females.

B. *CALLITHRICIDAE*

In contrast to what we now know about infant care in prosimians, infant care in the marmosets and tamarins is displayed as much by males as it is by females.

None of the marmosets and tamarins build nests. In the case of the pygmy marmoset (*Cebuella pygmaea*) the parents carry their infants at all times, until their combined weight (usually twins) equal their own (Jolly, 1972).

As we know, the social structure of the marmoset and tamarin is a family group with a monogamous bond between an adult male and an adult female. The infants in most species are weaned at about six months of age and reach sexual maturity at a very early age for a primate (fourteen months). Grown up juveniles are driven away by the parent of the same sex (cf. Mitchell, 1969).

All members of a monogamous group—males, females, and juveniles— will carry infants; however, the male usually does more carrying than do the other group members. In the pygmy marmoset (*Cebuella pygmaea*), male care is particularly obvious. The infant nurses on the mother but otherwise remains clinging to the father. The female, on the other hand, appears to be the more dominant individual (Mitchell, 1969).

Golden lion marmosets (*Leontopithecus spp.*) also have twins. These infant twins cling to the mother for the first week of their lives after which they are carried by the father. As early as the second day of life, however, both parents are actively involved in rearing the baby. Subsequently, the male assumes the majority of the infant care. Males and females appear to have equal dominance (Snyder, 1974).

Actually, marmoset groups in the wild vary considerably in size and composition (Epple, 1975a). In *Saguinus fuscicollis,* all members of all groups carry infants. Males carry them the most, followed by mothers, then juveniles. In one group, a male juvenile carried the infant more than did his twin sister. The adult male of a group usually assumes most of the infant care whether he has sired the infant or not (Epple, 1975b). In addition, castration of the male apparently has no dramatic effect on his parental performance. Despite this male predominance in infant care, in some groups the adult female carries the infants more than does the male (Epple, 1975b; also see Mitchell, 1969). In *Saguinus geoffroyi* infants are carried by fathers and mothers equally and a single female can rear twins successfully in captivity even in the absence of a male (Moynihan, 1970).

The marmoset genus most often studied is *Callithrix.* Immature females of *C. jacchus* are attracted to infants. Juveniles, adolescents, and subadults tend to engage the infant in play, whereas grooming interactions tend to center on the adult male and female (Ingram, 1976). The male can begin infant care by assisting the female with both hands during delivery of the neonate (Langford, 1963); however, the father's reaction to the newborn is more variable than is that of the mother. Only healthy neonates are cared for by the male and wet neonates attract more male care than do dry ones. Nonviable neonates are neglected by all marmoset group members and they may be cannibalized unless they vocalize. Fathers show more interest in the first born than in subsequent infants (Rothe, 1974):

> The adult male's first interactions with the neonates appeared to occur much earlier (less than ½ hour after birth) than in other nonhuman primates. (Stevenson, 1976, P. 365)

If a male carries the neonate, he will chastise other group members as they approach (Stevenson, 1976).

C. *CEBIDAE*

Females care for infants more than do males in the *Cebus* genus. However, adult male capuchins occasionally carry infants both dorsally and ventrally; and, adult males sometimes play vigorously with infants. The adult male is extremely tolerant of infants (cf. Mitchell, 1969).

Apparently, while adult male howler or howling monkeys (*Alouatta palliata*) sometimes attack or even kill infants, on other occasions they are seen retrieving fallen infants and playing with them. In addition, they often make a bridge with their bodies to help juveniles cross from one tree to another (cf. Mitchell, 1969). Some males become attached to specific infants and allow them to hold onto their ventral surfaces and may carry them in this position for short distances.

In the mantled howler (*Alouatta villosa*), "aunting" behaviors by adult females and attempted infant adoptions by females have been seen. Juvenile female mantled howlers show more interest in infants than do juvenile males. However, Glander (1975) saw baby-sitting in both juveniles and adult males.

Males of the species *Alouatta seniculus* tend to be more loosely attached to troops than do the females. The males sleep away from the matrifocal groupings; and, grooming data indicate that relations between females and infants are of great importance for group bonding, especially the relations involving *female* infants (Neville, 1972). Males usually show very little interest in infants.

Among the other nonmonogamous cebids, the adult male wooly monkey (*Lagothrix lagothrica*) shows interest in a birth, stays close to the delivering mother (even hugging her), licks the baby, and appears to be quite protective of the pair (Williams, 1967).

Squirrel monkey males (*Samiri sciureus*) sometimes have friendly interactions with infants but are usually indifferent towards them. An infant is also sometimes carried dorsally by a male or may be wrestled with; however, the infant is usually just tolerated (cf. Mitchell, 1969).

Unlike the case for the sex difference which we will repeatedly see in our section on Old World monkeys and apes, for squirrel monkeys there is disagreement in the literature concerning which sex of infant becomes independent of its mother the earliest. Kaplan (1970) claims that the female infant becomes independent earlier than the male infant, while Rosenblum (1974) reports that male squirrel monkeys move away from their mothers at a younger age than do females. Rosenblum (1974) also feels that there is greater infant-activated autonomy in the male. As adults, females are more active in affiliative behavior than are males (Strayer, Taylor, and Yanciw, 1975).

The monogamous species of cebids show much the same pattern of infant care as do the monogamous marmosets and tamarins. With monogamy, as we have already seen, there is little physical sexual dimorphism, less sexual behavior, delayed sexual maturation of the young in the presence of the parents so that only the adults breed, aid from juveniles in infant care, and large increases in male care of infants including carrying, feeding, defending, and socializing (Kleiman, 1977). We have already seen that marmosets and tamarins fit this description. Other New World species which can surely be included in this group of animals are the night monkey (*Aotus trivirgatus*) and the titi monkey (*Callicebus spp.*). *Pithecia spp.* (the saki monkeys) may also fit into the category.

Apparently, the larger the infant at birth, and the smaller the mother, the more immediate is the male attention to infant care. In those monogamous species having large infants, the male starts carrying the infant sooner. In any case, there is a large parental investment for males in monogamous New World species. The males do not get very involved in play with the young, but they do carry, groom, and care for the infants much of the time. More monogamy leads to less male energy expenditure, both in play and in sexual behavior, but more expenditure in infant care, grooming, and affiliation (Kleiman, 1977).

In the titi monkey (*Callicebus torquatus torquatus*), the adult male looks after the infant while the adult female leads the group throughout its daily ranging pattern (Kinzey et al., 1977). This particular species of titi monkey (the yellow-handed titi) has a much larger home range (forty times larger) than does another species of titi (e.g., *Callicebus moloch*). Despite this smaller range, the *C. moloch* male also handles most of the infant care (Mason, 1966).

Night monkeys or owl monkeys (*Aotus spp.*) also live in monogamous groups and both parents carry and care for the infant. The owl monkey, however, is nocturnal. For the first few days following parturition, the infant is carried on the mother's ventral surface but is then carried on her back. After the infant is about nine days old it is carried by the adult male, except when being nursed by the mother (cf. Mitchell, 1969; Leibrecht and Kelley, 1977).

D. THE RHESUS MONKEY (*MACACA MULATTA*) (SEE MITCHELL, 1971; MITCHELL, 1977)

1. Females

We already know that there are definite sex differences in infancy in the rhesus macaque.   Males display more social threats, play initiation, rough play, and sexual mounts. These differences, as we know, are related to differential amounts of prenatal androgen.

Partly because male and female infant rhesus behave differently, mothers in turn also tend to treat the two sexes differently. Mothers have more passive contact with, and restrain and retrieve female infants more than they do male infants. On the other hand, mothers withdraw from, play with, and present to male infants more than female infants. It cannot escape our attention that these maternal behaviors are the reciprocals of (the opposites of, or the "releasors" for) the behaviors shown more frequently by males than by females. Male infants also bite their mothers more than do female infants. Perhaps in retaliation, mothers punish male infants more than they do female infants (see Mitchell, 1971; Mitchell, 1977).

As we can see, differences in infants and mothers play a role in prompting a greater independence and activity in the male rhesus infant. In the rhesus at least, these differential behaviors subtly support the prenatally determined sex differences (Mitchell, 1977).

As rhesus infants pass through the juvenile period and approach adolescence, they too begin to display parentlike behaviors toward younger animals. Female preadolescent rhesus are very attracted to young infants and display what has been called "play mothering." Female preadolescents, in laboratory tests, show four times as much positive social behavior toward tiny infants as do males, while the male preadolescent rhesus display ten times as much hostility toward the infants as do females. The little positive behavior that is displayed by the preadolescent males toward infants is awkward and rough and involves play more than nurturance. Male infants elicit more play than do female infants, but they also elicit more aggression (cf. Mitchell, 1977).

These same sex differences apparently persist past puberty in rhesus monkeys. We have already seen in Chapter 11 that adult female rhesus are more interested in infants at birth than are adult male rhesus. Some females are so interested that they kidnap the infant from its mother. One such infant kidnapped by a nonlactating "aunt" died as a result of the kidnapping (Quiatt, 1977). The preference is a mutual one; that is, neonates and infants also prefer adult females to adult males (Sackett, 1970). Kidnapping is more likely to occur in adult females who have been rendered incapable of conception by oviduct ligation. Such females will take infants from their real mothers and carry them until the infants die, presumably of starvation. One such female carried a dead infant's carcass for twenty-five days (Vessey and Marsden, 1975).

Apparently, however, some parts of the female maternal behavior patterns are learned. Some learning presumably occurs in infancy, some in preadolescent play mothering, and some in actually giving birth and rearing infants. Primiparous (inexperienced) females have longer, more difficult deliveries than do multiparous (experienced) females, and they are usually more awkward and anxious with their infants. Primiparous females stroke, pet, restrain, and protect their infants more than do the multiparous mothers. The experienced mothers, on the other hand, are more relaxed; they punish and reject their infants more spontaneously and at an earlier age. Sex differences in their offspring appear to be greater. Apparently, male infants become more malelike with a multiparous mother (cf. Mitchell, 1977). There is also less behavioral variability among older experienced mothers than among young, inexperienced mothers. It is as if they had refined their maternal techniques to that most appropriate for their species (Mitchell, 1977).

Adult female rhesus, especially mothers but also nonmothers, respond to infant vocalizations more than do males. They become more active and respond by vocalizing themselves (Mitchell, 1977).

2. Males

In rhesus macaques, as in the primate order as a whole, there is great variability in male care of infants. In group situations, adult males usually show little interest in newborn infants. However, occasionally in the laboratory, an adult male rhesus shows a great deal of interest in the delivery of an infant (see Chapter 11).

There is little positive social interaction between free-ranging rhesus males and infants. The males are usually indifferent toward, rarely associate peacefully with, and sometimes attack or threaten infants. There is potential for very strong and warm emotional attachments between adult male and infant rhesus monkeys, however.

In our laboratory at the University of California at Davis, infant rhesus were removed from their mothers at less than one month of age and given to adult males who successfully raised them. The males differed from rhesus mothers, however, in at least four ways: (1) they allowed much less ventral contact; (2) their attachment to the infants increased over time; (3) they protected their infants by moving *toward* the threatening source, while the mothers retrieved the infants and withdrew; and (4) they played with their infants much more often and more intensely than did mothers. They raised physically and socially healthy male and female infants (Mitchell, 1977).

That these males and their infants were very attached to each other is seen in their responses to being separated from one another. The adult males reacted violently by threatening the experimenters, attempting to break the plastic barrier between the infant and themselves, or by biting themselves out of frustration. The infants protested by emitting distress calls, by assuming depressed postures, or by repeatedly throwing themselves against the separation barrier. Obviously these animals had strong affection for one another (Mitchell, 1977).

The most interesting aspect of the adult male-infant dyads was their play behavior; they played frequently and intensely. Prenatal androgens help determine sex differences in rhesus play in infancy. These male tendencies apparently carry over into "parental" behaviors, since adult male rhesus play more with infants than do adult female rhesus (cf. Mitchell, 1977).

In the wild, as already stated, adult and juvenile rhesus males usually avoid infants. However, Breuggeman (1973) has suggested that the sex differences in care-giving she observed were more quantitative than qualitative. For example, she described a cuddlinglike mount of adult males onto young males. She also noted that older rhesus males and castrates displayed more infant care than did other male rhesus, and that males tended to give more care to related individuals like infant sisters. There seemed to be a tendency for adult males to give much more care to male young than to female young, especially during the mating season. Apparently, male–male bonds can form in rhesus via "parental"care (Breuggeman, 1973). In our laboratory we also sensed that there were stronger bonds in the like-sexed male-infant dyads than in those involving female infants (cf. Redican and Mitchell, 1973).

In a laboratory environment where both adult males and females could be contacted by infants, adult male rhesus housed with mother-infant dyads provided infants with contact comfort when mothers rejected the infants, they reciprocated play, and they acted as defenders for their infants. In this so-called enriched "nuclear family" laboratory setting, "male infants played more with older siblings and peers and adult males; females played more with younger siblings and infants." (Suomi, 1972, P. 173.) Infants developed a preference for their own "father" over unfamiliar adult males (Suomi, 1972).

Additional primary sources on sex differences in rhesus monkey infant care can be found in our other reviews on this topic (Mitchell, 1969; Mitchell, 1971; Mitchell, 1976; Mitchell, 1977).

E. JAPANESE MACAQUES (cf. MITCHELL, 1969)

Adult male Japanese macaques (*Macaca fuscata*) take care of infants in the wild more than do rhesus males. New habits acquired by infants are first passed onto the adult males who care for them. There are some Japanese macaque troops in which there is no male care of infants. In these troops, none of the adult males pick up new food-eating habits (cf. Mitchell, 1969). Adult male leader Japanese macaques look after females and infants in the center of the troop. When infants reach ten months of age, such leaders and subleaders take over the care of the infants, particularly the infants of the dominant females. Some infant males in turn succeed in becoming leaders themselves, partly because they are able to interact with leaders when they are young.

Particularly during the birth season, adult male Japanese monkeys protect one- and two-year-old infants by hugging them, walking with them, or carrying them ventrally. Among one-year-old infants, the sex of the infant is of no consequence to male care; however, among two-year-old infants, female immatures receive more male care than do males (Mitchell, 1969).

Japanese macaque males whose social positions are rising most rapidly show the most male care of infants. Infant care by males is seen as being primarily cultural rather than hormonal in nature (cf. Mitchell, 1969).

The Japanese macaque infants become very attached to their adult male caretakers. Several days after one such paternalistic male had died, Japanese researchers found infants and juveniles gathering around the dead male's body and playing (Mitchell, 1969).

Despite such obvious male-infant attachments occurring in free-ranging Japanese macaques, Alexander (1970) feels that male-young affiliative and play interactions are not likely to contribute importantly to the socialization of the young. Alexander (1970) notes that the overwhelming majority of a young monkey's play and affiliative interactions are with his mother, siblings, and age-mates. The only adult male behaviors he feels to be of importance are aggressive ones. Adult males *are* responsible for virtually all of the attacks on a neonate under the age of three months. He feels that this alone may have some effect on the development of the young Japanese macaques (Alexander, 1970).

A particular male's "paternal" behavior can be fleeting, however. The paternal behavior may appear in one birth season, disappear in the mating season, and then not reappear the following birth season (Gouzoules, 1977).

Leader males may show infant care by defending infants against predators (Gouzoules, Fedigan, and Fedigan, 1975) and by mediating quarrels (Imanishi, 1963). Imanishi (1965) believes that female infants learn much by imitating the mother but that, since the male goes away from the mother earlier, the male must imitate an adult male after the first year of life. Since high-ranking mothers keep their male infants near the center of the troop where there are dominant males, their sons are likely to imitate dominant behaviors.

On the other hand, there *are* Japanese monkey troops in which there is no adult male care of infants; in fact, there are occasionally Japanese monkey troops in which there are no dominant adult males. In such a troop, adult males appear during the breeding season and then leave. The troop is run by females and there is a matriarchal dominance order based upon kinship which transcends generations (Kawamura, 1965). In such systems, the "principle of youngest ascendency" applies. This principle is that, among sisters over four years old, the youngest sister ranks just below the mother and holds the second rank among all the relatives. Brothers also tend toward youngest ascendency up to two to three years of age, but become peripheralized and eventually solitarized and this, of course, complicates male dominance (Koyama, 1967). There is some laboratory evidence that a Japanese monkey matriarchal society produces infant sex roles different from those seen in patriarchal societies (Lorinc and Candland, 1977). In matriarchal societies, a female infant of a dominant female manifests more dominant behaviors than does her male counterpart (a son of a dominant female).

F. CRAB-EATING MACAQUES

Crab-eating macaques (*M. fascicularis*) behave, in most ways, very much like rhesus monkeys. Adult male crab-eaters will kill infants in the laboratory if they are not familiar with the infant's mother. On the other hand, in stable groups in the laboratory, crab-eater adult males and even subadult males are tolerant of infant advances and often play actively with infants (cf. Mitchell, 1977; Brandt, Irons, and Mitchell, 1970). In free-ranging crab-eater monkeys, with blind infants as well as with normal infants, the adult male crab-eaters develop enduring social relationships involving retrieving, cradling, playing, grooming, and carrying of infants in the middle of the infant's first year (Berkson, 1977).

G. STUMPTAILED MACAQUES

In the stumptailed macaque (*M. arctoides*), adult males in captivity lip-smack to babies and even to pictures of babies (Blurton-Jones and Trollope, 1968). As in Japanese macaques, offspring generally come to hold a dominance rank immediately below that of their mothers. The direction and quantity of male care of infants are influenced by these dominance patterns. Male care is directed toward infants in an inverse relationship with the infant's age (Estrada, 1976).

In the stumptail macaque, a male will sometimes use a baby to regulate his own relationship with another monkey. In order to avoid being involved in a fight, for example, a male may pick up a baby. This is known as "agonistic buffering" (cf. Gouzoules, 1975).

Adult males not only lip-smack to infants but also show interest in them by giving soft "grunt-purrs" and by presenting to them (Trollope and Blurton-Jones, 1975).

H. BARBARY MACAQUES

Adult Barbary macaques (*M. sylvana*) of both sexes are interested in infants and dominant adult males take an active role in infant care. The infant Barbary macaque spends about 8 percent of its time with dominant males during its first three months of life. Barbary infants are extensively socialized by the entire group, not by a small subgroup of individuals as is true of rhesus. Very young (one to two months old) infants show little interest in adult males, whereas older infants (three to four months old) take a more active interest (cf. Mitchell, 1969).

Despite the absence of any overt interest on the part of the very young infant, "since the leader male takes the neonate from as early as the first day of life, it is the leader male who is the predominant influence in socialization of the infant until it is approximately two weeks of age." (Burton, 1972, P. 57.) According to Burton, the leader male reinforces the infant's mouth-sucking movements until these movements become socially meaningful communicative movements. The male also encourages the infant Barbary macaque in locomotor skills and rides the infant dorsally. The leader reorients the infant away from the mother by permitting subadult males to take the infant away to greater distances.

Burton's subadult Barbary macaque females, in contrast, play a small role in the infant's socialization. These females are peripheralized and rebuffed by all adults and subadult males and thus do not get to handle infants. This is *very* different from the pattern seen in other macaques (Burton, 1972). However, in 1940 there was evidence that this same population of Barbary macaques had a greater role for subadult females in infant socialization. Evidently there has been, in Burton's words, a "tradition drift" in infant care, with males becoming more and more involved and females less and less involved (Burton, 1972).

Barbary macaque males, like stumptails, often use infants to buffer agonistic interactions. Even dead infants are picked up and used by males to avert aggressive interactions (Merz, 1976). The typical pattern of agonistic buffering encounter, however, involves a male, the initiator, who invites a live infant to ride dorsally. The initiator then approaches another male with the infant in a dorsal position, hesitates, and then moves into the proximity of the new male. The new male then reaches out to hold the first male and all three animals then lip-smack. Females, as well as males, show agonistic buffering, but far less frequently than do males. In one study only 6.5 percent of such interactions involved females (Peffer and Smith, 1977). Apparently, this kind of behavior reduces the likelihood of aggression (Whiten and Rumsey, 1973).[1]

[1] After this book had cleared the galley proof stage Dr. David Taub presented evidence at the *American Society of Primatologists* Meeting in Atlanta, Ga., September 1978 which contradicts this view.

I. PIGTAIL MACAQUES

Both male and female pigtail infants (*M. nemestrina*) show a clear preference for one of their mother's nipples by the time they are one month old (Erwin, Anderson, and Bunger, 1975). Males, however, develop independence from their mothers earlier and to a greater extent than do female infants. The mothers play a major role in instigating the male's greater independence. Mothers bite male infants more than they do female infants and there is greater mutual antagonism between mothers and male infants (Jensen, 1966).

As in rhesus macaques, primiparous or inexperienced pigtail mothers are more anxious, and restrain and retrieve their infants more than do multiparous mothers (Kuyk, Dazey, and Erwin, in press).

In pigtail macaques, by the end of the first six months of life, males begin to show a more rapid decay in contact and orientation toward the mother than do females. In this way, they become peripheralized as they mature (Rosenblum, 1974).

In comparisons of adult male and female pigtail macaques, only *mother* pigtail monkeys respond regularly and consistently to recordings of infant vocalizations. They respond to them with increased pacing and calling (Simons, Bobbitt, and Jensen, 1968).

However, a depressive reaction in five-month-old pigtailed infants who are seperated from their mothers does elicit paternalistic protection from an adult male pigtail macaque. Also, pigtail adult males show some tolerance for the bouncing play of infants. Although typically they are indifferent to infants, adult males will protect infants when they are attacked (cf. Mitchell, 1977).

J. BONNET MACAQUES

There is a sex difference in preference for their own mothers in the bonnet macaque (*M. radiata*). After three to ten weeks of age, female infants show a preference for their own mother over a strange female while male infants do not (Rosenblum, 1974b). If the environment is complex, males also show lower levels of contact with the mother than do female infants, but this is not true if the environment is simple (Rosenblum, 1974).

In bonnet macaques:

Females develop their network primarily through grooming and remain closely associated with their mothers. Males rely more heavily on the play group, which includes participating subadult and adult males. (Simonds, 1974, P. 151)

Despite this similarity to the rhesus macaque pattern, bonnet subadult males are not peripheralized as much as are the rhesus. Bonnet macaques live in more altruistic, less competitive groups than do rhesus. There is less male migration among bonnet troops than among rhesus. There is also more infant sharing and infant care by animals other than the mother among bonnets. Even male bonnets hold and share infants (Wade, 1977).

There is a greater tolerance of adult male for adult male among bonnets. Adult males do not force subadult males out of the group. There is a "central hierarchy" of males who cooperate, move together, and provide a focal point for infants and females. Males do not touch the infant at birth; later they wrestle with infants but do not threaten in play as do juveniles and infants (Simonds, 1965). Even in the more accepting, more social bonnet macaques, however, females display more infant care than do males.

We will continue our coverage of sex differences in infant care in nonhuman primates in the next chapter in which other Old World monkeys, lesser apes and great apes are compared.

REFERENCES

Alexander, B. K. Parental behaviour of adult male Japanese monkeys. *Behaviour,* 1970, **36** (4), 270-285.

Bekoff, M. The development of social interaction, play, and meta communication in mammals: An ethological perspective. *Quarterly Review of Biology,* 1972, **47** (4), 412-434.

Berkson, G. The social ecology of defects in primates. In Chevalier-Skolnikoff, S. and Poirier, F. (Eds.) *Primate Biosocial Development.* New York: Garland, 1977, pp. 189-204.

Blurton-Jones, N. G. and Trollope, J. Social behaviour of stumptailed macaques in captivity. *Primates,* 1968, **9**, 365-394.

Brandt, E. M., Irons, R. and Mitchell, G. Paternalistic behavior in four species of macaques. *Brain, Behavior and Evolution,* 1970, **3**, 415-420.

Brandt, E. M. and Mitchell, G. Parturition in primates: Behavior related to birth. In Rosenblum, L. A. (Ed.) *Primate Behavior* (Vol. 2). New York: Academic Press, 1971, pp. 177-223.

Breuggeman, J. A. Parental care in a group of free-ranging rhesus monkeys (*Macaca mulatta*). *Folia Primatologica,* 1973, **20**, 178-210.

Buettner-Janusch, J. The breeding of Galagos in captivity and some notes on their behavior. *Folia Primatologica,* 1964, **2**, 93-110;

Burton, F. D. The integration of biology and behavior in the socialization of *Macaca sylvana* of Gibraltar. In Poirier, F. E. (Ed.) *Primate Socialization.* New York: Random House, 1972, pp. 29-62.

Doyle, G. A. The behaviour of the lesser bushbaby. In Martin, R. D., Doyle, G. A. and Walker, A. C. (Eds.) *Prosimian Biology.* London: Duckworth, 1974, pp. 213-231.

Epple, G. The behavior of marmoset monkeys (Callithricidae). In Rosenblum, L. A. (Ed.) *Primate Behavior* (Vol. 4). New York: Academic Press, 1975, pp. 195-239. (a)

Epple, G. Parental behavior in *Saguinus fuscicolliss ssp.* (Callithricidae). *Folia Primatologica,* 1975, 24, 221-238. (b)

Erwin, J., Anderson, B. and Bunger, D. Nursing behavior of infant pigtail monkeys (*Macaca nemestrina*): Preferences for nipples. *Perceptual and Motor Skills,* 1975, 40, 592-594.

Estrada, A. Social relations in a free-ranging troop of *Macaca arctoides.* Paper presented at the *International Primatological Society* Meeting, Cambridge, England, August, 1976.

Glander, K. E. Baby-sitting, infant sharing, and adoptive behavior in mantled howling monkeys. *American Journal of Physical Anthropology,* 1975, 41, 482.

Goldman, P. S. Maturation of the mammalian nervous system and the ontogeny of behavior. *Advances in the Study of Behavior,* 1976, 7, 1-90.

Gouzoules, H. Maternal rank and early social interactions of infant stumptail macaques, *Macaca arctoides. Primates,* 1975, 16, 405-418.

Gouzoules, H. Allopaternal behavior and paternity in a troop of *Macaca fuscata.* Paper presented at the *Animal Behavior Society* Meeting, University Park, Pennsylvania, June, 1977.

Gouzoules, H., Fedigan, L. M. and Fedigan, L. Responses of a transplanted troop of Japanese macaques (*Macaca fuscata*) to bobcat (*Lynx rufus*) predation. *Primates,* 1975, 16, 335-349.

Gucwinska, H. and Gucwinski, A. Breeding the Zanzibar galago at Wroclaw Zoo. *International Zoo Yearbook,* 1968, 8, 111-114.

Hamburg, D. A. Evolution of emotional responses: Evidence from recent research on nonhuman primates. *Science and Psychoanalysis,* 1968, 12, 39-54.

Imanishi, K. Social behavior in Japanese monkeys, *Macaca fuscata.* In Southwick, C. H. (Ed.) *Primate Social Behavior.* Princeton, N.J.: Van Nostrand, 1963, pp. 68-81.

Imanishi, K. Identification: A process of socialization in the subhuman society of *Macaca fuscata.* In Imanishi, K. and Altmann, S. (Eds.) *Japanese Monkeys.* Atlanta: Altmann, 1965, pp. 30-51.

Ingram, J. C. Social interactions within the marmoset family group (*C. jacchus*). Paper presented at the *International Primatological Society* Meeting, Cambridge, England, August. 1976.

Jensen, G. D. Sex differences in developmental trends of mother-infant monkey behavior (*M. nemestrina*). *Primates,* 1966, 7, 403.

Jolly, A. *The Evolution of Primate Behavior.* New York: Macmillan, 1972.

Kaplan, J. The effects of separation and reunion on the behavior of mother and infant squirrel monkeys. *Developmental Psychobiology,* 1970, **3**, 43-52.

Kawamura, S. Matriarchal social ranks in the Minoo-B troop. In Imanishi, K. and Altmann, S. (Eds.) *Japanese Monkeys.* Atlanta: Altmann, 1965, pp. 105-112.

Keefer, D. A. and Stumpf, W. E. Atlas of estrogen-concentrating cells in the central nervous system of the squirrel monkey. *Journal of Comparative Neurology,* 1975, **160**, 419-441.

Kinzey, W. G., Rosenberger, A. L., Heisler, P. S., Prowse, D. L. and Trilling, J. S. A preliminary field investigation of the yellow handed titi monkey, *Callicebus torquatus torquatus* in Northern Peru. *Primates,* 1977, **18**, 159-181.

Kleiman, D. G. Monogamy in mammals. *The Quarterly Review of Biology,* 1977, **52**, 39-69.

Kling, A. Differential effects of amygdalectomy in male and female nonhuman primates. *Archives of Sexual Behavior,* 1974, **3**, 129-134.

Kling, A. and Dunne, K. Social-environmental factors affecting behavior and plasma testosterone in normal and amygdala lesioned *Macaca speciosa. Primates,* 1976, **17**, 23-42.

Klopfer, P. H. Patterns of maternal care in lemurs: II. Effects of group size and early separation. *Zeitschrift für Tierpsychologie,* 1976, **30**, 277-296.

Klopfer, P. H. and Dugard, J. Patterns of maternal care in lemurs: III. *Lemur variegatus. Zeitschrift für Tierpsychologie,* 1976, **30**, 277-296.

Koyama, N. On dominance rank and kinship of a wild Japanese monkey troop in Arashiyama. *Primates,* 1967, **8**, 189-216.

Kress, J. H. Socialization patterns of *Lemur variegatus.* Paper presented at the *American Association of Physical Anthropologists* Meeting, Denver, Colorado, April, 1975.

Kuyk, K., Dazey, J. and Erwin, J. Primiparous and multiparous pigtail monkey mothers (*Macaca nemestrina*): Restraint and retrieval of female infants. *Journal of Biological Psychology,* in press.

Langford, J. B. Breeding behavior of *Hapale jacchus* (common marmoset). *South African Journal of Science,* 1963, **59**, 299-300.

Leibrecht, B. C. and Kelley, S. T. Some observations of behavior in breeding pairs of owl monkeys. Paper presented at the *American Society of Primatologists* Meeting, Seattle, Washington, April, 1977.

Lorinc, G. A. and Candland, D. K. The primate mother-infant dyad: Intergenerational transmission of dominance in *Macaca fuscata.* Paper presented at the *Animal Behavior Society* Meeting, University Park, Pennsylvania, June, 1977.

Martin, R. D. Tree shrews: Unique reproductive mechanism of systematic importance. *Science*, 1966, **152**, 1402-1404.

Mason, W. A. Social organization of the South American monkey, *Callicebus moloch:* A preliminary report. *Tulane Studies in Zoology*, 1966, **13**, 23-28.

Merz, E. Male-male interactions with dead infants in *Macaca sylvana*. Paper presented at the *International Primatological Society* Meeting, Cambridge, England, August, 1976.

Mitchell, G. Paternalistic behavior in primates. *Psychological Bulletin*, 1969, **71**, 399-417.

Mitchell, G. Parental and infant behavior. In Hafez, E. S. E. (Ed.) *Comparative Reproduction of Laboratory Primates* (Chapter 14). Springfield, Ill.: Charles C Thomas, 1971, pp. 382-402.

Mitchell, G. Attachment potential in rhesus macaque dyads (*Macaca mulatta*): A sabbatical report. *Catalog of Selected Documents in Psychology*, 1976, **6**, 97 pp.

Mitchell, G. Parental behavior in nonhuman primates. In Money, J. and Musaph, H. (Eds.) *Handbook of Sexology*. Amsterdam: Elsevier/North-Holland Biomedical Press, 1977, pp. 749-759.

Mitchell, G. and Brandt, E. M. Paternal behavior in primates. In Poirier, F. E. (Ed.) *Primate Socialization*. New York: Random House, 1972, pp. 173-206.

Moynihan, M. Some behavior patterns of platyrrhine monkeys: II. *Saguinus geoffroyi* and some other tamarins. *Smithsonian Contributions to Zoology*, 1970, **28**, 1-77.

Neville, M. K. Social relations within a troop of red howler monkeys (*Alouatta seniculus*). *Folia Primatologica*, 1972, **18**, 47-77.

Peffer, P. G. and Smith, E. O. "Agonistic buffering" in a captive Barbary macaque (*Macaca sylvana*) group. Paper presented at the *American Society of Primatologists* Meeting, Seattle, Washington, April, 1977.

Quiatt, D. Aunts and mothers and selection and adaptation. Paper presented at the *American Society of Primatologists* Meeting, Seattle, Washington, April, 1977.

Raphael, D. Uncle rhesus, auntie pachyderm, and mom: All sorts and kinds of mothering. *Perspectives in Biology and Medicine*, 1969, **12**, 290-297.

Redican, W. K. Adult male-infant interactions in nonhuman primates. In Lamb, M. E. (Ed.) *The Role of the Father in Child Development*. New York: Wiley, 1976, pp. 345-385.

Redican, W. K. and Mitchell, G. The social behavior of adult male-infant pairs of rhesus monkeys in a laboratory environment. *American Journal of Physcial Anthropology*, 1973, **38**, 523-526.

Rosenblum, L. A. Sex differences, environmental complexity and mother-infant relations. *Archives of Sexual Behavior*, 1974, **3**, 117-128. (a)

Rosenblum, L. A. Sex differences in mother-infant attachment in monkeys. In Friedman, R. C., Richart, R. M. and Van de Wiele, R. L. (Eds.) *Sex Differences in Behavior.* New York: Wiley, 1974, pp. 123-145. (b)

Rosenson, L. M. Interactions between infant greater bushbabies (*Galago crassicaudatus crassicaudatus*) and adults other than their mothers under experimental conditions. *Zeitschrift fur Tierpsychologie,* 1972, **31**, 240-269.

Rothe, H. Influence of new born marmosets (*Callithrix jacchus*) behaviour on expression and efficiency of maternal and paternal care. *Symposia of the Fifth Congress of the International Primatological Society.* Nagoya, Japan: Japan Science Press, 1974, pp. 315-320.

Rowell, T. E. Contrasting different adult male roles in different species of nonhuman primates. *Archives of Sexual Behavior,* 1974, **3**, 143-149.

Sackett, G. P. Unlearned responses, differential rearing experiences, and the development of social attachments by rhesus monkeys. In Rosenblum, L. A. (Ed.) *Primate Behavior* (Vol. 1). New York: Academic Press, 1970, pp. 111-140.

Sauer, E. G. F. and Sauer, E. M. The South West African bushbaby of the *Galago senegalensis* group. *The Journal of the South West African Scientific Society.* 1962, **16**, 5-36.

Schreiber, G. R. A note on keeping and breeding the Philippine tarsier (*Tarsius syrichta*) at Brookfield Zoo Chicago. *International Zoo Yearbook,* 1968, **8**, 114-115.

Simonds, P. E. The bonnet macaque in South India. In DeVore, I. (Ed.) *Primate Behavior.* New York: Holt, Rinehart and Winston, 1965.

Simonds, P. E. Sex differences in bonnet macaque networks and social structure. *Archives of Sexual Behavior,* 1974, **3**, 151-165.

Simons, R. C., Bobbitt, R. A. and Jensen, G. D. Mother monkeys (*Macaca nemestrina*) responses to infant vocalizations. *Perceptual and Motor Skills,* 1968, **27**, 3-10.

Snyder, P. A. Behavior of *Leontopithecus rosalia* (Golden lion marmoset) and related species: A review. *Journal of Human Evolution,* 1974, **3**, 109-122.

Sorenson, M. W. Behavior of tree shrews. In Rosenblum, L. A. (Ed.) *Primate Behavior* (Vol. 1). New York: Academic Press, 1970, pp. 141-194.

Sorenson, M. W. A review of aggressive behavior in the tree shrews. In Holloway, R. L. (Ed.) *Primate Aggression, Territoriality, and Xenophobia: A Comparative Perspective.* New York: Academic Press, 1974, pp. 13-30.

Sorenson, M. W. and Conaway, C. H. Observations on the social behavior of tree shrews in captivity. *Folia Primatologica,* 1966, **4**, 124-145.

Spencer-Booth, Y. The relationships between mammalian young and conspecifics other than mothers and peers: A review. In Lehrman, D. S., Hinde, R. A. and Shaw, E. (Eds.) *Advances in the Study of Behavior* (Vol. 3). New York: Academic Press, 1970, pp. 119-194.

Stevenson, M. F. Birth and perinatal behaviour in family groups of the common marmoset (*Callithrix jacchus jacchus*) compared to other primates. *Journal of Human Evolution*, 1976, **5**, 365-381.

Strayer, F. F., Taylor, M. and Yanciw, P. Group composition effects on social behaviour of captive squirrel monkeys (*Saimiri sciureus*). *Primates*, 1975, **16** (3), 253-260.

Suomi, S. J. Social development of rhesus monkeys reared in an enriched laboratory environment. *Abstract Guide of the 20th International Congress of Psychology*. Tokyo, Japan: Japan Science Press, 1972, pp. 173-174.

Trollope, J. and Blurton-Jones, N. G. Aspects of reproduction and reproductive behaviour in *Macaca arctoides*. *Primates*, 1975, **16**, 191-205.

Vessey, S. H. and Marsden, H. M. Oviduct ligation in rhesus monkeys causes maladaptive epimeletic (care-giving) behavior. In Kondo, S., Kawai, M. and Ehara, A. (Eds.) *Contemporary Primatology*. Basel: Karger, 1975, pp. 321-325.

Wade, T. Inbreeding and kin selection in the evolution of macaque social behavior. Paper presented at the *American Society of Primatologists* Meeting, Seattle, Washington, April, 1977.

Whiten, A. and Rumsey, T. J. Agonistic buffering in the wild Barbary macaque. *Primates*, 1973, **14**, 421-425.

Williams, L. Breeding Humboldt's woolly monkey (*Lagothrix lagothrica*) at Murrayton woolly monkey sanctuary. *International Zoo Yearbook*, 1967, **7**, 86-89.

13.

Infant Care: Other Old World Monkeys, Apes, and Humans

In our last chapter we discussed sex differences in infant care in the prosimians, New World monkeys, and macaques. We will continue our review of this topic by describing sex differences in infant care in other Old World monkeys, lesser apes, great apes, and people.

A. OTHER OLD WORLD MONKEYS

1. Baboons

Olive (*Papio anubis*) and yellow (*P. cynocephalus*) baboons do not live in one-male groups and they do not form harems. These species live in troops containing many adult males, females, and young. Adult male olive baboons approach newborn infants, lip-smack, touch the infants with their hands and mouth, and groom its mother. Both adult and subadult males are completely tolerant of the infant as it plays near them. They may sometimes touch the infant and will protect it and its mother against others in the troop. A few adult males carry infants on their ventral surfaces. Male baboons stand guard over play groups and prevent injuries from fighting within the troop. They also protect infants against predation. Some males adopt sick or orphaned infants who may become their constant companions (cf. Mitchell, 1969 for primary references).

Adult male olive baboons form a cooperative central hierarchy, and mothers with small infants stay near the males of the central hierarchy. Adult females other than mothers begin to rebuff infants not their own after the infants reach six months of age, but adult males do not. They remain permissive and protective toward these older infants so that, by ten months of age, the infant is spending more time in a play group of peers near a dominant male than it spends with its mother. In a crisis, an infant of this age is more likely to seek out an adult male than its mother (cf. Mitchell, 1969, for primary references).

At about thirty months of age, however, a juvenile olive baboon no longer receives preferential treatment from adult males. At this point he or she becomes part of the adult dominance hierarchy (Mitchell, 1969).

Mother–infant relations in yellow baboons and olive baboons have been described by DeVore (1963). Mothers and infants are the center of interest for other adult females when the infants are tiny. Mothers carrying black infants (black is the natal coat color for olive baboons) will not enter or leave a potentially dangerous clearing unless an adult male enters or leaves the clearing first. Black infants stay near adult males when the troop pauses in a clearing (Rhine and Owens, 1972).

According to Rowell, Din, and Omar (1968) adult males treat male and female infants differently. They apparently take far less interest in the female infants. However, Young (1977) found that this differential interest produced no sex differences in infants in the first three months. He detected no sex differences at this early age whether or not adult males were present in a group. Gilmore (1977) has described what appears to be "agonistic buffering" (males using infants to safely get close to more dominant males) in olive baboons.

In chacma baboons (*Papio ursinus*), there is a very obvious hierarchical organization among the males of a multimale troop. Idle males of this species act as sentinals to warn of approaching danger. Females show a strong interest in caring for infants and this interest appears to be developed as early as six months of age (Bolwig, 1959). Males behave much like olive and yellow baboons.

Hamadryas baboons (*P. hamadryas*) do not form a closed multimale troop, but live in parties made up of highly constant one-male harems. These one-male groups, including one adult male and one to nine females with young, never split up. Abandoned or orphaned infants may be adopted by the hamadryas male. Play groups form around him or around the oldest subadult male in his harem. Although the males rarely participate in infant play, frightened infants frequently run into their arms whereupon they threaten the immature aggressor. Thus, the male appears to baby-sit, but not for just one or two infants. He protects all of them and the infants appear to compete for his attention.

As the young mature, their relations with males change. The two-year-old female presents to the male. At an earlier age than two she gradually transfers her attachment from her mother to the male so that he protects her under stress. Between one and two years of age, she gradually enters into a kind of consortship, usually with a subadult male in the harem group. This young male later becomes her group leader in a new harem (cf. Mitchell, 1969 for primary references).

The two-year-old hamadryas male assumes a role of "mother" toward smaller infants, particularly female infants. When frightened he grabs an infant and approaches the dominant male with it ("agonistic buffering"). Female two-year-olds never pick up infants. If they are frightened they present or embrace a male. Thus, male hamadryas baboons become very "maternal" and may carry infants for up to half an hour. Motherless infants are invariably adopted by such young males (cf. Mitchell, 1969).

Gelada baboons (*Theropithecus gelada*) also live in one-male groups. Adult and subadult males sleep at the outside of a group in the most exposed positions, thus protecting more vulnerable members of the troop. There is a special role for "bachelor" males (subadults) in the one-male gelada group. They begin interacting with infants in the second year of life. Their rate of interaction with juveniles is greater than is that of the juveniles' mothers. They become the focus for immature animal activities. In fact, because of this "bachelor" male activity in geladas, there is actually more "paternal" behavior than "aunt" behavior in a gelada group (Bernstein, 1975). When infants are very young, however, females tend to be more interested in them than are the males (Dunbar and Dunbar, 1974).

2. Mangabeys (*Cercocebus spp.*)

Little is known about sex differences in mangabey infant care. Infants of *Cercocebus atys* (the sooty mangabey) certainly appear to receive special nonaggressive attention, however. The alpha male and adult females often try to pick the infant up and only juvenile male mangabeys are aggressive toward the infant (Bernstein, 1971). The majority of guardian behavior for the infant is performed by young nulliparous females and by relatives of the infant (Hunter, 1977).

Among *Cercocebus albigena,* adult females without infants appear to associate less with mothers than do females with infants. Some adult males form close ties to infants. In one instance reported by Chalmers (1968) an infant actually showed a preference for an adult male over its own mother.

3. Vervets (*Cercopithecus aethiops*)

Among the terrestrial vervets, immature females devote a considerable amount of attention to newborn infants (Struhsaker, 1971). They engage in much more infant caretaking than do juvenile males (Raleigh, 1977). These sex differences in infant care behavior observed in free-ranging vervets are also seen in captivity (Raleigh, 1977).

In vervets, it is the *matrifocal* core that provides the primate group with stability and continuity. The male wanders. An adult female does not change groups but she starts new ones and all of her offspring go with her. The older female vervets hold all of the troops valuable knowledge. Female coalitions dominate males in order to protect infants. The male role in infant care is very small (Lancaster, 1973).

The juvenile female vervet practices the maternal role in "play mothering" infants. This also occurs in juvenile males, but very infrequently (Lancaster, 1971). Adult females other than the mother handle vervet infants. They sometimes remove the infants from their mothers by force even when they have infants of their own (Krige and Lucas, 1974). Adult females also reinforce the play of young immature monkeys whereas adult males engage in rough play only with older male juveniles. Younger males, however (three years old) sometimes stimulate play in young infants (Fedigan, 1972). Males are not interested in neonates (Gartlan, 1969).

4. Other *Cercopithecus* Species

The other species of the genus *Cercopithecus* are not terrestrial but arboreal. In Lowe's guenon (*C. campbelli lowei*) adult males do not show interest in infants while females do. The adult male, for example, does not allow a group of infants to play close to him. On the other hand, males will come to the aid of infants when they are in danger (Bourliere, Bertrand, and Hunkeler, 1969).

Adult females of this species show "aunting" behavior. Males do not "mother" infants although occasionally they will play with them (Hunkeler, Bourliere, and Bertrand, 1972).

Among DeBrazza monkey groups (*Cercopithecus neglectus*), infants frequently play with the adult male, but he rarely initiates the play (Stevenson, 1973). The redtail monkeys of Uganda (*Cercopithecus ascanius schmidti*) live in one-male groups. In one study group, the leader male was replaced by a new adult male who immediately killed two newborn infants and partially consumed one of them. Following this he copulated with several females in the group. This infanticide is probably a regular feature accompanying group takeover among redtails. The behavior probably hastens the onset of estrus (Struhsaker, 1976).

Syke's monkey (*C. albogularis*) subadult males play frequently with juvenile males. Juvenile females play infrequently (Dolan, 1976). Removal of the adult male in a Syke's monkey group causes a decrease in the time spent by the infant away from the mother and increases the mother's proximity to her infant and time holding her infant. Removing an adult female does not produce such a change. Obviously the presence of the male makes both mother and infant more confident (Chalmers, 1971).

5. *Erythrocebus* and *Miopithecus*

Removing an adult male from a patas monkey (*Erythrocebus patas*) group has no effect on the mother–infant dyad. This is because the male patas monkey does not interact with or protect the mother–infant dyad (Chalmers, 1971).

The adult male patas also shows no interest in the newborn (Hall and Mayer, 1967).

Talapoin males (*Miopithecus talapoin*) are very unlike patas males. The adult male talapoin wrestles with infant males and supervises squabbles. Still, however, the female talapoin is more infant-oriented than is the male (Segre, 1970).

6. Hanuman Langur (*Presbytis entellus*)

Like the redtail monkey males discussed above, hanuman langur males will often kill newborn infants when they take over a troop (Sugiyama, 1966; Hrdy, 1977a). Dolhinow (1977a) has suggested that these langur infanticides are the result of abnormal stressed and crowded conditions; however, Hrdy (1977b) has replied that langurs in India have always lived in crowded and stressed environments and that the infanticide could not be a response to modern conditions:

> . . . vulnerability of a troop to male takeovers, not human disturbance of the habitat, appears to determine the differences in infant survivorship between troops. (Hrdy, 1977b, Letter)

It should be noted that new langur males do not attack infants of mothers they remember. In addition, alien infants kidnapped from other troops but held by familiar females are also not killed (Hrdy, 1977a). Infanticide apparently increases the new male's reproductive success by advancing the estrus of the females. Males are thus able to shortcut two or three year birth intervals (Hrdy, 1974; Chapman and Hausfater, 1977). On the other hand, there is *some* evidence that sufficiently low population density, and adequate home range size might be a factor in permitting the introduction of new males into a troop without disasterous results to troop infants. What we are suggesting is that the complete explanation of the infanticide phenomenon may be a combination of the crowding and troop vulnerability-reproductive success hypotheses (cf. Boggess, 1977).

Assuming infant langurs survive the neonatal period, the male infant has no contact with the adult male until the infant is ten months old. At this time he may run toward an adult or subadult male, mount him, and then embrace him. Female infants do not do this but instead become involved in mutual grooming with adult females. Because infant males are more active in play, adults threaten and chase them more than twice as often as they do female infants. In the wild, adult male langurs seldom play and adult females almost never play (Dolhinow and Bishop, 1970).

The newborn hanuman langur, as we know, is passed among and handled by many female caretakers from the day of birth (Dolhinow, 1977b). Adult male langurs show no interest in the newborn infants, however (Jay, 1963). A newborn may spend up to 48 percent of its first day on females other than its own mother (allomothers) (Hrdy and Hrdy, 1977). Apparently, the male's most important contribution to infant survival is troop protection (Hrdy, 1976).

7. Other Langurs

Among Nilgiri langurs (*P. johnii*), juveniles and subadults are often attracted to maternal groups but primarily to the mothers. Infant sharing of young babies by adult females is common and babies are passed from female to female. Males show little interest in the newborn or responsibility for the infant, but probably still show more interest than do some other langurs (Poirier, 1969; Poirier, 1970).

The spectacled langur (*P. obscurus*) behaves very differently from the langurs discussed thus far. Adult male spectacled langurs carry infants (although often clumsily) and kiss the infants backs as an adult female would do. Males have also been seen doing this to fifteen-month-old juveniles. Males also perform clasp greetings with juveniles, involving grins, embraces, and vocalizations with the mouth open sounding very much like human laughter. An adult male may sleep with a female juvenile while clasping her, carry an infant when it whimpers, and may be tolerant of infants climbing on him. In apparent sexual "training," the adult male actively pushes the young infant into a present posture (Horwich, 1974). As in other langurs, however, females also show interest in infants. Within hours after birth, babies are passed between adult females; and, while the male shows great interest, he is not allowed to handle the infant at this time (Badham, 1967).

In *P. cristatus* the male role appears to be much the same as that of *P. johnii* (Poirier, 1970). There is only one control male per troop who defends a territory against other lutong (*P. cristatus*) males. Adult males rarely carry juveniles and, when they do, it is usually the immature animal who initiates the interaction. In many instances a juvenile runs to the adult male while squealing and touches or even slaps at the male. Juvenile males sometimes play with subadult and adult males (Bernstein, 1968).

Infanticide and exclusion of immature animals from a group result when there is a male replacement or takeover in the purple-faced langurs (*Presbytis senex senex*) (Rudran, 1973).

8. Other Colobidae

A year old full sibling male of the species *Colobus polykomos* was seen holding its newborn younger brother soon after birth. While the adult male, on occasion, takes no interest in neonates, there have been reports that both males and females share infants freely. A female of *Pygathrix nemaeus* (douc langur) was seen holding another female's neonate. Among douc langurs, juvenile males may participate in infant sharing more than has been realized. There is also infant sharing in *Nasalis larvatus* (the proboscis monkey) but the sharing occurs at a later age. The male may also be involved (Hill, 1972; also personal observation).

There is infant sharing by females in *Colobus guereza* as well as in *Colobus polykomos* (see above). Male-infant interactions follow at four to nine weeks of age, after the peak of the female transfers has passed. After four to nine weeks, however, the male shows the same behaviors as the females. While the adult male grooms the infant, juvenile males do not. Two males never exchange an infant but they do give it up to an adult female. Apparently the male's interest functions primarily in defense of the infant (Horwich and Manski, 1975). Many of the social behaviors which increase in a group of *Colobus guereza* monkeys following a birth are maternal in nature but also appear infantile at times. Birth has a socially cohesive effect on the group (Horwich and Wurman, 1977).

B. LESSER APES

As we know, all lesser apes live in monogamous groups. Berkson (1966) has observed a captive gibbon group in which a male was put with a mother and infant after the infant reached three months of age. The male played with the infant even though the infant was afraid. Though the infant did not seem to like it, he was also carried ventrally by the male. By the time the infant was five months old, however, he initiated play with the adult male, and by nine to ten months he was initiating most if not all of the contact, which the male tolerated. According to Ellefson, white-handed gibbons (*Hylobates lar*) of both sexes become intolerant of immatures when the juveniles reach about 2½ years of age. From then on a gradual peripheralization of the immature gibbon takes place up to four years of age (Ellefson, 1966).

Within a siamang group (*Symphalangus syndactylus*), at night, the juvenile sleeps with the male and the infant with the female:

> When the infant is weaned from the female, the male takes over its daily care until it is fully independent. The subadult male is slowly peripheralized by the male. (Chivers, 1970, P. 21)

The departure of the oldest offspring of the group occurs on the day the mother gives birth to a new baby. In the siamang, the adult male carries the offspring in its second year and the young juvenile sleeps with the male at night if it has a younger sibling (Chivers and Chivers, 1975; see also Fairbanks, 1977).

C. GREAT APES

1. Chimpanzees (*Pan troglodytes*)

Among common chimpanzees, a mother continues to have close social relationships with her offspring after they reach adulthood. Daughters remain closer than sons to their mothers. Orphaned infants may be adopted by older siblings but family units are matriarchal; the father is not identifiable (cited in Bramblett, 1976).

Adolescents of both sexes may move about for several days on their own; and groups, comprised of females and young, often have no adult males in them. A mother's dominance status seems to be influenced by the standing of her adolescent son among the male status order (cf. Bramblett, 1976 for primary references).

Adult females and infants travel less extensively than do adult or adolescent males. Males often visit when a female is in estrus. A copulating adult pair may be harassed by juveniles and infants. The males are tolerant of harassing infants (cf. Bramblett, 1976).

Infants under five months old normally interact only with their mothers and siblings. The mother continues to assist in transportation until the infant is over four years old; weaning is not completed until the fifth or sixth year (cf. Bramblett, 1976).

Relative to adult males, females with infants were silent, wary, and shy. Except for the ties between a mother and her offspring, social bonds are loose. There are continuously shifting subgroups (cf. Bramblett, 1976).

In the common chimpanzee, mothers play with infants more than do adult males, but males also play with them occasionally. Infants also wrestle with adolescents of both sexes and with other mothers (Bingham, 1927; Goodall, 1965). Females, however, are much more interested in infants than are males. Kortlandt (1967) even observed female chimps showing a great deal of interest in a captive infant monkey and attempting to release it from its restraining device. One chimpanzee female who had lost her own child even accepted and nursed a five-week-old orphan that had been hand-reared by humans (Palthe and van Hooff, 1975).

But even though females show more infant care than males, males *do* show a great deal of potential for infant care. One adult male raised in captivity, even though socially deprived in early life, still showed infant care by carrying a young infant (Kollar, Beckwith, and Edgerton, 1968).

Clark (1977) has found what might be a sex difference in mother-infant weaning. When her two young male chimpanzees wanted their mothers to groom them, they would tug at the mother's hand until she responded. When they wanted her to travel they would shove her. On one occasion an infant male even pushed his mother over onto her back so that he could suckle. Clark's two females

Plate 1

1. Chimpanzee (*Pan troglodytes*) and orang-utan (*Pongo pygmaeus*) juveniles. (*Photo by T. Maple*) Chapter 2
2. A mature foot-clasp mount in the rhesus monkey (*Macaca mulatta*). (*Photo by J. Erwin*) Chapter 4

Plate 2

3. Sexual swelling in a female olive baboon. (*Papio anubis*). (*Photo by T. Maple*) Chapter 6

4. Sexual swelling in a female hamadryas baboon (*Papio hamadryas*). (*Photo by T. Maple*) Chapter 8

5. Adult rhesus monkey (*Macaca mulatta*) heterosexual posturing. (*Photo by W. K. Redican*) Chapter 9

6. An adult male yellow baboon (*Papio cynocephalus*) in the wild attempts to mount a female. (*Photo by T. Maple*) Chapter 10

Plate 3

7. A very pregnant female patas monkey (*Erythrocebus patas*). (*Photo by Evan Zucker, courtesy of Carribean Primate Research Center*) Chapter 11

8. An infant patas monkey (*Erythrocebus patas*) guarded by its mother. (*Photo by Evan Zucker, courtesy of the Carribean Primate Research Center*) Chapter 13

Plate 4

9. Rhesus monkeys (*Macaca mulatta*) sitting close together on food container on La Parguera. (*Photo by Evan Zucker, courtesy of the Carribean Primate Research Center*) Chapter 14

10. Spacing in the yellow baboon (*Papio cynocephalus*). (*Photo by T. Maple*) Chapter 15

Plate 5

11. A dominant male rhesus (*Macaca mulatta*) watches over thirsty infants. (*Photo by Evan Zucker, courtesy of the Carribean Primate Research Center*) Chapter 16
12. Dominance assertion via threat in the patas monkey (*Erythrocebus patas*). (*Photo by Evan Zucker, courtesy of the Carribean Primate Research Center*) Chapter 17

Plate 6

13. An example of an alliance in the patas monkey (*Erythrocebus patas*). (*Photo by Evan Zucker, courtesy of the Carribean Primate Research Center*) Chapter 18
14. Vigilance in the patas monkey (*Erythrocebus patas*). (*Photo by Evan Zucker, courtesy of the Carribean Primate Research Center*) Chapter 19

Plate 7

15. Feeding behavior in *Macaca mulatta* on La Parguera Island. (*Photo by Evan Zucker, courtesy of the Carribean Primate Research Center*) Chapter 20

16. Grooming in the patas monkey (*Erythrocebus patas*). (*Photo by Evan Zucker, courtesy of the Carribean Primate Research Center*) Chapter 21

Plate 8

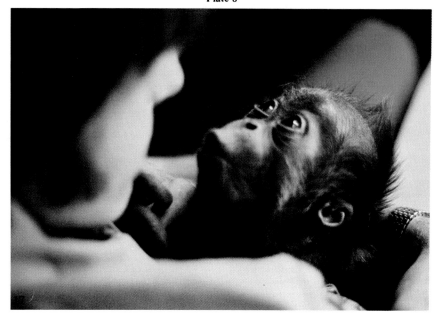

17. **An infant orang-utan (*Pongo pygmaeus*) vocalizing.** (*Photo by T. Maple*) Chapter 22
18. **Yawning in *Mandrillus sphinx*.** (*Photo by T. Maple*) Chapter 23

Plate 9

19. Rhesus monkey (*Macaca mulatta*) aggression. (*Photo by G. Mitchell*) Chapter 24
20. Aggression in the patas monkey (*Erythrocebus patas*). (*Photo by Evan Zucker, courtesy of the Carribean Primate Research Center*) Chapter 25

Plate 10

21. A vervet (*Cercopithecus aethiops*) displays a bizarre salute behavior. (*Photo by T. Maple*)
Chapter 26
22. A fear grimace in the rhesus monkey (*Macaca mulatta*). (*Photo by W. K. Redican*)
Chapter 27

Plate 11

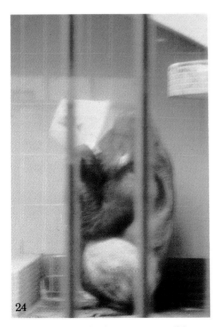

23. An adult male rhesus monkey (*Macaca mulatta*) with a facial wound. Chapter 28
24. A gorilla male uses a bucket to drink water. (*Photo by Evan Zucker*) Chapter 29
25. A chimpanzee female uses a towel for a shawl. (*Photo by T. Maple*) Chapter 29

26.a. and b. An orang-utan and a chimpanzee behave unpredictably and intelligently. (*Photos by T. Maple*) Chapter 32

27. *Homo sapiens* is the most self-aware of all the primates. (*Photo of Evan Zucker*) Chapter 33

never shoved or tugged at their mothers. Only the male infants were aggressive toward their mothers; the mothers occasionally responded in kind. Mothers of male infants also had to initiate grooming bouts with their infants more than did mothers of females. Nicolson (1977) noted that a male infant became independent of his mother before two female infants did.

As we already know, male chimpanzees leave their mothers and travel without them earlier than do females. In fact, females continue to travel with their mothers until they become sexually receptive and, even after this, they return to their mothers when not in estrus. Female infants also remain closer to the mother when with her, groom her more, and interact with younger siblings more than do males (Pusey, 1976).

There is evidence, however, that early experience with adults is not necessary for the development of sexually dimorphic and species-typical behavior patterns in chimpanzees. Experience with peers can facilitate sex differences (Nadler and Braggio, 1974).

Group play in chimpanzee infants, however, usually centers around the oldest male infant. Mothers play more with other infants than they do with their own. Mothers play with infants before infants play with each other; in fact, mothers actually encourage infants to play with each other by gently shoving them toward each other (Savage and Malick, 1977).

In a small cage environment, an adolescent male chimpanzee was seen to be always close to a mother and her infant. He frequently touched the tiny infant, sometimes stroking the infant's head (Nicolson, 1977).

Adult male chimpanzees can be very gentle with infants but occasionally a juvenile gets cuffed gently on the head when begging for food from a male. By the time a juvenile of either sex has reached adolescence, it has associated often with mature males. Males will not attack a female if she has an infant on her back. A full grown male chimpanzee will also protect an infant by charging straight toward a potential predator to pick up the infant. Kinships are definitely a factor in adult male chimpanzee paternalistic behaviors (cf. Mitchell, 1969).

Still, males are *rougher* with infants than are females. Juvenile males, for example, sometimes harass infants by pinching or biting them. There is also probably more sexual play in male juveniles and this sexual play is directed either at other infants or at mothers (Savage and Malick, 1977).

Despite this tendency for males to be rougher with infants than are females, there are some who believe chimpanzee females select males as mates on the basis of the male's caretaking abilities (Tutin, 1974). van Lawick-Goodall (1973) has found that adult males certainly do differ on such qualities as Tutin suggests:

In the chimpanzee, individuality is a very pronounced characteristic, probably rivalled only by personality differences in man. Thus, in a given social context, each of four different adult males may behave in a somewhat different and distinctive manner. (van Lawick-Goodall, 1973, P. 162)

2. Gorillas

The dominant silver-backed adult male gorilla occasionally grooms infants; and, although juveniles rarely interact with silver-backs, infants are attracted to them and often play on and around them. Adult males, however, usually carry infants only for short distances. Mothers carry the infants for much longer distances (cf. Mitchell, 1969). There has been an adult silver-backed male gorilla who has taken a large infant, left the main group, and remained alone with the infant for nearly a month (cited in Mason, 1964).

Baby gorillas travel on their mothers' backs until they are 2½ years old. Females and babies sit near the dominant male and the male sometimes accepts baby-sitting duties, tolerating as many as four infants clambering over him (Emlen and Schaller, 1963). Females answer the infants' calls more often than do males (Fossey, 1972).

Unlike chimpanzees, gorilla adults do not have sexual behavior with immature animals very often. Moreover, sexual behavior between immatures is also rare (Harcourt and Stewart, 1976). There may be sex differences in play and in infant independence, with males playing more and achieving independence earlier than female infants (Hoff, Nadler, and Maple, 1977); however, these early differences are only partially reflected in mature differences in peripheralization and group change. While subordinate males *are* found at the periphery of gorilla troops (Schaller, 1963), most *female* gorillas transfer from their parent group before maturity (Harcourt and Stewart, 1976).

At birth, gorilla males show interest in the newborn but remain at a distance. The presence of a male makes a female gorilla more attentive and a better mother. Infants show interest in adult males by the time they are seven months of age. When a ten to eleven-month-old infant was separated from its mother in captivity and then returned, the mother carried the infant to the adult male and allowed him to touch it (Nadler and Green, 1975).

In captivity, one silver-back male gorilla frequently took, or attempted to take, an infant from its mother. When he succeeded, according to Wilson et al. (1977), he carried the infant away awkwardly with the infant whimpering and the mother following. The infant was usually retrieved by the mother after a short time. The male adult gorilla seemed to prefer to take male infants.

3. Orangutans

According to Wilson et al. (1977), there are differences between gorilla and orangutan males in captivity which are largely qualitative. When an adult female with infant and an adult male are proximate, the female orangutan occasionally "presents" the infant to the male. The infant initiates most of the contact, including pulling on the male's hair. The adult male is tolerant of this. The orangutan infant does not appear to be as distressed in the male's care as does the gorilla infant (Wilson et al., 1977). At the Portland Zoo there was increasing paternal-infant interaction (both infant and male initiated) correlated with increasing infant independence from the mother (Lee, 1975). Extensive play between an adult male orangutan and his four-year-old juvenile offspring has been reported (cf. Zucker, Mitchell, and Maple, 1978).

In the wild (in Borneo or Sumatra) the only long-term social unit is the adult female and her dependent young. Adult males forage alone. Juveniles forage with increasing independence. Juvenile females stay with their mother for several years while juvenile males range further away at an earlier age than females. Solitary juvenile males are sometimes found in the wild (Horr, 1972). Okano (1971) describes such a young male playing with an adult female.

Play is seen more often in young males than in young females, but it *is* seen even in adult females. Mother and infant orangutans are closely attached. Okano(1971) speculates that at three to four years of age, juveniles gather to make play groups. At five to eight years of age males and females probably gather to make sexual groups. Okano believes that *both* males and females may live temporary solitary existences.

Horr (1977) has provided us with the most complete description of orangutan behavioral ontogeny in the wild. Most parental care comes almost completely from the mother, a situation that is unique among primates. Infants learn many things from the mother, such as nest building. In Borneo orangutans, adult males are unimportant in the defense of the young against predators; but, in Sumatra, several species prey on orangutans and the males stay closer to the females and young. In Borneo, nearby neighbors are most likely to be older female siblings of the mother or of the infant. Occasionally, they interact but certainly not as frequently as do "aunts" in other great ape species (Horr, 1977).

Mother orangutans share food with their infants and may actively feed them food or water. Older female siblings sometimes hold an infant but do not carry it. Mothers play idly with their infants for brief periods and so do older juvenile siblings. Juveniles very infrequently groom infants. "The burden is on the offspring imitating its mother rather than on any active socialization by the adult female." (Horr, 1977, P. 311.)

Contact with adult males for an infant orangutan in Borneo is practically nonexistent. As the infant becomes a juvenile, however, and if the mother has no younger infant, a male may copulate with the mother. Adult males either ignore or attack juveniles at this time in order to copulate with their mothers. Juvenile and subadult orangutan males, in particular, may have only negative experiences with adult males in Borneo (Horr, 1977).

D. THOUGHTS ABOUT HUMAN INFANT CARE

The fact that Borneo orangutans grow up having little or no contact with the adult male may remind us of some human individuals. In fact, Elias (1977) has made a direct comparison of human and orangutan mothers and their offspring. In both species, as infants grow older, proximity is maintained increasingly by the infant more than by the mother (primarily by vocal, facial, gestural, and locomotory means). There are many other similarities as well.

Of course, there *are* those that believe the adult male human is quite capable of providing good infant care. The fact that even orangutan males show gentle infant care in zoos and laboratories suggests that the allegedly more intelligent human male may be able to do even better. There have been several major publications on the role of the human father (for example, Biller, 1971; Lynn, 1974; Malinowski, 1927), including research reports on the human adult male's response to birth (Tanzer, 1973), and reports on the father's response to neonates and to infants in the first year of life (Lamb, 1977; Freudenberg and Driscoll, 1976; Parke and O'Leary, 1975; Sternglanz, Gray, and Murakami, 1977).

There has been much speculation about the biological or evolutionary origins of the human family. At one extreme is an explanation of sex roles for infant care in terms of what appear to be straightforward Darwinian rules. Tiger (1970), for example, has suggested that people might have evolved as large, dominant males, who form strong bonds with other males to help in organizing and managing the community, and as females oriented towards the propagation and stabilization of social structures. On the other hand, there are those who caution us that Old World monkeys evolved along with (in parallel with) the hominid line, that the outstanding behavioral characteristic of primates is variability, and that roles are individual behaviors not just age and sex characteristics (Burton, 1977).

Imanishi (1965) has speculated that human roles may have developed out of a need for incest avoidance. Many nonhuman primates accomplish incest avoidance by forming matriarchal mother–daughter kin groups and by the peripheralization and group changing of males. The females were left with the stable group to care for infants.

But human males *do* stay with the family. There may therefore have been some development of parental "drives" in human males; or, perhaps the elimination of estrus and substitution of diffuse receptivity in females kept the males near. On the other hand, perhaps the development of cultural transmission allowed people to make it traditional for males to help in infant care (cf. Etkin, 1963).

Much has been made of the fact that men hunt and that men hunt in groups. This orientation has drawn evolutionary theorists away from the male's role as a potential caretaker of infants. Tanner and Zihlman (1976), however, have attempted to reorient thought about the evolution of people by emphasizing the role of women in evolution. They have concluded that the role of females in human evolution has been critical because of:

1. the mother's *innovative* economic role, *gathering* with tools and *sharing* food with her offspring;
2. the mother's *social centrality* (she is the primary socializer and essential for kin selection, sexual selection, and *cultural transmission*); and
3. the mother's long investment in offspring, leading to her choice of a mate she considers to be most suitable for sexual interaction and infant care (most importantly *nonaggressive males*).

In human evolution, according to Zihlman's ideas, males developed small canine teeth and *less* aggressiveness than did male apes and most monkeys. Human males became closely integrated into their kin groups, a first step toward social "fatherhood." Males were thus both biologically *selected for* nonaggression and nurturance and *culturally* institutionalized (through cultural transmission or tradition) as parents (see Zihlman, in press).

But what actually occurs cross-culturally in *Homo sapiens*? Men are found less often with children than are women, even when men are free to be available to children. Groups of men who are with children tend to be smaller than groups of women who are with children. As children mature they tend to be found less often in groups of different gender. *Older* boys are found in men-only groups more than are younger boys (Mackey, in press; Mackey, 1977).

In studies of caretaking patterns for Kibbutz infants, father's visits were long visits especially to see their older boys (Gewirtz and Gewirtz, 1968). Using pupil dilation as an index of preference, adult females respond more than do men to infant faces (Sternglanz, Gray, and Murakami, 1977). Mothers of newborns smile at human infants more than do their fathers (Parke and O'Leary, 1975).

On the other hand in these same studies referred to above, men *are* seen with children (Mackey, 1977); men have the same preferences for type of infant face as do women (small chin, large forehead, large eyes); there are no significant differences in America in baby-sitting experiences for males and females up to the age twenty years (Sternglanz, Gray, and Murakami, 1977); human infants show no preference for either parent up through thirteen months of age; human infants respond *more* positively to play with the father than to play with the mother; women more often perform caretaking functions in holding infants whereas men more often have a play function (Lamb, 1977); in a hospital rooming-in setting, fathers actually hold and rock the newborn more than do mothers; and, if the mother is not present, the father is more likely to interact with the infant (Parke and O'Leary, 1975). In addition to all of this, there is a growing trend (of a cultural or institutionalized nature) for the modern father to play a greater part in child rearing. There is evidence that:

> . . . conditions *external* to the family are intimately involved in the pattern of parental behavior that determines personality and characterological outcomes for children. (Kemper and Reichler, 1976, P. 218).

This is particularly true in the development of gender differences.

At the other extreme, with regard to violence against children, a high proportion of households headed by women but with no biological father are involved in child abuse (Gil, 1970); and, more mothers than fathers kill their children. In a twenty-five year study of homocides involving preadolescent victems, nearly half of the children were killed by their mothers, the other half by their fathers and/or other males (Myers, 1967). This fact alone most assuredly makes *Homo sapiens* an extremely unique primate.

REFERENCES

Badham, M. A note on breeding the spectacled leaf monkey at Twycross Zoo. *International Zoo Yearbook,* 1967, **7**, 89.

Berkson, G. Development of an infant in a captive gibbon group. *Journal of Genetic Psychology,* 1966, **108**, 311-325.

Bernstein, I. S. The lutong of Kuala Selangor. *Behaviour,* 1968, **32**, 1-16.

Bernstein, I. S. The influence of introductory techniques on the formation of captive mangabey groups. *Primates,* 1971, **12**, 33-44.

Bernstein, I. S. Activity patterns in a gelada monkey group. *Folia Primatologica,* 1975, **23**, 50-71.

Biller, H. *Father, Child, and Sex Role.* Lexington, Mass.: Heath, 1971.

Bingham, H. C. Parental play of chimpanzees. *Journal of Mammalogy,* 1927, **8**, 77-89.

Boggess, J. Social change in a troop of langurs in Nepal. Paper presented at the *American Association of Physical Anthropologists* Meeting, Seattle, Washington, April, 1977.

Bolwig, N. A study of the behavior of the chacma baboon, *Papio ursinus. Behaviour,* 1959, **14**, 136-163.

Bourliére, F., Bertrand, M. and Hunkeler, C. L'ecologie de la mone de Lowe (*Cercopithecus campbelli Lowei*) en cote d'ivoire. *Extrait de la Terre et la Vie,* 1969, **2**, 135-163.

Bramblett, C. *Patterns of Primate Behavior.* Palo Alto, California: Mayfield, 1976.

Burton, F. D. Ethology and the development of sex and gender identity in non-human primates. *Acta Biotheoretica,* 1977, **26**, 1-18.

Chalmers, N. R. The social behaviour of free living mangabeys in Uganda. *Folia Primatologica,* 1968, **8**, 263-281.

Chalmers, N. R. Changes in mother-infant behavior following changes in group composition in Sykes and patas monkeys. *Proceedings of the Third Congress of the International Primatological Society* (Zurich, 1970), 1971, **3**, 116-120.

Chapman, M. and Hausfater, G. Infanticide in langurs, a mathematical analysis. Paper presented at the *American Society of Primatologists* Meeting, Seattle, Washington, April, 1977.

Chivers, D. J. Spatial relations within the siamang group. *Proceedings of the Third Congress of the International Primatological Society* (Zurich, 1970), 1971, **3**, 14-21.

Chivers, D. J. and Chivers, S. T. Events preceding and following the birth of a wild siamang. *Primates,* 1975, **16**, 227-230.

Clark, C. B. A preliminary report on weaning among chimpanzees of the Gombe National Park, Tanzania. In Chevalier-Skolnikoff, S. and Poirier, F. E. (Eds.) *Primate Bio-Social Development: Biological, Social, and Ecological Determinants.* New York: Garland, 1977, pp. 235-260.

DeVore, I. Mother-infant relations in free-ranging baboons. In Reingold, H. L. (Ed.) *Maternal Behavior in Mammals.* New York: Wiley, 1963, pp. 305-335.

Dolan, K. J. Meta communication in the play of a captive group of Sykes monkeys. Paper presented at the *American Association of Physical Anthropologists* Meeting, St. Louis, Missouri, April, 1976.

Dolhinow, P. J. Normal monkeys? *American Scientist,* 1977, **65**, letter to editor. (a)

Dolhinow, P. J. Caretaking patterns of the Indian langur monkey. Paper presented at the *American Association of Physical Anthropologists* Meeting, Seattle, Washington, April, 1977. (b)

Dolhinow, P. J. and Bishop, N. The development of motor skills and social relationships among primates through play. In Hill, J. P. (Ed.) *Minnesota Symposium of Child Psychology* (Vol. 4). Minneapolis, Minn.: University of Minnesota Press, 1970, pp. 141-198.

Dunbar, R. I. M. and Dunbar, P. Behavior related to birth in wild Gelada baboons (*Theropithecus gelada*). *Behaviour,* 1974, **50**, 185-191.

Elias, M. E. Proximity between mothers and their offspring: Human and orangutan (*Pongo pygmaeus*). Paper presented at the *Animal Behaviour Society* Meeting, University Park, Pennsylvania, June, 1977.

Ellefson, J. O. Group size and behavior in white-handed gibbons, *Hylobates lar*. Paper presented at the *American Association of Physical Anthropologists* Meeting, Berkeley, California, April, 1966.

Emlen, J. T. and Schaller, G. B. In the home of the mountain gorilla. In Southwick, C. H. (Ed.) *Primate Social Behavior*. Princeton, N.J.: Van Nostrand, 1963, pp. 124-135.

Etkin, W. Social behavioral factors in the emergence of man. *Human Biology*, 1963, **35**, 299-310.

Fairbanks, L. A. Animal and human behavior: Guidelines for generalization across species. In McGuire, M. T. and Fairbanks, L. A. (Eds.) *Ethological Psychiatry: Psychopathology in the Context of Evolution Biology*, 1977, pp. 87-110.

Fedigan, L. Social and solitary play in a colony of vervet monkeys, *Cercopithecus aethiops. Primates*, 1972, **13**, 347-364.

Fossey, Dian. Vocalizations of the mountain gorilla (*Gorilla gorilla beringei*). *Animal Behaviour*, 1972, **20**, 36-53.

Freudenberg, R. and Driscoll, J. W. Reactions of adult humans to cries of normal and abnormal infants. Paper presented at the *Animal Behavior Society* Meeting, Boulder, Colorado, June, 1976.

Gartlan, J. S. Sexual and maternal behavior of the vervet monkey, *Cercopithecus aethiops. Journal of Reproduction and Fertility* (Supplement), 1969, **6**, 137-150.

Gewirtz, H. B. and Gewirtz, J. L. Visiting and caretaking patterns for Kibbutz infants: Age and sex trends. *American Journal of Orthopsychiatry*, 1968, **38**, 427-443.

Gil, D. G. *Violence Against Children.* Cambridge, Mass.: Harvard University Press, 1970.

Gilmore, H. A. The evolution of "agonistic buffering" in baboons and macaques. Paper presented at the *American Association of Physical Anthropologists* Meeting, Seattle, Washington, April, 1977.

Goodall, J. Chimpanzees of the Gombe Stream Reserve. In DeVore, I. (Ed.) *Primate Behavior.* New York: Holt, Rinehart and Winston, 1965, pp. 425-473.

Hall, K. R. L. and Mayer, B. Social interactions in a group of captive patas monkeys (*Erythrocebus patas*). *Folia Primatologica*, 1967, **5**, 213-236.

Harcourt, A. H. and Stewart, K. J. Sexual behaviour of wild mountain gorilla. Paper presented at the *International Primatological Society* Meeting, Cambridge, England, August, 1976.

Hill, C. A. Infant sharing in the family Colobidae emphasizing *Pygathrix. Primates*, 1972, **13**, 195-200.

Hoff, M. P., Nadler, R. D. and Maple, T. The development of infant social play in a captive group of gorillas. Paper presented at the *American Society of Primatologists* Meeting, Seattle, Washington, April, 1977.

Horr, D. A. The Borneo orang-utan. *Borneo Research Bulletin,* 1972, 4, 46-50.

Horr, D. A. Orang-utan maturation: Growing up in a female world. In Chevalier-Skolnikoff, S. and Poirier, F. E. (Eds.) *Primate Bio-Social Development: Biological, Social and Ecological Determinants.* New York: Garland, 1977, pp. 289-322.

Horwich, R. H. Development of behaviors in a male spectacled langur (*Presbytis obscurus*). *Primates,* 1974, 15, 151-178.

Horwich, R. H. and Manski, D. Maternal care and infant transfer in two species of *Colobus* monkeys. *Primates,* 1975, 16, 49-63.

Horwich, R. H. and Wurman, C. The ontogenetic implications of infant birth on troop sociality in *Colobus* monkeys. Paper presented at the *Animal Behavior Society* Meeting, University Park, Pennsylvania, June, 1977.

Hrdy, S. B. Male-male competition and infanticide among the langurs (*Presbytis entellus*) of Abu, Ragasthan. *Folia Primatologica,* 1974, 22, 19-58.

Hrdy, S. B. Care and exploitation of nonhuman primate infants by conspecifics other than the mother. In Rosenblatt, J., Hinde, R., Shaw, E. and Beer, C. (Eds.) *Advances in the Study of Behavior* (Vol. 6). New York: Academic Press, 1976, pp. 101-158.

Hrdy, S. B. Infanticide as a primate reproductive strategy. *American Scientist,* 1977, 65, 40-49. (a)

Hrdy. S. B. Reply to Dolhinow. *American Scientist,* 1977, 65, letter to editor. (b)

Hrdy, S. B. and Hrdy, D. B. Allomaternal choice of charges among Hanuman langurs. Paper presented at the *American Association of Physical Anthropologists* Meeting, Seattle, Washington, April, 1977.

Hunkeler, C., Bourliere, F. and Bertrand, M. Le comportement social de la Mone de Lowe (*Cercopithecus campbelli lowei*). *Folia Primatologica,* 1972, 17, 218-236.

Hunter, J. L. Guardian behavior in the sooty mangabey (*Cercocebus atys*). Paper presented at the *Animal Behavior Society* Meeting, University Park, Pennsylvania, June, 1977.

Imanishi, K. The origin of the human family: A primatological approach. In Imanishi, K. and Altmann, S. (Eds.) *Japanese Monkeys.* Atlanta, Georgia: Altmann, 1965, pp. 113-140.

Jay, P. The Indian langur (*Presbytis entellus*). In Southwick, C. H. (Ed.) *Primate Social Behavior.* Princeton, N.J.: Van Nostrand, 1963, pp. 114-123.

Kemper, T. D. and Reichler, M. L. Father's work integration and types and frequencies of rewards and punishments administered by fathers and mothers to adolescent sons and daughters. *Journal of Genetic Psychology,* 1976, **129**, 207-219.

Kollar, E. J., Beckwith, W. C. and Edgerton, R. B. Sexual behavior of the ARL colony of chimpanzees. *Journal of Nervous and Mental Disease.* 1968, **147**, 444-459.

Kortlandt, A. Experimentation with chimpanzees in the wild. In Stark, D., Schneider, R. and Kuhn, H. (Eds.) *Progress in Primatology.* Stuttgart: Gustav Fischer, 1967, pp. 208-224.

Krige, P. D. and Lucas, J. W. Aunting behaviour in an urban troop of *Cercopithecus aethiops. Journal of Behavioural Science,* 1974, **2**, 55-61.

Lamb, M. E. Father-infant and mother-infant interaction in the first year of life. *Child Development,* 1977, **48**, 167-181.

Lancaster, J. B. Play-mothering: The relations between juvenile females and young infants among free-ranging vervet monkeys (*Cercopithecus aethiops*). *Folia Primatologica,* 1971, **15**, 161-182.

Lancaster, J. B. Stimulus/response: In praise of the achieving female monkey. *Psychology Today,* 1973, **7**, 30-102.

Lee, M. Q. Paternal behavior in the orang-utan (*Pongo pygmaeus*). Paper presented at the *Oregon-Washington Psychology* Meeting, Spring, 1975.

Lynn, D. B. *The father: His role in child development.* Monterey, California: Brooks-Cole, 1974.

Mackey, W. Parameters of the man-child bond in an evolutionary perspective. *American Journal of Physical Anthropology,* 1977, in press.

Mackey, W. C. The adult male-child bond: A cross-cultural analysis. Paper presented at the *Animal Behavior Society* Meeting, University Park, Pennsylvania, June, 1977.

Malinowski, B. *The Father in Primitive Psychology.* New York: Norton, 1927.

Mason, W. A. Sociability and social organization in monkeys and apes. In Berkowitz, L. (Ed.) *Recent Advances in Experimental Social Psychology* (Vol. 1). New York: Academic Press, 1964, pp. 277-305.

Mitchell, G. Paternalistic behavior in primates. *Psychological Bulletin,* 1969, **71**, 339-417.

Myers, S. A. The child slayer: A 25 year study of homocides involving preadolescent victims. *Archives of General Psychiatry,* 1967, **17**, 211-213.

Nadler, R. D. and Braggio, J. T. Sex and species difference in captive-reared juvenile chimpanzees and orang-utans. *Journal of Human Evolution,* 1974, **3**, 541-550.

Nadler, R. D. and Green, S. Separation and reunion of the gorilla (*Gorilla gorilla gorilla*). *International Zoo Yearbook,* 1975, **15**, 198-201.

Nicolson, N. A. A comparison of early behavioral development in wild and captive chimpanzees. In Chevalier-Skolnikoff, F. E. (Ed.) *Primate Bio-Social Development.* New York: Garland, 1977, pp. 529-562.

Okano, J. A preliminary observation of orang-utans in the rehabilitation station in Sepilok, Sabah. *The Annual of Animal Physiology,* 1971, **21**, 55-67.

Palthe, T. V. W. and van Hooff, J. A. R. A. M. A case of the adoption of an infant chimpanzee by a suckling foster chimpanzee. *Primates,* 1975, **16**, 231-234.

Parke, R. D. and O'Leary, S. Father-mother-infant interaction in the newborn period: Some findings, some observations, and some unresolved issues. In Riegel, K. and Meacham, J. (Eds.) *The Developing Individual in a Changing World* (Vol. 2): *Social and Environmental Issues.* The Hague: Mouton, 1975, pp. 653-663.

Poirier, F. E. Behavioral flexibility and intertroop variation among Nilgiri langurs (*Presbytis johnii*) of South India. *Folia Primatologica,* 1969, **11**, 119-133.

Poirier, F. E. The Nilgiri langur (*Presbytis johnii*) of South India. In Rosenblum, L. A. (Ed.) *Primate Behavior.* New York: Academic Press, 1970, pp. 251-383.

Pusey, A. E. Sex differences in the persistence and quality of the mother offspring relationship after weaning in a population of wild chimpanzees. Paper presented at the *International Primatological Society* Meeting, Cambridge, England, August, 1976.

Raleigh, M. J. Sex differences in the social behavior of juvenile vervet monkeys. Paper presented at the *American Society of Primatologists* Meeting, Seattle, Washington, April, 1977.

Rhine, R. J. and Owens, N. W. The order of movement of adult male and black infant baboons (*Papio anubis*) entering and leaving a potentially dangerous clearing. *Folia Primatologica,* 1972, **18**, 276-283.

Rowell, T. E., Din, N. A. and Omar, A. The social development of baboons in their first three months. *Journal of Zoology* (London), 1968, **155**, 461-483.

Rudran, R. Adult male replacement in one-male troops of purple-faced langurs (*Presbytis senex senex*) and its effect on population structure. *Folia Primatologica,* 1973, **19**, 166-192.

Savage, E. S. and Malick, C. Play and sociosexual behavior in a captive chimpanzee (*Pan troglodytes*) group. *Behaviour,* 1977, **60**, 179-194.

Schaller, G. *The Mountain Gorilla.* Chicago: University of Chicago Press, 1963.

Segre, A. Talapoins. *Animal Kingdom,* 1970, **73** (3), 20-25.

Sternglanz, S. H., Gray, J. L. and Murakami, M. Adult preference for infantile facial features: An ethological approach. *Animal Behaviour,* 1977, **25**, 108-115.

Stevenson, M. F. Observation of maternal behaviour and infant development in DeBrazza monkey, *Cercopithecus neglectus. International Zoo Yearbook,* 1973, **13**, 179-184.

Struhsaker, T. T. Social behaviour of mother and infant vervet monkeys (*Cercopithecus aethiops*). *Animal Behaviour,* 1971, **19**, 233-250.

Struhsaker, T. T. Infanticide in the redtail monkey (*Cercopithecus ascanius schmidti*) of Uganda. Paper presented at the *International Primatological Society* Meeting, Cambridge, England, August, 1976.

Sugiyama, Y. An artificial social change in a Hanuman langur troop (*Presbytis entellus*). *Primates,* 1966, **7**, 41-72.

Tanner, N. and Zihlman, A. Women in evolution Part I: Innovation and selection in human origins. *Signs: Journal of Women in Culture and Society,* 1976, **1** (3), 585-608.

Tanzer, D. Natural childbirth: Pain or peak experience. In Tavris, C. (Ed.) *The Female Experience.* New York: Ziff-Davis, 1973.

Tiger, L. The possible biological origins of sexual discrimination. *Impact of Science on Society,* 1970, **20**, 29-44.

Tutin, C. E. G. Exceptions to promiscuity in a feral chimpanzee community. Paper presented at the *International Primatological Society* Meeting, Nagoya, Japan, August, 1974.

Ullrich, W. Geburt und naturliche Geburtschilfe beim Orang-utan. *Der Zoologishe Garten,* 1970, **39**, 284-289.

van Lawick-Goodall, J. Cultural elements in a chimpanzee community. In Menzel, E. W. (Ed.) *Symposia of the Fourth International Congress of Primatology* (Vol. 1): *Precultural Primate Behavior.* Basel: Karger, 1973, pp. 144-184.

Wilson, M. E., Maple, T., Nadler, R. D., Hoff, M and Zucker, E. Characteristics of paternal behavior in captive orangutans (*Pongo pygmaeus abelii*) and lowland gorillas (*Gorilla gorilla gorilla*). Paper presented at the *American Society of Primatologists* Meeting, Seattle, Washington, April, 1977.

Young, G. H. Paternal influences on the development of behavioral sex differences in infant baboons. Paper presented at the *American Association of Physical Anthropologists* Meeting, Seattle, Washington, April, 1977.

Zihlman, A. Motherhood in transition: From ape to human. In Miller, W. and Newman, L. (Eds.) *First Child and Family Formation.* Chapel Hill, North Carolina: University of North Carolina Press, in press.

Zucker, E. L., Mitchell, G. and Maple, T. Adult male-offspring play interactions within a captive group of orang-utans (*Pongo pygmaeus*). *Primates,* 1978, **19** (2), 279-384.

14.

The Role of Gender in Social Spacing and Structure: Prosimians, New World Monkeys, and Macaques

As we have seen in previous chapters, gender-related variables such as physical sexual dimorphism, hormonal levels, pubescence, breeding season, estrus, birth season, and ontogenetic parent–infant relations can affect the social spacing and group structure.   For example, physical sexual dimorphism is significantly related to socionomic sex ratio and shows some relationship to body size. That is, the larger the male of a species relative to the female, the more females there will be relative to males in any given group and the larger will be the overall body size of the species regardless of sex (cf. Clutton-Brock and Harvey, 1976).

Sexual dimorphism also correlates positively, in a general way, with group size and negatively with arboreality (Jorde and Spuhler, 1974).   Relatively monomorphic species like marmosets, tamarins, titi monkeys, night monkeys, gibbons, and siamangs tend to live in small family groups (Kleiman, 1977). Group movements in small monogamous groups have a "remarkable level of synchrony" (Kleiman, 1977).   In chimpanzees, by contrast, group size is the most variable of all the primates (Marler, 1968).

In general, in most species of primates, where monogamy is not the rule: Adult females are often more closely spaced than are adult males (e.g., rhesus) although some males who are attached to one another may remain close (Marler, 1968).   In many instances physical proximity is correlated with aggression, particularly in males (Marler, 1976).

In most primate species, as we have seen (Poirier, 1973):

> Males are soon forced from their mothers out to the peer group where they mature and become less dependent upon the females. (P. 29)

Such males often come to live at the periphery of the troop, and may even become solitary or join another troop in adulthood.

But not all primate groups have a social structure like that just described. In some species there are few signs of sex differences in proximity, in peripheralization, or in group change. Southwick and Siddiqi (1974) have published a summary of the many different kinds of primate organization. Kummer (1971) has also written about some very different kinds of primate societies.

Let us now examine the social structures of a few different species of prosimians, paying particular attention to gender as a factor in the spacing and structure *within* a group.

A. PROSIMIANS

Some species of tree shrews (*Tupaia spp.*) live in polygamous family groups with male dominance while others appear to be asocial (Sorenson, 1974). In the aye-aye (*Daubentonia*) one male and one female live in a home range of twelve acres but the two are never observed together. They are apparently strictly solitary animals having very little communication between individuals. Although the vocal call of one is always followed by the call of the other (up to 100 yards apart), the calls evidently serve to *avoid* contact (Petter and Petter, 1967).

In bushbabies (*Galago senegalensis*), animals are found alone, in pairs, or in large groups. The "natural" group is thought to be a family group which is more open than that of the New World titi monkey (Doyle, 1974). Little is known concerning sex differences in spacing. It is likely, however, that both males and females show some avoidance of strange animals of the same sex. Associated females are usually related and there are usually more females than males in a *Galago* group. In *Galago alleni* there are six to eight females to every male with a pair formation bond between each female and the male. The male "visits" each female in turn (Charles-Dominique, 1977).

It was once assumed that nocturnal prosimians, like the slow loris (*Nycticebus coucang*) were relatively solitary nonsocial animals, but this is not necessarily true (Ehrlich and Musicant, 1977). Slow lorises live in "spatial harems" in which there is no discernable dominance hierarchy. However, there are both central and peripheral males.

Among prosimians, *Daubentonia* and *Lepilemur* are solitary, *Microcebus* apparently form sleeping groups, while *Propithecus* and *Indrii* form family groups. Lemurs, on the other hand, often form very large groups. Having an adult male present increases the rate at which an infant lemur becomes independent (Klopfer, 1972).

But even among the lemurs there are twenty-six nocturnal, three nocturnal-crepuscular (active at twilight), and nine diurnal or diurnal-crepuscular species. The social structures are quite varied. Charles-Dominique (1976) has pointed out that there are even some solitary forms but that solitary does not mean asocial.

All of the solitary species he studied are social, although the social relations are based upon distance communication using auditory and, more importantly, olfactory messages. Sex differences in marking are sometimes found.

When *Microcebus murinus* are put into groups of around eight animals with an approximately equal number of males and females, only one of the males mates. The other males are peripheralized and live at the fringes of the group (Perret, 1977). Petter and Petter (1968) believe that this nocturnal species of lemur along with *Cheirogaleus medius,* is solitary. *Hapalemur griseus,* however, is diurnal and lives in groups of two to ten individuals.

Murrell and Young (1977) have reported on two crepuscular species of lemurs (*Lemur mongoz* and *Varecia vareigatus*). The *Lemur mongoz* species spend ten times as much time in direct contact as do *Varecia vareigatus* (ruffed lemurs). The contact includes higher levels of grooming and huddling both of which are almost absent in the ruffed lemur. Ruffed lemurs are solitary and come together only for the breeding season whereas *Lemur mongoz* (the mongoose lemurs) are apparently monogamous. Tattersall and Sussman (1975) also say that *L. mongoz* is monogamous; however, they claim that the species is nocturnal not crepuscular. Very little else is known about sex differences in social spacing and structure in the prosimian suborder.

B. NEW WORLD MONKEYS

Vessey, Mortenson, and Muckenhirn (1976) examined the size and structure of eight different species of New World monkeys in 109 groups. The bearded Saki (*Chiropotes satanus*) was found in groups of four to twenty with some groups containing more than one sexually mature male. The white-faced Saki (*Pithecia pithecia*) occurred in groups of one to five with only one adult male and one or two adult females with young. One group, however, contained two adult males and three adult females. Red-handed tamarins (*Saguinas midas*) were found in an average group size of 5.2 individuals, while squirrel monkeys (*Saimiri sciureus*) were found in groups averaging thirty-one animals.

1. Squirrel monkeys (*Saimiri spp.*)

At the Monkey Jungle in Florida, all age groups of animals in a squirrel monkey troop were attracted to adult females. Adult males were not involved in most of the troops' activities (except during the mating season) and they traveled at the periphery of the troop (Baldwin, 1968). *Saimiri* have long been known to form isosexual (like sex) subgroups (Alvarez, 1975; Mason and Epple, 1968) especially squirrel monkey females. If two groups of squirrel monkeys are combined, collective aggression (coalition) appears and the females play the most active part in the aggressive behavior which is directed primarily toward strange males (Castell, 1969). According to Kaplan (1970) female infants become independent of their mothers earlier than do male infants and this can be related to the adult behavior of the females.

Baldwin and Baldwin (1972) studied a different species of squirrel monkey (*S. oerstedi*) in a natural forest in western Panama and noted that the two full adult males in their troop traveled at the edge of the troop and that the one young adult male traveled even further at the periphery of the troop. The two adult males were chased out by the adult females, while the young adult male was chased out by the two adult males.

Returning to *Saimiri sciureus,* the higher the male's position or rank the greater his preference to remain alone and avoid females. Preference for females is evidently a characteristic of submissive males (Candland et. al., 1973). Adult females engage in more affiliative behavior in their isosexual groups than do adult males (Strayer, Taylor, and Yanciw, 1975). The like-sex preference of adult female squirrel monkeys is so great that they clearly prefer female *strangers* over male strangers and over their own males (Mason, 1975).

If female squirrel monkeys receive estradiol it increases their affiliative behavior with males; if they receive progesterone it decreases their contact with males. The seasonal changes in heterosexual contact are evidently hormonally mediated (Anderson and Mason, 1977). Most of the following of the mate is done by the male, however (Fragaszy, 1977a). Squirrel monkeys do not show distress when their sexual partner is separated from them (Fragaszy and Mason, 1978).

2. Titi Monkeys (*Callicebus spp.*)

Titi monkeys, as we know, live in monogamous pairs in which the male participates in infant care (Mason, 1966). There is a strong and enduring emotional bond between the male and female with evidence of jealousy, particularly in the male (Cubicciotti and Mason, 1977). However, most of the following is done by the female titi; and, it is the female that is primarily responsible for the close spatial coordination of the pair. Social calls for contact, however, are emitted more by the male (Fragaszy, 1977a). The spacing between male and female titi monkeys is *much* closer than it is between male and female squirrel monkeys (Fragaszy, 1977b). Sex differences in spacing and group structure are more pronounced in *Saimiri* than in *Callicebus* (Mason, 1975). While female squirrel monkeys in captivity sometimes prefer strangers to male cage-mates, titi monkeys always prefer their familiar mates (Mason, 1975).

Despite the close spatial arrangements of *Callicebus* monkeys, there is some variability within the genus with regard to the size of the home range. *Callicebus torquatus,* for example, has a home range that is forty times larger than the home range of *Callicebus moloch* (Kinzey et al., 1977).

3. Howler Monkeys (*Alouatta spp.*)

In Argentina, the sex ratio of the howler monkey (*Alouatta caraya*) is one to one. Males, however, are much larger than females (Pope, 1966). This sex ratio is much different than the sex ratio for *Alouatta palliata* where there is often a preponderance of females. Howler monkeys use urine-rubbing to promote intra-troop cohesion and adult male animals urine-rub more than do adult females (Milton, 1975). There are peripheral males in howler monkey groups that serve to establish and maintain the sex ratio most appropriate for the species. In mantled howlers (*A. palliata*) this means fewer males than females; hence, there is a tendency for some males to become isolated from the troop (Scott, Malmgren, and Glander, 1976).

In a third howler species (*Alouatta seniculus*) males are also more loosely attached to the troop than are females but the adult sex ratio is about even. Solitary males are occasionally seen and males in the troop sleep away from matrifocal groupings (Neville, 1972).

4. Spider Monkeys (*Ateles spp.*)

Arboreal spider monkeys show little sexual dimorphism and males of the same group are less aggressive toward each other than are males of many multimale terrestrial primates (Klein and Klein, 1971; Eisenberg, 1966). Specific adult males do not serve as a focus for social activity for females and subgroups involving several males are common within a troop (Eisenberg and Kuehn, 1966). There is more aggression between the sexes than within the sex and more positive associations between like-sexed subjects (see Rondinelli and Klein, 1976). Olfactory marking appears to be of some importance for social spacing.

There are some interesting ecological correlates of spacing in spider monkeys. As altitude increases, group size diminishes, and the relative numbers of adult females and subadult/juveniles per group decreases (Durham, 1971).

5. Other New World Monkeys

Among *Cebus capucinus,* young males may leave the group for several years and join new groups. Solitary males have been seen. Urine-marking is used and probably affects spacing and group structure (Oppenheimer, 1969). The sex ratio for *Cebus nigrivittatus* is one to one (Oppenheimer and Oppenheimer, 1972).

Woolly monkeys (*Lagothrix lagothrica*) form the largest groups of South American monkeys except for squirrel monkeys. Groups range from twenty to seventy individuals with a large number of adult males (Nishimura and Izawa, 1975). The number of males or the male-to-female ratio is large compared to other multimale groups of New World monkeys (Nishimura and Izawa, 1975).

C. RHESUS MACAQUES (*MACACA MULATTA*)

Vandenbergh (1967) has studied the development of social structure in free-ranging rhesus monkeys. Female rhesus experience less difficulty in forming stable social units than do males. Bands in the wild typically develop from a basic group of adult females dominated by an adult male. The basic group of adult females is the most stable portion of the band and determines the rank of the band relative to other bands, but the presence of a dominant adult male is essential for the long-term survival of the band as a social unit. Males, however, sometimes remain solitary or form all-male groups, some for as long as 2½ years. The absence of kinship ties to nearby groups may be important in the maintenance of such all-male groups.

As females become adult, they remain with the group and hold a rank just below their mothers; as males become adult, they first become peripheralized (at puberty), and may gain or lose rank or leave the natal troop (Sade, 1967). These tendencies for peripheralization are seen as early as thirty weeks of age (cf. Hinde and Spencer-Booth, 1967). Juvenile and subadult males interact the least frequently with other age/sex classes than do other groups of animals (Altmann, 1968). Males and infants tend to avoid each other.

In two captive groups of rhesus monkeys connected by a tunnel, peripheral and immature males were the most frequent participants in intergroup activity (Marsden, 1968). After removal of two peripheral males (from one group) who made most of the intergroup visits, major movement in the tunnel between the two troops reversed in direction, and the relative dominance of the two troops also reversed. Reintroduction of the two males reversed the dominance again (Marsden, 1969). Apparently intergroup contact initiated by peripheral males helps to determine relative dominance between two groups.

Gender preference for individual rhesus monkeys develops in a curious way in the laboratory. Prior to seven months of age, neither sex shows any preference for one gender or another. After seven months, animals show a preference for interacting with their own sex. There is then a subsequent shift, earlier in females, toward a preference for the other sex, and a final shift after 3½ years of age back to a preference for a like-sexed animal (Suomi, Sackett, and Harlow, 1970). It is interesting that it is at around 3½ years of age that males are peripheralized and interact in all-male groups.

When rhesus males are removed by people from their free-ranging troops for periods of up to 103 days and are then reintroduced to the same troop, about one-half of them fail to rejoin their original groups and some of them remain solitary. Only one of nine females separated from her troop failed to rejoin it (Vessey, 1971). Laboratory studies have also indicated that bonds between females are stronger than are those between males (Erwin, Mobaldi, and Mitchell, 1971). In addition, when preadolescent female rhesus are brought into the laboratory from the wild, the stress of breaking their kinship groups often results in retarded sexual development and aberrant menstrual cycles (Mahoney, 1975).

In the wild, adult females and subadult males chase off solitary adult males that attempt to join their stable troop if the troop contains no adult males. Such a troop shows increased mobility, however. Evidently the troop leaders emerge from within the troop's ranks in such a case (Neville, 1968).

Changes in group membership in free-ranging rhesus are made almost exclusively by males (Drickamer, 1974), particularly by three- and four-year old males. The changes occur most often during the fall breeding season. Peripheral all-male subgroups are preferred by males who change groups, perhaps because siblings or peer-group associates are to be found in those subgroups, i.e., they join a new group by forming an affectional relationship with another male (Boelkins and Wilson, 1972).

In free-ranging rhesus troop dominance, the actions of peripheral males are also important in effecting the reversal of the relative dominance of two troops (Gabow, 1973). This agrees well with the tunnel-connected group studies done in captivity (Marsden, 1969). The adult female core of each group, however, also plays a key role in determining the bands' dominance status and its patterns of space utilization (Drickamer, 1975).

Wade (1976) studied the effects of strangers on rhesus monkey groups and concluded that:

> . . . females tend to prefer as alliance partners, in order, males most, then familiar females, and unfamiliar females least, while males tend to prefer less familiar over more familiar females. The latter preference may be a reason for inter-troop transfer by males in free-living troops. There was a decreased tolerance among females, especially unfamiliar ones, in the presence of males. Such a mechanism might be a contributor to the stability of troop affiliation in free-living females. (P. 212)

At about three years of age, consort activity is high in both male and female rhesus. At four and five years, consort activity decreases in both males and females. Males join all-male peripheral groups, become solitary, or join nonnatal groups at four or five years of age; and, the birth of a female's first infant occurs at four or five years of age. The infant's presence may inhibit consort activity. If a male stays in his natal group, he rises in rank over his mother and may even form a consort with her (Missakian-Quinn and Varley, 1977).

D. OTHER MACAQUES

1. Japanese Macaques (*M. fuscata*)

Japanese monkeys have a social spacing and social structure similar to that of the rhesus monkey. The social distance, first from the matrifocal kinship group, then from the natal troop, gradually increases with age for Japanese macaque males (Kortmulder, 1974). However, in a comparison of the social structures of five natural groups of Japanese monkeys on Shodoshima Island, there were large differences in numbers of males and females and in the age distribution of males across the five troops (Yamada, 1971).

At Ryozemyama, nineteen out of twenty-three males left the troop at four or five years of age and three others left before the age of twelve years. Not one of the twenty-three males stayed in his mother's troop until senescence. Most of the subadult and young adult males did not desert the troop alone but rather in groups of two or three males. A few years of solitary life, or free movement with other males, follows until about age seven when the male approaches a strange troop during mating season, mates, and then either joins the new troop or leaves. If he joins, he is likely to leave again within a few years. Males usually die at about fifteen to seventeen years of age (Sugiyama and Ohsawa, 1974).

Agonistic behavior between six- to nine-year-old males increases during the breeding season. Five-year-old males usually have positive social contacts with older brothers outside of their own troop (Suguwara, 1976). The importance of variables like age, sex, kinship, and spacing on dominance status is considerable. Many males have what is called a "dependent rank." Their dominance rank depends on whom they know (Quick, 1977).

While troop shifts usually occur in the mating season, male solitarization occurs throughout the year. The longer a male stays in a troop, and the older he is, the greater his dominance (Norikoshi and Koyama, 1975). Nishida (1966) has published an extensive study of solitary male monkeys.

As part of the general trend toward the avoidance of incest, there is an avoidance of contacts between adult females and young males (Mori, 1975). Preference data in the laboratory in rhesus suggest that the same thing occurs in *M. mulatta* (Suomi, Sackett, and Harlow, 1970). Japanese macaque males between four and seven years old do not have the bright red or scarlet face and hips seen in adult males who are at least eight years old (Masui et al., 1973). It is, therefore, quite easy for an adult female the detect the nonred males and to avoid them.

In contrast to young males, the dominant central (and red) males move with the females and have a "group focus role" in relation to group cohesion (Kurokawa, 1975). The relative dominance between troops reflects the relative rankings of the leader males of the troops (Koyama, 1970). When sweet potatoes are thrown into a troop, females catch most of them, perhaps because it is not necessary for the dominant males to do so. Peripheral males are kept away by the leaders (Kawai, 1967).

Among females over four years of age, the "principle of youngest ascendency" applies. That is, the youngest sister ranks just below the mother and holds second rank among the kinship group. Brothers also tend toward youngest ascendency at two or three years of age but this trend reverses into an elder over younger trend soon thereafter as peripheralization complicates male dominance (Koyama, 1967).

As in the rhesus macaque, occasionally there are troops of Japanese monkeys that have no dominant males. The adult males appear during the breeding season and then leave. The troop is run by adult females and the social structure is based on kinship and on a matriarchal order which transcends generations (Kawamura, 1965).

Frequently a solitary male may stay near a troop without joining it. Day and night he may remain between 20 and 200 meters of the troop and show no strong antagonism toward it (Kawai et al., 1968).

The female dominance system is of as much importance to social spacing and structure as is the male dominance system. Males of high-ranking mothers, for example, leave the troop at a much wider range of ages than do males of low-ranking mothers. Of course, the dominance hierarchy of the males themselves also determines who leaves and when (Itoigawa, 1974). Females, however, change troops only when there is troop fission (Itani, 1972). Thus, most of the transmission of genetic material from troop to troop is done by males. With regard to the acquisition and propagation of new food habits and other new behaviors, the transmission of these protocultural abilities from one troop to another is also done primarily by solitary and group-changing males (Itani, 1965).

In troop movements, young males both lead the procession and follow the procession (Imanishi, 1963). Dominant males remain at the core of the troop near females of high rank (Fujii, 1975).

2. Bonnet Macaques (*M. radiata*)

The sex ratio of free-ranging groups of bonnet macaques is close to one male for every female. In the seven groups studied by Rahaman and Parthasarathy (1967), there were slightly more males in one group and slightly more females in six groups. No exclusively male groups were seen, nor were there any solitary males. However, the clasping-while-sleeping position was seen infrequently in adult males; that is, they often slept alone (Koyama, 1973). Also, early in life, males venture further from their mothers than do females (Simonds, 1974), take longer to show a preference for their own mothers, and never display as strong a preference for their own mothers as do female infants (Rosenblum, 1975). In adulthood, however, there are a few central adult males around which females with new infants often cluster (Judge and Rodman, 1976).

There are stronger male–male bonds in bonnets than in rhesus and Japanese macaques. While isolation and solitarization are common in rhesus and *M. fuscata,* they are extremely rare in bonnets (Parthasarathy, 1976). The rank of a male in a bonnet troop is as evident from his nonagonistic tactile contact as from his agonism (Simonds, 1977).

3. Pigtail Macaques (*M. nemestrina*)

There is no sex difference in maternal preference in pigtail macaque infants (Rosenblum, 1975); however, males become independent of their mothers sooner and bite their mothers more than do females. In addition, mothers bite their male infants more than their female infants (Jensen, Bobbitt, and Gordon, 1966). Throughout life, female pigtail monkeys remain in proximity more than do males (Bernstein, 1972). Breaking up the natural family groups of female preadolescent pigtails can lead to aberrant menstrual cycles just as in rhesus (Mahoney, 1975).

4. Other Macaques (*Macaca spp.*)

In *M. fascicularis,* the crab-eating macaque, the sex ratio is initially slightly in favor of the males, but their mortality is higher (Angst, 1976). In the toque monkey (*M. sinica*), the sex ratio at birth is also in favor of males but senile females are seen more frequently than are senile males. Solitary males are never found, although single males who return to a troop in a few hours are found. Apparently such males attempt to mate in other troops by going on short excursions. Eventually they do not return. A male that lives to an old age probably changes troops five to six times in a lifetime (Dittus, 1975a; Dittus, 1975b). Most of the migration occurs in the breeding season. Females never migrate and migration occurs most often in young adolescent or low-ranking males. *All* males eventually leave their natal troops (Dittus, 1975a; Dittus, 1975b).

Although there is a considerable amount of variability in social structure from macaque species to macaque species, in all species males seem to be more loosely attached to their troops than are females. As we shall see in the next chapter, this is also true of many other Old World monkey species. However, we will see some interesting exceptions to the general rule that only males change groups when we discuss some of the great apes.

E. REFERENCES FOR CHAPTER 14

Altmann, S. A. Sociobiology of rhesus monkeys III. The basic communication network. *Behaviour,* 1968, **32**, 17-32.

Alvarez, F. Conditions of observation and social distance in groups of squirrel monkeys. *Primates,* 1975, **16** (4), 465-470.

Anderson, C. O. and Mason, W. A. Hormones and social behavior of squirrel monkeys (*Saimiri sciureus*) I. Effects of endocrine status of females on behavior within heterosexual pairs. *Hormones and Behavior,* 1977, **8**, 100-106.

Angst, W. Breeding statistics of *Macaca fascicularis* in Basel Zoo. Paper presented at the *International Primatological Society* Meeting, Cambridge, England, August, 1976.

Baldwin, J. D. The social behavior of adult male squirrel monkeys (*Saimiri sciureus*) in a seminatural environment. *Folia Primatologica,* 1968, **9**, 281-314.

Baldwin, J. D. and Baldwin, J. The ecology and behavior of squirrel monkeys (*Saimiri oerstedi*) in a natural forest in Western Panama. *Folia Primatologica,* 1972, **18**, 161-184.

Bernstein, I. S. Daily activity cycles and weather influences in a pigtail monkey group. *Folia Primatologica,* 1972, **18**, 390-415.

Boelkins, R. C. and Wilson, A. P. Intergroup social dynamics of the Cayo Santiago rhesus (*Macaca mulatta*) with special reference to changes in group membership by males. *Primates,* 1972, **13** (2), 125-140.

Candland, D. K., Tyrrell, D. S., Wagner, D. S. and Wagner, N. M. Social preferences of the squirrel monkey (*Saimiri sciureus*). *Folia Primatologica,* 1973, **19**, 437-449.

Castell, R. Communication during initial contact: A comparison of squirrel and rhesus monkeys. *Folia Primatologica,* 1969, **11**, 206-214.

Charles-Dominique, P. Social structure of nocturnal lemurs: From solitary to gregarious. Paper presented at the *International Primatological Society* Meeting, Cambridge, England, August, 1976.

Charles-Dominique, P. Urine marking and territoriality in *Galago alleni* (Waterhouse, 1837—Lorisoidea, Primates): A field study by radio telemetry. *Zeitschrift für Tierpsychologie,* 1977, **43**, 113-138.

Clutton-Brock, T. H. and Harvey, P. H. A statistical analysis of some aspects of primate ecology and social organization. Paper presented at the *International Primatological Society* Meeting, Cambridge, England, August, 1976.

Cubicciotti, D. D., III and Mason, W. A. A comparison of heterosexual jealousy responses in *Callicebus* and *Saimiri*. Paper presented at the *American Society of Primatologists* Meeting, Seattle, Washington, April, 1977.

Dittus, W. P. J. The ecology and behavior of the toque monkeys, *Macaca sinica.* *Dissertation Abstracts,* 1975, B35/09, 4725. (a)

Dittus, W. P. J. Population dynamics of the toque monkey, *Macaca sinica.* In Tuttle, R. H. (Ed.) *Socioecology and Psychology of Primates.* The Hague: Mouton, 1975, pp. 125-151. (b)

Doyle, G. A. The behaviour of the lesser bushbaby. In Martin, R. D., Doyle, G. A. and Walker, A. C. (Eds.) *Prosimian Biology.* London: Duckworth, 1974, pp. 213-231.

Drickamer, L. C. A ten-year summary of reproductive data for free-ranging *Macaca mulatta. Folia Primatologica,* 1974, **21**, 61-80.

Drickamer, L. D. Patterns of space utilization and group interactions among free-ranging *Macaca mulatta. Primates,* 1975, **16** (1), 23-33.

Durham, N. M. Effects of altitude differences on group organization of wild black spider monkeys (*Ateles paniscus*). *Proceedings of the International Congresss of Primatology* (Zurich, 1970) (Vol. 3). Basel: Karger, 1971, pp. 32-40.

Ehrlich, A. and Musicant, A. Social and individual behaviors in captive slow lorises. *Behaviour,* 1977, **60**, 195-220.

Eisenberg, J. F. and Kuehn, R. E. The behavior of *Ateles geoffroyi* and related species. *Smithsonian Miscellaneous Collections,* 1966, **151** (8), 1-63.

Erwin, J., Mobaldi, J. and Mitchell, G. Separation of juvenile rhesus monkeys of the same age. *Journal of Abnormal Psychology,* 1971, **78** (2), 134-139.

Fragaszy, D. M. Behavior in a novel environment in *Saimiri* and *Callicebus.* Paper presented at the *Western Psychological Association* Meeting, Seattle, Washington, April, 1977. (a)

Fragaszy, D. M. Social context and response to novel objects in *Saimiri* and *Callicebus.* Paper presented at the *American Society of Primatologists* Meeting, Seattle, Washington, April, 1977. (b)

Fragaszy, D. M. and Mason, W. A. Response to novelty in *Saimiri* and *Callicebus:* Influences of social context. *Primates,* 1978, **19**, in press.

Fujii, H. A psychological study of the social structure of a free-ranging group of Japanese monkeys in Katsuyama. In Kondo, S., Kawai, M. and Ehara, A. (Eds.) *Contemporary Primatology.* Basel: Karger, 1975, pp. 428-436.

Gabow, S. L. Dominance order reversal between two groups of free-ranging rhesus monkeys. *Primates,* 1973, **14**, 215-223.

Hinde, R. A. and Spencer-Booth, Y. The behaviour of socially living rhesus monkeys in their first two and one half years. *Animal Behaviour,* 1967, **15**, 169-196.

Imanishi, K. Social behavior in Japanese monkeys, *Macaca fuscata.* In Southwick, C. H. (Ed.) *Primate Social Behavior.* Princeton, N.J.: Van Nostrand, 1963, pp. 68-81.

Itani, J. A. Preliminary essay on the relationship between social organization and incest avoidance in nonhuman primates. In Poirier, F. E. (Ed.) *Primate Socialization*. New York: Random House, 1972, pp. 165-171.

Itani, J. On the acquisition and propagation of a new food habit in the troop of Japanese monkeys at Takasakiyama. In Imanishi, K. and Altmann, S. (Eds.) *Japanese Monkeys*. Atlanta: Altmann, 1965, pp. 52-65.

Itoigawa, N. Variables in male leaving a group of Japanese macaques. *Symposia of the Fifth Congress of the International Primatological Society*. Nagoya, Japan: Japan Science Press, 1974, pp. 233-244.

Jensen, G. D., Bobbitt, R. A. and Gordon, B. N. Sex differences in social interaction between infant monkeys and their mothers. *Recent Advances in Biological Psychiatry*, 1966, **9**, 283-293.

Jorde, L. B. and Spuhler, J. N. A statistical analysis of selected aspects of primate demography, ecology, and social behaviour. *Journal of Anthropological Research*, 1974, **30**, 199-224.

Judge, D. S. and Rodman, P. S. *Macaca radiata:* Intragroup relations and reproductive status of females. *Primates*, 1976, **17** (4), 535-539.

Kaplan, J. The effects of separation and reunion on the behavior of mother and infant squirrel monkeys. *Developmental Psychobiology*, 1970, **3**, 43-52.

Kawai, M. Catching behavior observed in the Koshima troop. A case of newly acquired behavior. *Primates*, 1967, **8**, 181-186.

Kawai, M., Yoshiba, K., Ando, S. and Azuma, S. Some observations on the solitary male among Japanese monkeys: A pilot report for a sociotelemetrical study. *Primates*, 1968, **9**, 1-12.

Kawamura, S. Matriarchal social ranks in the Minoo-B troop. In Imanishi, K. and Altmann, S. (Eds.) *Japanese Monkeys*. Atlanta: Altmann, 1965, pp. 105-112.

Kinzey, W. G., Rosenberger, A. L., Heisler, P. S., Prowse, D. L. and Trilling, J. S. A preliminary field investigation of the yellow handed titi monkey, *Callicebus torquatus torquatus,* in Northern Peru. *Primates*, 1977, **18**, 159-181.

Kleiman, D. G. Monogamy in mammals. *The Quarterly Review of Biology*, 1977, **52**, 39-69.

Klein, L. and Klein, D. Aspects of social behaviour in a colony of spider monkeys. *International Zoo Yearbook*, 1971, **11**, 175-181.

Klopfer, P. H. Patterns of maternal care in lemurs: II. Effects of group size and early separation. *Zeitschrift für Tierpsychologie*, 1972, **30**, 277-296.

Kortmulder, K. On ethology and human behaviour. *Acta Biotheoretica*, 1974, **23**, 55-78.

Koyama, N. On dominance rank and kinship of a wild Japanese monkey troop in Arashiyama. *Primates*, 1967, **8**, 189-216.

Koyama, N. Changes in dominance rank and division of a wild Japanese monkey troop of Arashiyama. *Primates*, 1970, **11**, 335-390.

Koyama, N. Dominance, grooming, and clasped-sleeping relationships among bonnet monkeys in India. *Primates,* 1973, **14**, 225–244.

Kummer, H. *Primate Societies.* New York: Aldine, 1971.

Kurokawa, T. An experimental field study of cohesion in Katsuyama group of Japanese monkeys. In Kondo, S., Kawai, M. and Ehara, A. (Eds.) *Contemporary Primatology.* Basel: Karger, 1975, pp. 437–444.

Mahoney, C. J. Aberrant menstrual cycles in *Macaca mulatta* and *Macaca fascicularis. Laboratory Animal Handbook,* 1975, **6**, 243–255.

Marler, P. Aggregation and dispersal: Two functions in primate communication. In Jay, P. C. (Ed.) *Primates: Studies in Adaptation and Variability.* New York: Holt, Rinehart and Winston, 1968, pp. 420–438.

Marler, P. On animal aggression: The roles of strangeness and familiarity. *American Psychologist,* 1976, **31**, 239–246.

Marsden, H. M. Behavior between two social groups of rhesus monkeys within two tunnel-connected enclosures. *Folia Primatoloigca,* 1968, **8**, 240–246.

Marsden, H. M. Dominance order reversal of two groups of rhesus monkeys in tunnel-connected enclosures. *Proceedings of the Second International Congress of Primatology.* Basel: Karger, 1969, pp. 52–58.

Mason, W. A. Social organization of the South American monkey, *Callicebus moloch:* A preliminary report. *Tulane Studies in Zoology,* 1966, **13**, 23–28.

Mason, W. A. Comparative studies of social behavior in *Callicebus* and *Saimiri:* Strength and specificity of attraction between male-female cagemates. *Folia Primatologica,* 1975, **23**, 113–123.

Mason, W. A. and Epple, G. Social organization in experimental groups of *Saimiri* and *Callicebus. Proceedings of the Second International Congress of Primatology.* Basel: Karger, 1968, pp. 59–65.

Masui, K., Nishimura, A., Ohsawa, H. and Sugiyama, Y. Population study of Japanese monkeys at Takasakiyama. *Journal of the Anthropological Society of Nippon,* 1973, **81** (4), 236–248.

Milton, K. Urine-rubbing behavior in the mantled howler monkey, *Alouatta palliata. Folia Primatologica,* 1975, **23**, 105–112.

Missakian-Quinn, E. A. and Varley, M. A. Consort activity in groups of free-ranging rhesus monkeys on Cayo Santiago. Paper presented at the *American Society of Primatologists* Meeting, Seattle, Washington, April, 1977.

Mori, A. Intratroop spacing mechanism of the wild Japanese monkeys of the Koshima troop. In Kondo, S., Kawai, M. and Ehara, A. (Eds.) *Contemporary Primatology.* Basel: Karger, 1975, pp. 423–427.

Murrell, S. and Young. D. Comparative study of activity cycles and social behavior of western black and white ruffed lemurs (*Varecia variegatus editorum*) and mongoose lemurs (*Lemur mongoz mongoz*). Paper presented at the *American Society of Primatologists* Meeting, Seattle, Washington, April, 1977.

Neville, M. K. A free-ranging rhesus monkey troop lacking adult males. *Journal of Mammalogy,* 1968, **49**, 771-773.

Neville, M. K. Social relations within troops of red howler monkeys (*Alouatta seniculus*). *Folia Primatologica,* 1972, **18**, 47-77.

Nishida, T. A sociological study of solitary male monkeys. *Primates,* 1966, **7**, 141-204.

Nishimura, A. and Izawa, K. The group characteristics of woolly monkeys (*Lagothrix lagothrica*) in the upper Amazonian basin. In Kondo, S., Kawai, M. and Ehara, A. (Eds.) *Contemporary Primatology.* Basel: Karger, 1975, pp. 351-357.

Norikoshi, K. and Koyama, N. Group shifting and social organization among Japanese monkeys. *Symposia of the Fifth Congress of the International Primatological Society.* Tokyo: Japan Science Press, 1975, pp. 43-61.

Oppenheimer, J. R. Behavior and ecology of the white-faced monkey, *Cebus capucinus,* on Barro Colorado Island, C.Z. *Dissertation Abstracts,* 1969, **30**, 811.

Oppenheimer, J. R. and Oppenheimer, E. C. Behavioral observations of *Cebus nigrivittatus* on the Venezuelan llanos. *American Zoologist,* 1972, **12**, 69.

Parthasarathy, M. D. Social ostracism among bonnet monkeys. Paper presented at the *International Primatological Society* Meeting, Cambridge, England, August, 1976.

Perret, M. Influence du groupement social sur l'activation sexuelle saisonniere chez le ♂ de *Microcebus murinus* (Miller, 1777). *Zeitschrift für Tierpsychologie,* 1977, **43**, 159-179.

Petter, J. J. and Petter, A. The aye-aye of Madagascar. In Altmann, S. A. (Ed.) *Social Communication Among Primates.* Chicago: University of Chicago Press, 1967, pp. 195-205.

Petter, A. and Petter, J. J. Primates, Prosimii, Lemuriformes. In Master, J. (Ed.) *Smithsonian Institution's Preliminary Identification Manual for African Mammals #26.* Washington, D.C.: Smithsonian, 1968, pp. 1-16.

Poirier, F. E. Socialization and learning among nonhuman primates. In Kimball, S. T. and Burnett, J. H. (Eds.) *Learning and Culture.* Seattle, Washington: University of Washington Press, 1973, pp. 3-41.

Pope, B. L. The population characteristics of howler monkeys (*Alouatta caraya*) in Northern Argentina. *American Journal of Physical Anthropology,* 1966, **24**, 361-370.

Quick, L. B. The social environment of behavior of Japanese macaques. Paper presented at the *American Association of Physical Anthropologists* Meeting, Seattle, Washington, April, 1977.

Rahaman, H. and Parthasarathy, M. D. A population survey of the bonnet monkey, *Macaca radiata* (Geoffroy) in Bangalore, South India. *Journal of the Bombay Natural History Society,* 1967, **64** (2), 251-255.

Rondinelli, R. and Klein, L. L. An analysis of adult social spacing tendencies and related social interactions in a colony of spider monkeys (*Ateles geoffroyi*) at the San Francisco Zoo. *Folia Primatologica*, 1976, 21, 12-31.

Rosenblum, L. A. Studies of three species of monkeys for clues to behavior. *Downstate Reporter*, 1975, 6, 8-12.

Sade, D. S. Determinants of dominance in a group of free-ranging rhesus monkeys. In Altmann, S. A. (Ed.) *Social Communication Among Primates*. Chicago, Ill.: University of Chicago Press, 1967, pp. 99-114.

Scott, N. J., Malmgren, L. A. and Glander, K. E. Grouping behaviour and sex ratio in mantled howling monkeys (*Alouatta palliata*). Paper presented at the *International Primatological Society* Meeting, Cambridge, England, August, 1976.

Simonds, P. E. Sex differences in bonnet macaque networks and social structure. *Archives of Sexual Behavior*, 1974, 3, 151-165.

Simonds, P. E. Aggression and social bonds in an urban troop of bonnet macaques (*Macaca radiata*). Paper presented at the *American Association of Physical Anthropologists* Meeting, Seattle, Washington, April, 1977.

Sorenson, M. W. A review of aggressive behavior in the tree shrews. In Holloway, R. H. (Ed.) *Primate Aggression, Territoriality, and Xenophobia: A Comparative Perspective*. New York: Academic Press, 1974, pp. 13-30.

Southwick, C. H. and Siddiqi, M. F. Contrasts in primate social behavior. *Bioscience*, 1974, 24 (7), 389-406.

Stott, K. Sakis: Imps of the rain forest. *Zoonooz*, 1976, 49 (4), 4-8.

Strayer, F. F., Taylor, M. and Yanciw, P. Group composition effects on social behavior of captive squirrel monkeys (*Saimiri sciureus*). *Primates*, 1975, 16 (3), 253-260.

Sugiyama, Y. and Ohsawa, H. Life history of male Japanese macaques at Ryozenyama. *Proceedings of the Fifth International Congress of Primatology*. Nagoya, Japan: Japan Science Press, 1974, pp. 407-410.

Suguwara, K. Analysis of the social relations among adolescent males of Japanese monkeys (*Macaca fuscata fuscata*) at Koshima Islet. *Journal of the Anthropological Society of Nippon*, 1976, 83 (4), 330-354.

Suomi, S. J., Sackett, G. P. and Harlow, H. F. Development of sex preference in rhesus monkeys. *Developmental Psychology*, 1970, 3 (3), 326-336.

Tattersall, I. and Sussman, R. W. Observations on the ecology and behavior of the mongoose lemur, *Lemur mongoz mongoz*, Linnaeus (Primates, Lemuriformes) at Ampijora, Madagascar. *Anthropological Papers of the American Museum of Natural History*, 1975, 52 (4), 195-216.

Vandenbergh, J. G. The development of social structure in free-ranging rhesus monkeys. *Behaviour*, 1967, 29, 179-194.

Vessey, S. H. Free-ranging rhesus monkeys: Behavioral effects of removal, separation, and reintroduction of group members. *Behaviour,* 1971, **34,** 216-227.

Vessey, S. H., Mortenson, B. K. and Muckenhirn, N. A. Size and structure of primate groups in Guyana, South America. Paper presented at the *International Primatological Society* Meeting, Cambridge, England, August, 1976.

Wade, T. D. The effects of strangers on rhesus monkey groups. *Behaviour,* 1976, **56,** 194-214.

Yamada, M. Five natural troops of Japanese monkeys on Shodoshima Island: II. A comparison of social structure. *Primates,* 1971, **12** (2), 125-150.

15.

The Role of Gender in Social Spacing and Structure: Old World Monkeys, Apes, and Implications for Humans

In Chapter 14 we saw that, in most macaques, males become independent of their mothers earlier, become more peripheralized in the troop, become solitary more or join isosexual groups more, and change troops more than do females. In the present chapter we will see to what degree these same sex differences apply to other Old World primates, including both monkeys and apes.

A. OTHER OLD WORLD MONKEYS

1. Hamadryas Baboons (*Papio hamadryas*)

As we already know, hamadryas baboons live in one-male harems. The hamadryas male "herds" his females by biting the napes of their necks to keep them near him (Kummer and Kurt, 1963). Other baboons, for example *P. anubis,* live in multimale groups. There is evidence from mixed group studies and female hybrids (*P. hamadryas* and *P. anubis*) that the females modify their behavior according to the social pattern imposed upon them by the male's behavior (Nagel, 1970). Male *P. hamadryas* by *P. anubis* hybrids follow and groom females but do not herd them. Thus, the two different baboon social spacing structures can be determined by both genetics and by tradition. The hybrid males display considerably more behavioral variation than do the hybrid females (Nagel, 1970).

Kummer et al. (1976) have published laboratory data indicating that hamadryas females are more intimate with one another than are hamadryas males. For example, inhibition of taking a food container seen in the possession of a like-sexed partner is less frequent in female dyads than in male dyads. Intimacy is characterized by short interindividual distance, a high frequency of friendly tactile interactions, and low frequencies of objectless fighting. Thus, females are usually spaced closer together than are males.

2. Gelada Baboons (*Theropithecus gelada*)

Gelada baboons also live in one-male groups in which the main role of the single adult male is enforcing the presence of his females and preventing the takeover of the females by other males. Females are ranked in a dominance order maintained by orientation to the male (Crook, 1968). Each male may lead from one to twenty-four females (Mori and Kawai, 1975).

Among gelada baboons, lone males are sometimes seen. They are not solitary but rather emigrate from herd to herd. A herd is a gathering of many one-male groups. There are also all-male groups that move from herd to herd. Evidently females make the choice of whom should become the leader and each male must live through a freelance or all-male existence prior to becoming acceptable as a leader (Mori and Kawai, 1975). Kummer (1974) has published data showing that female like-sex dyads are more compatible than are male like-sex dyads. The female gelada apparently plays the primary cohesive role. Neck-biting and herding by a male are not necessary (Fischer, 1977). Gelada females stay within five to six feet of the dominant male (Emory, 1976).

Also in contrast to other Old World monkeys, estrus does not seem to affect social spacing in geladas (Dunbar, 1976). Males tend to sleep at the periphery of a gelada harem at night (Crook, 1966), but the one-male unit is extremely cohesive. The females of a specific one-male group form coalitions against animals of other one-male groups (Bramblett, 1970). Although the male is conspicuous, he is peripheral to the females. If the level of excitement is high, for example, the females seem to react as a class to chase the male. The female's structure results from long-term relationships with other females. Even the group movements of one-male groups are frequently initiated by females (Bramblett, 1970); however, all-male groups are found at the head of a herd (cited in Rhine, 1975).

3. Mandrills (*Papio sphinx*)

Mandrills also live in one-male groups but they are semiarboreal rather than completely terrestrial like geladas. The female mandrills do not stay quite as close to their male as do the female geladas. Solitary mandrill males are sometimes seen (Emory, 1976), and, when traveling, it is the male that stays near the rear of the troop (Jouventin, 1975).

4. Other Baboons (*P. anubis, P. cynocephalus, P. ursinus, P. papio*)

All the other baboons live in multimale troops. Males seem to be more loosely attached to the troops than are females. In some areas, male baboons have been known to leave a troop for three days at a time and travel up to two miles in order to hunt. Females do not do this (Harding and Strum, 1976). Because of peripheralization in multimale baboon species, males interact more frequently with extragroup conditions or events (Rhine, 1975).

In *P. papio* (the Guinea baboon), the adult sex ratio is one adult male to twelve to twenty-two adult females (Masure and Bourliére, 1971). Only adult male or subadult male guinea baboons sleep alone. In sleeping parties, only one male per party is the rule, and during the day, there are no all-female parties. In traveling, adult males bring up the rear (Dunbar and Nathan, 1972). In chacma baboons (*P. ursinus*), adult males initiate movement and chase others back into the troop (Buskirk, Buskirk, and Hamilton, 1974).

5. Mangabeys (*Cercocebus spp.*)

In the three troops of *Cercocebus galeritus* observed by Quris (1975), there was only one adult male who functioned as the leader in each troop. Solitary males were also seen. In white-cheeked mangabeys (*Cercocebus albigena*), males and females sleep in different trees and do not travel together for about one-half of the daytime hours (Mizuno, Kawai, and Ando, 1976). In this species, presenting is used almost exclusively for sex and only by females. However, females seem to space themselves closer together than do males (Chalmers and Rowell, 1971). In two groups of mangabeys observed by Chalmers (1968), there were approximately equal numbers of males and females. Mangabeys, of course, are arboreal primates.

6. Patas Monkeys (*Erythrocebus patas*)

Patas monkeys are terrestrial and live in one-male groups. However, the male is very much more peripheral to the group than is the male of baboon one-male groups. Harem males of the patas species drive off other males. All-male groups of patas are seen frequently (Struhsaker and Gartlan, 1970). The male harem leader remains aloof from his females and their young (Hall and Mayer, 1967). Females, however, do not attempt to move from one group to another, whereas males almost always leave their natal group. Single or solitary males are common but, without a group of females, such males are very tolerant of each other and readily form all-male groups. Males travel from group to group and over a much larger range than do females. Membership in an all-male group is a normal developmental stage for males, some starting as young as eighteen months of age. However, only full adult males are solitary. A male is originally driven from his natal group by the harem leader (Gartlan, 1975). If a leader male is removed from a patas group in captivity it has no effect on the spatial relations of the remaining animals in the group (Chalmers, 1971).

7.   *Cercopithecus spp.*

In the many arboreal species of the genus *Cercopithecus,* the oldest progeny are often expelled from the group (cf. Morike, 1976). Many species (e.g., *C. ascanis* and *C. mitis*) have vocalizations specific to males which are used to rally the group (Marler, 1973) and which obviously affect spacing. There is a spontaneous departure from the group of maturing subadult males in *C. campbelli,* but these males frequently take some females with them. The alpha male of this species is more often alone than not (Hunkeler, Bourliére, and Bertrand, 1972).

In the one terrestrial species of the genus *Cercopithecus,* females space themselves closer together than do males. Even male juveniles are more solitary than are female juveniles (Raleigh, 1977). It is the female matrifocal core that provides vervets with stability and continuity because, as we have said before, the male is more loosely attached to this group. An adult female does not change groups but does start new ones by taking all of her relations away with her. Female coalititons occur frequently and, in coalition, they can dominate males. Only one male per group is necessary (Lancaster, 1973).

8.   *Miopithecus talapoin*

In the talapoin monkey, as we know, juvenile sex differences in many behaviors are not correlated with sex differences in adulthood. Juvenile females, however, do show more proximity and contact than do juvenile males; and, although the reasons for the sex differences are probably different at different ages, they also show closer spacing and proximity in adulthood. Adult male talapoins are relatively withdrawn and inactive. They remain apart from the group somewhat as do squirrel monkey adult males (Wolfheim, 1977). Outside of the breeding season, there is a tendency for talapoins to live in isosexual subgroups. During the breeding season the females leave their infants with the female group and join male subgroups (Rowell and Dixson, 1975).

9.   Langurs *(Presbytis spp.)*

Hanuman langurs *(Presbytis entellus)* live in both one-male and multimale groups. Permanently isolated males as well as all-male groups are found (but see Curtin, 1977). Occasionally, langur females may leave a troop or form temporary alliances with other females within a troop as a counterstrategy against a male who is attempting a group takeover with infanticide (Hrdy, 1974).

Among sixteen groups of hanuman langurs observed by Tiwari and Mukherjee (1976), twelve were bisexual and two were all-male groups. Two-thirds of the bisexual groups had only one adult male, while one-third of the bisexual groups had from three to eleven males. The one-male groups appeared to be the most stable, most relaxed, and least aggressive (see Vogel, 1970).

According to Weber and Vogel (1970):

... we found about 74% of all adult and subadult males living in only-male groups. The majority of males therefore spend some portion of their lives within only-male groups. (P. 79)

Apparently, in one-male groups, the one male controls the group for about three years (Itani, 1972).

The Nilgiri langur (*P. johnii*) society is female focal (Poirier, 1972) with females showing more proximity and contact with one another than the males. In *P. senex,* adult male replacements in one-male troops lead to infant mortality and exclusion of immature animals from the group (Rudran, 1973). Golden langurs (*P. geei*) also live in one-male, bisexual, and all-male groups (Mukherjee and Saha, 1974).

10. Colobidae

In the Angola Colobus monkey (*C. angolensis*) group of Tanzania there is usually only one male who performs a distinctive jump-roar display to maintain spacing and intragroup cohesion. The male is peripheral to the females and young (Groves, 1973; see also Waser, 1977, for similar evidence of the use of adult male displays for intragroup cohesion). Most species of the genus *Colobus* show peripheralization of males relative to females.

The proboscis monkeys (*Nasalis larvatus*) live in multimale troops but little else is known about their group spacing and structure in the wild.

B. LESSER APES

The lesser apes live in monogamous family groups; however, in seminatural situations, juvenile groups have been formed. If a three-year-old-juvenile-male gibbon (*Hylobates lar*) is placed with two juvenile females, he will select one of the females and form a pair. This tendency toward pair formation occurs despite rearing in captivity (Paluck, Lieff, and Esser, 1970).

Adult gibbons become intolerant of immatures of 2½ years of age; and, at this age, the immatures are gradually peripheralized from the monogamous group. The process of peripheralization lasts from about 2½ to 4 years of age (Ellefson, 1966). The vocalizations of female Kloss gibbons (*Hylobates klossi*) are louder than those of males. When she vocalizes her offspring wait at a distance from her. The great risk of these loud female vocalizations in terms of predation may be why males accompany females and why the infants remain away from the vocalizing mothers (Tenaza, 1976). Male songs probably serve in mate attraction.

Siamangs (*Symphalangus syndactylus*) show behavioral similarity to the gibbons in regard to the peripheralization of maturing subadults (Fox, 1972). Solitary males are found occasionally (Chivers, 1975; Chivers, 1976). At night, the juvenile sleeps with the male and the infant sleeps with the female. (Chivers, 1970, P. 20.) The subadult male is slowly peripheralized by the adult male. The drier the weather, the more dispersed is the siamang group. The usual traveling order is female, juvenile, male and weaned juvenile, and, finally, the subadult male (if not yet peripheralized) (Chivers, 1970).

C. GREAT APES

1. Common Chimpanzees (*Pan troglodytes*)

Male chimpanzees leave their mothers earlier than do female chimpanzees. They therefore become more peripheralized from the rather open chimpanzee group than do females (van Lawick-Goodall, 1973a). The female chimpanzee may stay with her mother for life except when in estrus (van Lawick-Goodall, 1973b).

When males reach adolescence, they start to sleep alone unless a female is in estrus, in which case he sleeps with her (Riss and Goodall, 1976). Mothers sometimes follow their sons when they leave to associate with other groups (Pusey, 1976). Females are sometimes solitary. Among chimpanzees, 28 percent of the groups seen in the wild are all-male groups or males on their own (cf. Mason, 1970). Bauer (1976) has also discussed sex differences in chimpanzee aggregation.

2. Gorillas (*Gorilla gorilla*)

Gorilla groups are much more permanent than are chimpanzee groups. Gorilla males, however, like chimpanzee males, cover more ground than do females. Solitary males are common among gorillas, whereas solitary females are not. Gorillas are quieter than are chimpanzees and they use more individual rather than communal displays in social spacing (Reynolds, 1965).

Gorilla infants reach independence a little earlier than do chimpanzee infants. One large infant left its main group with an adult silver-backed male and remained alone with him for nearly a month (Mason, 1964). The dominant male gorilla determines troop movement. The females stay in the center of the troop whereas the subordinate males remain on the periphery (Schaller, 1963). However, most gorilla *females* transfer out of their natal troops before maturity (Harcourt and Stewart, 1976).

3. Orangutans (*Pongo pygmaeus*)

As we know, orangutans are found in open individualistic societies with adults minimizing social contacts. There is a tendency toward greater interindividual distance in males than in females (Rijksen, 1975). It is probably impossible to keep two adult male orangutans together in a zoo. This is not true for females (Pitcairn, 1974).

Okano (1971) has guessed that, at three to four years of age, juvenile orangutans, males or females, come together to make play groups; and, that at five to eight years of age, males and females make sexual groups. Okano also theorized that both adult males and adult females are solitary at some time in their lives.

Whether or not the Okano pattern is true, the only long-term social unit is the adult female and her dependent offspring. Adult males live a solitary existence almost all of the time. Juvenile males leave the mother at a younger age than do juvenile females and probably at a younger age than do chimpanzee males. Solitary juvenile males are occasionally found. Males travel further than do females (Horr, 1972).

Orangutans have been assumed to be arboreal; however, males descend to the ground to forage, to travel, and even to nap. The huge males also apparently disappear from their home area for up to two years (Galdikas-Brindamour, 1975).

As has been noted by Cohen (1975), males and adolescents appear by themselves much more often than expected, whereas females and juveniles appear alone much less often than expected. Adult heterosexual pairs also appear much less often than expected. There is no other species of primate that has a spacing or a social structure like that of the orangutan (Cohen, 1975).

Rodman (1976) has surmised that the strong sexual dimorphism of the orangutan forces the segregation of the sexes. The large males must travel and feed more. Rodman feels that this explanation holds true only for frugivorous animals, which orangutans are.

D. THOUGHTS ABOUT *HOMO SAPIENS*

There have been many recent books and articles written on social spacing and social structure in humans (cf. Sommer, 1969; Weitz, 1974, for examples). However, what can we find in the human literature to compare with the almost universal tendency toward peripheralization of males seen in nonhuman primates?

In studies of preschool children (three to five year olds) it is known that boys and older children play outdoors more than do girls and younger children (Sanders and Harper, 1976). Among second graders, affiliation measures show that greater distances prevail between different-sexed than between same-sexed children (Travis and Fontenelle, 1977).

In college students, shy young males remain away from females but do not maintain great distances from other males (Carducci, Webber, and Brandt, 1977). In studies of the relative age/sex groupings for people in public, there is a greater relative frequency of exclusively subadult male groups compared to female groups of the same age and a lower relative frequency of subadult males in the company of people of other age/sex groups. There is a tendency, even in humans, for females to be gradually integrated into the adult group as they reach reproductive age but for males to remain apart associating largely with same-sex peers (Lockard and Adams, 1977).

REFERENCES

Bauer, H. R. Sex differences in aggregation and sexual selection in Gombe chimpanzees. *American Zoologist,* 1976, **16**, 209.

Bramblett, C. A. Coalitions among gelada baboons. *Primates,* 1970, **11**, 327–333.

Buskirk, W. H., Buskirk, R. E. and Hamilton, W. J. III. Troop mobilizing behavior of adult male chacma baboons. *Folia Primatologica,* 1974, **22**, 9–18.

Carducci, B., Webber, A. W. and Brandt, C. Shyness, personal space, and helping behavior. Paper presented at the *Western Psychological Association* Meeting, Seattle, Washington, April, 1977.

Chalmers, N. R. Group composition, ecology, and daily activities of free-living mangabeys in Uganda. *Folia Primatologica,* 1968, **8**, 247–262.

Chalmers, N. R. Changes in mother-infant behavior following changes in group composition in Sykes and patas monkeys. *Proceedings of the Third International Congress of Primatology* (Zurich, 1970), 1971, **3**, 116–120.

Chalmers, N. R. and Rowell, T. E. Behaviour and female reproductive cycles in a captive group of mangabeys. *Folia Primatologica,* 1971, **14**, 1–14.

Chivers, D. J. Spatial ralations within the siamang group. *Proceedings of the Third International Congress of Primatology* (Zurich, 1970), 1970, **3**, 14–21.

Chivers, D. J. Communication within and between family groups of siamang (*Symphalangus syndactylus*). *Behaviour,* 1976, **57**, 116–135.

Chivers, D. J., Raemaekers, J. J. and Aldrich-Blake, F. P. G. Long-term observations of siamang behavior. *Folia Primatologica,* 1975, **23**, 1–49.

Cohen, J. E. The size and demographic composition of social groups of wild orangutans. *Animal Behavior,* 1975, **23**, 543–550.

Crook, J. H. Gelada baboon herd structure and movement. *Symposia of the Zoological Society of London,* 1966, **18**, 237–258.

Crook, J. H. Primate societies and individual behavior. *Journal of Psychosomatic Research,* 1968, **12**, 11–19.

Curtin, R. A. Socioecology of langurs in the Nepal Himalaya. Paper presented at the *American Association of Physical Anthropologists* Meeting, Seattle, Washington, April, 1977.

Dunbar, R. I. M. Oestrous behaviour and social relations among gelada baboons. Paper presented at the *International Primatological Society* Meeting, Cambridge, England, August, 1976.

Dunbar, R. I. M. and Nathan, J. F. Social organization of the Guinea baboon, *Papio papio. Folia Primatologica,* 1972, **17**, 321-334.

Ellefson, J. O. Group size and behavior in white-handed gibbons, *Hylobates lar.* Paper presented at the *American Association of Physical Anthropologists* Meeting, Berkeley, California, April, 1966.

Emory, G. Social structure in mandrills and gelada baboons. *Bios,* 1976, **1**, 1-9.

Fischer, R. B. Sociometrics of captive gelada baboons. Paper presented at the *Animal Behavior Society* Meeting, University Park, Pennsylvania, June, 1977.

Fox. G. J. Some comparisons between Siamang and gibbon behaviours. *Folia Primatologica,* 1972, **18**, 122-139.

Galdikas-Brindamour, B. Orangutans, Indonesia's people of the Forest. *National Geographic,* 1975, **148** (4), 444-472.

Gartlan, J. S. Adaptive aspects of social structure in *Erythrocebus patas. Proceedings of the Fifth Congress of the International Primatological Society* (1974). Tokyo: Japan Science Press, 1975, pp. 161-171.

Groves, C. P. Notes on the ecology and behaviour of the Angola colobus (*Colobus angolensis*) in N.E. Tanzania. *Folia Primatologica,* 1973, **20**, 12-26.

Hall, K. R. L. and Mayer, B. Social interactions in a group of captive patas monkeys (*Erythrocebus patas*). *Folia Primatologica,* 1967, **5**, 213-236.

Harcourt, A. H. and Stewart, K. J. Sexual behaviour of wild mountain gorilla. Paper presented at the *International Primatological Society* Meeting, Cambridge, England, August, 1976.

Harding, R. S. O. and Strum, S. C. The predatory baboons of Kekopey. *Natural History* (N.Y.), 1976, **85**, 46-53.

Horr, D. A. The Borneo orangutan, *Borneo Research Bulletin,* 1972, **4**, 46-50.

Hrdy, S. B. Male-male competition and infanticide among the langurs (*Presbytis entellus*) of Abu, Rajasthan. *Folia Primatologica,* 1974, **22**, 19-58.

Hunkeler, C., Bourliére, F. and Bertrand, M. Le comportement social de la Mone de Lowe (*Cercopithecus campbelli lowei*). *Folia Primatologica,* 1972, **17**, 218-236.

Itani, J. A preliminary essay on the relationship between social organization and incest avoidance in nonhuman primates. In Poirier, F. E. (Ed.) *Primate Socialization.* New York: Random House, 1972, pp. 165-171.

Jouventin, P. Observations sur la socio-ecologie du Mandrill. *Extrait de la Terre et la Vie, Reune d' Ecologie Appliquee,* 1975, **29**, 493-532.

Kawabe, M. and Mano, T. Ecology and behavior of the wild proboscis monkey *Nasalis larvatus* (Wurmb) in Sabah, Malaysia. *Primates,* 1972, **13**, 213-228.

Kummer, H. Rules of dyad and group formation among captive gelada baboons (*Theropithecus gelada*). *Proceedings from the Symposia of the Fifth Congress of the International Primatological Society.* Nagoya, Japan: Japan Science Press, 1974, pp. 129-159.

Kummer, H., Abegglen, H., Bachmann, C., Falett, J. and Sigg, H. Intimacy and fear of aggression in baboon social relationships. Paper presented at the *International Primatological Society* Meeting, Cambridge, England, August, 1976.

Kummer, H. and Kurt, F. Social units of a free-living population of hamadryas baboons. *Folia Primatologica,* 1963, **1-2**, 4-18.

Lancaster, J. B. Stimulus/Response: In praise of the achieving female monkey. *Psychology Today,* 1973, **7**, 30-102.

Lockard, J. S. and Adams, R. M. Peripheral males: A primate model for a human subgroup. Paper presented at the *Animal Behavior Society* Meeting, University Park, Pennsylvania, June, 1977.

Marler, P. A comparison of vocalizations of red-tailed monkeys and blue monkeys, *Cercopithecus ascanis* and *C. Mitis,* in Uganda. *Zeitschrift für Tierpsychologie,* 1973, **33**, 223-247.

Mason, W. A. Sociability and social organization in monkeys and apes. In Berkowitz, L. (Ed.) *Recent Advances in Experimental Social Psychology* (Vol. 1). New York: Academic Press, 1964, pp. 277-305.

Mason, W. A. Chimpanzee social behavior. In Bourne, G. (Ed.) *The Chimpanzee* (Vol. 2). New York: Karger, 1970, pp. 265-288.

Masure, A. M. and Bourliére, F. Surpeuplement, fecondite, mortalite et aggressivite dans une population captive de *Papio papio*. *Extrait de la Terre et la Vie,* 1971, **25**, 491-505.

Mizuno, A., Kawai, M. and Ando, S. Ecological studies of forest-living monkeys in the Kibale forest of Uganda. In Tani, Y. (Ed.) *Kyoto University African Studies,* 1976, **10**, 1-35.

Mori, U. and Kawai, M. Social relations and behavior of gelada baboons. *Contemporary Primatology.* Basel: Karger, 1975, pp. 470-474.

Mörike, Doris. Verhalten einer gruppe von Brazzameerkatzen (*Cercopithecus neglectus*) im Heidelberger Zoo. *Primates,* 1976, **17** (4), 475-512.

Mukherjee, R. P. and Saha, S. S. The golden langurs (*Presbytis geei*) of Assam. *Primates,* 1974, **15**, 327-340.

Nagel, U. Social organization in a baboon hybrid zone. *Proceedings of the Third International Congress of Primatology,* 1970, **3**, 48-57.

Okano, J. A preliminaty observation of orang-utans in the rehabilitation station in Sepilok, Sabah. *The Annual of Animal Physiology,* 1971, **21**, 55-67.

Paluck, R. J., Lieff, J. D. and Esser, A. H. Formation and displacement of a group of juvenile *Hylobates lar*. *Primates,* 1970, **11**, 185-194.

Pitcairn, T. K. Aggression in natural groups of pongids. In Holloway, R. L. (Ed.) *Primate Aggression, Territoriality and Xenophobia.* New York: Academic Press, 1974, pp. 241-272.

Poirier, F. E. Nilgiri langur behavior and social organization. In Voget, F. W. and Stephenson, R. L. (Eds.) *For the Chief: Essays in Honor of Luther S. Cressman.* Eugene, Oregon: University of Oregon Anthropology Paper No. 4, 1972, pp. 119-134.

Pusey, A. E. Sex differences in the persistence of quality of the mother off-spring relationships after weaning in a population of wild chimpanzees. Paper presented at the *International Primatological Society* Meeting, Cambridge, England, August, 1976.

Quris, R. Ecologie et organisation sociale de *Cercocebus galeritus agilis,* dans la nord-est du gabon. *Terre et la Vie, Revue d'Ecologie Apliquee,* 1975, **29**, 337-398.

Raleigh, M. J. Sex differences in the social behavior of juvenile vervet monkeys. Paper presented at the *American Society of Primatologists* Meeting, Seattle, Washington, April, 1977.

Reynolds, V. Some behavioral comparisons between the chimpanzee and the mountain gorilla in the wild. *American Anthropologist,* 1965, **67**, 693-706.

Rhine, R. J. The order of movement of yellow baboons (*Papio cynocephalus*). *Folia Primatologica,* 1975, **23**, 72-104.

Rijksen, H. D. Social structure in a wild orang-utan population in Sumatra. In Kondo, S., Kawai, M. and Ehara, A. (Eds.) *Contemporary Primatology.* Basel: Karger, 1975, pp. 373-379.

Riss, D. and Goodall, J. Sleeping behavior and associations in a group of captive chimpanzees. *Folia Primatologica,* 1976, **25**, 1-11.

Rodman, P. S. Sexual dimorphism, diet, and the social systems of hominoids. Paper presented at the *American Association of Physical Anthropologists* Meeting, St. Louis, Missouri, April, 1976.

Rowell, T. E. and Dixson, A. F. Changes in social organization during the breeding seasons of wild talapoin monkeys. *Journal of Reproduction and Fertility,* 1975, **43**, 419-434.

Rudran, R. Adult male replacement in one-male troops of purple-faced langurs (*Presbytis senex senex*) and its effect on population structure. *Folia Primatologica,* 1973, **19**, 166-192.

Sanders, K. M. and Harper, L. V. Free-play fantasy behavior in preschool children: Relations among gender, age, season, and location. *Child Development,* 1976, **47**, 1182-1185.

Schaller, G. *The Mountain Gorilla.* Chicago: University of Chicago Press, 1963.

Sommer, R. *Personal Space,* Englewood Cliffs, N.J.: Prentice-Hall, 1969.

Struhsaker, T. T. and Gartlan, J. S. Observations on the behavior and ecology of the patas monkey (*Erythrocebus patas*) in the Waza reserve. *Cameroon Journal of Zoology, London,* 1970, **161**, 49-63.

Tenaza, R. R. Songs, choruses and countersinging of the Kloss' gibbons (*Hylobates klossi*) in Siberut, Indonesia. *Zeitschrift für Tierpsychologie,* 1976, **40**, 37-52.

Tiwari, K. K. and Mukherjee, R. P. Social structure of hanuman langur populations around Ramtek, Nagpur District, India. Paper presented at the *International Primatological Society* Meeting, Cambridge, England, August, 1976.

Travis, Cheryl and Fontenelle, G. Nonverbal correlates of affiliation in children. Paper presented at the *Animal Behavior Society* Meeting, University Park, Pennsylvania, June, 1977.

van Lawick-Goodall, J. The behavior of chimpanzees in their natural habitat. *American Journal of Psychiatry,* 1973, **130**, 1-11. (a)

van Lawick-Goodall, J. Cultural elements in a chimpanzee community. In Menzel, E. W. (Ed.) *Symposia of the Fourth International Congress of Primatology,* (Vol. 1). *Precultural Primate Behavior.* Basel: Karger, 1973, pp. 144-184. (b)

Vogel, C. Behavioral differences of *Presbytis entellus* in two different habitats. *Proceedings of the Third International Congress of Primatology,* 1970, **3**, 41-47.

Waser, P. M. Individual recognition, intragroup cohesion and intergroup spacing: Evidence from sound playback to forest monkeys. *Behaviour,* 1977, **60**, 28-74.

Weber, I. and Vogel, C. Sozialverhalten in ein — und zweigeschlechtigen Langurengruppen. *Homo,* 1970, **21**, 73-80.

Weitz, S. (Ed.) *Nonverbal Communication.* New York: Oxford, 1974.

Wolfheim, J. H. Sex differences in behavior in a group of captive juvenile talapoin monkeys (*Miopithecus talapoin*), *Behaviour,* 1977, **63**, 110-128.

16.

Dominance: Prosimians, New World Monkeys, Rhesus, and Japanese Macaques

Dominance rank or status in a group or troop is an intervening variable. It is useful when the different ways in which it is assessed agree with each other (Hinde, 1974). Individual dominance has to be differentiated from aggressiveness, leadership, and from the derived dominance we frequently see resulting from alliances or in the offspring of a dominant mother.

Aggressive animals are not necessarily dominant, nor are dominant animals always aggressive. In addition, dominant animals are not necessarily leaders, nor are leaders always dominant. Just as an animal may be a leader in one situation but not in another, so might it be dominant in one situation but not in another. Thus, while dominance may be an individual characteristic it does not have to be characteristic of that individual in all situations.

In primate research, dominance is usually defined as priority of access to some desired incentive. This incentive may be food, water, a sitting place, an estrous female, or the opportunity to groom or be groomed. No one kind of behavior is correlated with dominance and not all groups of primates even have obvious dominance hierarchies (Kolata, 1976).

At times when dominance is not obvious, the removal of a specific animal may produce social instability in the group. Such an animal, when present, evidently keeps "control" and maintains group stability (cf. Boelkins, 1967).

"Dominance displays" are often referred to in the primate literature. These include such behaviors as tree or branch shaking, jumping and roaring, or chest hitting. Aggression is not necessarily a part of the display. In fact, an animal's position in the rank order of its group may be better characterized by these almost playlike displays or performances than by aggressive behavior (Kortmulder, 1974).

It is also often assumed that the more dominant males always have better access to sexual partners, and particularly to those most ready to conceive.

However, this is not always the case for subordinate males often impregnate females by sneaking in when the more dominant males are not looking or are busy elsewhere (Bernstein, 1976). Thus, we cannot assume that the animals producing the most offspring are necessarily the most dominant. Dominance means several things, certainly more than this (Syme, 1974).

Within-sex interactions often differ from between-sex interactions particularly with regard to dominance and status (Mazur, 1973; Mazur, 1976). It is often assumed that males and females have separate and qualitatively different kinds of dominance hierarchies.[1] As we already know, sex hormones and most notably testosterone, have been implicated in dominance attainment. However, the hormone itself does not cause a rise in status; the effect of the hormone depends upon social and situational cues (Mazur, 1976).

There are apparently many sexual signals among primates that have come to have meaning in the context of dominance and subordination. For example, a sexual mount may actually be a dominance mount, a sexual present may be a show of subordination, and showing an erect penis is commonly believed to be a dominance display for the squirrel monkey (cf. Wickler, 1967).

Dominance has been assumed to be learned by animals as they grow up. In sexually dimorphic species it has often been said that males learn how to be dominant in their play groups. But often infant males of dominant mothers become dominant themselves after having little interaction with peers. In addition, there is very little evidence that the function of play is to produce or to learn stable dominance hierarchies (cf. Symons, 1977).

Finally, because a dominant animal has priority of access to social and nonsocial objects does not necessarily mean that he will always take advantage of having priority. Dominant adult males, for example, are rarely the first to explore new objects. In fact, they appear to be very conservative and seem to have less need to test new possibilities. E. W. Menzel (1968) has described this situation as follows:

> I have seen many cases in which adults — especially high ranking males — seem to actually wait for a subordinate or a juvenile to test out a situation and then immediately displace him (if food is involved) or follow suit. (P. 182)

Thus, dominance is not a unidimensional characteristic but is, rather, a quite complex one. Admitting this, we can now begin our survey of sex differences in dominance in the primate order.

A. PROSIMIANS

As has been the case with just about every other group of prosimian behaviors surveyed in this book, dominance in the prosimians has not been studied as thoroughly as it has been in other groups of primates.

---

[1]Of course, this point of view may have been somewhat influenced by sexism in science.

Some species of tree shrews that apparently live in polygamous family groups have male dominance (Sorenson, 1974); however, little else is known concerning dominance in these primitive species. It is known, however, that male bush-babies (*Galago senegalensis*) are usually dominant over females. The males emit a call which is also occasionally heard in adult females when they become dominant in a cage in captivity. An adult male who is in his home territory is usually dominant over one that is not (Bearder and Doyle, 1974). Dominance in male bushbabies is more benign than in females. If a female becomes dominant, she becomes highly aggressive (Doyle, 1974). Apparently a male weight advantage and male dominance is necessary for reproduction in *Microcebus murinus* (Perret, 1977). Male dominance is apparently the rule among prosimians.

B. NEW WORLD MONKEYS

1. Squirrel monkeys (*Saimiri sciureus*)

In squirrel monkeys, genital display is a dominance display and occurs more frequently in males than in females. Within the squirrel monkey group there are two separate hierarchies, one for males and one for females. However, there is an "omega" male role to connect the male hierarchy with the female hierarchy. There is often also a female "scapegoat" role, where a passive female receives all kinds of attention from most of the others (Alvarez, 1975).

When a group of squirrel monkeys is established in captivity a male soon becomes the dominant member, regardless of how the group is formed (Anschel and Talmage-Riggs, 1976).

Female squirrel monkeys receive significantly more penile displays than do males (Candland, 1971). All animals have a fairly strong preference for the high-ranking male. The higher the male's position, however, the greater is his preference to remain alone and avoid females. Preference to be with females, at least among squirrel monkeys, is a characteristic of submissive males (Candland et al., 1973).

In feeding situations, male squirrel monkeys are dominant over females, but initial possession of a single item of food is not necessarily a male's (Fragaszy, 1976). In the squirrel monkey, in fact, there is a small positive correlation between dominance and testosterone levels. There is also evidence that, in agonistic situations, dominant squirrel monkeys have a greater choice of signals to employ and show greater variability than do nondominant animals (Maurus, Hartmann, and Kuhlmorgen, 1974). Lolling, sprawling, and back-rolling are shown only by dominant males (Maurus, et al., 1975). Interestingly, however, when direct brain stimulation is used, dominance gestures are elicited many more times from brain sites in subordinate females than from brain sites in dominant males (Maurus and Ploog, 1971).

Low-ranked squirrel monkey females never engage in mouth-to-mouth behavior related to food sharing. Dominant males frequently display this behavior (Peters, 1970). Females living together without males for one to two years eventually come to display dominance hierarchies similar to heterosexual ones (Talmage-Riggs and Anschel, 1973). Clark and Dillon (1974) have also discussed squirrel monkey dominance.

2. Monogamous New World Monkeys

Among many New World monogamous species, male dominance is not as clear as it is in the squirrel monkey. In fact, there is often female dominance and aggression by the female toward the male (Kleiman, 1977). Kleiman has noted that, in these species, as an individual's relative dominance goes down, his or her grooming initiates go up.

In the golden lion marmoset (*Leontopithecus rosalia*) there is apparently equal dominance between the male and the female (Snyder, 1974). Dawson (1976) sees dominance in marmosets as being one male–one female dominance systems rather than a characteristic of an individual. In captivity, however, one dominant female inhibits all other females in her group from reproducing (Epple, 1975). In *Cebuella pygmaea,* a female usually seems to be the dominant individual; and, among *Saguinous fuscicollis,* dominance interactions between females are of a more serious nature (involving vicious fighting) than are those between males. Dominant animals do more "scent-marking" than do subordinates. Apparently it is a "triumph ceremony" (Epple, 1975).

In *Callithrix jacchus,* only the dominant males react to alterations of the environment with increased scent-marking, but both sexes scent-mark as adults in response to another adult of their own sex, and marking has an important function in demonstrating social dominance in both males and females (Epple, 1970).

In titi monkeys, which are not marmosets but which are monogamous and monomorphic, males initiate more social interactions at feeding and are more aggressive than females. Although female titis attempt to steal food more than do males, they have less success. Initial possession of a single food item is usually the male's (Fragaszy, 1976).

3. Other New World Monkeys

In *Cebus nigrivittatus,* adult males are dominant over adult females (Oppenheimer and Oppenheimer, 1972). This is also true of howling monkeys; that is, adult males are dominant over adult females (Glander, 1975; Carpenter, 1934). In the black howler, *Alouatta caraya,* "the males maintain a certain extent of control over females and often intervene when conflicts arise." (Benton, 1976, P. 151.)

Among spider monkeys, ritualized presenting and mounting as forms of dominance are lacking. Dominance is not strongly dimorphic (Eisenberg and Kuehn, 1966). Olfactory marking correlates highly with dominance in the spider monkey. Social dominance cannot be predicted solely on the basis of sex and age but it might be predicted on the basis of marking. In spider monkeys, association, popularity, or affiliation are concentrated in the upper ranks more than is individual power (Dare, 1977). As we know, the spider monkey is not strongly dimorphic.

C. RHESUS AND JAPANESE MACAQUES (MACACA SPP.)

1. Rhesus monkeys (Macaca mulatta)

There is good evidence for a role for the male sex hormone testosterone in the dominance of rhesus monkeys. Even *infant female* monkeys who are given androgen show an increase in dominance and the dominance persists for a year after the last hormone injection (Joslyn, 1973).

When adult male and female rhesus are given DHTP (dihydrotestosterone proprionate) for six weeks their dominance rank rises to the highest levels in their groups and these changes in dominance persist (Cochran and Perachio, 1977). With a sudden and decisive defeat, on the other hand, the plasma testosterone in the defeated male decreases sharply (Rose, Gordon, and Bernstein, 1972). Apparently then, not only do hormonal changes result in social changes but social changes also lead to hormonal alterations.

When rhesus groups fight, males in the defeated group show depressed testosterone levels while the dominant male of the victors shows a brief threefold rise in testosterone (Bernstein, Gordon, and Rose, 1973).

In a study of the behavior of free-ranging castrated rhesus monkeys, six of ten castrated males fell in dominance after several years but two of them fought and dominated males twice their weight (Wilson and Vessey, 1968). Hormone levels alone do not explain all there is to know about status.

In rhesus, it may be that testosterone is directly related to behavior aimed at achieving and/or maintaining dominance but not to dominance status per se (Mazur, 1976). In any case, there is at least as good evidence for a relationship between testosterone and dominance as there is for a relationship between testosterone and sexual activity (Gordon, Rose, and Bernstein, 1977). In fact, the established dominance hierarchies of both males and females can control spontaneous and even electrically evoked (through brain stimulation) sexual and aggressive behavior (Perachio, 1973; Alexander and Perachio, 1973).

Aside from the role of testosterone in rhesus male dominance, it is the male rhesus that branch shakes, yawns, mounts, struts, chews and gnashes, emits warning barks, and moves with tail raised high, all of which are correlated with dominance (Marler, 1968; Hinde and Rowell, 1962; Møller, Harlow, and Mitchell, 1968).

In the establishment of dominance in dyads in the laboratory, male–male rhesus dyads fight more than do female–female rhesus dyads. Females are much less aggressive than males when dominance is to be established, at least in a dyad (Angermeier et al., 1968).

According to Bernstein and Sharpe (1966), the role of the dominant male is the most sharply differentiated of all roles in a rhesus troop, but the adult female core of each group plays the key role in determining the *band's* dominance status relative to other bands as well as its patterns of space utilization (Drickamer, 1975). Peripheral males do help in reversing the relative dominance of two troops (Gabow, 1973; Marsden, 1969), however.

In rhesus, females experience less trouble in forming stable groups than do males, but bands in the wild typically develop from a basic group of adult females dominated by an adult male (Vandenbergh, 1967); and, there is evidence from laboratory research that infants reared by both adult females and adult males are more active, aggressive, and more cooperative than are those raised by mothers alone (Suomi, 1974). The infants reared by both adult females and adult males tend to operate as teams and form coalitions or alliances more easily.

Dominance in the rhesus depends upon more than size. Alliances and coalitions referred to above are quite important to rhesus status (Varley and Symmes, 1966). This is especially true for females. The status of individuals in grooming networks correlates with dominance in female rhesus but not in male rhesus (Sade, 1972). As females become adults, they rank just below their mothers. Males do also, until puberty. At puberty males either gain or lose in rank or leave the troop (Sade, 1967). Brother and sister dominance depends upon age until the male reaches three or four years of age, at which time he automatically becomes dominant over his sister if he remains in the troop (Missakian, 1972). A female almost always ranks below her mother; however, if she acts independently of her mother, she may sometimes achieve a rank higher than her mother when the mother's rank falls (Marsden, 1968). Also, submissive females, even in groups of unfamiliar animals, can often gain status by eliciting the aid of dominant others (Masserman, Wechkin, and Woolf, 1968). Males may even mate with their mothers if they become dominant over them. Most of them, however, do not (Missakian, 1973).

The most likely time for a change in the male membership and hence dominance hierarchy of a troop of rhesus monkeys is during the breeding season. Physical strength and aggressiveness do not guarantee immediate success, however (Neville, 1968). Adult females and subadult males can chase off a solitary adult male that attempts to join their stable band. Old females can dominate a troop in this way (Neville, 1968).

According to Missakian-Quinn and Varley (1977), high-ranking males (top one-third of the hierarchy) account for 49 percent of all consort activity, the second one-third account for 21 percent, and the lowest one-third in rank account for 30 percent of all consort activity. The presence of a dominant male completely inhibits the occurrence of mounting in a subordinate male. Despite this, numbers of ejaculations do not differ from the top third to the bottom third, and similar conclusions are reported for female dominance. However, despite apparent ejaculation equivalence, low-ranked females in captivity do not get pregnant or remain pregnant as much as do high-ranked females, apparently because of high stress levels (Goo and Sassenrath, 1977).

Female choice undoubtedly plays an important role in mating, and preferences of females for certain males are related to factors other than male dominance (Lindburg, 1975). In a large group of rhesus monkeys undergoing a division on Cayo Santiago, off Puerto Rico, the dominant male has a low probability of fathering infants, central dominant and subordinate peripheral males cannot be differentiated by probabilities of fathering infants, and *individual* middle-ranking and peripheral males have high probabilities of siring infants. It is probable, however, that under more stable conditions dominant males may do better (Sade, Chepko-Sade, and Schneider, 1977). Nonetheless, it is true that while dominance in the rhesus may lead to power and control, it does not necessarily lead to reproductive success.

2. Japanese Monkeys (*Macaca fuscata*)

Among Japanese macaques, dominant individuals, usually males, have priority of access to such things as food. However, once food is in the hand it is rarely robbed even by a superior. This phenomenon is apparently sustained by a leader's supervision (Torii, 1975).

While testosterone levels have been correlated with the achieving or maintenance of dominance in rhesus macaques, there is little evidence, pro or con, for a role for hormones in dominance in Japanese macaques. Eaton and Resko (1974) found no correlation between testosterone levels and dominance in laboratory Japanese macaques and concluded that the adult level of androgen was less important than social factors. However, they did find that, in a confined troop, dominance displays occurred more frequently in males than in females and that individual displays were significantly correlated with ejaculation rank. It is notable, however, that the significant increase in the frequency of male displays during mating season was not correlated with individual dominance rank but only with ejaculation rank. Modahl and Eaton (1977) conclude that the displays themselves function to influence the female's choice of males more than to effect changes in dominance.

On the other hand, it has been reported (Stephenson, 1974) that higher class males mate with higher class females and that higher status males mate with females who have not as yet conceived whereas lower class males mate with lower class and already pregnant females. High-ranking Japanese macaque males are also known to intrude in the courtship activities of lower ranking males and particularly when the female in question has not yet conceived. In Japanese macaque males, age and rank are correlated.

The effect that other individuals may have on the rank of a given Japanese macaque has been studied extensively. In many cases, an animal's status is based on "dependent rank" which in turn depends upon other variables, as we shall see (Quick, 1977).

Among daughters, as we have discussed in our chapters on infant care and spacing, the "principle of youngest ascendency" applies. That is, among sisters over four years of age, the youngest sister ranks just below the mother. Among sons, youngest ascendency works only up to two or three years of age after which peripheralization complicates the issue so that, in general, older males dominate younger males (Koyama, 1967). As we know, almost all males eventually leave the natal troop, those remaining the longest being those closest to their mothers and those not peripheralized (Itoigawa, 1974). Age and length of time in the troop are positively correlated with dominance rank (Norikoshi and Koyama, 1975). Also, males who remain with a troop after troop fission are higher in rank than those who shift troops (Koyama, 1970).

When females are assigned to particular males by proximity to these males, many subgroups can be seen with the most dominant males being at the center of the subgroups containing the most females (Fujii, 1975). This gives one the distinct impression that the dominant central males move with the females, but that peripheral males move independently. It also helps to account for the "group focus role" of the dominant male in regard to group cohesion (Kurokawa, 1975).

Japanese monkey young males both lead the procession and follow the procession. Occasionally, dominant males go out to associate with them, but females with infants never do (Imanishi, 1963). There is apparently a linear rank system of dominance only among the males. If a male is to learn dominance, he must be the son of a dominant female who associates with dominant males (Imanishi, 1965), if there are dominant males.

The Bucknell colony of Japanese macaques is female dominated and controlled. In this colony, a female infant, offspring of the dominant female, displayed more dominant behaviors than either of two male infants that were present. Lorinc and Candland (1977) suggest that a matriarchal society produces infant sex roles different from those of a patriarchal society. It is definitely known that ". . . canines are not essential for either the attainment or maintenance of high rank" (Alexander and Hughes, 1971, P. 91) in the Japanese macaque.

Depending upon which of these two macaques we select for our studies of dominance we may get the impression that testosterone is either important or unimportant in dominance. However, it is quite clear from our survey that males of both species are usually but not necessarily dominant and that dominance may be, but is not necessarily, related to reproductive success. In the next chapter, we will examine some other Old World monkeys and the apes to see if the literature on sex differences in dominance on these species can tell us anything else about gender and dominance.

REFERENCES

Alexander, B. K. and Hughes, J. Canine teeth and rank in Japanese monkeys (*Macaca fuscata*). *Primates,* 1971, **12**, 91-93.

Alexander, M. and Perachio, A. A. An influence of social dominance on sexual behavior in rhesus monkeys. *American Zoologist,* 1970, **10**, 224.

Alexander, M. and Perachio, A. A. The influence of target sex and dominance in evoked attack in rhesus monkeys. *American Journal of Physical Anthropology,* 1973, **33** (2), 543-547.

Alvarez, F. Social hierarchy under different criteria in groups of squirrel monkeys, *Saimiri sciureus. Primates,* 1975, **16** (4), 437-455.

Angermeier, W. F., Phelps, J. B., Murray, S. and Howanstine, J. Dominance in monkeys: Sex differences. *Psychonomic Science,* 1968, **12** (7), 344.

Anschel, S. and Talmage-Riggs, G. Social structure dynamics in small groups of captive squirrel monkeys. Paper presented at the *International Primatological Society* Meeting, Cambridge, England, August, 1976.

Bearder, S. K. and Doyle, G. A. Field and laboratory studies of social organization in bushbabies (*Galago senegalensis*). *Journal of Human Evolution,* 1974, **3**, 37-50.

Benton, L. The establishment and husbandry of a black howler (*Alouatta caraya*) colony at Columbia Zoo. *International Zoo Yearbook,* 1976, **16**, 149-152.

Bernstein, I. S. Dominance, aggression, and reproduction in primate societies. *Journal of Theoretical Biology,* 1976, **60**, 459-472.

Bernstein, I. S., Gordon, T. P. and Rose, R. M. Plasma testosterone changes following merger of two rhesus monkey groups. *American Zoologist,* 1973, **13**, 1267.

Bernstein, I. S. and Sharpe, L. G. Social roles in a rhesus monkey group. *Behaviour,* 1966, **26**, 91-104.

Boelkins, R. C. Determination of dominance hierarchies in monkeys. *Psychonomic Science,* 1967, **7**, 317-318.

Candland, D. K. Changes in heartrate during social organization of the squirrel monkey (*Saimiri sciureus*). *Proceedings of the Third International Congress of Primatology* (Zurich 1970), 1971, **3**, 172-179.

Candland, D. K., Tyrell, D. S., Wagner, D. S. and Wagner, N. M. Social preferences of the squirrel monkey (*Saimiri sciureus*). *Folia Primatologica*, 1973, **19**, 437-449.

Carpenter, C. R. A field study of the behavior and social relations of howler monkeys (*Alouatta palliata*). *Comparative Psychology Monographs*, 1934, **10**, 168.

Clark, D. L. and Dillon, J. E. Social dominance relationships between previously unacquainted male and female squirrel monkeys. *Behaviour*, 1974, **50**, 217-231.

Cochran, C. A. and Perachio, A. A. Dihydrotestosterone proprionate effects on dominance and sexual behaviors in gonadectomized male and female rhesus monkeys. *Hormones and Behavior*, 1977, **8**, 175-187.

Dare, R. Power, affiliation, and status in primates: Analysis with a new set of indices. Paper presented at the *American Association of Physical Anthropologists* Meeting, Seattle, Washington, April, 1977.

Dawson, G. A. Behavioral ecology of the Panamanian tamarin, *Saguinus oedipus* (Callitrichidae, Primates). *Dissertation Abstracts International*, 1976, **B37**, 645-646.

Doyle, G. A. The behaviour of the lesser bushbaby. In Martin, R. D., Doyle, G. A. and Walker, A. C. (Eds.) *Prosimian Biology*. London: Duckworth, 1974, pp. 213-231.

Drickamer, L. D. Patterns of space utilization and group interactions among free-ranging *Macaca mulatta*. *Primates*, 1975, **16**, 23-33.

Eaton, G. G. and Resko, J. A. Plasma testosterone and male dominance in a Japanese macaque (*Macaca fuscata*) troop compared with repeated measures of testosterone in laboratory males. *Hormones and Behavior*, 1974, **5**, 251-259.

Eisenberg, J. F. and Kuehn, R. E. The behavior of *Ateles geoffroyi* and related species. *Smithsonian Miscellaneous Collections*, 1966, **151** (8), 1-63.

Epple, G. Quantitative studies on scent marking in the marmoset (*Callithrix jacchus*). *Folia Primatologica*, 1970, **13**, 48-62.

Epple, G. The behavior of marmoset monkeys (Callithricidae). In Rosenblum, L. A. (Ed.) *Primate Behavior* (Vol. 4). New York: Academic Press, 1975, pp. 195-239.

Fragaszy, D. M. Contrasts in feeding behavior in captive pairs of *Saimiri* and *Callicebus*. Paper presented at the *Animal Behavior Society* Meeting, Denver, Colorado, March, 1976.

Fujii, H. A psychological study of the social structure of a free-ranging group of Japanese monkeys in Katsuyama. In Kondo, S., Kawai, M. and Ehara, A. (Eds.) *Contemporary Primatology*. Basel: Karger, 1975, pp. 428-436.

Gabow, S. L. Dominance order reversal between two groups of free-ranging rhesus monkeys. *Primates*, 1973, **14**, 215-223.

Glander, K. E. Dominance in mantled howling monkeys. *American Journal of Physical Anthropology*, 1975, **42**, 303.

Goo, G. P. and Sassenrath, E. N. Dominance rank and reproduction in captive rhesus monkeys. Paper presented at the *American Society of Primatologists* Meeting, Seattle, Washington, April, 1977.

Gordon, T. P., Rose, R. M. and Bernstein, I. S. Social and hormonal influences on sexual behavior in the rhesus monkey. Paper presented at the *American Society of Primatologists* Meeting, Seattle, Washington, April, 1977.

Hinde, R. A. *Biological Bases of Human Social Behavior*. New York: McGraw-Hill, 1974.

Hinde, R. A. and Rowell, T. E. Communication by postures and facial expressions in the rhesus monkey (*Macaca mulatta*). *Proceedings of the Zoological Society of London*, 1962, **138**, 1-21.

Imanishi, K. Social behavior in Japanese monkeys, *Macaca fuscata*. In Southwick, C. H. (Ed.) *Primate Social Behavior*. Princeton, N.J.: Van Nostrand, 1963, pp. 68-81.

Imanishi, K. Identification: A process of socialization in the subhuman society of *Macaca fuscata*. In Imanishi, K. and Altmann, S. (Eds.) *Japanese Monkeys*. Atlanta: Altmann, 1965, pp. 30-51.

Itoigawa, N. Variables in male leaving a group of Japanese macaques. *Symposia of the Fifth Congress of the International Primatological Society* (Nagoya, Japan). Tokyo: Japan Science Press, 1974, pp. 233-244.

Joslyn, W. D. Androgen-induced social dominance in infant female rhesus monkeys. *Journal of Child Psychology and Psychiatry*, 1973, **14**, 137-145.

Kleiman, D. G. Monogamy in mammals. *The Quarterly Review of Biology*, 1977, **52**, 39-69.

Kortmulder, K. On ethology and human behaviour. *Acta Biotheoretica*, 1974, **23**, 55-78.

Koyama, N. On dominance rank and kinship of a wild Japanese monkey troop in Arashiyama. *Primates*, 1967, **11**, 189-216.

Koyama, N. Changes in dominance rank and division of a wild Japanese monkey troop of Arashiyama. *Primates*, 1970, **11**, 335-390.

Kurokawa, T. An experimental field study of cohesion in Katsuyama group of Japanese monkeys. In Kondo, S., Kawai, M. and Ehara, A. (Eds.) *Contemporary Primatology*. Basel: Karger, 1975, pp. 437-444.

Lindburg, D. G. Mate selection in the rhesus monkey, *Macaca mulatta*. *American Journal of Physical Anthropology*, 1975, **42**, 315.

Lorinc, G. A. and Candland, D. K. The primate mother-infant dyad: Intergenerational transmission of dominance in *Macaca fuscata*. Paper presented at the *Animal Behavior Society* Meeting, University Park, Pennsylvania, June, 1977.

Marler, P. Aggregation and dispersal: Two functions in primate communication. In Jay., P. D. (Ed.) *Primates: Studies in Adaptation and Variability.* New York: Holt, Rinehart and Winston, 1968, pp. 420-438.

Marsden, H. M. Agonistic behaviour of young rhesus monkeys after changes induced in social rank of their mothers. *Animal Behaviour,* 1968, **16**, 38-44.

Marsden, H. M. Dominance order reversal of two groups of rhesus monkeys in tunnel-connected enclosures. *Proceedings of the Second International Congress of Primatology* (Atlanta, Georgia, 1968). Basel: Karger, 1969, pp. 52-58.

Masserman, J. H., Wechkin, S. and Woolf, M. Social relationships and aggression in rhesus monkeys. *Archives of General Psychiatry,* 1968, **18**, 210-213.

Maurus, M., Hartmann, E. and Kühlmorgen, B. Invariant qualities in communication processes of squirrel monkeys. *Primates,* 1974, **15**, 179-192.

Maurus, M., Kühlmorgen, B., Hartmann-Wiesner, E. and Pruscha, H. An approach to the interpretation of the communicative meaning of visual signals in agonistic behavior of squirrel monkeys. *Folia Primatologica,* 1975, **23**, 208-226.

Maurus, M. and Ploog, D. Social signals in squirrel monkeys: Analysis by cerebral radio stimulation. *Experimental Brain Research,* 1971, **12**, 171-183.

Mazur, A. A cross-species comparison of status in small established groups. *American Sociological Review,* 1973, **26**, 214-226.

Menzel, E. W., Jr. Primate naturalistic research and problems of early experience. *Developmental Psychobiology,* 1968, **1** (3), 175-184.

Missakian, E. A. Genealogical and cross-genealogical dominance relations in a group of free-ranging rhesus monkeys (*Macaca mulatta*) on Cayo Santiago. *Primates,* 1972, **13**, 169-180.

Missakian, E. A. Genealogical mating activity in free-ranging groups of rhesus monkeys (*Macaca mulatta*) on Cayo Santiago. *Behaviour,* 1973, **45**, 225-241.

Missakian-Quinn, E. A. and Varley, M. A. Consort activity in groups of free-ranging rhesus monkeys on Cayo Santiago. Paper presented at the *American Society of Primatologists* Meeting, Seattle, Washington, April, 1977.

Modahl, K. B. and Eaton, G. G. Display behaviour in a confined troop of Japanese macaques (*Macaca fuscata*). Paper presented at the *American Society of Primatologists* Meeting, Seattle, Washington, April, 1977.

Møller, G. W., Harlow, H. F. and Mitchell, G. Factors affecting agonistic communication in rhesus monkeys (*Macaca mulatta*). *Behaviour,* 1968, **31**, 339-357.

Neville, M. K. A free-ranging rhesus monkey troop lacking adult males. *Journal of Mammalogy,* 1968, **49**, 771-773.

Neville, M. K. Male leadership change in a free-ranging troop of Indian rhesus monkeys (*Macaca mulatta*). *Primates,* 1968, **9**, 13-27.

Norikoshi, K. and Koyama, N. Group shifting and social organization among Japanese monkeys. *Symposia of the Fifth Congress of the International Primatological Society.* Tokyo: Japan Science Press, 1975, pp. 43-61.

Oppenheimer, J. R. and Oppenheimer, E. C. Behavioral observations of *Cebus nigrivittatus* on the Venezuelan llanos. *American Zoologist,* 1972, **12**, 69.

Perachio, A. A., Alexander, M. and Marr, L. D. Hormonal and social factors affecting evoked sexual behavior in rhesus monkeys. *American Journal of Physical Anthropology,* 1973, **38**, 227-232.

Perret, M. Influence du groupement social sur l'activation sexuelle saisonniere chez le ♂ de *Microcebus murinus* (Miller, 1777). *Zeitschrift für Teirpsychologie,* 1977, **43**, 159-179.

Peters, M. Mouth to mouth contact in squirrel monkeys (*Saimiri sciureus*). *Zeitschrift für Tierpsychologie,* 1970, **27**, 1009-1010.

Quick, L. B. The social environment behavior of Japanese macaques. Paper presented at the *American Association of Physical Anthropologists* Meeting, Seattle, Washington, April, 1977.

Rose, R. M., Gordon, T. P. and Bernstein, I. S. Plasma testosterone levels in the male rhesus: Influences of sexual and social stimuli. *Science,* 1972, **178**, 643-645.

Sade, D. S. Determinants of dominance in a group of free-ranging rhesus monkeys. In Altmann, S. A. (Ed.) *Social Communication Among Primates.* Chicago: University of Chicago Press, 1967, pp. 99-114.

Sade, D. S. Sociometrics of *Macaca mulatta.* I: Linkages and cliques in grooming matrices. *Folia Primatologica,* 1972, **18**, 196-223.

Sade, D. S., Chepko-Sade, B. D. and Schneider, J. Paternal exclusions among free-ranging rhesus monkeys on Cayo Santiago. Paper presented at the *Animal Behavior Society* Meeting, University Park, Pennsylvania, June, 1977.

Snyder, P. A. Behavior of *Leontopithecus rosalia* (Golden lion marmoset) and related species: A review. *Journal of Human Evolution,* 1974, **3**, 109-122.

Sorenson, M. W. A review of aggressive behavior in the tree shrews. In Holloway, R. L. (Ed.) *Primate Aggression, Territoriality and Xenophobia: A Comparative Perspective.* New York: Academic Press, 1974, pp. 13-30.

Stephenson, G. R. Social structure of mating activity in Japanese macaques. *Symposia of the Fifth Congress of the International Primatological Society* (Nagoya, 1974). Tokyo: Japan Science Press, 1974, pp. 63-115.

Suomi, S. J. Social interactions of monkeys reared in a nuclear family environment versus monkeys reared with mothers and peers. *Primates,* 1974, **15**, 311-320.

Syme, G. J. Competitive orders as measures of social dominance. *Animal Behaviour,* 1974, **22**, 931-940.

Symons, D. The question of function: Dominance and play. Paper presented at the *Animal Behavior Society* Meeting, University Park, Pennsylvania, June, 1977.

Talmage-Riggs, G. and Anschel, S. Homosexual behavior and dominance hierarchy in a group of captive female squirrel monkeys (*Saimiri sciureus*). *Folia Primatologica*, 1973, **19**, 61-72.

Torii, M. Possession by non-human primates. In Kondo, S., Kawai, M. and Ehara, A. (Eds.) *Contemporary Primatology*. Basel: Karger, 1975, pp. 310-314.

Varley, M. and Symmes, D. The hierarchy of dominance in a group of macaques. *Behaviour*, 1966, **27**, 54-75.

Wickler, W. Socio-sexual signals and their intraspecific imitation among primates. In Morris, D. (Ed.) *Primate Ethology*. Chicago: Aldine, 1967, pp. 69-147.

Wilson, A. P. and Vessey, S. H. Behavior of free-ranging castrated rhesus monkeys. *Folia Primatologica*, 1968, **9**, 1-14.

17.

Dominance: Other Old World Monkeys, Apes, and Implications for Humans

As we have seen in Chapter 16, males are usually dominant over females in physically dimorphic species but are usually not dominant over females in relatively monomorphic species. Dominance, however, is also affected by hormonal levels, age, kinship, coalition, group change, and it is situationally dependent. In the present chapter we will continue our review of sex differences in dominance, starting with Old World monkeys other than *Macaca mulatta* and *Macaca fuscata*.

A. OTHER MACAQUES

1. Pigtail Macaques (*M. nemestrina*)

Among pigtail monkeys, the most dominant animal in a group is usually a male. Once a group has been together for some time, the pigtail status hierarchy remains remarkably stable if a dominant male is present. Even a normal replacement of the alpha male under natural conditions does not change the rank order structure very significantly (Bernstein, 1969*). However, if a pigtail group is artificially formed in captivity, there is actually more disruption than there is in capuchin or rhesus monkey group formation (Bernstein, 1969[†]).

The dominant male pigtail displays what has been referred to as a "control role." He defends the group and prevents internal disturbance (Smith, 1973). Groups not having a control male show increased aggressiveness in captivity (Oswald and Erwin, 1976). Female dominance is important as far as sexual behavior is concerned. Females low in social dominance apparently show low frequencies of sexual behavior (even in a preovulatory condition)

*Folia Primatologica.
[†]*Animal Behaviour.*

and they rarely conceive (Goldfoot, 1971). Even the most dominant females are subordinate to the alpha male, however. Male dominance in the pigtail monkey also extends to situations where pigtail and bonnet (*M. radiata*) macaques live in heterospecific groups. The males of both these species are dominant over the females of both species (Stynes, Kaufman, and Reiser, 1975).

2. Bonnet Macaques (*M. radiata*)

Despite the fact that bonnets show friendlier behavior toward conspecifics than do pigtails, the male bonnet is also dominant over the female bonnet (and, as mentioned above, even over the female pigtail). There is a straight linear ranking among bonnet monkeys and this hierarchy is more rigid in males than in females (Koyama, 1973). During the mating season, social tensions can be quite high and as the relative ranks of the males change, their tactile relationships also change (Simonds, 1977).

Despite the rather clear and even rigid hierarchy of the bonnet males, however, the bonnet ranking is still less clear-cut than is the case for the Japanese monkey. Bonnet males seem to have a greater tolerance for one another. On occasion they even embrace one another. (This embracing is never seen in Japanese macaque males.) In the bonnet macaque, there may be several males that play the "control role" by defending the group and mediating quarrels (Sugiyama, 1971). The toque macaque (*M. sinica*) displays a dominance system quite similar to that of the bonnet (cf. Dittus, 1975).

3. The Stumptail Macaque (*M. arctoides*)

A linear dominance hierarchy can also be detected in free-ranging stumptail groups. As in rhesus and Japanese macaques, offspring often rank immediately below their mothers. Dominant individuals are attractive to others and there is usually an alpha male who is the focus of attention for immatures. In stumptails, individuals who hold adjacent ranks tend to interact with each other more than with those of distant rank (Estrada, 1976). Maternal rank influences the quality and quantity of infant experience. If a mother is dominant, her infant often receives more submissive behavior and more social contact (including grooming) than do other infants. Her infant is also harassed less frequently (Gouzoules, 1975).

Plasma testosterone levels are related to social rank among both male and female stumptail monkeys (Kling and Dunne, 1976). Female dominance also increases after the female delivers an infant (unless the infant is stillborn) (Weisband and Goy, 1976).

4. The Crab-Eating Macaque (*M. fascicularis*)

Male rhesus establish dominance by direct physical encounter whereas females tend to establish dominance by withdrawal of the subordinate animal before physical interaction takes place. In the closely related crab-eating macaque (*M. fascicularis*), *females* apparently engage in more physical aggression when establishing dominance (Hendricks, Seay, and Barnes, 1975). When the most dominant male and female are removed from small captive groups of crab-eaters, the social behavior of the remaining group members changes drastically. Nineteen out of forty behavior categories changed in one study (Hendricks, Seay, and Barnes, 1975).

Even though changes in dominance often come about by physical means in this species, aggression is not inevitable, and the change in rank can be gradual despite physical wounding. In one captive group of crab-eaters, an old alpha male was injured by a bite from his adult son. Tension between the males was evident and the original leader showed submission. Following the physical confrontation, aggression increased in the rest of the group, particularly aggression by the males toward others. But the older dominant male began to show appeals for aggressive help from others and *especially* from the dominant son that had wounded him. He received the aid he "asked" for, and the two males helped each other in restoring and maintaining the social order during the transition period from old leader to new leader. DeWaal (1975) interprets these data as indicating that there is some amount of "respect for aging leaders" in crab-eating macaques.

There are roles for dominant macaque males other than group defense and internal control. In some species (e.g., *M. sylvana*), dominant males play an active role in infant care and infant socialization (Lahiri and Southwick, 1966).

B. OTHER OLD WORLD MONKEYS

1. Talapoin Monkeys (*Miopithecus talapoin*)

The pattern of sex differences in dominance seen in macaques does not apply to talapoin monkeys. Female talapoins rank above males (Dixson and Herbert, 1974). The status of females relative to males also apparently increases during the breeding season (Rowell and Dixson, 1975). As infants and juveniles, males are more assertive than are females, but as adults, the reverse is true. Female talapoins will frequently "gang up" on males, at least in captivity (Wolfheim, 1977). In two captive groups studied by Dixson, Herbert, and Rudd (1972), females dominated males in both groups. The status of females was not as dependent upon gonadal hormones as was the status of males. Although males are slightly heavier than females, they are of lower rank and are sometimes mounted by high-ranking females. Females also mount other females, although dominant females sometimes prefer to *be* mounted by others of their own sex (Dixson, Scruton, and Herbert, 1975). Despite the fact that males are subordinate to females they prefer the most dominant females as sexual partners (Keverne, 1976).

2. Baboons

Among Old World monkeys other than talapoins, male dominance appears to be the rule. Dominance ranks are usually more obvious in males, and it is an adult male that usually has the highest rank. This is particularly so in those species showing pronounced sexual dimorphism (Bernstein, 1970). Males seem to be more status conscious than do females (Nagel and Kummer, 1974). Despite the apparent rigidity of many Old World monkey male hierarchies, "control male" behaviors are sometimes actively displayed by individuals other than an alpha male. In addition, "control male" behavior is not dependent on the presence of a well-formed dominance hierarchy (Bernstein, 1974). However, if there is some instability in the male status hierarchy, troop fission is likely to follow (Nash, 1976).

Even in the baboons, where sexual dimorphism is very pronounced, males change ranks frequently and no one kind of social behavior is correlated with dominance (Kolata, 1976). In *Papio cynocephalus* (the yellow baboon), branch-shaking and prancing *are* performed primarily by adult and subadult males, but these behaviors are also occasionally displayed by adult females (Anthoney, 1969). (This is also true of guinea baboons, *Papio papio*. See Dunbar and Nathan, 1972). Vertical tail carriage, a behavior thought to be related to dominance in macaques, apparently shows no relationship to dominance in the yellow baboon (Hausfater, 1977).

Estrus does not affect female dominance in *P. cynocephalus,* but the presence of estrous females does produce dominance changes in the males (Hausfater, 1975b) as they compete for females. In terms of reproductive success, those males having longevity in a group and a high rank at the start of their reproductive careers hold an edge over other males (Saunders and Hausfater, 1977). Hausfater (1975a) has written an extensive monograph on dominance and reproduction in the yellow baboon. Among olive baboons (*P. anubis*), estrus is not a major disruptive factor, but low-ranking females in estrus change more than do high-ranking females (Rowell, 1969).

The chacma baboon (*P. ursinus*) female has a higher rank when she is in estrus although males are still dominant over her (Bolwig, 1959). The most aggressive male may be dominant in individual encounters; but, in one troop, "the male ranking lowest in individual aggressive encounters initiated significantly more troop progressions and was most sexually active." (Saayman, 1971, P. 36.) This was an old male.

The hamadryas (*P. hamadryas*) male is dominant over the females of his harem. A female asserts her dominance over other females in the harem by getting close to the male and then threatening other females (Crook, 1968). Males may also use others in dominance interactions. The "agonistic buffering" techniques described in previous chapters are employed by subadult hamadryas males (Gilmore, 1977). Old harem males tend to lose control of their females to a rival male who grows up in his own harem. In fact, he loses his harem to the male with whom he is most intimate (Kummer et al., 1976).

In a different genus of baboons (*Mandrillus*), coloration of the male is extremely important in maintaining dominance. In *M. sphinx,* the posterior coloration of the dominant male, which is almost luminous, shows up well in the dark forest undergrowth. His penile coloration (red) also has a dominance function, and the elaborate facial coloration has apparently evolved through rivalry between adult males for females (Jouventin, 1975).

As do hamadryas baboon males, gelada baboon (*Theropithecus gelada*) males use "agonistic buffering" in dominance interactions (Gilmore, 1977). In their one-male dominated groups, females have a hierarchy which is maintained by each female's orientation to the male (as in hamadryas). Estrus has no effect on the dominance behavior of the male and little effect on the females, although there is a tendency for estrous females to be harassed by the more dominant females in their one-male units (Dunbar, 1976).

3. *Cercopithecus spp.*

In the terrestrial African vervet (*Cercopithecus aethiops*) dominance gestures like canine display, body bounce, and tail curve are limited to males and the head-bob and erect posture occur significantly more frequently in males than in females. Eye flashes and gapes are also seen more often in males than in females whereas females display more submissive crouching (Durham, 1969). In *C. aethiops,* as in some macaques, there is evidence that an animal's rank depends upon the rank of its mother. The relative ranking among males is unstable, whereas the dominance hierarchy of females is very stable. While males are usually dominant over females, female coalitions can dominate males (Lancaster, 1973).

In the more arboreal species of the genus *Cercopithecus* (e.g., *C. campbelli lowei*), males are also dominant over females. While the alpha male acts as a watchdog and control animal, he does not exert a strong dominance and is seen more often alone than not. On the other hand, he is groomed by adult females, so the male of this species is not quite as isolated as is the patas male (Hunkeler, Bourliere, and Bertrand, 1972).

4. Langurs (Presbytis spp.)

According to Dolhinow and co-workers, *P. entellus* females do not have marked or linear dominance hierarchies. Age and reproductive success among females are not related to dominance based upon displacement. When a female gives birth, however, her dominance status does increase (Vonder Haar Laws, Dolhinow, and McKenna, 1977). Others disagree with these conclusions concerning female dominance. Hrdy and Hrdy (1976), for example, believe that the female langur dominance hierarchy has a lot to do with reproductive success.

Unlike the females, the dominance hierarchy of adult males is rigid (Jay, 1963) and males are dominant over females. Even subadult males are dominant over adult females (Dolhinow and Bishop, 1970).

Among Nilgiri langurs (*P. johnii*), males are again dominant over females but when adult females do have dominance encounters, these encounters seem to create more tension and excitement than do male dominance interactions. There are more loud vocalizations in female dominance squabbles. A dominant male, on the other hand, does not need to vocalize to emphasize itself or call attention to a subordinate, although he does occasionally emit a dominance grunt or grind his canines (Poirier, 1970). But these rules regarding male and female langur dominance are not hard and fast ones. As Poirier (1970) has pointed out, dominance hierarchies vary not only with the species but also with the troop, with the "personalities" of the animals, and especially with the "personality" of the alpha male. The same thing could be said about any genus of Old World monkeys.

C. APES

1. Chimpanzees (Pan troglodytes)

Among chimpanzees, mature males are generally dominant over adult females, although there may be a reversal in dominance between the male and female when a female is at the height of her sexual swelling. This occurs even in female–female dyads. The administration of estrogen increases female dominance (Birch and Clark, 1950) and the estrogen-treated female tends to receive a lot of grooming. In fact, chimpanzees in general tend to direct more grooming toward a dominant partner than toward a subordinate one (Mason, 1970). Dominant-acting *strangers,* however, are treated with caution and wariness (Juno, Miller, and Nadler, 1977).

As in monkeys, dominance and leadership are not always equivalent. In one group of chimpanzees studied by Menzel (1973), the *least* preferred leader was the most dominant male unless the activity involved was defense against a potential predator.

Male chimpanzees are, in general, more assertive than females in social interactions. Males tend to initiate social interactions more frequently than do females, particularly in the juvenile period (Nadler and Braggio, 1974).

2. Gorillas

The silver-backed male gorilla (*Gorilla gorilla*) is the most dominant animal in a gorilla group, but he is not necessarily the leader in every activity (Menzel, 1973). Chest beating and many vocalizations are associated with dominance in gorilla males (Fossey, 1972).

As subadult and adult females begin to show the slight sexual swelling typical of female gorillas they also begin to show chest beating. Thus, as in female chimpanzees, female dominance is related to estrus in the gorilla (Keiter, 1977; see also Hess, 1973; Nadler, 1976).

3. Orangutan

Adult male orangutans are dominant over females and are more assertive in social interactions (Nadler and Braggio, 1974). There is an obvious physical sexual dimorphism in orangutans, so much so that it is possible for the male to take the female sexually by force (Rijksen, 1975). (However, gorillas also show marked sexual dimorphism but the males do not take the females by force.) Adult male orangutans show an impressive display accompanied by a "long call" vocalization. The reaction of the receiver to such a call depends upon his dominance rank (Rijksen, 1975). Among immature animals, however, dominance appears to depend on physical size more than on gender (Okano, 1971).

D. IMPLICATIONS FOR HUMANS

1. Human Children

Braggio et al. (1976) have directly compared chimpanzees, orangutans, and children and have found that, in all three species, males are more *assertive* than are females. In children aged seven to forty-five months reared at Synanon (an institution where children are peer-reared) however, both boys and girls appear at high, middle, and low dominance ranks (Missakian, 1976). The authors interpreted this finding as an indication that peer-rearing in humans reduces gender differences (see also Milch and Missakian-Quinn, 1977). In addition, Jeffers (1977) has shown that dominance in preschool children is situation specific. Even when a preschool child has previously been subordinate, he or she may be dominant in his or her own home. Preschool boys, however, show higher levels of aggression at home and away from home, whether peer-reared or parent-reared.

In mixed-sex classrooms of preschool children, boys tend to be at the top of the dominance hierarchy with females at the bottom, but with substantial overlap. In single-sexed preschool classrooms, however, it is as important for a girl to be tough as it is for a boy and the hierarchies which develop are very similar (Omark and Brenner, 1976).

In preadolescent and adolescent children, females submit more readily when competing against males than when competing against other females. In contrast, even boys who submit readily in all-boy games are seen to compete more actively in mixed-sex games (Cronin, Callaghan, and Weisfeld, 1977). Despite this difference, it is interesting that in middle childhood, both sexes of children already *perceive* the same-sex parent as being the most dominant and most punitive (Goldin, 1969). Apparently this is an indication of already well-formed like-sex or same-sex dominance hierarchies.

In high school boys, dominance reflects athletic ability and physical attractiveness rather than intelligence. The important social traits are shaped six or more years prior to high school (Weisfeld, Omark, and Cronin, 1977).

2. Human Adults

In terms of what could be predicted on the basis of degree of sexual dimorphism, people are intermediate between species like baboons, where males may weigh two to three times as much as females, and gibbons, in which males and females are nearly equal in size. In people, however, there is much individual variation within sex and a great deal of overlap between the sexes. Thus, there is a fairly high frequency of female dominance over males (cf. Scott, 1974).

In within-sex male human dominance hierarchies, there is often real and fantasized sexual inhibition of the subordinate male in the presence of the higher ranking male just as we have seen in some of the macaques (cf. Jensen, 1976). Whether mixed-sex groups of adults consistently show male dominace over females or not, the kinds of advertising for men's products seen on television and the products themselves for men suggest male dominance over females. For example, men's watches show more prestige, more accuracy, and greater size than do women's watches (Wagner, 1975). There is evidence, however, that a large share of these kinds of differences in people are the result of differential rearing and that they are often "sexist" and not based upon real *biological* group differences. As our modern society becomes less and less a subsistence society and more and more one which requires less sexual differentiation in roles, these differences in dominance, or perceived dominance, become less necessary. Russell and Russell (1971) have guessed that, among primates, the more dangerous the environment, the greater the sexual differentiation, and the more male dominance is seen. Whether this applies to people (or to other primates for that matter) is open to question.

With regard to the implications of dominance hierarchies in people there are those who feel that, regardless of sex differences, competition and dominance are unhealthy for modern life. Assuming that human males have dominance over human females, Price (1967) has noted that:

> Depression may be commoner in females, and this accords with the fact that in animal groups the status of a female depends not only on the status of her male, but also on her status within her male's harem. In the females of some species a reduction of status occurs every month between ovulation and menstruation, as she becomes sexually less attractive: It may not be fortuitous that depression is common in our own species at this time (P. 5)

The importance of dominance, competition, and assertiveness to people is also evident in sports. As Maple and Howard (1977) have noted:

> In victory, the individual assumes an erect posture, leaps, clasps hands, shakes arms and fists, and "dances". . . . In defeat, subjects respond in a manner which appears to be the antithesis of victory: shoulders hunched, head down, movements slow . . . . Female athletes emit the same range of behaviors as do males, but they seem to be somewhat less exuberant. (Abstract)

However, studies of human dominance and assertiveness do not always report greater male assertiveness. In answers given on the College Self-Expression Scale, for instance, female students were more assertive than were male students. As is often found in children, each sex was significantly more assertive toward members of the same than toward members of the opposite sex (Stebbins et al., 1977).

Among college student females, there is evidence that the status of a male affects the evaluations by others (especially by other females) of a female in his company. Females evaluated a female, in the company of an attractive male having high status, as being intelligent, self-confident, and friendly. Evaluations by men of the same female were unaffected by the status of her male partner (Sheposh, Deming, and Young, 1977). There is other evidence that women respond more intensely to authority figures than do men, and that females show extreme responses to female authority figures (that is, women in authority generate more extreme reactions than do males in authority and particularly so in other women) (Wright, 1976).

This tendency to rate a woman according to the nature of the man in her company is probably related to a similar phenomenon often seen following marriage. Nonmarried and nonrelated men and women are usually rated stereotypically regardless of the roles they play whereas:

> . . . marriage facilitates the expression of stereotypically "masculine" charactersitics in women and of stereotypically 'feminine' characteristics in men. (Gerber and Balkin, 1977, P. 9)

There are some who even feel that the hormone testosterone, apparently implicated in dominance in some species of Old World monkeys, is also related to activities of achieving or maintaining status in people. The testosterone model, however, is not limited to human males. As Mazur (unpublished) has said:

> Needless to say, the involvement of a "male hormone" in status processes provides no moral justification for the privileged position of males in most societies. (P. 13)

In contrast to the emphasis on hormones advanced by Mazur and the assumptions of human male dominance seen in many studies, Maccoby and Jacklin (1974), in their book on the psychology of sex differences, make the following statement after an extensive review of the literature:

> We doubt, however, that dominance and leadership are inevitably linked to aggression. . . . aggression is certainly not the method most usually employed for leadership among mature human beings. (P. 368)

They conclude that whether or not male humans are more dominant than females is an open question. They say that there is too little evidence or that the findings are ambiguous, but that:

> Among adult mixed pairs or groups, formal leadership tends to go to males in the initial phases of interaction, but the direction of influence becomes more equal the longer the relationship lasts . . . . (P. 353)

This is in agreement with the results related to marriage reported above.

REFERENCES

Anthony, T. R. Threat activity in wild and captive groups of savannah baboons. *Proceedings of the Second International Congress of Primatology.* Basel/New York: Karger, 1969, pp. 108-113.

Bernstein, I. S. Introductory techniques in the formation of pigtail monkey troops. *Folia Primatologica,* 1969, **19**, 1-19. (a)

Bernstein, I. S. Stability of the status hierarchy in a pigtail monkey group (*Macaca nemestrina*). *Animal Behaviour,* 1969, **17**, 452-458. (b)

Bernstein, I. S. Primate status hierarchies. In Rosenblum, L. A. (Ed.) *Primate Behavior* (Vol. 1). New York: Academic Press, 1970, pp. 71-109.

Bernstein, I. S. Principles of primate group organization. In Chiarelli, A. B. (Ed.) *Perspectives in Primate Biology.* New York: Plenum, 1974, pp. 283-298.

Birch, H. G. and Clark, G. Hormonal modification of social behavior. IV: The mechanism of estrogen-induced dominance in chimpanzees. *Journal of Comparative Physiological Psychology,* 1950, **43**, 181-193.

Bolwig, N. A study of the behavior of the chacma baboon, *Papio ursinus. Behaviour,* 1959, **14**, 136-163.

Braggio, J. T., Nadler, R. D., Lance, J. and Myseko, D. Sex differences in apes and children. Paper presented at the *International Primatological Society* Meeting, Cambridge, England, August, 1976.

Cronin, C. L., Callaghan, J. W. and Weisfeld, G. E. Sex differences in competitive behavior in children. Paper presented at the *Animal Behavior Society* Meeting, University Park, Pennsylvania, June, 1977.

Crook, J. H. Primate societies and individual behaviour. *Journal of Psychosomatic Research,* 1968, **12**, 11-19.

DeWaal, F. B. M. The wounded leader: A spontaneous temporary change in the structure of agonistic relations among captive Java monkeys (*Macaca fascicularis*). *Netherlands Journal of Zoology,* 1975, **25** (4), 529-549.

Dittus, W. P. J. Dynamics of migration in *Macaca sinica* (Linnaeus, 1771) and its relation to male dominance, reproduction, and mortality. *American Journal of Physical Anthropology,* 1975, **42**, 298.

Dixson, A. F. and Herbert, J. Gonadal hormones and aggressive behaviour in captive groups of talapoin monkeys (*Miopithecus talapoin*). *Journal of Endocrinology,* 1974, **61**, 46.

Dixson, A. F., Herbert, J. and Rudd, B. T. Gonadal hormones and behaviour in captive groups of talapoin monkeys (*Miopithecus talapoin*). *Journal of Endocrinology,* 1972, **57**, 41.

Dixson, A. F., Scruton, D. M. and Herbert, J. Behaviour of the talapoin monkey (*Miopithecus talapoin*) studied in groups in the laboratory. *Journal of Zoology* (London), 1975, **176**, 177-210.

Dolhinow, P. J. and Bishop, N. The development of motor skills and social relationships among primates through play. In Hill, J. P. (Ed.) *Minnesota Symposia on Child Psychology* (Vol. 4). Minneapolis, Minnesota: University of Minnesota Press, 1970, pp. 141-198.

Dunbar, R. I. M. Oestrous behaviour and social relations among gelada baboons. Paper presented at the *International Primatological Society* Meeting, Cambridge, England, August, 1976.

Dunbar, R. I. M. and Nathan, M. F. Social organization of the Guinea baboon, *Papio papio. Folia Primatologica*, 1972, **17**, 321-334.

Durham, N. M. Sex differences in visual threat displays of West African vervets. *Primates*, 1969, **10**, 91-95.

Estrada, A. Social relations in a free-ranging troop of *Macaca arctoides*. Paper presented at the *International Primatological Society* Meeting, Cambridge, England, August, 1976.

Fossey, D. Vocalizations of the mountain gorilla (*Gorilla gorilla beringei*). *Animal Behaviour*, 1972, **20** (1), 36-53.

Gerber, G. L. and Balkin, J. Sex-role stereotypes as a function of marital status and role. *The Journal of Psychology*, 1977, **95**, 9-16.

Gilmore, H. A. The evolution of "agonistic buffering" in baboons and macaques. Paper presented at the *American Association of Physical Anthropologists* Meeting, Seattle, Washington, April, 1977.

Goldfoot, D. A. Hormonal and social determinants of sexual behavior in the pigtail monkey (*Macaca nemestrina*). In Stoelinga, G. B. A. and van der Werff ten Bosch, J. J. (Eds.) *Normal and Abnormal Development in Brain and Behavior.* Leiden: University of Leiden Press, 1971, pp. 325-342.

Goldin, P. G. A review of children's reports of parent behaviors. *Psychological Bulletin*, 1969, **71** (3), 222-236.

Gouzoules, H. Maternal rank and early social interactions of infant stumptail macaques, *Macaca arctoides. Primates*, 1975, **16** (4), 405-418.

Hausfater, G. Dominance and reproduction in baboons (*Papio cynocephalus*). *Contributions to Primatology*, 1975, **7**, 1-150. (a)

Hausfater, G. Estrous females: Their effects on the social organization of the baboon group. *Proceedings of the Symposia of the Fifth International Congress of Primatology.* Tokyo: Japan Science Press, 1975, pp. 117-127. (b)

Hausfater, G. Tail carriage in baboons (*Papio cynocephalus*): Relationships to dominance rank and age. *Folia Primatologica*, 1977, **27**, 41-59.

Hendricks, D. E., Seay, B. M. and Barnes, B. The effects of the removal of dominant animals in a small group of *Macaca fascicularis. Journal of General Psychology*, 1975, **92**, 157-168.

Hess, J. P. Some observations on the sexual behavior of captive lowland gorillas, *Gorilla g. gorilla* (Savage and Wyman). In Michael, R. P. and Crook, J. H. (Eds.) *Comparative Ecology and Behavior of Primates.* New York: Academic Press, 1973, pp. 508-581.

Hrdy, S. B. and Hrdy, D. B. Hierarchal relations among female hanuman langurs (Primates: Colobinae, *Presbytis entellus*). *Science*, 1976, **193**, 913-915.

Hunkeler, C., Bourliére, F. and Bertrand, M. Le comportement social de la Mone de Lowe (*Cercopithecus campbelli lowei*). *Folia Primatologica,* 1972, **17**, 218-236.

Jay, P. The Indian langur monkey (*Presbytis entellus*). In Southwick, C. H. (Ed.) *Primate Social Behavior.* Princeton, N.J.: Van Nostrand, 1963, pp. 114-123.

Jeffers, V. Situation specificity of social dominance in preschool children. Paper presented at the *Animal Behavior Society* Meeting, University Park, Pennsylvania, June, 1977.

Jensen, G. D. Comparisons of sexuality of chimpanzees and humans. Paper presented at the *International Primatological Society* Meeting, Cambridge, England, August, 1976.

Jouventin, P. Les rôles des colorations du mandrill (*Mandrillus sphinx*). *Zeitschrift für Tierpsychologie,* 1975, **39**, 445-462.

Juno, C. J., Miller, L. C. and Nadler, R. D. Response to strangers in young chimpanzees. Paper presented at the *Animal Behavior Society* Meeting, University Park, Pennsylvania, June, 1977.

Keiter, M. D. Reproductive behavior in subadult captive lowland gorillas (*Gorilla gorilla gorilla*). Paper presented at the *American Society of Primatologists* Meeting, Seattle, Washington, April, 1977.

Keverne, E. B. Dominance, aggression, and sexual behaviour in social groups of talapoin monkeys. Paper presented at the *International Primatological Society* Meeting, Cambridge, England, August, 1976.

Kling, A. and Dunne, K. Social-environmental factors affecting behavior and plasma testosterone in normal and amygdala lesioned *M. speciosa. Primates,* 1976, **17** (a), 23-42.

Kolata, G. B. Primate behavior: Sex and the dominant male. *Science,* 1976, **191**, 55-56.

Koyama, N. Dominance, grooming, and clasped-sleeping relationships among bonnet monkeys in India. *Primates,* 1973, **14** (2-3), 225-244.

Kummer, H., Abegglen, H., Bachman, C., Falett, J. and Sigg, H. Intimacy and fear of aggression in baboon social relationships. Paper presented at the *International Primatological Society* Meeting, Cambridge, England, August, 1976.

Lahiri, R. K. and Southwick, C. H. Parental care in *Macaca sylvana. Folia Primatologica,* 1966, **4**, 257-264.

Lancaster, J. B. Stimulus/response: In praise of the achieving female monkey. *Psychology Today,* 1973, **7**, 30-102.

Maccoby, E. E. and Jacklin, C. *The Psychology of Sex Differences.* Stanford, California: Stanford University press, 1974.

Maple, T. and Howard, S. The thrill of victory and the agony of defeat: Ethological correlates. Paper presented at the *American Society of Primatologists* Meeting, Seattle, Washington, April, 1977.

Mason, W. A. Chimpanzee social behavior. In Bourne, G. (Ed.) *The Chimpanzee* (Vol. 2). New York: Karger, 1970, pp. 265-288.

Mazur, A. Effects of testosterone on status in primate groups. *Folia Primatologica*, 1976, **26**, 214-226.

Menzel, E. W., Jr. Leadership and communication in young chimpanzees. *Symposia of the Fourth International Congress of Primatology*. Basel: Karger, 1973, pp. 192-225.

Milch, K. H. and Missakian-Quinn, E. Longitudinal study of social play in peer-raised children. Paper presented at the *Animal Behavior Society* Meeting, University Park, Pennsylvania, June, 1977.

Missakian, E. A. Gender differences in agonistic behavior and dominance relations of communally reared children. Paper presented at the *Animal Behavior Society* Meeting, Durham, N.C., June, 1976.

Nadler, R. D. and Braggio, J. T. Sex and species differences in captive-reared juvenile chimpanzees and orangutans. *Journal of Human Evolution*, 1974, **3**, 541-550.

Nadler, R. D. Sexual behavior of captive lowland gorillas. *Archives of Sexual Behavior*, 1976, **5**, 487-502.

Nagel, U. and Kummer, H. Variation in cercopithecoid aggressive behavior. In Holloway, R. L. (Ed.) *Primate Aggression, Territoriality and Xenophobia*. New York: Academic Press, 1974, pp. 159-184.

Nash, L. T. Troop fission in free-ranging baboons in the Gombe Stream National Park, Tanzania. *American Journal of Physical Anthropology*, 1976, **44** (1), 63-77.

Okano, J. A preliminary observation of orangutans in the rehabilitation station in Sepilok, Sabah. *The Annual of Animal Physiology*, 1971, **21**, 55-67.

Omark, D. R. and Brenner, M. Institutional ecological factors affecting cognitive and social behaviors. Paper presented at the *Animal Behavior Society* Meeting, Boulder, Colorado, June, 1976.

Oswald, M. and Erwin, J. Control of intragroup aggression by male pigtail monkeys (*Macaca nemestrina*). *Nature*, 1976, **262**, 686-688.

Poirier, F. E. Dominance structure of the Nilgiri langur (*Presbytis johnii*) of South India. *Folia Primatologica*, 1970, **12**, 161-186.

Price, J. The dominance hierarchy and the evolution of mental illness. *The Lancet*, 1967, July, 243-246.

Rijksen, H. D. Social structure in a wild orangutan population in Sumatra. In Kondo, S., Kawai, M. and Ehara, A. (Eds.) *Contemporary Primatology*. Basel: Karger, 1975, pp. 373-379.

Rowell, T. E. Intra-sexual behaviour and female reproductive cycles of baboons (*Papio anubis*). *Animal Behaviour*, 1969, **17**, 159-167.

Rowell, T. E. and Dixson, A. F.   Changes in social organization during the breeding season of wild talapoin monkeys. *Journal of Reproduction and Fertility,* 1975, **43**, 419–434.

Russell, C. and Russell, W. M. S.   Primate male behaviour and its human analogue. *Impact of Science on Society,* 1971, **21**, 63–74.

Saayman, G. S.   Behaviour of the adult males in a troop of free-ranging chacma baboons. *Folia Primatologica,* 1971, **15**, 36–57.

Saunders, C. D. and Hausfater, G.   Demographic influences on sexual selection in baboons: A computer simulation. Paper presented at the *American Society of Primatologists* Meeting, Seattle, Washington, April, 1977.

Scott, J. P.   Agonistic behavior of primates: A comparative perspective. In Holloway, R. L. (Ed.) *Primate Aggression, Territoriality and Xenophobia.* New York: Academic Press, 1974, pp. 417–434.

Sheposh, J. P., Deming, M. and Young, L. E. The radiating effects of status and attractiveness of a male upon evaluations of his female partner. Paper presented at the *Western Psychological Association* Meeting, Seattle, Washington, April, 1977.

Simonds, P. E.   Aggression and social bonds in an urban troop of bonnet macaques (*Macaca radiata*).   Paper presented at the *American Association of Physical Anthropologists* Meeting, Seattle, Washington, April, 1977.

Smith, E. O.   A further description of the control role in pigtail macaques, *Macaca nemestrina. Primates,* 1973, **14** (4), 413–419.

Stebbins, C. A., Kelly, B. R., Tolor, A. and Power, M. E.   Sex differences in assertiveness in college students. *The Journal of Psychology,* 1977, **95**, 309–315.

Stynes, A. J., Kaufman, I. C. and Reiser, S. M. Social behavior in heterospecific groups of young macaque monkeys (*M. nemestrina* and *M. mulatta*): Preliminary observations. *American Zoologist,* 1975, **15**, 768.

Sugiyama, Y.   Characteristics of the social life of bonnet macaques. *Primates,* 1971, **12**, 247–266.

Vonder Haar Laws, J., Dolhinow, P. J. and McKenna, J. J. Female hanuman langur monkey rank and reproduction. Paper presented at the *American Society of Primatologists* Meeting, Seattle, Washington, April, 1977.

Vandenbergh, J. G. The development of social structure in free-ranging rhesus monkeys, *Behaviour,* 1967, **29**, 179–194.

Wagner, J. The sex of time-keeping. *International Journal of Symbology,* 1975, **6** (3), 23–30.

Weisband, C. and Goy, R. W. Effect of parturition and group composition on competitive drinking order in stumptail macaques (*Macaca arctoides*). *Folia Primatologica,* 1976, **25**, 95–121.

Weisfeld, G. E., Omark, D. R. and Cronin, C. L. Brains or brawn? A longitudinal study of dominance in high school boys. Paper presented at the *Animal Behavior Society* Meeting, University Park, Pennsylvania, June, 1977.

Wolfheim, J. H. Sex differences in behavior in a group of captive juvenile talapoin monkeys (*Miopithecus talapoin*). *Behaviour*, 1977, **63**, 110-128.

Wright, F. The effects of style and sex of consultants and sex of members in self-study groups. *Small Group Behavior*, 1976, **7** (4), 433-456.

18.

Leadership and Alliances

At many different points in our review of sex differences in dominance in the last two chapters, we emphasized the facts that: (1) dominance and leadership were not one and the same thing; and (2) an alliance between two or more individuals that might individually be subordinate can often lead to dominance over a dominant individual. A chapter on these two topics — leadership and alliances — is therefore a natural supplement to our two chapters on dominance. We will now cover each topic in turn using, as usual, a taxonomic framework.

A. LEADERSHIP AND GROUP CONTROL

1. General Comments

Leadership is apparent in specific behaviors more than in specific individuals. That is, different individuals can be leaders at different things. In the howler monkey, for example, the male will very often initiate group movement by giving a series of deep, coarse, clucking sounds (Carpenter, 1934); and, in the hamadryas baboon, a male starts group progression from the sleeping rock by displaying a particular surging gait (cf. Marler, 1968). Rowell (1972) has pointed out that these kinds of behaviors are also often displayed by female forest baboons. The dominance hierarchy is, to a large extent, irrelevant as far as many leadership behaviors are concerned. As we have seen, even the existence of the "control animal role" is not dependent on the presence of a dominance hierarchy; and, even in the presence of a rigid hierarchy, the control role may be displayed by individuals other than the alpha male (cf. Bernstein, 1974).

2. South American Monkeys (*Alouatta spp.*)

Carpenter (1934), in his early studies of the howler monkey (*Alouatta palliata*), pointed out that mature males lead the group when moving. In squirrel monkeys, the male also does most of the leading in exploratory movement (Fragaszy, 1977).

In the monogamous New World species *Callicebus moloch,* on the other hand, the adult female leads in exploratory movement and she also leads the family group throughout its daily ranging pattern (Kinzey et al., 1977). However, the *male* "coordinates" the activities of the group in both *C. moloch* and *C. torquatus.*

At present there are too few studies of leadership activities in New World monkeys to come to any conclusions regarding the presence or absence of sex differences. However, it is likely that, as in many behavioral characteristics, the monomorphic and monogamous species differ in leadership roles from those species which are not monomorphic and monogamous.

3. Rhesus Macaques (*M. mulatta*)

We already know that male rhesus monkeys play a control role which maintains internal group order (cf. Harlow and Lauersdorf, 1974). However, the basic group of adult females is the most stable portion of the band and determines the rank of the band relative to other bands (Vandenbergh, 1967).

In one free-ranging group of rhesus, females were more active interferers than males in intragroup fights (Kaplan, 1975). Neville (1968) even reported on the existence of a free-ranging rhesus monkey troop which lacked adult males. The two eldest females in the troop ran the troop and chased off adult males that tried to join it. In addition, it is thought that females have a large say in mating associations, that is, female choice plays an important role in mating (Lindburg, 1975).

4. Other Macaques (*Macaca spp.*)

Pigtail macaques (*M. nemestrina*) have been studied more than any other macaques regarding the control role of dominant males. The control male pigtail defends the group and controls internal disturbance and aggression (Smith, 1973; Oswald and Erwin, 1976; Sackett, Oswald, and Erwin, 1975; Erwin, 1976). The control role is limited to control of recognized group members, not to just any animals or any specific age or sex groups (Bernstein, 1966).

In the Japanese monkey (*M. fuscata*), the dominant male has a "group focus role" and is the center for group cohesion (Kurokawa, 1975). However, just as in the rhesus monkey, there are troops having no dominant males. The troop control role is then assumed by adult females (Kawamura, 1965). Leadership roles are frequently difficult to study in captive Japanese monkey groups since a fence around the troop "deprives the troop leader (male) of many of his normal functions such as guiding troop migrations, leading in confrontations with neighboring troops, and guarding against predators." (Alexander and Bowers, 1969, P. 241.)

Bonnet monkey males also display a control role (Sugiyama, 1971), but both sexes of bonnets will defend their home area. During movement, however, it is usually a male that moves first and there is usually a male at the rear of a progression. When leaving the feeding area for a resting place, adult males display a vocalization and others follow them (Rahaman and Parthasarathy, 1969).

In crab-eaters, we have already mentioned that in a transition from one dominant male to another there is "respect for the aging leader." (DeWaal, 1975).

5. Langurs (*Presbytis spp.*)

In many troops of many different species of langurs (*Presbytis spp.*), there is only one control male per troop (Bernstein, 1968); however, there are also multimale and all-male troops in which subadult and adult males must get along with one another. In *P. cristatus,* it is fairly clear that the dominant adult male controls or leads troop movement with immature males following a few feet behind him (Bernstein, 1966).

6. *Cercopithecus spp.* and Related Species

In the terrestrial vervet (*C. aethiops*), males usually bring up the rear during troop progression (Dunbar, 1974). The leader male initiates male displays and is able to inhibit the activity of other animals (Fairbanks and McGuire, 1977). Lancaster (1973), however, has emphasized the role of the female in the vervet group. She claims that field studies of all primates have suffered from an overemphasis on the behavior of adult males simply because they are large and conspicuous. While the adult female does not change groups, she is the leader in starting new groups, and in maintaining the integrity of the old groups. It is the old females who hold the troop's valuable knowledge, and the decision of where to go comes from the female core. In addition, Lancaster reminds us that dominance rank originally depends upon the rank of the mother and that, in the last analysis, female coalitions can dominate even the largest males. The female hierarchy is also more stable than is that of the males. These facts make females more valuable to the troop as leaders and in many more activities than males (Lancaster, 1973). In arboreal species of *Cercopithecus* (e.g., *C. campbelli*), the male also *appears* to lead the group, but Lancaster's criticisms of emphasis on males may apply here as well (Bourliére, Bertrand, and Hunkeler, 1969).

In the patas monkey (*Erythrocebus patas*), the one male in the group serves solely as a lookout and as a decoy. He could hardly be considered a leader in any other sense. Adult females certainly initiate and coordinate group movement (Gartlan, 1975). Similarly, in the talapoin monkey (*Miopithecus talapoin*), the males act as sentinels but the females are the leaders and are also dominant over the males (Segre, 1970).

7. Baboons

Among yellow baboons (*Papio cynocephalus*) there is a striking tendency for adult males to occupy front or rear positions in movement (Rhine, 1975). This is also true for *Papio papio* (Bert et al., 1967).

In chacma baboons (*P. ursinus*) adult males take an active role in initiating group movement and in keeping the group together. They even chase stray animals back into the group. Troop mobilizing behavior such as this is never seen in females (Buskirk, Buskirk, and Hamilton, 1974). The males thus regulate the social behavior of the troop (Saayman, 1971).

Adult male olive baboons (*P. anubis*) enter a forest clearing first (Rhine and Owens, 1972), but Rowell (1972) has described other situations where forest baboon females lead group movements.

Gelada baboons (*Theropithecus gelada*) males are found at the head of a herd; and, gelada males who are old take up the rear (Rhine, 1975). Although the herding of the males in one-male groups makes them conspicuous as leaders, in many ways they are peripheral to the more central female core. Females sometimes act as a group to chase the dominant male; and, group movements are frequently initiated by females (Bramblett, 1970). The male that becomes the leader of a one-male group depends on the choice of the females (Mori and Kawai, 1975).

Mandrill (*Mandrillus sphinx*) baboons are more arboreal than are hamadryas and gelada baboons. Mandrill males emit a rally call for group movement and tend to stay near the rear of the troop when traveling. If the troop is frightened, however, males lead the troop in traveling (Jouventin, 1975).

8. Apes

The travel order for the lesser ape siamang (*Symphalangus syndactylus*) is as follows: adult female, juvenile, adult male, weaned infant, and subadult males. The movement of these monogamous animals is very synchronized. The female also leads group movement in Kloss' gibbon (*Hylobates klossii*). There is a tendency, however, for male and female gibbons and siamangs to be more equal in leading and in coordinating group activities than is the case for nonmonogamous primates (cf. Schaller, 1963).

Among chimpanzees (*Pan troglodytes*) leadership is extremely variable. In one young group of chimpanzees studied by Menzel (1973), any of five different chimpanzees in a group of eight would serve as a leader in terms of being able to induce others to follow him or her. Also, no single individual seems to serve as *the* leader in gorilla (*Gorilla gorilla*) groups; however, the silver-back male does tend to lead movements (Emlen and Schaller, 1963).

9. Implications for Humans

As we can see from this new look at group control from the point of view of leadership, rather than of dominance, it is not altogether clear that males are more often leaders than are females. With dominance hierarchies put aside, in fact, it appears that in many cases the female matrifocal core of a group has more leadership qualities than does the peripheral male group or even the dominant male. Just because, in some species, large, dominant males evolved and formed strong bonds with other males does not mean that the males are the most important in organizing and managing the "community" (but see Tiger, 1970).

We have seen that dominant males are frequently the center of attention because they are conspicuous, at least to the human observers doing the field studies but presumably also to the other animals in a troop. However, this conspicuousness may often be more related to the aggressive potential of the male than to his leadership qualities.

Among human children of six and seven years of age, there is no doubt that men appear to be larger and more powerful to them. However, children of this age perform much better on learning tasks after observing female models than they do after observing male models. This is true even when the task involves a so-called male activity like a visual-spatial skill (building with wooden blocks) (Bartlett, 1977).

We have already pointed out that women respond more intensely to authority figures and particularly to female authority figures than do men (Wright, 1976). However, what we have not said is that:

> Female leaders described themselves as performing more consideration and tolerance of uncertainty leader behaviors and as being more satisfied with co-workers than male leaders. (Bartol and Wortman, 1976)

Overall, however, among human beings, there are very few "job-related" differences between male and female leaders. There are probably fewer sex differences in leadership than in dominance.

Aggressiveness does not appear to be as important a characteristic for a leader of today as it has been in the past. As Maccoby and Jacklin (1974) note:

But the day of the iron-fisted tycoon appears to be waning . . . leaders must be supportive toward the people with whom they work, and more skilled in guiding a group toward consensus than in imposing their own wills. (P. 368)

B. ALLIANCES AND COALITIONS

Very little is known about possible alliances and coalitions in the prosimians. It is known, however, that some female prosimians will protect another female from a male when the second female is giving birth (see Chapter 11). It is probable that other kinds of alliances occur in these primitive primates.

1. New World Monkeys

Among squirrel monkeys (*Saimiri sciureus*) females engage in more affiliative behavior within their unisexual subgroups than do males (Strayer, Taylor, and Yanciw, 1975). They show greater social cohesiveness than do the males (Alvarez, 1975). In the Monkey Jungle in Florida, all age and sex groups appear to be attracted to adult females (Baldwin, 1968). Adult males are occasionally chased out of the group's core by adult females (Baldwin and Baldwin, 1972). The extremely stable female subgroup retains its integrity by refusing to associate with nongroup females and by actively rejecting males (Fairbanks, 1974).

In summary, there are definitely more alliances and coalitions among female squirrel monkeys than among males; and, two females, in coalition, can definitely overcome the individual dominance of a male.

Spider monkeys (*Ateles spp.*), on the other hand, which do not show strong dominance for either sex, appear to show a sex difference in coalition somewhat different from that of the squirrel monkey. Males are rarely seen to assist females in an attack on other females or males, but males do assist other males in attacks on either males or females. In contrast, no single female or group of female spider monkeys is ever seen to threaten a male. Hence, coalitions and/or alliances are more frequent in male spider monkeys than they are in females (Klein and Klein, 1971), and most of the coalitions are alliances of males against females.

In monogamous species of nonhuman primates, alliances and coalitions are almost always across sex. That is, one adult male and one adult female form a monogamous coalition against others of the same species. Male and female owl monkeys (*Aotus spp.*), for example, tend to stay close together and mutually defend their family group (cf. Leibrecht and Kelley, 1977).

2. Rhesus Monkeys (*Macaca mulatta*)

Among rhesus monkeys, females have higher proximity, huddle, and contact scores than do males (Bernstein and Draper, 1964). As Bernstein and Sharpe (1966) note:

> Adult females may be regarded as a focus of group organization in that they interact at multiple levels with all other animals. (P. 100)

Subadult males do form peripheral subgroups, but these males are relatively loosely attached to their groups. The coalitions or alliances which result from these all-male peripheral subgroups are of more use in group change or in forming all-male groups than they are in alliances for greater dominance within the natal troop. Peripheral males, however, are frequently able to join a new group by forming an alliance with another male (Boelkins and Wilson, 1972). In one colony of rhesus studied by Chance (1956), however, there was a persistent affinity of three adult males for each other which contrasted sharply with the lack of constant association in the females.

But the Chance colony seems to be the exception rather than the rule. Even in groups of unfamiliar animals, submissive females are more adept at gaining status by eliciting the aid of dominant others than are submissive males (Masserman, Wechkin, and Woolf, 1968).

Female rhesus, in general, establish stable status hierarchies and close associations by grooming. Males do not (Sade, 1972). Just how this status hierarchy develops ontogenetically is open to question. However, it is known that rhesus of both sexes come to prefer to interact with members of their own sex by the time they are seven months of age. Between seven months of age and three years, they begin to prefer opposite-sexed animals again, but by the time the animals are three to 3½ years of age they again prefer to associate with their own sex (Suomi, Sackett, and Harlow, 1970). Alliances, therefore, depend as much upon the age of the animals as on the sex of the animals. In any case, among rhesus, dominance depends upon more than size. Alliances, particularly like-sex alliances, greatly affect the dominance hierarchy (Varley and Symmes, 1966).

In stable groups, female rhesus prefer a female–female coalition, but among strangers, females prefer as alliance partners, in order, males most, then familiar females. Females, in general, will not form alliances with strange females (Wade, 1976). Males, on the other hand, will form alliances with strange females more than they will with familiar ones. The outcome of these kinds of coalitions is that females tend to have closed within-group ties, whereas males tend to have open between-group ties. Mitchell (1976), in a review of dyad formations in the laboratory, has come to the same conclusion.

It is interesting that, regardless of sex, rhesus monkeys raised in more complex social environments are more capable of forming alliances or coalitions than those reared in less complex environments (Suomi, 1974; Anderson and Mason, 1974).

3. Other Macaques (*Macaca spp.*)

According to Berkson (1977), crab-eater macaque males (*M. fascicularis*) stay together in permanent like-sex subgroupings less often than do females. We might therefore expect more like-sex coalitions among females than among males.

Tactile relationships are of great importance to bonnet monkey (*M. radiata*) groups (Simonds, 1977). Monkeys of the same sex tend to sleep with one another and they form coalitions on the basis of these close affectional bonds (Koyama, 1973). The sex differences in like-sex coalitions, however, are probably not as great in bonnets as they are in the rhesus or crab-eater.[1]

4. Other Old World Monkeys

Mangabeys (*Cercocebus spp.*) show stronger like-sex bonds between females than between males (Chalmers, 1968). Females even embrace one another more than do males (Chalmers and Rowell, 1971).

Langur females, but not males, form like-sex alliances against the other sex. In *Presbytis entellus,* adult females sometimes even show temporary alliances against a male who is attempting infanticide (Hrdy, 1974).

The patas monkey (*Erythrocebus patas*) male, in a one-male group, is often chased by allied females in the group (Struhsaker and Gartlan, 1970); and, as we know, talapoin females (*Miopithecus talapoin*) gang up on talapoin males. In captivity, such female coalitions sometimes even kill males (Wolfheim, 1977).

Thus, in most Old World monkeys, female alliance is common, male alliance less common. Females in coalition are fully capable of dominating the leader male or males; and, in some species, they may even be able to kill them.

5. Apes

Female–female alliance, at least in chimpanzees, is stronger than is male–male alliance. Many of the like-sex female coalitions are based upon mother–daughter associations of many years (Pusey, 1976). As Fox (1975) has noted:

> With females . . . the more prominent the status of the kinship group, the tighter its unity. With males, however, this fact does not always apply (P. 16)

---

[1]Some recent unpublished research by Nancy Caine may force us to come to a different conclusion on this.

6. Implications for Humans

Among human children of three to five years of age, cooperative role taking is more frequent in girls than in boys (Sanders and Harper, 1976). By the time children reach the second grade, they, like many primates, begin to show a preference for same-sex peers. Children at this age are more likely to be in the proximity of, smile at, and to touch a preferred same-sex peer than a preferred opposite-sex peer (Travis and Fontenelle, 1977). At five to twelve years of age, girls obtain higher cooperation and altruism scores than do boys (Skarin and Moely, 1976).

Among college students and older adults, females are more deliberate than males in selecting a friend. Their friend must more closely approximate their ideal as far as intelligence is concerned than must a male's friend (Bailey, DiGiacomo, and Zinser, 1976).

There are many human ethological studies, however, which suggest that adult human female–female friendships develop more easily than do male–male friendships. In measurements as simple as eye-to-eye contacts, both sexes prefer the silent stares of the opposite sex, but this is *more* true of men than of women. On all indices of visual social behavior, women are more active than men. In task-oriented groups, for example, females have like-sex eye contact twice as often as do males. The sex difference in eye-to-eye visual contact appear at around the fourth grade or at nine years of age (Exline, 1971). It is interesting, incidentally, that even in monkey–human eye-to-eye contact, female monkeys (*Macaca mulatta*) make more eye contact with a human observer than do male monkeys (Thomsen, 1974).

An interesting general reference on cooperation, alliances, or coalitions in animals has been published by Crook (1971). Maccoby and Jacklin (1974), however, in their review of *human* sex differences, say that girls are not more "social" than boys and that young boys are highly oriented toward a peer group and congregate in larger groups than do girls. Girls, on the other hand, tend to associate in pairs or small groups of age-mates.

In each of seven cultures studied by Whiting and Pope (1974), girls were more likely to attempt to control the behavior of another person in the interests of some social value or in the interests of that other person's welfare. Boys, on the other hand, were more "egoistic" and individualistic in their dominance interactions. According to Maccoby and Jacklin (1974):

> By college age, there is some reason to believe that male leaders employ more authoritarian methods of leadership and control within their own groups than women do within their groups. (P. 263)

To the degree that *Homo sapiens* is a monogamous species:

. . . any man-woman pair usually forms a coalition in which, in the interest of maintaining the mutually rewarding aspects of the relationship, aggression is deliberately minimized (Maccoby and Jacklin, 1974, P. 265)

In such a monogamous pair (married couple), just as in monomorphic and monogamous New World monkeys and lesser apes, it becomes difficult to identify the dominant partner:

. . . wives become relatively more dominant (or more equal in dominance) the longer the marriage continues. (Maccoby and Jacklin, 1974, P. 262)

REFERENCES

Alexander, B. K. and Bowers, J. M. Social organization of a troop of Japanese monkeys in a two-acre enclosure. *Folia Primatologica,* 1969, **10**, 230-242.

Alvarez , F. Conditions of observation and social distance in groups of squirrel monkeys. *Primates,* 1975, **16** (4), 465-470.

Anderson, C. L. and Mason, W. A. Early experience and complexity of social organization in groups of young rhesus monkeys (*Macaca mulatta*). *Journal of Comparative and Physiological Psychology,* 1974, **87** (4), 681-690.

Bailey, R. C., DiGiacomo, R. J. and Zinser, O. Length of male and female friendship and perceived intelligence in self and friend. *Journal of Personality Assessment,* 1976, **40**, 635-640.

Baldwin, J. D. The social behavior of adult male squirrel monkeys (*Saimiri sciureus*) in a seminatural environment. *Folia Primatologica,* 1968, **9**, 281-314.

Baldwin, J. D. and Baldwin, J. The ecology and behavior of squirrel monkeys (*Saimiri oerstedi*) in a natural forest in Western Panama. *Folia Primatologica,* 1972, **18**, 161-184.

Bartlett, L. The effects of sex of model on task performance in young children. Paper presented at the *Western Psychological Association* Meeting, Seattle, Washington, April, 1977.

Bartol, K. M. and Wortman, M. S., Jr. Sex effects in leader behavior self-descriptions and job satisfaction. *Journal of Psychology,* 1976, **94**, 177-183.

Berkson, G. The social ecology of defects in primates. In Chevalier-Skolnikoff, S. and Poirier, F. (Eds.) *Primate Biosocial Development.* New York: Garland, 1977, pp. 189-204.

Bernstein, I. S. A field study of the male role in *Presbytis cristatus* (lutong). *Primates,* 1966, **7** (3), 402-403.

Bernstein, I. S. An investigation of the organization of pigtail monkey groups through the use of challenges. *Primates,* 1966, **7** (4), 471-480.

Bernstein, I. S. The lutong of Kuala Selangor. *Behaviour,* 1968, **32**, 1-16.

Bernstein, I. S. Daily activity cycles and weather influences in a pigtail monkey group. *Folia Primatologica,* 1972, **18**, 390-415.

Bernstein, I. S. Principles of primate group organization. In Chiarelli, A. B. (Ed.) *Perspectives in Primate Biology.* New York: Plenum, 1974, pp. 283-298.

Bernstein, I. S. and Draper, W. A. The behaviour of juvenile rhesus monkeys in groups. *Animal Behaviour,* 1964, **12**, 84-91.

Bernstein, I. S. and Sharpe, L. G. Social roles in a rhesus monkey group. *Behaviour,* 1966, **26**, 91-104.

Bert, J., Ayats, H., Martino, A. and Collumb, H. Note sur l'organisation de las vigilance sociale chez le babouin *Papio papio* dans l'est senegalais. *Folia Primatologica,* 1967, **6**, 44-47.

Boelkins, R. C. and Wilson, A. P. Intergroup social dynamics of the Cayo Santiago rhesus (*Macaca mulatta*) with special reference to changes in group membership by males. *Primates,* 1972, **13** (2), 125-140.

Bourliére, F., Bertrand, M. and Hunkeler, C. L'ecologie de la mone de Lowe (*Cercopithecus campbelli Lowei*) en cote d'ivoire. *Extrait de la Terre et la Vie,* 1969, **2**, 135-163.

Bramblett, C. A. Coalitions among gelada baboons. *Primates,* 1970, **11**, 327-333.

Buskirk, W. H., Buskirk, R. E. and Hamilton, W. J., III. Troop mobilizing behavior of adult male chacma baboons. *Folia Primatologica,* 1974, **22**, 9-18.

Carpenter, C. R. A field study of the behavior and social relations of howler monkeys (*Alouatta palliata*). *Comparative Psychology Monograph,* 1934, **10** (2), 168.

Chalmers, N. R. The social behaviour of free living mangabeys in Uganda. *Folia Primatologica,* 1968, **8**, 263-281.

Chalmers, N. R. and Rowell, T. E. Behaviour and female reproductive cycles in a captive group of mangabeys. *Folia Primatologica,* 1971, **14**, 1-14.

Chance, M. R. A. Social structure of a colony of *Macaca mulatta. British Journal of Animal Behavior,* 1956, **4**, 1-13.

Chivers, D. J. Spatial relations within the siamang group. *Proceedings of the Third International Congress of Primatology* (Zurich), 1970, **3**, 14-21.

Crook, J. H. Sources of cooperation in animals and man. In *Man and Beast: Comparative Social Behavior.* Washington: Smithsonian Institution Press, 1971.

DeWaal, F. B. M. The wounded leader: A spontaneous temporary change in the structure of agonistic relations among captive Java monkeys (*Macaca fascicularis*). *Netherlands Journal of Zoology,* 1975, **23** (4), 529-549.

Dunbar, R. I. M. Observations on the ecology and social organization of the green monkey *Cercopithecus sabaeus,* in Senegal. *Primates,* 1974, **15** (4), 341-350.

Emlen, J. T. and Schaller, G. B. In the home of the mountain gorilla. In Southwick, C. H. (Ed.) *Primate Social Behavior.* Princeton, N.J.: Van Nostrand, 1963, pp. 124-125.

Erwin, J. Aggressive behavior of captive pigtail macaques: Spatial conditions and social controls. *Laboratory Primate Newsletter,* 1976, **15** (2), 1-10.

Exline, R. V. Visual interaction: The glances of power and preference. *Nebraska Symposium on Motivation,* 1971, **19**, 163-206.

Fairbanks, L. An analysis of sub-group structure and process in a captive squirrel monkey (*Saimiri sciureus*) colony. *Folia Primatologica,* 1974, **21**, 209-224.

Fairbanks, L. A. and McGuire, M. T. Inhibition of control male behavior in vervet monkeys. Paper presented at the *American Society of Primatologists* Meeting, Seattle, Washington, April, 1977.

Fox, R. Primate kin and human kinship. In Fox, R. (Ed.) *Biosocial Anthropology.* New York: Wiley, 1975, pp. 9-35.

Fragaszy, D. M. Behavior in a novel environment in *Saimiri* and *Callicebus.* Paper presented at the *Western Psychological Association* Meeting, Seattle, Washington, April, 1977.

Gartlan, J. S. Adaptive aspects of social structure in *Erythrocebus patas. Proceedings of the Symposia of the Fifth Congress of the International Primatological Society.* Tokyo: Japan Science Press, 1975, pp. 161-171.

Harlow, H. F. and Lauersdorf, H. E. Sex differences in passion and play. *Perspectives in Biology and Medicine,* 1974, **17**, 348-360.

Hrdy, S. B. Male-male competition and infanticide among the langurs (*Presbytis entellus*) of Abu, Rajasthan. *Folia Primatologica,* 1974, **22**, 19-58.

Jouventin, P. Observations sur la socio-ecologie du Mandrill. *Extrait de la Terre et ala Vie,* 1975, **29**, 493-532.

Kaplan, J. R. Interference in fights and the control of aggression in a group of free-ranging rhesus monkeys. *American Journal of Physical Anthropology,* 1975, **44**, 189.

Kawabe, M. and Mano, T. Ecology and behavior of the wild proboscis monkey *Nasalis larvatus* (Wurmb) in Sabah, Malaysia. *Primates,* 1972, **13**, 213-228.

Kawamura, S. Matriarchal social ranks in the Minoo-B troop. In Imanishi, K. and Altmann, S. (Eds.) *Japanese Monkeys.* Atlanta, Georgia: Altmann, 1965, pp. 105-112.

Kinzey, W. G., Rosenberger, A. L., Heister, P. S., Prowse, D. L. and Trilling, J. S. A preliminary field investigation of the yellow handed titi monkey, *Callicebus torquatus torquatus,* in Northern Peru. *Primates,* 1977, **18**, 159-181.

Klein, L. and Klein, D. Aspects of social behaviour in a colony of spider monkeys. *International Zoo Yearbook,* 1971, **11**, 175-181.

Koyama, N. Dominance, grooming, and clasped-sleeping relationships among bonnet monkeys in India. *Primates,* 1973, **14**, 225-244.

Kurokawa, T. An experimental field study of cohesion in Katsuyama group of Japanese monkeys. In Kondo, S., Kawai, M. and Ehara, A. (Eds.) *Contemporary Primatology.* Basel: Karger, 1975, pp. 437-444.

Lancaster, J. B. In praise of the achieving female monkey. *Psychology Today,* 1973, **7**, 30-102.

Leibrecht, B. C. and Kelley, S. T. Some observations of behavior in breeding pairs of owl monkeys. Paper presented at the *American Society of Primatologists* Meeting, Seattle, Washington, April, 1977.

Lindburg, D. G. Mate selection in the rhesus monkey, *Macaca mulatta. American Journal of Physical Anthropology,* 1975, **42**, 315.

Maccoby, E. E. and Jacklin, C. N. *The Psychology of Sex Differences.* Stanford, California: Stanford University Press, 1974.

Marler, P. Aggregation and dispersal: Two functions in primate communication. In Jay, P. C. (Ed.) *Primates: Studies in Adaptation and Variability.* New York: Holt, Rinehart and Winston, 1968, pp. 420-438.

Masserman, J. H., Wechkin, S. and Woolf, M. Social relationships and aggression in rhesus monkeys. *Archives of General Psychiatry,* 1968, **18**, 210-213.

Menzel, E. W., Jr. Leadership and communication in young chimpanzees. *Symposia of the Fourth International Congress of Primatology.* Basel: Karger, 1973, pp. 192-225.

Mitchell, G. Attachment potential in rhesus macaque dyads (*Macaca mulatta*): A sabbatical report. *JSAS Catalogue of Selected Documents in Psychology,* 1976, **6**, 97 pp. Ms. 1177.

Mori, U. and Kawai, M. Social relations and behavior of gelada baboons. *Contemporary Primatology.* Basel: Karger, 1975, pp. 470-474.

Neville, M. K. A free-ranging rhesus monkey troop lacking adult males. *Journal of Mammalogy,* 1968, **49**, 771-773.

Oswald, M. and Erwin, J. Control of intragroup aggression by male pigtail monkeys (*Macaca nemestrina*). *Nature,* 1976, **262**, 686-688.

Pusey, A. E. Sex differences in the persistence and quality of the mother offspring relationship after weaning in a population of wild chimpanzees. Paper presented at the *International Primatological Society* Meeting, Cambridge, England, August, 1976.

Quris, R. Ecologie et organisation sociale de *Cercocebus galeritus agilis* dan le nord-est du Gabon. *Terre et la Vie, Revue d'Ecologie Appliquee,* 1975, **29**, 337-398.

Rahaman, H. and Parthasarathy, M. D. The home range, roosting places, and the day ranges of the bonnet macaque (*Macaca radiata*). *Journal of Zoology* (London), 1969, **157**, 267-276.

Rhine, R. J. The order of movement of yellow baboons (*Papio cynocephalus*). *Folia Primatologica,* 1975, **23**, 72-104.

Rhine, R. J. and Owens, N. W. The order of movement of adult male and black infant baboons (*Papio anubis*) entering and leaving a potentially dangerous clearing. *Folia Primatologica,* 1972, **18**, 276-283.

Rowell, T. E. *The Social Behaviour of Monkeys.* Baltimore: Penguin, 1972.

Saayman, G. S. Aggressive behavior in free-ranging chacma baboons (*Papio ursinus*). *Journal of Behavioural Science,* 1971, **1** (3), 77-83.

Sackett, D. P., Oswald, M. and Erwin, J. Aggression among captive female pigtail monkeys in all-female and harem groups. *Journal of Biological Psychology,* 1975, **17**, 17-29.

Sade, D. S. Sociometrics of *Macaca mulatta.* I: Linkages and cliques in grooming matrics. *Folia Primatologica,* 1972, **18**, 196-223.

Sanders, K. M. and Harper, L. V. Free-play fantasy behavior in preschool children: Relations among gender, age, season and location. *Child Development,* 1976, **47**, 1182-1185.

Schaller, G. B. Behavioral comparisons of the apes. In DeVore, I. (Ed.) *Primate Behavior.* New York: Holt, Rinehart and Winston, 1963, pp. 474-484.

Segre, A. Talapoins. *Animal Kingdom,* 1970, **73**, 20-25.

Simonds, P. E. Aggression and social bonds in an urban troop of bonnet macaques (*Macaca radiata*). Paper presented at the *American Association of Physical Anthropologists* Meeting, Seattle, Washington, April, 1977.

Skarin, K. and Moely, B. E. Altruistic behavior: An analysis of age and sex differences. *Child Development,* 1976, **47**, 1159-1165.

Smith, E. O. A further description of the control role in pigtail macaques, *Macaca nemestrina. Primates,* 1973, **14** (4), 413-419.

Strayer, F. F., Taylor, M. and Yanciw, P. Group composition effects on social behaviour of captive squirrel monkeys (*Saimiri sciureus*). *Primates,* 1975, **16** (3), 253-260.

Struhsaker, T. T. and Gartlan, J. S. Observations on the behaviour and ecology of the patas monkey (*Erythrocebus patas*) in the Waza Reserve, Cameroon. *Journal of Zoology* (London), 1970, **161**, 49-63.

Sugiyama, Y. Characteristics of the social life of bonnet macaques. *Primates,* 1971, **12**, 247-266.

Suomi, S. J. Social interactions of monkeys reared in a nuclear family environment versus monkeys reared with mothers and peers. *Primates,* 1974, **15**, 311-320.

Suomi, S. J., Sackett, G. P. and Harlow, H. F. Development of sex preference in rhesus monkeys. *Developmental Psychology,* 1970, **3**, 326-336.

Tenaza, R. R. I. Monogamy, territory and song among Kloss' gibbons (*Hylobates klossii*) in Siberut Island, Indonesia. II. Kloss' gibbon sleeping trees relative to human predation: Implications for the sociobiology of forest-dwelling primates. Doctoral Dissertation, University of California, Davis, 1974.

Thomsen, C. E. Eye contact by non-human primates toward a human observer. *Animal Behaviour,* 1974, **22**, 144-149.

Tiger, L. The possible biological origins of sexual discrimination. *Impact of Science on Society,* 1970, **20**, 29-44.

Travis, C. B. and Fontenelle, G. Nonverbal signals of affiliation in children. Paper presented at the *Animal Behavior Society* Meeting, University Park, Pennsylvania, June, 1977.

Vandenbergh, J. G. The development of social structure in free-ranging rhesus monkeys. *Behaviour,* 1967, **29**, 179-194.

Varley, M. and Symmes, D. The hierarchy of dominance in a group of macaques. *Behaviour,* 1966, **27**, 54-75.

Wade, T. D. The effects of strangers in rhesus monkey groups. *Behaviour,* 1976, **56**, 194-214.

Whiting, B. and Pope, C. A cross-cultural analysis of sex differences in the behavior of children aged three to eleven. *Journal of Social Psychology,* 1973, **91**, 171-188.

Wolfheim, J. H. Sex differences in behaviour in a group of captive juvenile talapoin monkeys (*Miopithecus talapoin*). *Behaviour,* 1977, **63**, 110-128.

Wright, F. The effects of style and sex of consultants and sex of members in self-study groups. *Small Group Behavior,* 1976, **7**, 433-456.

19.
Extratroop Behavior: Vigilance, Protection, Territoriality, and Intertroop Behavior

As we have seen in the chapters on dominance, leadership, and alliances, the roles of males and females in nonhuman primate groups are to some degree overlapping. While males in multimale heterosexual groups are usually dominant over females this is not always the case. In addition, dominant males are not always the leaders and if they lead the troop in one activity this does not necessarily mean that they will lead the troop in other activities as well. Both male–male and female–female alliances are common in nonhuman primates, particularly in those species which are nonmonogamous. In monogamous species, of course, cross-sex alliances between adults are more common than are same-sex alliances.

In the previous two or three chapters we have repeatedly referred to what has been termed the "control role" in a group of primates. The "control role" is a role typically taken by an adult male in a nonmonogamous species of primate. The control role includes two major activities, troop or group protection and the settling of internal strife. The former activity usually deals with extratroop or intertroop threats to the home band. The latter activity involves within-group leadership and control. We have dealt with, up to now, the internal organization, spacing, structure, and stability of primate groups and the role of focal animals like adult males and females in that organization. In short, we have primarily discussed those behaviors, seen in animals playing a control role, which deal with the internal disturbance or internal stability of the group. In the current chapter, however, we will concentrate on the other major half of the control role which includes vigilance for possible danger from outside the group, group protection, territoriality, and various and sundry intertroop behaviors. In the chapter which follows this one, we will discuss feeding habits and, in particular, predation which is also, in a sense, an extratroop activity. In the present chapter we will, in part, deal with primates as prey; in the chapter which follows this one, we will deal primarily with primates as predators.

A. GENERAL COMMENTS ON PROTECTION

According to Hrdy (1976), protection is the male's most important contribution to primate survival. *Males* of many species are said to "learn" assertiveness and aggressiveness to protect the group from external harm (Poirier, 1973). Rowell (1974) has also seen the most common role feature for males to be defense of females and young. Some males attack as all others retreat, while other males (like the patas male) perform diversionary displays as the core group of females and young escape danger (Rowell, 1974). In short, in most species of primates, the males rather than the females are the group protectors.

Let us now examine sex differences in vigilance, protection, territoriality, and other intertroop behaviors in a taxonomic framework to see to what extent this statement can be generalized.

B. PROSIMIANS

Some species of tree shrews (*Tupaia spp.*) are territorial while others are not (Sorenson, 1974). Most tree shrews scent-mark so that others of their own species know they have been present. Castration reduces marking in males (von Holst and Buergel-Goodwin, 1975).

In terms of potential lookout behavior or vigilance behavior, in slow lorises (*Nycticebus coucang*) it is known that males climb twice as much as do females (Tenaza et al., 1969). It is possible that this sex difference in climbing could reflect differences in vigilance or group protection.

Bushbaby males (*Galago senegalensis*) are known to protect infants when danger threatens (Gucwinska and Gucwinski, 1968). Bushbabies are territorial. There are often direct confrontations between males controlling females in adjacent territories and between males with females or with neighboring young males not having fixed territories. Females have smaller territories within the larger territory of the one male who controls them. They, too, sometimes compete for territories with each other; however, adjacent females are usually related (Charles-Dominique, 1977). The adult male, however, is the dominant figure in territoriality and only he emits territorial calls (Doyle, 1974). Alarm calls are also more frequent in males than in females (Doyle, 1974).

In captive lemurs (*Lemur fulvus*) the male scent-marks more than does the female but when alarmed by human observers, *females* mark more and emit more alarm vocalizations than do males (Harrington, 1977). Lemurs of the nocturnal species *Lemur mongoz,* however, do not use marking to specify home range limits and do not defend their home ranges (Tattersall and Sussman, 1975).

Thus, within the prosimian suborder, it is not always the male that is the most vigilant, protective, or territorial but it is probably the male more than the female in most prosimian species.

C. NEW WORLD MONKEYS

1. Monogamous Species

In monogamous primates there is a more equal responsibility in the two sexes for guarding the joint territory, scent-marking, and for aggression toward conspecific intruders. Usually each animal in a monogamous pair defends the joint territory against adults of their same sex (Kleiman, 1977). However, even in the monogamous marmoset (*Callithrix jacchus*), only the male reacts to alterations of the environment with increased scent-marking. Hence, even in a monogamous bond the male may still take the leading role in territorial behavior (Epple, 1970). In another South American genus suspected of being monogamous, *Pithecia spp.* (Kleiman, 1977), only the male emits the loud bark which is believed to be an expression of territoriality.

2. Nonmonogamous Species

In a captive colony of black howler monkeys (*Alouatta caraya*) at the Columbia Zoo, "the only aggression displayed by the male has been directed at the keeper who enters the enclosure to remove food pans, and on these occasions consists of threat gestures only." (Benton, 1976, P. 151.) In mantled howler monkeys (*A. palliata*), adult males urine-rub more than do adult females, but this is supposedly for intratroop cohesion (Milton, 1975).

*Intra*group aggression between spider monkey males is rare but *inter*group aggression *is* seen (Klein and Klein, 1971). Male spider monkeys (*Ateles geoffroyi*) definitely express hostile behavior toward intruders (Eisenberg and Kuehn, 1966).

Also, among white-faced monkeys (*Cebus capucinus*), males over eight years of age protect the group from predators (Oppenheimer, 1969). It appears, therefore, that in most nonmonogamous New World monkeys, males are more vigilant, protective, and territorial than are females, even in relatively monomorphic species like spider monkeys (*Ateles spp.*).

D. MACAQUES (*MACACA SPP.*)

1. Rhesus Monkeys (*Macaca mulatta*)

Four-year-old male rhesus are the most frequent participants in intergroup interactions of all age and sex groups of rhesus monkeys (Hausfater, 1972).

Intergroup change and conflict are most frequent in the breeding season (Lindburg, 1969), and males are wounded more often than females in intergroup conflict. Male rhesus monkeys definitely protect the troop from external enemies (Harlow and Lauersdorf, 1974).

During introductions of new animals into captive groups of rhesus, an intruder affects males more than females. Host groups containing males respond more intensely to intruders and these responses are specific to *male* intruders (Bernstein, Gordon, and Rose, 1974). This like-sex aggression, however, is also seen among females:

> The age and/or sex class corresponding to that of the introduced monkeys was the one which initiated most of the threat and attack behavior. (Southwick, 1967, P. 208)

The actions of peripheral males are particularly important in dominance interactions between two groups of rhesus monkeys (Gabow, 1973).

Adult male and female rhesus show different but very special responses to infants. In the first place, if an infant is introduced to their group neither sex shows intolerance, xenophobia, or aggressiveness toward the infant (Southwick and Siddiqi, 1972). In the second place, if an infant is threatened from outside the group, both males and females protect the infant. The adult female picks up the infant, retreats from the source of danger and threatens the intruder. The male, on the other hand, moves directly toward the intruder, places himself between the infant and the intruder, and either attacks or threatens (Mitchell, 1977).

2. Other Macaques

Among pigtail monkeys (*M. nemestrina*), males tend to climb to the highest locations and spend more time passively alert or vigilant than do females (Bernstein, 1972). One of the control roles of the dominant pigtail male is, as we know, defense of the group (Smith, 1973).

The stumptail macaque adult male is extremely protective of infants (Rhine and Kronenwetter, 1972):

> ... the presence of infants in the group increased the aggressive boldness of adult males vis-a-vis humans. (P. 23)

Male stumptails generally perform social vigilance, and they alert the group to outside danger; they also act as group protectors (Chevalier-Skolnikoff, 1972).

One entire troop of Japanese macaques (*Macaca fuscata*) has been transferred from Japan to south Texas. The transplanted troop initially suffered from predation by bobcats (*Lynx rufus*). According to Gouzoules, Fedigan, and Fedigan (1975):

> . . . while there may be a tendency for adult males to be more active in driving off the bobcat and in coming to the source of an alarm—all age-sex classes participated in the "mobbing" of the predator. (P. 345)

In free-ranging Japanese troops of *M. fuscata,* leader males definitely protect the troop (Imanishi, 1963). A corral in captivity deprives a leader male of many of his normal functions such as guarding against predators (Alexander and Bowers, 1969).

Bonnet monkey (*M. radiata*) females who have newborn or very young infants tend to space themselves closer to a central adult male than do other females, undoubtedly for added protection (Judge and Rodman, 1976). Adult male bonnets drop to the rear during flight from danger to see that the rest of the troop runs safely to the tree tops (Rahaman and Parthasarathy, 1969a) but both sexes defend the area. Dominant males, however, keep a watchful look, shake branches, and make aggressive noises toward strangers (Rahaman and Parthasarathy, 1969b). Leader bonnets also emit alarm calls and protect the troop from predators (Sugiyama, 1971).

The lion-tailed macaque (*M. silenus*) is an especially interesting species since it is the most arboreal of the macaques. There is much vocal intimidation between neighboring troops involving a loud whooping display. Sex differences have not been reported (Sugiyama, 1968).

E. OTHER OLD WORLD MONKEYS

Among most Old World monkeys, adult males act as sentinels, give alarm calls, and show defensive behaviors (Bramblett, 1973).

1. *Cercopithecus spp.*

Lookout behavior and group protection is seen in adult male vervets (*C. aethiops*) and in adult male guenons (*C. campbelli*) (Chevalier-Skolnikoff, 1972). The leader male of *C. campbelli* is usually found at a greater height and is the only male to emit loud calls of warning and to show the jumping display. The adult male of this species is also often seen near a new mother (Hunkeler, Bourliére, and Bertrand, 1972). Adult males of this species even react to the distress calls of an infant of an alien species (*C. petaurista*) and come to rescue it if it is in danger (Bourliére, Bertrand, and Hunkeler, 1969). *Cercopithecus sabaeus* males also sit higher in the trees than do others, evidently serving as sentries or lookouts (Dunbar, 1974).

In *Cercopithecus aethiops,* intertroop aggressive encounters almost always involve young peripheral males and not the leader who is the sentinel and the protector against predators (Struhsaker, 1967).

Even among the two closely related genera which are usually very unlike the genus *Cercopithecus* in other behavioral sex differences (*Miopithecus* and *Erythrocebus*), the usual sex differences in vigilance and protection hold up. Adult male talapoins (*Miopithecus*) in the wild are definitely more vigilant than are females (Wolfheim, personal communication), and male talapoins in captivity sit higher in the branches than do females as they survey activities around them like watchdogs (Segre, 1970). Patas (*Erythrocebus*) males also serve as lookouts and perform diversionary displays when near potential predators (Struhsaker and Gartlan, 1970).

2. Langurs (*Presbytis spp.*)

According to Naomi Bishop (1975) adult male langurs who live at high altitudes do not emit all of the alarm barks. In adult male golden langurs (*P. geei*) there is a complete absence of territorial whoop calls in adult males. However, golden langur males emit two different kinds of alarm calls and males show the greatest reaction to outside disturbance (Mukherjee and Saha, 1974). In contrast, *only* Nilgiri langur males (*P. johnii*) emit the territorial whoop vocalizations. Nilgiri langur males are vigilant and also emit the alarm calls that evoke the strongest responses from the rest of the troop (Poirier, 1968; Poirier, 1970).

Among hanuman langurs, all-male and multimale groups show more aggression toward intruders than do one-male groups (Tiwari and Mukherjee, 1976). Ceylon gray langur (*P. entellus*) males maintain an observational tonus in which they are more alive to more distant conditions and events. This behavior is referred to by Ripley (1967) as "dominance vigilance." As in many hanuman langur troops, there is only one control male per troop in the lutong of Kuala Selangor (*P. cristatus*). He defends the territory against other males and shows hostility toward other troops (Bernstein, 1968).

3. Baboons (*Papio spp.*)

Savannah baboon (*P. cynocephalus*) males protect infants and females from outside disturbances (Hall and DeVore, 1965). Adult males protect the troop from predators (Hamburg, 1971). In encounters with lions, however, baboons of all ages and both sexes emit an alarm bark as all animals withdraw (Saayman, 1971). The males, in particular, climb to the top of trees to look at the lions from a distance. According to Rhine (1975), nature and experience combine to yield bolder behavior in adult males than in others. Even preadolescent males, because of peripheralization, behave more confidently to extratroop conditions or events than do females. Males interpose themselves between the troop and potential danger. In *Papio anubis,* black infants stay near males when the troop passes through potentially dangerous territory (Rhine and Owens, 1972).

In *Papio papio,* the guinea baboon, the "watchdogs" are the juvenile males and not the adult males. Defense by branch-shaking, however, is most typical of adult males (Dunbar and Nathan, 1972). Thus, males play the role of protectors while juveniles play the prominent role in vigilance (Bert et al., 1967).

In *Papio ursinus,* the chacma baboon, the system is different. Males act as sentinels (Bolwig, 1959). Males also predominate in intertroop conflicts (Buskirk, Hamilton, and Buskirk, 1975).

Chacma baboons are particularly inventive about the manner in which they defend themselves against intruders. Hamilton, Buskirk, and Buskirk (1975a) describe instances of defensive stoning by baboons in which both sexes and all age groups older than two year olds participate. The most frequent hurlers of rocks, however, are four-year-old (and older) adult males. Pettet (1975) describes one situation quite explicitly:

. . . the largest adult male started dropping stones down the slope towards me and was later joined by two, sometimes three, of the older baboons. (P. 549)

During intertroop encounters, males also take the initiative. There are often violent chases by adult males while adult females withdraw or are chased from encounter sites (Hamilton, Buskirk, and Buskirk, 1975b).

4. Miscellaneous Old World Monkeys

In the mangabey, *Cercocebus galeritus agilis,* there is usually only one control male per group who acts as the defender (Quris, 1975). In *Colobus spp.,* the males also function primarily in defense of females and infants, primarily the latter (Horwich and Manski, 1975). Territorial aggressiveness is most pronounced in males and, in most cases, territorial clashes are started by the younger, subordinate males (Schenkel and Schenkel-Hulliger, 1967). Proboscis monkey (*Nasalis larvatus*) adult males are leaders who attack, threaten, sound warnings, and serve as watchdogs for the group (Kawabe and Mano, 1972).

F. APES

1. Lesser Apes

In the monogamous gibbons (*Hylobates spp.*), males and females are more equal in guarding the family group (Schaller, 1963), but males still surpass females.

Intergroup conflicts last an hour or so in *Hylobates lar*. The main protagonists are the adult males (one per group, of course). The males may continue their vocal squabble as others in the group return to feeding (Ellefson, 1966). Females also engage in less chasing of extragroup intruders than do males (Brockelman, Ross, and Pantuwatana, 1974).

There is a tendency in *H. klossii* for adults to defend the territory against an animal of the same sex. Adult males, however, guard against predators, establish the territories, and remain further from each other than do adult females (Tenaza, 1974). Males also emit more territorial vocalizations than do females (Tenaza, 1976).

2. Great Apes

An adult female chimpanzee (*Pan troglodytes*) will charge a human being to protect her infant (Reynolds and Reynolds, 1965); but usually, in such activities as mobbing a potential predator, the most dominant male is the leader (Menzel, 1973). Highly aggressive charging displays are seen more frequently in males than in females (Pitcairn, 1974). Communal male displays are common (Reynolds, 1965). Pant-hooting is heard more frequently from adult male chimpanzees than from adult females (Marler and Hobbett, 1975). Chimps do not defend territories (Reynolds, 1965).

Bluff charges are made much more often by gorilla (*Gorilla gorilla*) males than by gorilla females. The two- or three-toned roar and hooting are also emitted only by adult male gorillas, especially during the chest-beating display. Such displays are not communal but individual in gorillas. Gorillas also do not defend territories (Reynolds, 1965).

Jorge Sabater Pi has described several gorilla attacks on humans. Actually, aggressive acts by gorillas toward people are rare. They often involve a response by the gorilla to its own injury, a response to being taken by surprise, or a response to protect an infant or infants. Most of the attacks are by dominant adult males either in defense of a group, particularly infants, or when taken by surprise when alone. However, as if to show that it can happen, one attack was by an adult female gorilla without offspring (Sabater Pi, 1966).

Adult male orangutans (*Pongo pygmaeus*) show threat displays when approaching on the forest floor or when in the trees. They also have a "long-call" which may be used in spacing or territoriality to "signal a disagreeable mood" (Pitcairn, 1974). Since there are few predator threats, it is not clear whether or not males serve to protect the female and young, especially since, in Borneo at least, the males do not remain close to the females with offspring (Horr, 1972; Horr, 1977).

G. HUMAN CONSIDERATIONS

Homo sapiens is, of course, to a degree, a territorial species. People will defend their homes, their towns, and even their place in line (Sommer, 1969). That there are human sex differences in these behaviors appears to be certain to us. Males appear to be the major protagonists when it comes to defense of one's own family (or possessions for that matter). People of both sexes "personalize" their places of abode. The walls above the beds in college dormitories are especially often used for "territorial marking." In this case, it is known that males use larger but less personal items on their dormitory room walls than do college women (Vinsel et al., 1977).

Does a man tend to come to the aid of a woman or a child? He will if he knows them, or if he is related to them. He also will come to the aid of a woman or child if he perceives that the woman and/or child are not in the company of another man. In studies of bystander responses to assault, when a man attacks a woman, the bystander is more likely to help if the assaulting man appears to be a stranger to the woman he attacks. If the two are perceived by the bystander as being a married couple, the bystander is *much* less likely to help the woman (Shotland and Straw, 1976).

We began this chapter with the intention of examining four different classes of behavior: vigilance, protection, territoriality, and intertroop activity. We found that, in all four types of behavior, males usually surpassed females. This sex difference was found even in the monogamous New World monkeys and the monogamous lesser apes and there were very few exceptions. This is the most consistent sex difference in nonhuman primate behavior discussed in this book up to this point. It would be very surprising to us if human males and females did not differ in a similar direction.

It is interesting, in passing; that Maccoby and Jacklin (1974), in their volume on human sex differences, did not discuss protection, vigilance, territoriality, or "intertroop" behavior. Apparently, there have been very few studies of human sex differences in these characteristics. Perhaps there are some extremely important roles for human males that are being overlooked by social scientists.

REFERENCES

Alexander, B. K. and Bowers, J. M. Social organization of a troop of Japanese monkeys in a two-acre enclosure. *Folia Primatologica,* 1969, **10**, 230-242.

Benton, L. The establishment and husbandry of a black howler (*Alouatta caraya*) colony at Columbia Zoo. *International Zoo Yearbook,* 1976, **16**, 149-152.

Bernstein, I. S. The lutong of Kuala Selangor. *Behaviour,* 1968, **32**, 1–16.

Bernstein, I. S. Daily activity cycles and weather influences on a pigtail monkey group. *Folia Primatologica,* 1972, **18**, 390-415.

Bernstein, I. S., Gordon, T. P. and Rose, R. M. Factors influencing the expression of aggression during introductions to rhesus monkey groups. In Holloway, R. L. (Ed.) *Primate Aggression, Territoriality, and Xenophobia: A Comparative Perspective.* San Francisco: Academic Press, 1974, pp. 211-240.

Bert, J., Ayats, H., Martino, A. and Collomb, H. Note sure l'organisation de la vigilance sociale chez le babouin *Papio papio* dans l'est senegalais. *Folia Primatologica,* 1967, **6**, 44-47.

Bishop, N. Vocal behavior of adult male langurs in a high altitude environment. *American Journal of Physical Anthropology,* 1975, **42**, 291 (abstract).

Bolwig, N. A study of the behavior of the chacma baboon, *Papio ursinus. Behaviour,* 1959, **14**, 136-163.

Bourliére, F., Bertrand, M. and Hunkeler, C. L'ecologie de la mone de Lowe (*Cercopithecus campbelli Lowei*) en cote d'ivoire. *Extrait de la Terre et la Vie,* 1969, **2**, 135-163.

Bramblett, C. A. Social organization as an expression of role behavior among old world monkeys. *Primates,* 1973, **14** (1), 101-112.

Brockelman, W. Y., Ross, B. A. and Pantuwatana, S. Social interaction of adult gibbons (*Hylobates lar*) in an experimental colony. In Rumbaugh, D. M. (Ed.) *Gibbon and Siamang.* Basel: Karger, 1974, pp. 137-156.

Buskirk, R. E., Hamilton, W. J., III and Buskirk, W. H. Defense of space by baboon troops. *American Journal of Physical Anthropology,* 1975, **42**, 293.

Charles-Dominique, P. Urine marking and territoriality in *Galago alleni* (Waterhouse, 1837 Lorisoidea, Primates). A field study by radio-telemetry. *Zeitschrift für Tierpsychologie,* 1977, **43**, 113-138.

Chevalier-Skolnikoff, S. Sex role differentiation in nonhuman primates, with implications for man. *Medical Anthropology Newsletter,* 1972, **3** (3), 9-13.

Doyle, G. A. The behaviour of the lesser bushbaby. In Martin, R. D., Doyle, G. A. and Walker, A. C. (Eds.) *Prosimian Biology.* London: Duckworth, 1974, pp. 213-231.

Dunbar, R. I. M. Observations on the ecology and social organization of the green monkey (*Cercopithecus sabaeus*) in Senegal. *Primates,* 1974, **15** (4), 341-350.

Dunbar, R. I. M. and Nathan, M. F. Social organization of the guinea baboon. *Papio papio. Folia Primatologica,* 1972, **17**, 321-334.

Eisenberg, J. F. and Kuehn, R. E. The behavior of *Ateles geoffroyi* and related species. *Smithsonian Miscellaneous Collections,* 1966, **151** (8), 1-63.

Ellefson, J. O. Group size and behavior in white-handed gibbons, *Hylobates lar.* Paper presented at the *American Physical Anthropologists* Meeting, Berkeley, California, April, 1966.

Epple, G. Quantitative studies on scent marking in the marmoset (*Callethrix jacchus*). *Folia Primatologica,* 1970, **13**, 48-62.

Gabow, S. L.   Dominance order reversal between two groups of free-ranging rhesus monkeys. *Primates,* 1973, **14** (2-3), 215-223.

Gouzoules, H., Fedigan, L. M. and Fedigan, L.   Reponses of a transplanted troop of Japanese macaques (*Macaca fuscata*) to bobcat (*Lynx rufus*) predation. *Primates,* 1975, **16** (3), 335-349.

Gucwinska, H. and Gucwinski, A.   Breeding the Zanzibar galago at Wroclaw Zoo. *International Zoo Yearbook,* 1968, **8**, 111-114.

Hall, K. R. L. and DeVore, I.   Baboon social behavior. In DeVore, I. (Ed.) *Primate Behavior.* New York: Holt, Rinehart and Winston, 1965, pp. 53-110.

Hamburg, D. A.   Aggressive behavior of chimpanzees and baboons in natural habitats. *Journal of Psychiatric Research,* 1971, **8**, 385-398.

Hamilton, W. J., III, Buskirk, R. E. and Buskirk, W. II.   Defensive stoning by baboons. *Nature,* 1975, **256**, 488-489. (a)

Hamilton, W. J., III, Buskirk, R. E. and Buskirk, W. H.   Chacma baboon tactics during intertroop encounters. *Journal of Mammalogy,* 1975, **56**, 857-870. (b)

Harlow, H. F. and Lauersdorf, H. E.   Sex differences in passion and play. *Perspectives in Biology and Medicine,* 1974, **17**, 348-360.

Harrington, J. E.   Discrimination between males and females by scent in *Lemur fulvus. Animal Behaviour,* 1977, **25**, 147-151.

Hausfater, G.   Intergroup behavior of free-ranging rhesus monkeys (*Macaca mulatta*). *Folia Primatologica,* 1972, **18**, 78-107.

Horr, D. A.   The Borneo orangutan. *Borneo Research Bulletin,* 1972, **4**, 46-50.

Horr, D. A.   Orangutan maturation: Growing up in a female world. In Chevalier-Skolnikoff, S. and Poirier, F. E. (Eds.) *Primate Bio-Social Development.* New York: Garland, 1977, pp. 289-322.

Horwich, R. H. and Manski, D.   Maternal care and infant transfer in two species of Colobus monkeys. *Primates,* 1975, **16** (1), 49-73.

Hrdy, S. B.   Care and exploitation of nonhuman primate infants by conspecifics other than the mother. *Advances in the Study of Behavior,* 1976, **6**, 101-158.

Hunkeler, C., Bourliére, F. and Bertrand, M.   Le comportement social de la Mone de Lowe (*Cercopithecus campbelli Lowei*). *Folia Primatologica,* 1972, **27**, 218-236.

Imanishi, K.   Social behavior in Japanese monkeys, *Macaca fuscata.* In Southwick, C. H. (Ed.) *Primate Social Behavior.* Princeton, N.J.: Van Nostrand, 1963, pp. 68-81.

Judge, D. S. and Rodman, P. S.   *Macaca radiata:* Intragroup relations and reproductive status of females. *Primates,* 1976, **17** (4), 535-539.

Kawabe, M. and Mano, T.   Ecology and behavior of the wild proboscis monkey *Nasalis larvatus* (Wurmb) in Sabah, Malaysia. *Primates,* 1972, **13**, 213-228.

Kleiman, D. G.   Monogamy in mammals. *The Quarterly Review of Biology,* 1977, **52**, 39-69.

Klein, L. and Klein, D. Aspects of social behaviour in a colony of spider monkeys. *International Zoo Yearbook,* 1971, **11**, 175-181.

Lindburg, D. G. Rhesus monkeys: Mating season mobility of adult males. *Science,* 1969, **166**, 1176-1178.

Maccoby, E. E. and Jacklin, C. N. *The Psychology of Sex Differences.* Stanford, California: Stanford University Press, 1974.

Marler, P. and Hobbett, L. Individuality in long-range vocalizations of wild chimpanzees. *Zeitschrift für Tierpsychologie,* 1975, **38**, 97-107.

Menzel, E. W., Jr. Leadership and communication in young chimpanzees. *Symposium of the Fourth International Congress of Primatology.* Basel: Karger, 1973, pp. 192-225.

Milton, K. Urine-rubbing behavior in the mantled howler monkey, *Alouatta palliata. Folia Primatologica,* 1975, **23**, 105-112.

Mitchell, G. Parental behavior in nonhuman primates. In Money, J. and Musaph, H. (Eds.) *Handbook of Sexology.* Amsterdam: Elsevier/North Holland Biomedical Press, 1977, pp. 749-759.

Mukherjee, R. P. and Saha, S. S. The golden langurs (*Presbytis geei*) of Assam. *Primates,* 1974, **15**, 327-340.

Oppenheimer, J. R. Behavior and ecology of the white-faced monkey, *Cebus capucinus,* on Barro Colorado Island, C.Z. *Dissertation Abstracts,* 1969, **30**, Ord. No. 69-10, 811.

Pettet, A. Defensive stoning by baboons. *Nature,* 1975, **258**, 549.

Pitcairn, T. K. Aggression in natural groups of pongids. In Holloway, R. L. (Ed.) *Primate Aggression, Territoriality, and Xenophobia: A Comparative Perspective.* New York: Academic Press, 1974, pp. 241-272.

Poirier, F. E. Nilgiri langur: Territorial behavior. *Proceedings of the Second International Congress of Primatology,* 1968, **1**, 31-35.

Poirier, F. E. The communication matrix of the Nilgiri langur (*Presbytis johnii*) of South India. *Folia Primatologica,* 1970, **13**, 92-136.

Poirier, F. E. Socialization and learning among nonhuman primates. In Kimball, S. T. and Burnett, J. H. (Eds.) *Learning and Culture.* Seattle: University of Washington Press, 1973, pp. 3-41.

Quris, R. Ecologie et organisation sociale de *Cercocebus galeritus agilis* dans le nord-est du gabon. *Terre et la Vie, Revue d'Ecologie Appliqueé,* 1975, **29**, 337-398.

Rahaman, H. and Parthasarathy, M. D. Studies on the social behaviour of bonnet monkeys. *Primates,* 1969, **10**, 149-162. (a)

Rahaman, H. and Parthasarathy, M. D. The home range, roosting places, and the day ranges of the bonnet macaque (*Macaca radiata*). *Journal of Zoology* (London), 1969, **157**, 267-276. (b)

Reynolds, V. Some behavioral comparisons between the chimpanzee and the mountain gorilla in the wild. *American Anthropologist,* 1965, **67**, 693-706.

Reynolds, V. and Reynolds, F. Chimpanzees of the Budongo forest. In DeVore, I. (Ed.) *Primate Behavior.* New York: Holt, Rinehart and Winston, 1965, pp. 368-374.

Rhine, R. J. The order of movement of yellow baboons (*Papio cynocephalus*). *Folia Primatologica,* 1975, **23**, 72-104.

Rhine, R. J. and Kronenwetter, C. Interaction patterns of two newly formed groups of stumptail macaques (*Macaca arctoides*). *Primates,* 1972, **13** (1), 19-33.

Rhine, R. J. and Owens, N. W. The order of movement of adult male and black infant baboons (*Papio anubis*) entering and leaving a potentially dangerous clearing. *Folia Primatologica,* 1972, **18**, 276-283.

Ripley, S. Intertroop encounters among Ceylon gray langurs (*Presbytis entellus*). In Altmann, S. (Ed.) *Social Communication Among Primates.* Chicago: University of Chicago Press, 1967, pp. 237-254.

Rowell, T. E. Contrasting different adult male roles in different species on non-human primates. *Archives of Sexual Behavior,* 1974, **3**, 143-149.

Saayman, G. S. Baboons' responses to predators. *African Wild Life,* 1971, **25**, 46-49.

Sabater Pi, J. Gorilla attacks against humans in Rio Muni, West Africa. *Journal of Mammalogy,* 1966, **47**, 123-124.

Schaller, G. B. Behavioral comparisons of the apes. In DeVore, I. (Ed.) *Primate Behavior.* New York: Holt, Rinehart and Winston, 1963, pp. 474-484.

Schenkel, R. and Schenkel-Hulliger, L. On the sociology of free-ranging Colobus (*Colobus guereza caudatus,* Thomas 1885). *First Congress of the International Primatological Society* (Frankfurt, 1966). Published in Starck, D., Schneider, R., and Kuhn, H. J. (Eds.) *Neue ergebnisse der primatologie.* Stuttgart: Gustav Fischer, 1967, pp. 183-194.

Segre, A. Talapoins. *Animal Kingdom,* 1970, **73** (3), 20-25.

Shotland, R. L. and Straw, M. K. Bystander response to an assault: When a man attacks a woman. *Journal of Personality and Social Psychology,* 1976, **34**, 990-999.

Smith, E. O. A further description of the control role in pigtail macaques, *Macaca nemestrina. Primates,* 1973, **14** (4), 413-419.

Sommer, R. *Personal Space.* Englewood Cliffs, N.J.: Prentice Hall, 1969.

Sorenson, M. W. A review of aggressive behavior in the tree shrews. In Holloway, R. L. (Ed.) *Primate Aggression, Territoriality, and Xenophobia: A Comparative Perspective.* New York: Academic Press, 1974, pp. 13-30.

Southwick, C. H. An experimental study of intragroup agonistic behavior in rhesus monkeys (*Macaca mulatta*). *Behaviour,* 1967, **28**, 182-209.

Southwick, C. H. and Siddiqi, M. F. Experimental studies on social intolerance in wild rhesus groups. *American Zoologist,* 1972, **12** (4), 651-652.

Stott, K. Sakis: Imps of the rain forest. *Zoonooz,* 1976, **49** (4), 4-8.

Struhsaker, T. T. *Behavior of Vervet Monkey (Cercopithecus aethiops).* Berkeley: University of California Press, 1967.

Struhsaker, T. T. and Gartlan, J. S. Observations on the behavior and ecology of the patas monkey (*Erythrocebus patas*) in the Waza Reserve, Cameroon. *Journal of Zoology, London,* 1970, **161**, 49-63.

Sugiyama, Y. The ecology of the lion-tailed macaque (*Macaca silenus*) (Linnaeus): A pilot study. *Journal of the Bombay Natural History Society,* 1968, **65** (2), 283-292.

Sugiyama, Y. Characteristics of the social life of bonnet macaques. *Primates,* 1971, **12** (3-4), 247-266.

Tattersall, I. and Sussman, R. W. Observations on the ecology and behavior of the mongoose lemur *Lemur mongoz mongoz* Linnaeus (Primates, Lemuriformes) at Ampijora, Madagascar. *Anthropological Papers of the American Museum of Natural History,* 1975, **52** (4), 195-216.

Tenaza, R. R. I. Monogamy, territory and song among Kloss' gibbons (*Hylobates klossii*) in Siberut Island, Indonesia. II. Kloss' gibbon sleeping trees relative to human predation: Implications for the sociobiology of forest dwelling primates. Doctoral Dissertation, University of California, Davis, 1974.

Tenaza, R. R. Songs, choruses and countersinging of Kloss' gibbons (*Hylobates klossii*) in Siberut Island, Indonesia. *Zeitschrift für Tierpsychologie,* 1976, **40**, 37-52.

Tenaza, R. R., Ross, B. A., Tanticharoenyos, P. and Berkson, G. Individual behavior and activity rhythms of captive slow lorises (*Nycticebus coucang*). *Animal Behaviour,* 1969, **17**, 664-669.

Tiwari, K. K. and Mukherjee, R. P. Social structure of hanuman langur populations around Ramtek, Nagpur District, India. Paper presented at the *International Primatological Society* Meeting, Cambridge, England, August, 1976.

Vinsel, A., Wilson, J., Brown, B. B. and Altman, I. Personalization in dormitories: Sex differences. Paper presented at the *Western Psychological Association* Meeting, Seattle, Washington, April, 1977.

Von Holst, D. and Buergel-Goodwin, U. The influence of sex hormones on chinning by male *Tupaia belangeri. Journal of Comparative Physiology,* 1975, **103**, 123-151.

20.

Feeding and Predation

The different roles of males and females in intragroup and intergroup activities affect their behaviors with regard to feeding and predation. Dominance, for example, is related to priority of access to food. In fact, priority of access to food is a measure of dominance. Knowledge of travel routes to previously abundant food sources is also related to leadership. Often the older adult females of a troop retain information of this sort.

Feeding in nonhuman primates involves relatively complicated and variable behavior. Primates are omnivorous, which is to say that they eat almost anything, including other animals. As we have seen in the last chapter, interest in activities outside the troop is primarily a male proclivity. Predation on other animals requires this kind of extratroop vigilance and attention. We will discuss sex differences in predation in the second half of this short chapter. For now, however, we will examine sex differences in feeding behavior.

A. GENERAL COMMENTS ON FEEDING

As might be expected, feeding behavior is related to just about every characteristic of a given species or individual. If the day range of a particular species is large, for example, it is likely that the amount of food available is low. If the amount of food available is low but concentrated in one area, then it is likely that aggressive behavior will be frequent. Paradoxically, there is also a relatively high positive correlation across species between a low amount of food and frequent grooming behavior. In general, arboreal species have more to eat and eat more than do terrestrial ones (Jorde and Spuhler, 1974).

While dominance is often defined by who gets food first, if the food is a *new* food, the dominant male may actually wait for a more exploratory subordinate animal or a juvenile to test out the situation, after which he may immediately move in to take the rest of the food (Menzel, 1968).

B. FEEDING IN SOUTH AMERICAN MONKEYS

Among black howler monkeys (*Alouatta caraya*), feeding, especially in captivity, is directly related to aggression, and apparently there are sex differences in the degree of competition over food. Benton (1976) has reported that:

> Such aggressive interactions as there are take place only between females, during feeding, and have easily been resolved by increasing the quantities of the items in dispute and by presenting the feed in two additional containers. The only aggression displayed by the male has been directed at the keeper who enters the enclosure to remove food pans. (P. 151)

We can readily see the female intratroop versus male extratroop aggressive orientations with respect to food in the above quotation. But this is not always the case for New World monkeys.

When food is given to primates in captivity, there are rarely any sex differences in regards to whether or not the animals accept food. For example, there are no sex differences in accepting food in spider monkeys (*Ateles spp.*) (Pechtel, Masserman, and Aarons, 1961). Although sex differences are not apparent, affiliation in the relatively monomorphic spider monkey is related to feeding behavior (Dare, 1977).

In many New World monkeys there is food sharing as well as competition for food. It may be that feeding together is important for dyadic social-emotional attachment. Mouth-to-mouth behavior in the squirrel monkey (*Saimiri sciureus*), for example, is related to food sharing. Adult male squirrel monkeys are both active and passive participants in mouth-to-mouth contacts, whereas adult females are rarely seen as *active* participants. Lower ranked females almost never display even passive oral-oral contact (Peters, 1970).

While male squirrel monkeys share food, they also aggressively compete for it, especially against females. Around a food dish in captive groups, aggressive interactions are seen mostly in males and particularly in males *toward* females. Initial possession of an item of food in a food dish, however, is *not* male dominated in the squirrel monkey (Fragaszy, 1976).

In monogamous New World monkeys, we often see a different pattern. Male titi monkeys (*Callicebus moloch*) in captivity *do* initiate more social interactions at a food dish and *are* more aggressive than are females, but the sex difference is not as great as it is in the squirrel monkey. The male titi usually gains initial possession of a single item of food and male titi monkeys also eat more than do female titi monkeys. [In fact, in most species of primates and even in *human* infants, children, and adults, males eat longer and more than do females (Gewirtz and Gewirtz, 1968).] A female titi monkey will attempt to steal food from her mate more than he will from her. She is not as successful as he is, however (Fragaszy, 1976). Titi monkeys, while omnivorous, are about 65 percent frugivorous (Kinzey et al., 1977).

Food sharing occurs in some monogamous species. The lion tamarin male (*Leontopithecus rosalia*), for example, presents food to his offspring (Kleiman, 1977).

C. FEEDING IN OLD WORLD MONKEYS

Food, once in the possession of an Old World monkey, is rarely robbed even by a superior. In fact, this inhibition to steal is even sustained by a leader's supervision (Torii, 1975).

As in spider monkeys, there are no sex differences in accepting food in rhesus monkeys (*Macaca mulatta*) (Pechtel, Masserman, and Aarons, 1961). However, in competition for food, males outcompete females (Wise, Zimmerman, and Strobel, 1973).

There are no sex differences in food acceptance in crab-eating macaques (*M. fascicularis*) (Pechtel, Masserman, and Aarons, 1961). Males consume more calories than do females, even in the first sixty days of life (Willes, Kressler, and Truelove, 1977). Premature pigtail males, however, are especially susceptible to retardation in intake rate (Sackett et al., 1974).

Giving food to another individual appeared spontaneously in a home-reared pigtail monkey (*M. nemestrina*) female, but she gave only what was of no value to her. Interestingly, she had also been seen "cheating" when giving food. She gave symbolically; and, she "bartered" (cf. Bertrand, 1976).

Among Japanese macaques (*M. fuscata*), females catch more sweet potatoes than do males when the sweet potatoes are thrown into a troop. Dominant males do not have to catch them, however, since they keep peripheral males away from them (Kawai, 1967). If a new food, like candy, is given to a Japanese macaque troop, the two to three year olds eat it first. Adult males are the last to acquire new food-eating habits unless they directly care for infants and juveniles (Kawamura, 1963).

In food-getting situations among Japanese macaques, threat responses are seen no more frequently in males than in females; however, there is a difference in food acceptance in this species. Refusing food rarely occurs in females but in males it does (Norikoshi, 1971).

In bonnet macaques (*M. radiata*), males usually lead the troop to the feeding grounds (Rahaman and Parthasarathy, 1969). There is probably less competition among males for food in bonnets, toque macaques, and lion-tailed macaques (cf. Sugiyama, 1968) than in rhesus and pigtails.

There are no sex differences in food acceptance in vervets (*Cercopithecus aethiops*) or in mangabeys (*Cercocebus spp.*) (Pechtel, Masserman, and Aarons, 1961). Male vervets, however, feed more than do female vervets (Raleigh, 1977). In patas monkeys, there are *group* differences of relevance to gender. Heterosexual groups, for example, have priority over all-male groups at water holes (Gartlan, 1975); and, in times of climatic catastrophe, all-male units are the first to succumb (Gartlan, 1976).

Food sharing is seen in douc langurs (*Pygathrix nemaeus nemaeus*). Active behavior of breaking off and giving another douc a piece of food occurs, but there are no sex differences in this behavior (Kavanagh, 1972).

Among yellow baboons (*Papio cynocephalus*), there are statistically significant differences between the diets of males and females (Post, 1977). Adult males sit in one place longer and obtain more food per sitting than do females, but there is no sex difference in rate of eating (Rhine and Westlund, 1977). Rose (1976) has noted that female baboons of the species *Papio anubis* use bipedalism in feeding more than do males.

D. APE FEEDING

Feeding tolerance in the siamang (*Symphalangus syndactylus*) family is a reflection of the strong cohesiveness of the family group. According to Boyd, Page, and Schusterman (1977), bidirectional food transfer is most likely to occur between the adult female and the infant, although it also occurs between the adult male and a subadult or adult female.

In chimpanzees (*Pan troglodytes*), food stealing rarely occurs. Once food is in the hands of one animal another animal will not steal it. Torii (1975) considers this to be a sign of "manners" in chimpanzees. In chimpanzees, there is also food begging or "asking." An animal does not take food from another without "permission" (Torii, 1975). Adult females share food with all infants, but especially with their own infants (Savage and Malick, 1977). There is some indication that the degree of attachment and sexual possessiveness between a male and female chimpanzee can be seen in the amount of generosity displayed in food sharing. Tutin (1974), for example, in his description of temporary consort pairs or temporary monogamous courtships in chimpanzees, notes that degree of "monogamy" is correlated with the amount of time the male spends grooming the female and with his generosity in food sharing. Maple (personal communication) has observed food sharing in captive orangutans (*Pongo pygmaeus*). Field workers, however, have noted that gorillas do not share food (personal communications.)

E. SEX DIFFERENCES IN PREDATION[1]

Rowell (1974) has noted that consideration of defensive behavior rather than hunting as a role of nonhuman primate males "might be a more profitable approach to understanding behavioral changes in prehuman and early human societies." (P. 143.) We have seen that the most consistent sex differences thus far do indeed occur in group protection or group defense. However, as far as nonhuman primates are concerned, hunting has not been adequately considered as a male or female role. There have been numerous recent accounts of a surprising number of primate species hunting, killing, and eating meat. No one has really examined the literature *across species* to determine whether this is exclusively a male role, primarily a male role, or if it is a role in which both males and females participate.

Perhaps the most well-known studies on primate predation have involved baboons in Africa. Harding (1973) for example, saw a troop of olive baboons (*Papio anubis*) catch and eat forty-seven different animals in 1970 and 1971. He reports that adult males caught and ate all but three of these forty-seven animals. The adult females caught three hares; however, only once was a female able to keep and eat most of the animal she had caught. She usually lost her catch to males.

Strum (1975), working with Harding, has also reported predation in the olive baboon with definite hunting and *sharing* of food. "Starting as an adult male activity in the olive baboon troop, this tradition rapidly expanded to include capture and consumption of prey by adult females and juveniles of all ages and both sexes." (P. 755.)

Baboon males kill small gazelles, young lambs, quails, hares, and guinea fowl. According to Harding and Strum (1976), adult male olive baboons (*Papio anubis*) move *deliberately* through herds of grazing gazelles in a true hunting fashion. Initially, at least, capturing prey is a solitary activity; that is, animals do not cooperate, and there is only a single hunter.

In two short years (1972 and 1973), however, these same olive baboons studied by Harding and Strum (1976) doubled their predation to 100 animals. Moreover, females began to kill more frequently and even *infants* learned to eat meat from their mothers or from close males. Juveniles also learned to hunt and kill, and hunting became more systematic. The baboons learned to chase prey toward other baboons. A hunt for one gazelle could last for up to two hours or even up to three days and (only in the males) to as far as two miles away from the main troop. The baboons also started to share meat, even though they never shared any other food item. Apparently, this increase in hunting resulted from an antelope population explosion.

Another closely related species of baboon, the yellow baboon (*Papio cynocephalus*) kills hares, vervet monkeys, and gazelles in that order of frequency (Hausfater, 1976); Predation in the yellow baboon is higher during the dry season. Hausfater

[1] I wish to thank Roberta N. Chinn for help in the literature search on this topic.

(1976) believes that the predation results from a need for vitamin B_{12}. As in the case of the olive baboon, males eat more meat than do the females.

In summary, in baboons, males start most of the hunting, kill most of the prey, and eat most of the meat; however, females and even juveniles *do* hunt. Hunting is not *exclusively* a male role. Among macaques, a captive adult male lion-tailed macaque (*Macaca silenus*) was seen to capture and consume a young peacock (Maple, personal communication).[2]

In the common chimpanzee (*Pan troglodytes*), the picture is much the same as in the two baboon species examined above. There is more hunting done by males, but females *do* hunt (van Lawick-Goodall, 1973). In such activities as "mobbing a snake," which may be a defensive behavior more than a hunting behavior, the dominant male is also the leader (Menzel, 1973).

Chimpanzees eat other primates (as do baboons); in fact, chimpanzees eat baboons (Bygott, 1972). Not only do chimpanzees eat their "lower" primate relatives, they also practice cannibalism. Bygott (1972) has reported two instances of cannibalism in chimpanzees. In both instances, *adult males* stole an infant chimpanzee from an unfamiliar female and ate it alive. It is probable that the males did not recognize the infant as a member of the group. An adult female would not eat it, although her adult son did.

Chimpanzee prey (e.g., young baboons) is usually caught by adult males. Adult male chimpanzees certainly seem to eat much more of the meat than do the females (although the females do eat some).

Thus, in baboons and chimpanzees, the two nonhuman primates most often compared to man, males are the primary, though not the only, hunters. In macaques, however, females seem to kill and eat at least as much as do males. What about other nonhuman primates? Another nonhuman primate known to kill and eat meat is Goeldi's monkey (*Callimico goeldii*), which preys on eastern ribbon snakes. Adult males of this species kill and eat snakes. Capuchins (*Cebus spp.*) also kill and eat snakes (Lorenz, 1971), but a sex difference has not been reported in the literature. As we know, there are also reports of cannibalism in captivity in many species of nonhuman primates. Rothe (1974) saw marmosets (*Callithrix jacchus*) eat nonviable offspring; and *male* bushbabies (*Galago senegalensis*) also may eat the young of their captive groups (Buettner-Janusch, 1964). Although these are not examples of hunting behavior, they *are* examples of meat eating; and, males do the killing more than do the females.

Up to this point we have not discussed in detail the possibilities of sex differences in hunting behavior in *arboreal* as opposed to terrestrial primate species.

---

[2]Estrada and associates (1977, 1978) have recently reported predation in a free-ranging group of stumptail macaques (*M. arctoides*) in which the females killed and ate more prey than did males. In addition, a rhesus macaque female in a colony at Davis, CA. recently killed a bird and ate it (Caine, personal communication).

While marmosets, capuchins, Goeldi's monkeys, and galagos are all arboreal, the "prey" reportedly killed by them has been either a snake of a conspecific infant in captivity. As we shall now see, arboreal monkeys and apes also kill birds.

In southern Africa, large male Samango monkeys (*Cercopithecus mitis*) are mobbed by flycatchers, shrikes, and bulbuls. In the trees, thirty feet above the ground, they rob the nests of waxbills and mannikins. The monkeys completely ignore the diving birds while deliberately searching for birds' nests (Oatley, 1970).

The arboreal lesser ape, the gibbon (*Hylobates lar*), also hunts and eats birds. An adult *female* with her offspring will catch and eat house sparrows. The young also eat the birds (Blackwell, 1969). Newkirk (1973) notes that free-ranging *female* gibbons in Bermuda kill and attempt to eat speckled domestic white hens. The female gibbons also steal and eat the hens' eggs, although the gibbon males help to eat them as well. Gibbons also eat nestling birds and catch birds in midair. In the Berlin Zoo, gibbons attack guinea fowl and peacocks, and in the Cologne Zoo they attack sparrows, magpies, and ducks (Newkirk, 1973). Tenaza (1976) has reported that wild mynah birds mimic the calls of gibbons in Indonesia. This may possibly serve as a defense against their natural predators. However, a similar convergence with the calls of birds in marmosets has been interpreted as occurring because marmosets are small and encounter the same kinds of threats that birds face, for example the "monkey eagle" (Vencl, 1977).

In summary, arboreal as well as terrestrial species of primates forage and/or hunt for eggs or meat. Both males and females do the killing. In the monogamous gibbons, and in macaques, there is a possibility that the females may kill more than the males.

Integrating all of the studies reported above, it is probably fair to say that, in *most* primates, the adult male does *most* of the hunting, killing, and eating of meat. However, these roles are by no means exclusively male roles, and in some species, particularly in monogamous ones, these roles may actually occur more frequently in females. Even more surprising, however, is the fact that immature or juvenile nonhuman primates also learn to hunt; and, even infants (in baboons, chimps, and gibbons) have been seen eating meat.

As Teleki (1973) has pointed out, predatory behavior depends upon an awareness of the prey when it is near the troop or group. (See Table 20-1 for Teleki's model of predation.) As we know, in most species of primates, males show more vigilance than do females. It is likely, therefore, that males would be more aware of the potential prey than would females. In addition, hunting often requires traveling at various distances from the troop. As we know, males of many species of primates are more loosely attached to their groups than are females and are thus free to do more hunting. Another factor of possible importance in predation is the size of the predator relative to the prey. In many species of primates, males are larger than are females. In the recorded incidents

Table 20-1   Teleki's (1973) Model of Predation (After Chinn and Mitchell, unpublished)

INTEREST

STAGE ONE: Pursuit

(3 modes)

SEIZING: Involves explosive action by the predators in taking full advantage of exceptionally
promising circumstances that arise suddenly.
Usually dependent upon special situations: where
the predator and the prey are intermingling freely
in close proximity, where the prey individual and
other members of its community are inattentive
or distracted, and where visual predator-prey
contact can be easily maintained. More a matter of food collection than true predation.

CHASING: Time
elapsed and dis-
tance covered, as
well as energy ex-
pended, is less than
in pursuit. More
deliberate and con-
trolled than pur-
suit. Is tenacious
and persistent.

STALKING: Like chasing, as it is
more deliberate and controlled than
pursuit. Differs from chasing, however, mainly in degree. Is usually
longer in duration than other modes.
Appears to be the most deliberate and
controlled. Lacks the explosive high-
speed actions of other modes; entails
a stealthy, disciplined intent to cor-
ner prey without alerting it to danger.

STAGE TWO: CAPTURE

a)   ACQUISITION.
b)   KILLING. Highly individualistic behavior.
c)   DIVISION OF CARCASS.

STAGE THREE: CONSUMPTION

a)   DISMANTLING THE CARCASS.
b)   DISTRIBUTING THE MEAT.*
   —Recovery of meat.
   —Taking of meat.
   —Requesting of meat.

*All of the animal is consumed including internal organs, connective tissue and cartilage, bones and skin; in addition,
nails, teeth and hair are often consumed.

of primate predation, the prey is always relatively smaller than the predator. An example would be the size of yellow baboons (*P. anubis*) as compared to vervet monkeys (*C. aethiops*): the adult male baboon is 100 pounds, the adult female baboon is 30-66 pounds; the adult male vervet monkey is only 15 pounds (Rowell, 1974).

REFERENCES

Benton, L. The establishment and husbandry of a black howler (*Alouatta caraya*) colony at Columbia Zoo. *International Zoo Yearbook*, 1976, **16**, 149-152.

Bertrand, M. Acquisition by a pigtail macaque of behavior patterns beyond the natural repertoire of the species. *Zeitschrift für Tierpsychologie*, 1976, **42**, 139-169.

Blackwell, K. A brief note on bird-catching behavior. in the *Lar* gibbon. *International Zoo Yearbook*, 1969, **9**, 157.

Boyd, T. C., Page, C. and Schusterman, R. J. Activity patterns in a captive siamang family (*Symphalangus syndactylus*). Paper presented at the *American Society of Primatologists* Meeting, Seattle, Washington, April, 1977.

Buettner-Janusch, J. The breeding of *Galagos* in captivity and some notes on their behavior. *Folia Primatologica*, 1964, **2**, 93-110.

Bygott, J. D. Cannibalism among wild chimpanzees. *Nature*, 1972, **238**, 410-411.

Dare, R. Power, affiliation, and status in primates: Analysis with a new set of indices. Paper presented at the *American Association of Physical Anthropologists* Meeting, Seattle, Washington, April, 1977.

Estrada, A. and Estrada, R. Patterns of predation in a free-ranging troop of stumptail macaques (*Macaca arctoides*): Relations to the ecology II. *Primates*, 1977, **18** (3), 633-646.

Estrada, A., Sandoval, J. M. and Manzolillo, D. Further data on predation by free-ranging stumptail macaques (*Macaca arctoides*). *Primates*, 1978, **19** (2), 401-407.

Fragaszy, D. M. Contrasts in feeding behavior in captive pairs of *Saimiri* and *Callicebus*. Paper presented at the *Animal Behavior Society* Meeting, March, 1976.

Gartlan, J. S. Adaptive aspects of social structure in *Erythrocebus patas*. *Proceedings of the Symposium of the 5th Congress of the International Primatological Society*. Tokyo: Japan Science Press, 1975, pp. 161-171.

Gartlan, J. S. Ecology and behaviour of the patas monkey. Film presented at the *International Primatological Society* Meeting, Cambridge, England, August, 1976.

Gewirtz, H. B. and Gewirtz, J. L. Visiting and caretaking patterns for Kibbutz infants: Age and sex trends. *American Journal of Orthopsychiatry*, 1968, **38**, 427-443.

Harding, R. S. O. Predation by a troop of olive baboons (*Papio anubis*). *American Journal of Physical Anthropology*, 1973, **38**, 587-591.

Harding, R. S. O. and Strum, S. C. The predatory baboons of Kekopey. *Natural History*, 1976, **85**, 46-53.

Hausfater, G. Predatory behavior of yellow baboons. *Behaviour*, 1976, **56**, 44-68.

Jorde, L. B. and Spuhler, J. A. A statistical analysis of selected aspects of primate demography, ecology, and social behavior. *Journal of Anthropological Research*, 1974, **30**, 199-224.

Kavanagh, M. Food-sharing behavior within a group of Douc monkeys (*Pygathrix nemaeus nemaeus*). *Nature* (London), 1972, **239**, 406-407.

Kawai, M. Catching behavior observed in the Koshima troop. A case of newly acquired behavior. *Primates*, 1967, **8**, 181-186.

Kawamura, S. The process of sub-culture propagation among Japanese macaques. In Southwick, C. H. (Ed.) *Primate Social Behavior*. Princeton, N.J.: Van Nostrand, 1963, pp. 82-89.

Kinzey, W. G., Rosenberger, A. L., Heisler, P. S., Prowse, D. L. and Trilling, J. S. A preliminary field investigation of the yellow handed titi monkey, *Callicebus torquatus torquatus*, in Northern Peru. *Primates*, 1977, **18** (1), 159-181.

Kleiman, D. G. Monogamy in mammals. *The Quarterly Review of Biology*, 1977, **52**, 39-69.

Lorenz, R. Goeldi's monkey *Callimico goeldii* (Thomas 1904) preying on snakes. *Folia Primatologica*, 1971, **15**, 133-142.

Menzel, E. W., Jr. Primate naturalistic research and problems of early experience. *Developmental Psychobiology*, 1968, **1** (3), 175-184.

Menzel, E. W., Jr. Leadership and communication in young chimpanzees. *Symposium of the IVth International Congress of Primatology*. Basel: Karger, 1973, pp. 192-225.

Newkirk, J. B., III. A possible case of predation in the gibbon. *Primates*, 1973, **14** (2-3), 301-304.

Norikoshi, K. Tests to determine the responsiveness of free-ranging Japanese monkeys in food-getting situations. *Primates*, 1971, **12** (2), 113-124.

Oatley, T. B. Predatory behavior by the Samango monkey (*Cercopithecus mitis*). *Lammergcyer*, 1970, **11**, 000.

Pechtel, C., Masserman, J. H. and Aarons, L. Food acceptance in monkeys. *Recent Advances in Biological Psychiatry*, 1961, **3**, 228-235.

Peters, M. Mouth-to-mouth contact in squirrel monkeys (*Saimiri sciureus*). *Zeitschrift für Tierpsychologie*, 1970, **27**, 1009-1010.

Post, D. G. Baboon feeding behavior and the evolution of sexual dimorphism. Paper presented at the *American Association of Physical Anthropologists* Meeting, Seattle, Washington, April, 1977.

Rahaman, H. and Parthasarathy, M. D. The home range, roosting places, and day ranges of the bonnet macaque (*Macaca radiata*). *Journal of Zoology of London,* 1969, **157**, 267-276.

Raleigh, M. J. Sex differences in the social behavior of juvenile vervet monkeys. Paper presented at the *American Society of Primatologists* Meeting, Seattle, Washington, April, 1977.

Rhine, R. J. and Westlund, B. J. The nature and development of a primary feeding habit in yellow baboons (*Papio cynocephalus*). Paper presented at the *American Society of Primatologists* Meeting, Seattle, Washington, April, 1977.

Rose, M. D. Bipedal behavior of olive baboons (*Papio anubis*) and its relevance to an understanding of the evolution of human bipedalism. *American Journal of Physical Anthropology,* 1976, **44**, 247-261.

Rothe, H. Influences of newborn marmosets (*Callithrix jacchus*) behavior on expression and efficiency of maternal and paternal care. *Symposia of the Fifth International Congress of Primatology.* Nagoya, Japan, 1974, pp. 315-320.

Rowell, T. E. Contrasting different adult male roles in different species of non-human primates. *Archives of Sexual Behavior,* 1974, **3**, 143-149.

Sackett, G. P., Holm, R. A., Davis, A. E. and Fahrenbruch, C. E. Prematurity and low birth weight in pigtail macaques: Incidence, prediction, and effects on infant development. *Symposia of the Fifth International Congress of Primatology.* Nagoya, Japan, 1974, pp. 189-205.

Savage, E. S. and Malick, C. Play and socio-sexual behaviour in a captive chimpanzee (*Pan troglodytes*) group. *Behaviour,* 1977, **60**, 179-194.

Strum, S. C. Primate predation: Interim report on the development of a tradition in a troop of olive baboons. *Science,* 1975, **187**, 755-757.

Sugiyama, Y. The ecology of the lion-tailed macaque (*Macaca silenus*) (Linnaeus): A pilot study. *Journal of the Bombay Natural History Society,* 1968, **65** (2), 283-292.

Teleki, G. *Predatory Behavior in Wild Chimpanzees.* New Jersey: Associated University Presses, Inc., 1973.

Tenaza, R. R. Wild mynahs mimic wild primates. *Nature,* 1976, **259**, 561.

Torii, M. Possession by nonhuman primates. In Kondo, S., Kawai, M. and Ehara, A. (Eds.) *Contemporary Primatology.* Basel: Karger, 1975, pp. 310-314.

Tutin, C. E. G. Exceptions to promiscuity in a feral chimpanzee community. *Symposia of the Fifth International Congress of Primatology.* Nagoya, Japan, 1974, pp. 445-449.

Van Lawick-Goodall, J. Cultural elements in a chimpanzee community. In Menzel, E. W. (Ed.) *Symposia of the Fourth International Congress of Primatology* (Vol. 1): *Precultural Primate Behavior.* Basel: Karger, 1973, pp. 144-184.

Vencl, F. A case of convergence in vocal signals between marmosets and birds. *American Naturalist,* 1977, **111**, 777-782.

Willes, R. F., Kressler, P. L. and Truelove, J. F. Nursery rearing of infant monkeys (*Macaca fascicularis*) for toxicity studies. *Laboratory Animal Science,* 1977, **27**, 90-98.

Wise, L. A., Zimmerman, R. R. and Strobel, D. A. Dominance measurements of low and high protein reared rhesus macaques. *Behavioral Biology,* 1973, **9**, 77-84.

21.

Sex Differences in Grooming

Grooming is a behavior commonly seen in nonhuman primate social interaction and is often taken for granted. The present chapter compares male and female nonhuman primates regarding grooming behavior. At the outset, we should expect to find that, in general, female nonhuman priamtes groom more than do males. We will organize this chapter taxonomically as we have the others. A large share of this material will be based upon a previous review (Mitchell and Tokunaga, 1976).

A. PROSIMIANS

Among the Lorisidae, adult galagos (*Galago crassicaudatus*) often direct grooming to the other sex. Flinn and Nash (1975) reported that when grooming *was* seen between members of the same sex it was almost always between females. Roberts (1971) saw no grooming between males but much grooming between females. In infants, however, Rosenson (1972) observed that both same- and opposite-sexed infant galagos (seven days old) elicited interest and grooming from adult males as well as from adult females. The sex difference found among Lorisidae has also been seen in *Nycticebus coucang* (the slow loris) in which "females groomed 57 percent more than did males." (Tenaza et al., 1969, P. 669).

[1] This chapter is an expansion, revision and updating of the following, previously published article: Mitchell, G. and Tokunaga, D. H. Sex differences in non-human primate grooming. *Behavioural Processes,* 1976, **1**, 335-347 (© Elsevier Scientific Publishing Co., Amsterdam, The Netherlands).

B. SOUTH AMERICAN SPECIES

In the common marmoset (*Callithrix jacchus*), the amount of grooming done by the adult male and the adult female may depend upon the estrous condition of the female. At postpartum, the male marmoset makes more attempts to groom the female than the female does the male. Also at postpartum, the female refuses more invitations to groom (displayed by the male) than the male refuses when the female invites. Thus, at least during the time following the female's delivery of an infant, the male seems more interested in grooming than does the female (Rothe, 1974). The adult male and female groom more than do immatures (Ingram, 1976).

Kleiman (1977), in her review of monogamous primates, points out that male marmosets and tamarins groom quite a lot. Whereas in polygamous animals the females rest together and groom more than do males, in monogamous primates, males initiate grooming more than do females and there is more heterosexual grooming. In fact, Kleiman (1977) concludes the following:

> Thus, a basic characteristic of monogamy may be the simple fact that a male and a female sleep and groom together. (P. 59)

Leibrecht and Kelley (1977) have reported similar close heterosexual grooming between monogamous male and female night monkeys (*Aotus spp.*)

Among the Cebidae, there is variability from species to species as to which sex grooms the most. In spider monkeys (*Ateles spp.*) (Eisenberg and Kuehn, 1966), more dominant individuals groom more, whereas in *Cebus nigrivittatus* the adult females (which are subordinate to the adult males) do most of the grooming (Oppenheimer and Oppenheimer, 1972). Neville (1972) has reported that grooming patterns in the howler monkey (*Alouatta seniculus*) reveal the importance of mother–infant relations in group bonding. The female infants, in particular, continue to groom throughout life, remaining close to the mother and to other females. Males, in turn, tend to be more loosely attached to their troops.

In sum, sex differences in grooming among the New World primates apparently depend at least upon the individual genus being considered. In the sexually dimorphic howler and cebus monkeys as well as in others, females tend to be groomers, whereas in the more sexually monomorphic spider monkeys, sex differences in grooming are not as apparent. In the more primitive and monogamous marmosets, the male has been reported to groom more than the female, at least directly following parturition in the female. He also initiates more bouts of grooming.

C. OTHER OLD WORLD ARBOREAL MONKEYS

Since all South American monkeys are arboreal, it might be informative, at this time, to review the arboreal Old World monkeys (e.g., langurs, colobus, talapoins, *Cercopithecus spp.*). Before doing this, however, we should probably mention the one species group of *Cercopithecus* which is terrestrial (*C. aethiops*). Bernstein (1970) noted that in vervets (*Cercopithecus aethiops sabaeus*), males did less grooming than they received, whereas females both displayed and received more grooming than did the males. This difference was obtained with a group of vervets that did not have an adult male at the highest rank. However, Buckley (1976) also has noted this sex difference in the acquisition of grooming in captive vervets, and Raleigh (1977) has seen the same difference in juveniles.

In another *Cercopithecus* study (*Cercopithecus sabaeus*), adult females both gave and received more grooming than did males, and adult males were groomed more than they groomed. Males were never seen to groom other males. When males were groomed by females, most of the adult female grooming was directed toward one central male (Dunbar, 1974). Hunkeler, Bertrand, and Bertrand (1972) reported that the adult alpha male Campbell monkey (*Cercopithecus campbelli lowei*) was often groomed by adult females and that females groomed each other frequently.

Among talapoins (*Miopithecus talapoin*), there is no intersexual grooming, since the animals tend to stay in unisexual groups outside of the breeding season (Rowell and Dixson, 1975). Although little has been published on sex differences in intrasexual grooming, Segre (1970) has described a group of talapoins in which females groomed more and were more infant-oriented than were males. (Wolfheim, 1977) found that juvenile females groom more than do juvenile males.

Bernstein (1970) suggested that langurs (*Presbytis entellus*), like spider monkeys, may have grooming behavior that depends as much on status as on the sex of the animal. Even in the langurs, however, where dominance as well as gender seemed to be positively correlated with grooming, sex differences in development have been noted by Dolhinow and Bishop (1970). At ten months of age the male infant approaches, mounts, and embraces adult and subadult males (with no grooming), whereas the female infant has no contacts with adult males but instead grooms and is groomed by adult females. Poirier (1970, 1972) has reported that langur females groom more frequently and longer than do males.

Horwich and Manski (1975) report that in two species of *Colobus* (*C. guereza* and *C. polykomos*), adult males groomed four- to nine-week-old infants. In both red colobus and black and white colobus monkeys, however, females groom more than do males (Struhsaker and Oates, 1975).

Mangabey (*Cercocebus atys*) males do less grooming than they receive and females groom and receive more grooming more than do males. For another species of mangabey (*Cercocebus albigena johnstoni*), grooming is frequent between adult males and females, between adult females with infants, and between adult females without infants, but not between adult males (Chalmers, 1968; Bernstein, 1970).

In summary, Old World arboreal species, like those of the New World, present us with great variability from genus to genus. However, in the Old World arboreal monkeys, it is quite apparent overall that the adult females do more grooming than do the adult males.

D. OLD WORLD TERRESTRIAL MONKEYS

1. Macaques

In the rhesus macaque (*Macaca mulatta*) the female grooms more than does the male (cf. Bernstein, 1970). Even when strangers are introduced to unisexual and heterosexual groups of rhesus, the introduced females display and receive more grooming than do the introduced males, (Bernstein, Gordon, and Rose, 1974; Hansen, Harlow, and Dodsworth, 1966). Moreover, in rhesus monkeys housed in single cages in a laboratory, *self*-grooming occurs more frequently in females than in males, as does the invitation to groom directed toward a *human* observer (Cross and Harlow, 1965).

Mason, Green, and Posepanko (1960) have been among the few to report more invitations to groom in male rhesus than in female rhesus. Their animals were monitored in cages in captivity and the invitations to groom were also directed toward observers. There is no obvious reason for this different result. Perhaps differential histories in the laboratory prior to the observations are implicated.

In support of the sex difference reported by Cross and Harlow (1965), are data showing that even isolate-reared females display more self-grooming and invitations to groom than do their male counterparts (Cross and Harlow, 1965), although normally reared subjects show more grooming (particularly social grooming) overall (Hansen, Harlow, and Dodsworth, 1966). Mason (1960) has shown that early deprivation decreases grooming, particularly in male rhesus. Suomi, Harlow, and Kimball (1971) found that rhesus reared in social deprivation exhibited less grooming than did those that were socially reared. This also was particularly true of males. Ruppenthal et al. (1974), in studies of rhesus monkeys caged in what they called a "nuclear-family environment," reported that their rhesus females groomed more than did rhesus males. Even in familial or kinship subgroups of free-ranging rhesus, most of the kinship grooming involves mother–daughter grooming (Varley, 1977).

Ovarian hormones administered to a female will increase the amount of grooming by the male (Herbert, 1967). In addition, applications of extracts of vaginal secretions from estrous females to the sexual skin of nonestrous females is said to increase grooming of the painted female by the adult male (Keverne and Michael, 1971). Maple, Erwin, and Mitchell (1974) found that following a two-day separation from a specific male, upon return there was twice as much grooming displayed by estrous females as by menstruating females in laboratory heterosexual dyads. However, separating a familiar heterosexual dyad, and then reuniting after two days apart, increased the amount of grooming in *both* groups of females from preseparation to reunion. Prior to separation, however, menstruating females groomed the males more than did estrous females.

Not controlling for time together or familiarity, Michael and Herbert (1963) found that grooming of the male by the female reached a *minimum* at midcycle, whereas grooming of the female by the male reached a *maximum* at midcycle. While there seems to be little doubt that estrus affects grooming, *how* it affects grooming apparently depends upon the social context and whether there are social stresses. If, for example, a familiar pair of animals has just been reunited after separation, as in the Maple, Erwin, and Mitchell (1974) study, the estrous female may actually seek contact by grooming more than will the male. That is, a stress by sex interaction may be exacerbated by the estrous condition. Without the separation stress, grooming by the female is greater during menstruation rather than during midcycle. If this is borne out by data in the future, the combination of grooming, estrus, attachment, and sex differences would become an area of great interest indeed. The role of grooming and the role of estrus are both too important in the evolution of social behavior in primates *not* to be closely examined for potential interactions between the two. A summary of some relations of grooming to reproduction in the rhesus has been published by Agar and Mitchell (1975).

In an overview of socializing functions in many species of primates, not just rhesus monkeys, Poirier and Smith (1974) confidently state that females groom more than do males. Female grooming, in turn, is so much at the core of group behavior in free-ranging rhesus monkeys that Missakian (1973) has used cessation of grooming behavior between adult females as a definition of troop fission. It is interesting that she utilizes intra*female* grooming as the definition of whether or not a group even exists! According to Sade (1972), the status of individuals in grooming networks correlates with dominance only in females. In summary, for the rhesus monkey, there is clearly a sex difference in grooming and, more importantly, the difference is felt, by many, to be at the core of the social organization of the species.

Pigtail macaques (*Macaca nemestrina*) are larger than are rhesus and they are sexually dimorphic. Pigtail adult males do less grooming than do pigtail adult females. Males do less grooming than they receive, while females both groom and receive more grooming than do males (Bernstein, 1970; Bernstein, 1972). Thus, female pigtails spend a greater total time in social grooming than do males. Bernstein (1972) views this as one indication of their stronger social bonding. Rosenblum, Kaufman, and Stynes (1966) also reported that pigtail females groom more than do pigtail males; however, this is apparently not true of *self*-grooming.

Macaca fascicularis (the crab-eating macaque), is smaller than the rhesus, yet it too is sexually dimorphic and the grooming behavior it displays resembles that of the rhesus. Males do less grooming than they receive and females both groom more and receive more grooming than do males (Bernstein, 1970). In dyadic encounters, Thompson (1967) found a much higher frequency of grooming in female–female pairs than in male–male or even in male–female pairs.

The bonnet macaque (*Macaca radiata*) behaves in many ways quite differently from the three macaques discussed above (Rosenblum, Kaufman, and Stynes, 1966). Bonnets show higher levels of contact than do, for example, pigtails (Wade, 1977). Nevertheless, even in the bonnet monkey, grooming is most frequent in adult female–adult female dyads. All bonnet monkeys groom most with females. Even though there is greater tolerance of males by males in bonnets than in rhesus (Sugiyama, 1971), grooming is still much less frequent between male than between female bonnet macaques (Koyama, 1973). As in the pigtail macaque, females show more social grooming than do males, but not necessarily more self-grooming (Rosenblum, Kaufman, and Stynes, 1966). "Females develop their network primarily through grooming and remain closely associated with their mother. Males rely more heavily on the play group, which includes participating subadult and adult males." (Simonds, 1974, P. 151.)

The Barbary macaque male (*Macaca sylvana*) displays more care of infants than does the rhesus and is in other ways more "social" than is *M. mulatta*. In caring for an infant, the adult male Barbary macaque not only embraces it, holds it, and carries it, but also grooms it in ways similar to its mother (Lahiri and Southwick, 1966). There are no direct reports of sex differences in grooming, but it seems likely that the sex difference would be less in *Macaca sylvana* than in the rhesus, pigtail, or crab-eater macaques.

Stumptail macaques (*Macaca arctoides*) show extreme variability in social behavior, especially in sexual behavior (Chevalier-Skolnikoff, 1974). Based upon previous studies of this species (Bertrand, 1969) we would expect a smaller sex difference in grooming in the stumptail than in the rhesus. Males, for example, interact with infants more than do rhesus males (Gouzoules, 1975). Gouzoules (1975) has reported that the rank of an infant's mother correlates positively with the amount of grooming the infant receives. Here, as in the case of some of the New World and Old World arboreal species, status as well as gender affects grooming. It is interesting that, in the stumptail at least, females groom males *because* the males are dominant but males groom females for other reasons (i.e., sex). Grooming in the stumptail is correlated with dominance rank, and individuals holding adjacent ranks tend to interact with each other (Estrada, 1976). The males groom females more when the females are in estrus (Jones and Trollope, 1968). Male grooming is lowest just before menstruation, while female grooming declines sharply at midcycle (just after ovulation). These findings are similar to those reported for rhesus males in an unstressed situation. After ovariectomy, females increase their grooming and males increase their grooming of females (Slob et al., 1975).

Rhine and Kronenwetter (1972) directly compared male and female stumptail macaques for the amount of grooming displayed. Despite marked variability and basic behavioral differences from the rhesus and other macaques, stumptail macaque females groom more than do males.

Like the Barbary macaque males, Japanese macaque males have been observed interacting closely with infants (Itani, 1963). This has included grooming of the infants. Despite this tendency toward infant care in the males, females still groom infants more frequently than do males. Nonrelated females, in turn, groom less than do related females. Vocal sounds play an important role in grooming, but only in female–female grooming and not in mother–infant or male–female grooming. The rate of vocal sounds is low in adult males and so is the frequency of grooming (Mori, 1975). Baxter and Fedigan (1977) have shown the importance of grooming in Japanese macaque consort relations.

In summary, in most species of macaques, females do indeed groom more frequently than do their conspecific males. This is true even in those species of macaques where male–male tolerance is maximal. Although, for example, bonnet macaque males may groom more frequently than do rhesus macaque males, the males of both species groom less than do their female conspecifics. In addition, dominance status and estrus also affect male and female grooming behaviors (as does the presence of infants); and, grooming has been noted as a behavior which reduces tension within the troop (Terry, 1970). In fact, *female* grooming has been cited as *the* behavior which best defines which animals are members of a troop (Missakian, 1973).

2. Baboons

According to Bernstein (1970), among some baboons, females groom longer but not more frequently than do males (e.g., *Papio cynocephalus*). But Satnick (1977) has reported that female yellow baboons do a lot more grooming overall, whereas males tend to do most of their grooming with estrous females. In hamadryas baboons (*Papio hamadryas*), which live in harem groups, the male grooms the most. Although when an infant olive baboon (*Papio anubis*) is born only adult dominant males are allowed to handle it, and although the young have early and intimate contact with adult males, including grooming of the infant by the males (cf. Dolhinow and Bishop, 1970), females do more grooming than do males (cf. Hamburg and Lunde, 1966). There is, however, some degree of flexibility in the behavior of all baboons within the genus *Papio*. Female baboons, for example, modify their behavior according to the social pattern imposed on them by the much larger males. Hamadryas males rally and herd females in their harems whereas savannah baboon males (*Papio anubis*) tend to groom and follow females more. A hybrid male (half *P. hamadryas* and half *P. anubis*) follows and grooms females and does not herd them, yet the hamadryas females adjust to the hybrid male's behavioral pattern (Nagel, 1970). This finding is particularly enlightening regarding the degree of fixedness or species-specificity in grooming patterns. It shows that the system of social organization adopted by a species is determined by both genetics and social tradition. Apparently, grooming patterns are also determined by both.

Rowell (1968) has examined the role of estrus in baboon grooming behavior and has concluded that when female baboons are swollen they increase their grooming of males. She also noted, however, that males increase their grooming of females when the females are swollen. A summary of the relation of grooming to reproduction in baboons (as well as in macaques) has been published by Rowell (1972).

In chacma baboons (*Papio ursinus*), grooming is described as being typical of females (Bolwig, 1959). For the gelada baboon (*Theropithecus gelada*), Bernstein (1975) has reported that females are more active in grooming than are the males. Bramblett (1970) points out that coalitions among gelada baboons are expressions of social relationships but that such coalitions are less frequently displayed than is grooming. Females who frequently groom one another are also likely to form coalitions against others in the group, even against the harem leader.

E. APES

Just as in the monogamous New World monkeys, monogamous male gibbons and siamangs groom frequently. Free-ranging gibbons groom fifteen to thirty minutes per day, most of it heterosexual male–female grooming, whereas chimpanzees groom only ten minutes per day and gorillas even less (Kleiman, 1977). In monogamous primates, as we may remember, there is less sexual behavior, more grooming, and fewer sex differences in grooming. Gibbons and siamangs tend to groom with the opposite sex rather than with the same sex. A juvenile male is groomed less than is a juvenile female by an adult male siamang (Boyd, Page, and Schusterman, 1977).

According to Mason (1970), chimpanzees tend to direct more grooming toward a dominant partner than toward a subordinate but this is true only of female chimpanzees (administration of estrogen increases female chimpanzee dominance). Moreover, Van Hooff (1973) has reported that chimpanzee males both give and receive a great deal of grooming, whereas females tend to give a lot but not to receive as much. Schaller (1965), however, has said that chimpanzee grooming is primarily a within-sex *female* activity. On the other hand, the degree of male sexual possessiveness of a female (degree of monogamy) in a heterosexual consorting dyad correlates positively with the amount of time the male spends grooming the female (Tutin, 1974).

Reynolds (1965) has compared gorillas and chimpanzees in grooming. He reports that in the gorillas, females do most of the grooming, while in chimpanzees, both males and females groom a lot.

These reports on the great apes are not consistent from study to study. It is clear in chimpanzees, however, that as the chimpanzee matures, grooming takes over for play as a means of cementing bonds. It is also certain that mothers having infants who play with each other will groom each other as their infants play (Savage and Malick, 1977). Because male offspring tend to wander away sooner than do female infants, it is likely that most female infants receive and reciprocate more grooming from their mothers for longer than do most male infants (cf. Pusey, 1976).

In summary, there is probably a tendency for female chimpanzees and gorillas to groom more than do their male conspecifics; however, this grooming difference depends as much upon such things as group composition, dominance, kinship, and female estrus as it does upon gender.

Even in captivity, however, gorillas and orangutans exhibit very little grooming as compared to macaques, baboons, or even to gibbons and chimpanzees. Female orangutans, however, groom more than do males and they groom for longer bouts. In one captive heterosexual adult dyad 77 percent of the grooming was done by the female toward the adult male (Zucker et al., 1977; Maple, personal communication).

Our conclusions regarding sex differences in grooming are as follows: It is probably reasonably safe to assume that, in most primates, under most circumstances, females will have a greater predisposition to groom than will males. However, this may not be as true and, in fact, may be completely false for some monogamous arboreal species of monkeys or for some of those species which show low sexual dimorphism in general. Conversely, the sex difference generalization is probably most true for those terrestrial Old World monkeys which display strong sexual dimorphism. There are also changes in the grooming sex difference generalization with changes in estrus in the females, with changes in relative dominance, with separation from familiar partners, with changes in group composition, and with changes in familiarity, kinship, early experiences, and social tradition.

F. IMPLICATIONS FOR HUMANS

Maccoby and Jacklin (1974) listed as their very first "myth" concerning sex differences, the "myth" that girls are more "social" than boys. They pointed out that there was no evidence that girls were more likely than boys to be concerned with people, that girls were not more interested in social stimuli, that girls were not more responsive to social rewards such as praise from others. They insisted that any differences which do exist in socialness are more of kind than degree.

Admitting from the start that *Homo sapiens* is a far more plastic and intelligent species than the other primates, and granting that both males and females of our species are probably more capable of playing a broader range of roles, it is still conceivable that these grooming data may be of import to our own species.

Grooming has been likened to "socialness." Some researchers have defined the existence of the social group itself by the occurrence or nonoccurrence of grooming (Missakian, 1973). Grooming has been shown to be tension reducing (Terry, 1970) and to play a strong role in infant care and in sexual behavior (cf. Agar and Mitchell, 1975; Mitchell, 1976). Therefore, if there is *indeed* a general sex difference in grooming, there may well be a general sex difference in "socialness" (cf. Maccoby and Jacklin, 1974) within the *non*-human primates, with females being more "social" than males.

Questionnaire data on same-sex friendships in people indicate that women are much more likely to sit and talk than are men (Caldwell and Peplau, 1977). We already know that in decorating their dormitory walls, college women use more personal and social items than do college men (Vinsel et al., 1977). Even in cross-species interactions, women hold eye contact longer with rhesus monkeys than do men (especially with female rhesus monkeys) (Thomsen, 1974); and, in many studies conducted over the last ten years, it has been found that animal pets substitute for humans (and are about half as psychologically significant as humans to their owners), particularly for *female* humans. The proclivity for "socializing" with pets in human females is very great (Cameron, 1977). One could easily debate whether this is a liability or a benefit to society.

Whatever use the nonhuman grooming sex difference may have for humans, it must be remembered that: (1) the difference is *most* well established among Old World terrestrial monkeys, and (2) the difference is not so prewired even among these nonhuman primates as to make differential early and even adult experience ineffective in producing changes.

REFERENCES

Agar, M. E. and Mitchell, G. Behavior of free-ranging rhesus adults: A review. In Bourne, G. (Ed.) *The Rhesus Monkey* (Vol. 1, Chapter 8). New York: Academic Press, 1975, pp. 323–342.

Baxter, M. J. and Fedigan, L. M. A comparative analysis of grooming and consort partner selection in a troop of Japanese monkeys. Paper presented at the *American Society of Primatologists* Meeting, Seattle, Washington, April, 1977.

Bernstein, I. S. Primate status hierarchies. In Rosenblum, L. A. (Ed.) *Primate Behavior: Developments in Field and Laboratory Research* (Vol. 1). New York: Academic Press, 1970, pp. 71-109.

Bernstein, I. S. Daily activity cycles and weather influences on a pigtail monkey group. *Folia Primatologica,* 1972, **18**, 390-415.

Bernstein, I. S. Activity patterns in a gelada monkey group. *Folia Primatologica,* 1975, **23**, 50-71.

Bernstein, I. S., Gordon, T. P. and Rose, R. M. Factors influencing the expression of aggression during introduction to rhesus monkey groups. In Holloway, R. L. (Ed.) *Primate Aggression, Territoriality and Xenophobia: A Comparative Perspective.* San Francisco: Academic Press, 1974, pp. 211-240.

Bertrand, M. The behavioral repertoire of the stumptail macaques. *Bibliotheca Primatologica,* 1969, **11**, 1-273.

Bolwig, N. A study of the behavior of the chacma baboon, *Papio ursinus. Behaviour,* 1959, **14**, 136-163.

Boyd, T. C., Page, C. and Schusterman, R. J. Activity patterns in a captive siamang family (*Symphalangus syndactylus*). Paper presented at the *American Society of Primatologists* Meeting, Seattle, Washington, April, 1977.

Bramblett, C. A. Coalitions among gelada baboons. *Primates,* 1970, **11**, 327-333.

Buckley, J. S. Sexual differences in acquisition of grooming behaviors in captive vervet monkeys (*Cercopithecus aethiops*). *American Journal of Physical Anthropology,* 1976, **44**, 168.

Caldwell, M. A. and Peplau, L. A. Sex differences in friendship. Paper presented at the *Western Psychological Association* Meeting, Seattle, Washington, April, 1977.

Cameron, P. The pet threat. Paper presented at the *Western Psychological Association* Meeting, Seattle, Washington, April, 1977.

Chalmers, N. R. The social behavior of free living mangabeys in Uganda. *Folia Primatologica,* 1968, **8**, 263-281.

Chevalier-Skolnikoff, S. Male-female, female-female, and male-male sexual behavior in the stumptail monkey, with special attention to the female orgasm. *Archives of Sexual Behavior,* 1974, **3**, 95-116.

Cross, H. A. and Harlow, H. F. Prolonged and progressive effects of partial isolation on the behavior of macaque monkeys. *Journal of Experimental Research on Personality,* 1965, **1**, 39-49.

Dolhinow, P. J. and Bishop, N. The development of motor skills and social relationships among primates through play. In Hill, J. P. (Ed.) *Minnesota Symposia on Child Psychology* (Vol. IV). Minneapolis: University of Minnesota Press, 1970, pp. 141-198.

Dunbar, R. I. M. Observation on the ecology and social organization of the green monkey, *Cercopithecus sabaeus,* in Senegal. *Primates,* 1974, **15**, 341-350.

Eisenberg, J. F. and Kuehn, R. E. The behavior of *Ateles geoffroyi* and related species. *Smithsonian Miscellaneous Collections,* 1966, **151** (8), 1-63.

Estrada, A. Social relations in a free-ranging troop of *Macaca arctoides.* Paper presented at the *International Primatological Society* Meeting, Cambridge, England, August, 1976.

Flinn, L. and Nash, L. T. Group formation in recently captured lesser galagos. Paper presented at the *American Association of Physical Anthropologists* Meeting, Denver, Colorado, 1975.

Gouzoules, H. Maternal rank and early social interactions of infant stumptail macaques, *Macaca arctoides. Primates,* 1975, **16**, 405-418.

Hamburg, D. A. and Lunde, D. T. Sex hormones in the development of sex differences in human behavior. In Maccoby, E. E. (Ed.) *The Development of Sex Differences.* Stanford, California: Stanford University Press, 1966, 1-24.

Hansen, E. W., Harlow, H. F. and Dodsworth, R. O. Reactions of rhesus monkeys to familiar and unfamiliar peers. *Journal of Comparative Physiological Psychology,* 1966, **61**, 274-279.

Herbert, J. The social modification of sexual and other behaviour in the rhesus monkey. *Proceedings of the First International Congress of Primatology,* 1967, pp. 232-246.

Horwich, R. H. and Manski, D. Maternal care and infant transfer in two species of *Colobus* monkeys. *Primates,* 1975, **16**, 49-73.

Hunkeler, C., Bertrand, B. and Bertrand, M. Le comportement social de la Mone de Lowe *(Cercopithecus campbelli lowei). Folia Primatologica,* 1972, **17**, 218-236.

Ingram, J. C. Social interactions within marmoset family groups *(C. jacchus).* Paper presented at the *International Primatological Society* Meeting, Cambridge, England, August, 1976.

Itani, J. Paternal care in the wild Japanese monkey, *Macaca fuscata.* In Southwick, C. H. (Ed.) *Primate Social Behavior.* Princeton, N.J.: Van Nostrand, 1963, pp. 91-97.

Jones, N. G. B. and Trollope, J. Social behavior of stump-tailed macaques in captivity. *Primates,* 1968, **9**, 365-394.

Keverne, E. B. and Michael, R. P. Sex-attractant properties of ether extracts of vaginal secretions from rhesus monkeys. *Journal of Endocrinology,* 1971, **51**, 313-322.

Kleiman, D. G. Monogamy in mammals. *The Quarterly Review of Biology,* 1977, **52**, 39-69.

Koyama, N. Dominance, grooming, and clasped-sleeping relationships among bonnet monkeys in India. *Primates,* 1973, **14** (2/3), 225-244.

Lahiri, P. K. and Southwick, C. H. Parental care in *Macaca sylvana. Folia Primatologica,* 1966, **4**, 257-264.

Leibrecht, B. C. and Kelley, S. T. Some observations of behavior in breeding pairs of owl monkeys. Paper presented at the *American Society of Primatologists* Meeting, Seattle, Washington, April, 1977.

Maccoby, E. E. and Jacklin, C. N. *The Psychology of Sex Differences.* Stanford, California: Stanford University Press, 1974.

Maple, T. Personal communication, 1977.

Maple, T., Erwin, J. and Mitchell, G. Separation of adult heterosexual pairs of rhesus monkeys: The effect of female cycle phase. *Journal of Behavioral Science,* 1974, **2** (2), 81-86.

Mason, W. A. The effects of social restriction on the behavior of rhesus monkeys: I. Free social behavior. *Journal of Comparative and Physiological Psychology,* 1960, **53**, 582-589.

Mason, W. A. Chimpanzee social behavior. In Bourne, G. H. (Ed.) *The Chimpanzee* (Vol. 2). New York: Karger, 1970, pp. 265-288.

Mason, W. A., Green, P. C. and Posepanko, C. J. Sex differences in affective-social responses of rhesus monkeys. *Behaviour,* 1960, **16**, 74-83.

Michael, R. P. and Herbert, J. Menstrual cycle influences grooming behavior and sexual activity in the rhesus monkey. *Science,* 1963, **140**, 500-501.

Missakian, E. A. The timing of fission among free-ranging rhesus monkeys. *American Journal of Physical Anthropology,* 1973, **38**, 621-624.

Mitchell, G. Parental behavior in non-human primates. In Money, J. and Musaph, H. (Eds.) *Handbook of Sexology* (Chap. 54). Amsterdam: Elsevier, 1976, pp. 749-759.

Mitchell, G. and Tokunaga, D. H. Sex differences in nonhuman primate grooming. *Behavioural Processes,* 1976, **1**, 335-347.

Mori, A. Signals found in the grooming interactions of wild Japanese monkeys of the Koshima troop. *Primates,* 1975, **16**, 107-140.

Nagel, U. Social organization in a baboon hybrid zone. *Proceedings of the Third International Congress of Primatology* (Zurich), 1970, **3**, 48-57.

Neville, M. K. Social relations within troops of red howler monkeys (*Alouatta seniculus*). *Folia Primatologica,* 1972, **18**, 47-77.

Oppenheimer, J. R. and Oppenheimer, E. C. Behavioral observations of *Cebus nigrivittatus* on the Venezuelan llanos. *American Zoologist,* 1972, **12**, 69.

Poirier, F. E. Dominance structure of the Nilgiri langur (*Presbytis johnii*) of South India. *Folia Primatologica,* 1970, **12**, 161-186.

Poirier, F. E. Nilgiri langur behavior and social organization. In Voget, F. W. and Stephenson, R. L. (Eds.) *For the Chief: Essays in Honor of Luther S. Cressman.* Eugene, Oregon: Univeristy of Oregon Anthro-Papers No. 4, 1972, pp. 119-134.

Poirier, F. E. and Smith, E. O. Socializing functions of primate play. *American Zoologist,* 1974, **14**, 275-287.

Pusey, A. E. Sex differences in the persistence and quality of the mother off-spring relationship after weaning in a population of wild chimpanzees. Paper presented at the *International Primatological Society* Meeting, Cambridge, England, August, 1976.

Raleigh, M. J. Sex differences in the social behavior of juvenile vervet monkeys. Paper presented at the *American Society of Primatologists* Meeting, Seattle, Washington, April, 1977.

Reynolds, V. Some behavioral comparisons between the chimpanzees and the mountain gorilla in the wild. *American Anthropologist,* 1965, **67**, 693-706.

Rhine, R. J. and Kronenwetter, C. Interaction patterns of two newly formed groups of stumptailed macaques (*Macaca arctoides*). *Primates,* 1972, **13**, 19-33.

Roberts, P. Social interactions of *Galago crassicaudatus. Folia Primatologica,* 1971, **14**, 171-181.

Rosenblum, L. A., Kaufman, I. C. and Stynes, A. J. Some characteristics of adult social and autogrooming patterns in two species of macaque. *Folia Primatologica,* 1966, **4**, 438-451.

Rosenson, L. M. Interactions between infant greater bushbabies (*Galago crassicaudatus*) and adults other than their mothers under experimental conditions. *Zeitschrift für Tierpsychologie,* 1972, **31**, 240-269.

Rothe, H. Allogrooming by adult *Callithrix jacchus* in relation to postpartum oestrus. *Journal of Human Evolution,* 1974, **3**, 535-540.

Rowell, T. E. Grooming by adult baboons in relation to reproductive cycles. *Animal Behaviour,* 1968, **16**, 585-588.

Rowell, T. E. *The Social Behaviour of Monkeys.* Baltimore: Penguin, 1972.

Rowell, T. E. and Dixson, A. F. Changes in social organization during the breeding season of wild Talapoin monkeys. *Journal of Reproduction and Fertility,* 1975, **43**, 419-434.

Ruppenthal, G. C., Harlow, M. K., Eisele, C. D., Harlow, H. F. and Suomi, S. J. Development of peer interactions of monkeys reared in a nuclear-family environment. *Child Development,* 1974, **45**, 670-682.

Sade, S. S. Sociometrics of *Macaca mulatta.* I. Linkages and cliques in grooming matrices. *Folia Primatologica,* 1972, **18**, 196-223.

Satnick, R. Grooming relationships in yellow baboons, and their significance for other measures of social organization. Paper presented at the *American Association of Physical Anthropologists* Meeting, Seattle, Washington, April, 1977.

Savage, E. S. and Malick, C. Play and socio-sexual behaviour in a captive chimpanzee (*Pan troglodytes*) group. *Behaviour,* 1977, **60**, 179-194.

Schaller, G. B. Behavioral comparisons of the apes. In DeVore, I. (Ed.) *Primate Behavior.* New York: Holt, Rinehart and Winston, 1965, pp. 474-484.

Segre, A. Talapoins. *Animal Kingdom,* 1970, **73** (3), 20-25.

Simonds, P. E. Sex differences in bonnet macaque networks and social structure. *Archives of Sexual Behavior,* 1974, **3**, 151-165.

Slob, A. K., Goy, R. W., Wilgand, S. J. and Scheffler, G. Gonadal hormones and behaviour in the stumptail macaque (*Macaca arctoides*) under laboratory conditions: A preliminary report. *Journal of Endocrinology,* 1975, **64**, 38.

Struhsaker, T. T. and Oates, J. F. Comparison of the behavior and ecology of red colobus and black and white colobus monkeys in Uganda: A summary. In Tuttle, R. H. (Ed.) *Socioecology and Psychology of Primates.* The Hague: Mouton, 1975, pp. 103-123.

Sugiyama, Y. Characteristics of the social life of bonnet macaques. *Primates,* 1971, **12** (3/4), 247-266.

Suomi, S. J., Harlow, H. F. and Kimball, S. D. Behavioral effects of prolonged partial social isolation in the rhesus monkey. *Psychological Reports,* 1971, **29**, 1171-1177.

Tenaza, R. R., Ross, B. A., Tanticharvenyos, P. and Berkson, G. Individual behavior and activity rhythms of captive slow lorises (*Nycticebus coucang*). *Animal Behaviour,* 1969, **17**, 664-669.

Terry, R. L. Primate grooming as a tension reduction mechanism. *Journal of Psychology,* 1970, **76**, 129-136.

Thompson, N. S. Some variables affecting the behaviour of irus macaques in dyadic encounters. *Animal Behaviour,* 1967, **15**, 307-311.

Thomsen, C. E. Eye contact by non-human primates toward a human observer. *Animal Behaviour,* 1974, **22**, 144-149.

Tutin, C. E. G. Exceptions to promiscuity in a feral chimpanzee community. *Proceedings of the Fifth International Congress of Primatology,* 1974, pp. 445-449.

Van Hooff, J. A. R. A. M. A structural analysis of the social behaviour of a semi-captive group of chimpanzees. In von Cronach, M. and Vine, I. (Eds.) *Social Communication and Movement.* London: Academic Press, 1973, pp. 75-162.

Varley, M. A. Grooming relations in three groups of rhesus monkeys on Cayo Santiago. Abstract for the *Animal Behavior Society* Meeting, University Park, Pennsylvania, June, 1977.

Vinsel, A., Wilson, J., Brown, B. B. and Altman, I. Personalization in dormitories: Sex differences. Paper presented at the *Western Psychological Association* Meeting, Seattle, Washington, April, 1977.

Wade, T. Inbreeding and kin selection in the evolution of macaque social behavior. Paper presented at the *American Society of Primatologists* Meeting, Seattle, Washington, April, 1977.

Wolfheim J. H. Sex differences in behavior in a group of captive juvenile tala-poin monkeys (*Miopithecus talapoin*). *Behaviour,* 1977, **63**, 110–128.

Zucker, E. L., Wilson, M. E., Hoff, M. P., Nadler, R. D. and Maple, T. Grooming behaviors of orangutans and gorillas: Description and comparison. Paper presented at the *Animal Behavior Society* Meeting, University Park, Pennsylvania, June, 1977.

22.

Sex Differences in Vocalizations[1]

As we have seen in previous chapters, male and female primates differ considerably in the degree to which they are intragroup- or extragroup-oriented. For example, males are often peripheralized, more vigilant of extragroup events, change groups more frequently, and are more often involved in predation. Females, on the other hand, remain at the core of the troop and groom more often than do males. Given these very general differences in the behavioral characteristics of the two sexes, we might expect to find sex differences in loud or distant vocalizations as well as in the more quiet and intimate contact vocalizations. Let us examine the literature with regard to these two general types of vocalizations. Rather than employing a strict taxonomic format throughout, we will instead employ a format in which we first present these vocalizations where males surpass females and then the vocalizations where females surpass males.

A. GENERAL COMMENTS

A very extensive description of rhesus monkey (*Macaca mulatta*) vocalizations has been published by Rowell (1962; also Rowell and Hinde, 1962). More recent research on macaque vocalizations has been reviewed by Rowell (1972) and by Erwin (1975). Lindburg (1971) has described the use of vocalizations by rhesus monkeys in the wild.

Basically speaking, rhesus vocalizations have been divided into two major groups by Rowell (1962): harsh noises and clear calls. The harsh noises (i.e.,

[1]The present chapter is based in large part upon the following unpublished article: Mitchell, G. and Prassa, S. P. The effects of gender, experience, and other factors on the vocalizations of nonhuman primates. *Paper of the Harlow Symposium.* Tucson, Arizona, December, 1976.

bark, roar, growl, screech, etc.) are most often heard in agonistic contexts, whereas the clear calls (e.g., whoo) seldom occur in agonistic contexts. Of course, there are many gradations *within* each category, whether harsh or clear, and there are also gradations between the two categories. An example of the former is the rhesus "screech-bark." To the human ear this vocalization, which occurs during fighting, sounds like it is a screech and a bark occurring simultaneously. An example of between-category confusion is the "coo-screech." This vocalization, which sometimes occurs in the initial moments following separation from a familiar partner, has components of both fear noises and contact calls occurring at least serially and perhaps even simultaneously. It starts with a clear "coo" and increases in pitch and harshness until it becomes a screech.

Our goal in this chapter is not to look at the physical characteristics of primate sounds. We are instead concerned primarily with those sex differences in vocalizations which can be easily discriminated by human listeners. Since more is known in this regard about *Macaca mulatta* than about any other species of primate, let us begin with sex differences in rhesus harsh noises.

B. "MALE" VOCALIZATIONS

1. Rhesus Macaques

Altmann (1968) noted, in one of his often cited papers on the sociobiology of the rhesus, that in agonistic encounters, rhesus males more often assumed a strategy of "come closer or scream" than rhesus females who more often "hit or withdrew." In this description of male rhesus agonism, Altmann noted that the adult males "lunged and screeched" more than did females. We cannot, however, come to the general conclusion on the basis of the Altmann report that male rhesus monkeys of all ages emit a greater total of agonistic or harsh noises than do females. Mason and Berkson (1975) paired lab-born rhesus with wild-born rhesus and discovered that, in these pairings, males screeched (harsh noise associated with fear) *less* than did females but barked *more* than did females. Baysinger, Brandt, and Mitchell (1972), on the other hand, found that female *infant* rhesus (in the laboratory) vocalized more than did male rhesus, and that this was even noticeable in the "bark" vocalizations (a harsh call). Clearly, age and/or social or nonsocial environment have much to do with the nature of the sex difference observed. In one laboratory study (Maple, Risse, and Mitchell, 1973), *adult* females emitted more harsh bark vocalizations than did their male conspecifics; however, these vocalizations may not have been "typical" barks since they occurred following social separation.

As noted above, it is often difficult for the human ear to distinguish between harsh sounds since there are intermediate sounds between the various kinds.

Particularly in extreme agonistic contexts the screech and bark merge together. However, the animals in the noncorroborative Mason and Berkson (1975) study were all approximately 1½ years of age and they were not involved in extreme aggressive encounters.

2. Other Macaques

Not only wild rhesus males, but adult males of other wild macaques have been reported to emit the alarm cry or warning bark more frequently then do their adult female conspecifics. Among the other macaques which evince a gender difference in this regard are *Macaca fascicularis* and *Macaca fuscata*. In fact, this adult sex difference has been reported for many Old World species (Marler, 1968; Bramblett, 1973).

As in the case of the macaques discussed above, a gender difference in vocalizations involving intertroop conflict has been reported for the bonnet monkey (*M. radiata*). Adult male bonnets more often make aggressive noises near strangers than do females. When leaving feeding grounds for a resting place, the adult males, who keep a watchful eye on extratroop activity, display a vocalization which apparently induces others in the group to follow (Rahaman and Parthasarathy, 1969).

The more arboreal lion-tailed macaque (*Macaca silenus*) displays vocal intimidation between neighboring troops using whooping displays (rare in macaques) similar to those emitted by Nilgiri langurs (which, incidentally, live in the same area). The lion-tailed macaque troop has a higher male to female ratio than other macaques and has a more open social organization (Sugiyama, 1968). It would be interesting to note whether the gender differences in vocalization seen in other macaques are as marked in *Macaca silenus* as they are in the more terrestrial primates living in more closed societies.

In summary, macaques other than rhesus tend to display gender differences in vocalization similar to the rhesus, but in many of the macaques (e.g., *M. sylvana, M. fuscata, M. arctoides, M. radiata, M. silenus*) it is likely that the sex differences are not nearly as clear-cut and it is probable that the adult males utilize soft "social" sounds more often than do rhesus males.

Bramblett (1973) has made the general statement that most Old World monkey *adult* males give alarm calls (a harsh noise) more than do their adult female conspecifics. Marler (1968), referring to these alarm calls as warning barks, lists some Old World monkeys in which the adult male typically emits the warning bark. Among these are *Macaca mulatta, Macaca fascicularis,* and *Macaca fuscata.* On the whole, the following statement appears to be true for the macaques: Adult males emit more warning barks than do adult females.

3. Baboons

In the first three months of life, evidently baboon infant males vocalize more than do infant females (Young, 1977). However, the nature of these vocalizations is not clear.

Adult male chacma baboons (*Papio ursinus*) are regularly seen mobilizing the troop by giving loud vocalizations (Buskirk, Buskirk, and Hamilton, 1974). They do this by chasing troop members (but only in *one* troop out of five). This finding seems related to the report of a male "rally call" in bonnet macaques (Rahaman and Parthasarathy, 1969). Jouventin (1975) noted a sex difference in a rallying call in mandrills (*Mandrillus sphinx*). In this species, leader males are the only ones who emit rally calls while *both* males and females emit alarm calls. When savannah baboons encounter a group of lions, animals of all ages and both sexes bark (Saayman, 1971). In *Papio papio* (the Guinea baboon), on the other hand, males tend to bark and roar while females scream, screech, or squeal (Masure and Bourliére, 1971). If Rowell's (1952) categories can be applied to the Guinea baboon, one set of "harsh vocalizations" (barks and roars) has more to do with aggression, whereas the other type of vocalization (also harsh) is more often correlated with fear. The first set of vocalizations is male-related; the second type is female-related.

4. Langurs

In *Presbytis cristatus* (the silver leaf-monkey) troops, there is usually a single adult male. According to Bernstein (1966), it is the male's role to primarily defend the troop's territory from extratroop adult males, including the use of territorial vocalizations. In the hanuman langur (*Presbytis entellus*) studied by Bishop (1975), adult males were responsible for all vocal behavior *except* infant vocalizations and alarm barks; however, in Dolhinow's (1972) langurs (of the same species) adult males did not squeal or scream. In addition, her adult females did not emit the belch vocalization whereas the males did. This suggests that hanuman langur gender differences may depend upon the troop.

Male langurs usually perform the territorial whoop vocalizations (Marler, 1968); however, in the golden langurs of Assam (*Presbytis geei*), there is an *absence* of whoop calls in adult males and the alarm calls are also different from those of the hanuman langur. The golden langur male emits two different kinds of alarm calls (Mukherjee and Saha, 1974).

Among Nilgiri langurs (*Presbytis johnii*) vocalizations within dominance encounters vary with gender. There are more loud vocalizations in female dominance encounters than there are in the dominance encounters of males. Separation of troops is maintained by male vigilance and by the male whoop call as in *Presbytis entellus* (Poirier, 1968; Poirier, 1970a). It is interesting that adult male alarm calls in *Presbytis johnii* evoke stronger responses from the troop than do those of the adult female (Poirier, 1970b).

Wilson and Wilson (1975) have recommended the use of langur male vocalizations as taxonomic aids within the *Presbytis aygula-melalophos* group of langurs in Sumatra. They write:

> Most variable is general pelage color; next is general color pattern and lie of the hair on the head. Less variable in terms of geographical distribution represented by each is the loud call of the adult male. (P. 461)

And:

> . . . we believe that the loud call of the adult male is a valid and useful tool in elucidating phylogenetic relationships. (P. 462)

In summary, the vocal behavior of langurs varies greatly from species to species, so much so that variations in the loud call of males can be used to taxonomic advantage. In general, adult males probably emit more whoop, territorial, and alarm calls than do females; however, even in-this vocalization there is at least one species of langur in which the male does not whoop at all (the golden langur). Squeals and screams, on the other hand, are apparently more often emitted by females.

5. *Cercopithecus spp.*

In contrast to the langurs, it is the alarm call of the *female* vervet not of the male which brings the greatest response from the troop (Poirier, 1970b). In Lowe's guenon (*Cercopithecus campbelli*), however, the male *does* act as a sentinel and emits a warning bark (Bourliére, 1969). It is the alpha male which acts as a watchdog and a control animal and he is the *only* male to emit loud calls of warning (Hunkeler, Bourliére, and Bertrand, 1972).

There are interesting age differences in vocalizing in *C. aethiops*. Vocalizations represent a much larger portion of an infant's communicative repertoire than they do of an adult's communicative repertoire which is composed of many more visual facial signals (Struhsaker, 1967).

Marler (1968) makes the general statement that warning barks occur more frequently in male vervets than in female vervets. Like the loud calls of the langur males, the rally and territorial calls of *Cercopithecus* males (red-tailed and blue monkeys) have been selected for a high degree of specific distinctiveness when they function to maintain intergroup spacing or to rally the group.

The alarm calls, on the other hand, have a minimum of specific distinctiveness. Of seven blue monkey (*C. mitis*) vocalizations, three vocalizations were restricted to adult males; of five red-tailed monkey (*C. ascanis*) vocalizations, two vocalizations were restricted to adult males (Marler, 1973).

In summary, gender differences in vocalizations in *Cercopithecus* species follow along lines seen elsewhere. That is, males usually emit most of the warning barks as well as the rally and territorial calls.

6. Other Monkeys of the Old World

Kawabe and Mano (1972) have shown that adult *male* proboscis monkeys (*Nasalis larvatus*) emit growls and honks of warning (alarm) and threat, whereas adult females emit shrieks and screams of fear and surprise.

In the Angola colobus (*Colobus angolensis*), it is usually a single male that performs the distinctive jumping-*roaring* display. This display, including the vocalized roar, probably serves a diversionary function much like a similar display performed by the male patas monkey (*Erythrocebus patas*). It may also maintain spacing (Groves, 1973; see also Marler, 1968; Bramblett, 1973). Mangabeys (*Cercocebus galeritus*) males emit calls of strong intensity which appear only after sexual maturity and which depend on the male's sexual activity cycle (Quris, 1975).

Based upon the Old World monkey material included in this review to this point, vocalizations involving territorial, defense, warning, and alarm calls are male-related sounds, whereas sounds related to fear or flight are female sounds. Moreover, based upon the rhesus studies, it appears that females vocalize (overall) more frequently than do males, unless the males are involved in intense agonistic encounters. How widespread are these gender differences within the primate order? Do they extend outside the limits of Old World monkeys? Let us look at some other major groups within the primate order.

7. Prosimians

In tree shrews (*Tupaia spp.*), the most primitive of primates (if they are primates at all), there is a gender difference in vocalizations during aggressive encounters. Male aggression involves chasing and biting whereas female aggression involves threats and vocalizations (Sorenson, 1974). Thus, female aggression is more vocal than male aggression.

In the lesser bushbaby (*Galago crassicaudatus*), there is a specific *male* call associated with territorial spacing and urine washing. The natural group of this genus is sometimes somewhat like a monogamy for the female but not for the male. Sex differences are usually not as marked in monogamous and less dimorphic species; yet, the bushbaby shows a gender difference in territorial calling with males calling more than females (Doyle, 1974).

For the aye-ayes (*Daubentonia madagascariensis*), which are apparently solitary animals, there has been no mention of sex differences in vocalizations, yet the call of one of these animals is always followed by the call of another (up to 100 yards apart) and these calls apparently *do* function in territorial spacing (Petter and Petter, 1967).

When alarmed by people, *female* lemurs (*Lemur fulvus*) display more alarm vocalizations than do males (Harrington, 1977). In another lemur (*Lemur catta*), males emit more calls associated with low intensity and submissive situations, while females emit higher rates for increased intensity and aggressive situations (King and Fitch, 1977).

Thus, the Old World monkey gender differences apparently cannot be readily generalized (without qualification) to the prosimians, primarily because of the lemur data and because little has been published on sex differences in prosimians.

8. New World Monkeys

The marmosets and tamarins, as we know, are often treated like New World likenesses of Old World prosimians. Marmosets are monogamous and, in many of them (e.g., *Cebuella pygmaea*), females are dominant (Epple, 1975). There are few if any sex differences in vocalizations.

In the less primitive New World monkeys, gender differences more often resemble those seen in Old World monkeys. In the howler monkey (*Alouatta spp.*) which is *not* monogamous, as well as in the titi monkey (*Callicebus moloch*), which *is* monogamous, warning and territorial vocalizations are usually emitted by males. As Peters and Ploog (1973) point out, an assumed lack of sexual dimorphism in behavior may often be a function of the subtleness of the signals in question. In field observations, sex recognition is often aided by vocalizations in the context of different patterns of body movements and postures (e.g., for *Cebus capuchinus*).

In another New World primate, where physical dimorphism is not obvious (e.g., in *Pithecia pithecia*, a Saki), males emit a loud bark, *perhaps* as an expression of territoriality, whereas females do not (Stott, 1976).

In summary, the more primitive and monogamous marmosets and tamarins do not display gender-related warning and territorial vocalizations, whereas many of the more advanced New World monkeys, including some monogamous ones, do.

9. Lesser Apes

All lesser apes tend to live in monogamous family units. Using electrical brain stimulation to elicit vocalizations in *Hylobates lar* (the white-handed gibbon),

Apfelbach (1972) found no sex differences in elicited vocalizations. In the wild, *both* male and female gibbons of this species give loud "morning calls" for spacing (Marler, 1968), However, in Kloss' gibbon (*Hylobates klossii*), adults defend their territories only against members of the same sex, with males "singing" before sunrise and females "singing" after sunrise. Moreover, these morning territorial songs of Kloss' gibbon *are* sexually dimorphic (Tenaza, 1974). Kloss' female songs are of lower frequency, are louder, and carry further than those emitted by males. Despite the fact that their voices carry further, *females vocalize closer to one another than do males.* It is interesting that in this species (*Hylobates klossii*) males do more of the territorial singing than do females (Tenaza, 1976).

Among siamangs (*Symphalangus syndactylus*), calls are emitted less frequently (although louder) than in gibbons and no sex differences have been reported (cf. Chivers, 1975; Koyama, 1971).

In summary, among the lesser apes, those species that show the greatest sex differences in vocalization (e.g., Kloss' gibbon) tend toward the same gender differences seen in Old World monkeys, at least in terms of territorial calls. On the whole, however, it is probable that the monogamous lesser apes evince fewer sex differences than do the Old World monkeys.

10. Great Apes

Pitcairn (1974) has summarized some sex differences in vocalizations for the gorilla (*Gorilla gorilla*). The two- or three-toned roar is made only by the adult male gorilla, and "hooting" during the threatening chest-beating display is emitted only by large silver-backed males (see also Marler, 1968). The female gorilla, on the other hand, emits "the pant ho-ho" when chest beating. While there is some disagreement in the literature, the majority opinion is that, unlike the Old World monkeys, it is the *male* gorilla which has the largest repertoire of calls and it is the male which calls most frequently (Pitcairn, 1974; Fossey, 1972). Alarm barks are typical of adult silver-back males when observers are present. According to Fossey (1972), however, even when observers are not a factor, and the alarm barks are not given, the males vocalize more than do the females.

Schaller (1963) reported a different finding. He said that females emit the largest variety of sounds, with short harsh grunts being emitted by males and screams by females (Emlen and Schaller, 1963). It is possible that there is a real difference between the two study areas or the two groups of gorillas.

According to Fossey (1972), there are marked individual differences in the voices of gorillas. Relative to chimpanzees, however, gorillas are very quiet, and they are seldom in chorus. Most of their displays and vocalizations are emitted by individuals. Chimpanzees, on the other hand, are very vocal and noisy, are often in chorus, and often evince communal displays (Reynolds, 1965).

Among chimpanzees, there are consistent sex differences in long-range vocalizations. The "pant-hooting" of adult females differs from that of the adult males. Females have no "climax" section to their vocalizations and the average duration of "pant-hooting" is longer in females. Also, the pitch of the first harmonic is *deeper* in the loud calls of the females (Marler and Hobbett, 1975). The function of these calls is not understood.

Only the adult *male* orangutan has a "long-call" signalling a disagreeable mood (Pitcairn, 1974). There is an obvious sexual dimorphism in vocal display (particularly in the long-call) (Rijksen, 1975). The male does not stay with the female; he wanders and gives loud bellowing vocalizations apparently in competition for females (Horr, 1972).

In summary, sex differences in vocalizations have not been closely and thoroughly examined in the great apes. While there *are* definitely sex differences in great ape vocalizations, they do not neatly agree with the differences reported for Old World monkeys. Individual differences, and possibly even differences from area to area and from group to group, confound our attempts to find any consistency. It is apparent, however, that the adult males of the great ape species tend to emit warning calls and territorial calls. It is also strongly possible that females may emit more fear calls. There is still disagreement, however, as to which sex vocalizes most frequently overall.

C. "FEMALE" VOCALIZATIONS

It is interesting that most of the research done on clear calls and other softer vocalizations often associated with the female gender has been done in the laboratory; whereas, most of the research on harsh noises, territorial calls, and agonistic sounds has been done in the field.

In a laboratory study of both socially deprived and socially normal rhesus monkeys, Cross and Harlow (1965) noted that female rhesus, regardless of rearing or age, vocalize more (overall) than do male rhesus. (A similar finding is reported in Møller, Harlow, and Mitchell, 1968). If stimulated by a human observer, the sex differences in overall numbers of vocalizations becomes even more apparent (Cross and Harlow, 1965). However, there is a steady decrease in numbers of vocalizations in both sexes as they mature (from infancy through puberty). Maple, Brandt, and Mitchell (1975) report a similar age difference in frequency of vocalization following separation. Erwin and Mitchell (1973) note a decrease in rhesus vocalizations with *age*, but in this case it is in a specific vocalization, the coo vocalization (nonagonistic clear call). There is a consistent sex difference, with females cooing more frequently than males at all ages *except* at those ages when the females have already reached puberty, but the males have not. The onset of puberty in both sexes decreases coo vocalizations. The clear coo calls are usually monitored following social separation from another familiar animal; hence, the condition resembles Cross and Harlow's (1965) stimulated more than their passive condition.

While exceptions to the finding of more coo vocalizations in female infants have been reported (Hinde and Spencer-Booth, 1971), those reporting the exceptions *have* found the opposite result in another study (Rowell and Hinde, 1963). Moreover, sex differences in rates of rhesus coo vocalizations, with females cooing more, are found in separated rhesus infants as early as three months of age (Scollay, 1970). The only circumstances in which male rhesus appear to vocalize (overall) more than do females involve extreme aggressive encounters (cf. Altmann, 1962; Erwin, 1975).

Even among adults, females typically emit more clear calls and girning sounds (a sound of greeting) than do males (cf. Erwin, 1975; Maple, Erwin, and Mitchell, 1974). In studies involving the separation of adult rhesus heterosexual dyads (Maple, Risse, and Mitchell, 1973), adult females coo more frequently than do males. There is also evidence that adult rhesus females and other adult macaque females emit more coo vocalizations and other calls if separated when in estrus than if separated when not in estrus (Maple, Erwin, and Mitchell, 1974; Grimm, 1967; Itani, 1963). The coo in adulthood may have a different meaning (it seems to have a lower pitch) than it does in infancy. Even in preadolescence, females separated in cross-sexed dyads emit more "coos" than do those separated in like-sexed dyads (Maple, Brandt, and Mitchell, 1975). Perhaps a "call for contact" matures into a "call for sexual contact." Since "coo" vocalizations decrease at puberty in both sexes, yet a sex difference still exists after puberty, the relationship between hormonal state and calls for contact in the females is not a simple one.

The coo vocalization is a clear call which generally occurs in the rhesus monkey when there is a stimulus change. In teleological terms, it is often referred to as a "call for contact" (Møller, Harlow, and Mitchell, 1968). Separation of an animal from a familiar conspecific or from a familiar environment produces an increase in the frequency and intensity as well as a change in the quality of these calls. The calls occur less frequently in adults than in infants (cf. Møller, Harlow, and Mitchell, 1968). Both infant and adult isolation-reared rhesus emit the coo call less frequently than do socially reared conspecifics (cf. Maple, Brandt, and Mitchell, 1975). Isolates also display structural abnormalities in coo calls on sound spectrographs (Newman and Symmes, 1974). Perhaps, even more importantly, the isolates coo in inappropriate contexts (Brandt, Baysinger, and Mitchell, 1972). It is therefore quite apparent that the vocalization sex difference reported above must be qualified not only by the *age* of the animals but also by the *rearing experience* of the animals (Mitchell and Redican, 1972).

In studies of subadult and adult rhesus monkeys that have been reared in social isolation compared to normally reared animals, Mitchell (1968) found that while isolate-reared adult rhesus of both sexes emit fewer "coo" vocalizations than do their like-sex normal controls, female social isolates *still* "coo" more than do the isolate males. The normal adult females also coo more than do the normal adult males. However, despite the fact that the sex difference holds up, early deprivation decreases coo vocalizations so much in the isolates that the isolate females actually coo *less* than do the *normal* males. Thus, while there does appear to be a sex difference in this vocalization, early experiences can change the magnitude of the difference, and *differential* early experience can actually reverse the direction of the difference.

Not only is there a sex difference among rhesus subadults in emitting the coo vocalization, there is also a difference which depends upon the sex of the animal with whom a subadult rhesus is paired. If a subadult rhesus, isolation-reared *or* socially-reared, is paired with a normal *female* rhesus (of any age, but especially one not yet pubescent), that subadult, whether male or female, will coo more than if paired with a *male* rhesus (Mitchell et al., 1966). Thus, there appears to be social facilitation involved in emitting coo vocalizations. While males usually coo less than females, the number of coos emitted by a male may be increased by pairing him with a female. We should therefore add to: (1) *age* and (2) *social deprivation*, (3) the *sex* of the social partner as a factor which affects rhesus sex difference in clear call vocalization.

If the sex of the social partner makes a difference, and the age of the animal vocalizing makes a difference, what about the age of the social partner? Do rhesus monkeys coo more frequently in the presence of juveniles than they do in the presence of age-mates? The answer is yes, six to ten times more (Mitchell et al., 1966)! Thus, we must add a fourth factor (*age of social partner*) to the list of factors affecting the coo vocalization sex difference.

In the case of the emission of coo vocalizations, there is a steady decline with age in both sexes. If social facilitation of the coo is a powerful factor, that is, if being with an animal who coos more frequently induces our subject animal (a subadult rhesus) to coo more frequently, then we would *expect* coos in the presence of infants or juveniles to be more frequent than in the presence of age-mates, which would in turn be more frequent than in the presence of adults. Does this occur? Only if the animals are separated from each other physically but are allowed to hear each other.

If the animals are allowed to contact one another, we get a different pattern. While coo vocalizations occur more frequently with physical access to infants than with physical access to age-mates, the coos are emitted *less* frequently with age-mates than with adults. With physical access to *familiar* age-mates, rhesus subadults increase their affiliative behaviors (e.g., grooming) and have no need to increase their "calls for contact" (coos). With age-mates which are *strangers,* rhesus subadult monkeys increase their agonistic behaviors (fear and aggression). Since the coo is a nonagonistic call for contact, we should therefore *expect* (in our subadult rhesus) the *fewest* number of coo vocalizations with age-mates, the *most* with infants (because of social facilitation), and an intermediate number with adults. This is, in fact, the finding reported by Mitchell et al. (1966). Therefore, the age of the social partner works as follows when physical contact is permitted: the younger the social partner, the more frequent the number of coos in the presence of that partner *except* when agonistic and affiliative behaviors preclude the occurrence of "calls for social contact." If the dyads in the Mitchell et al. (1966) study had been *separated* from one another, but could still hear each other, there would undoubtedly have been a near perfect inverse relationship between the number of vocalizations and the age of the social partner from which the subject had been separated.

Based upon the above somewhat tortuous trip through factors affecting clear calls, we must conclude that sex differences in rhesus vocalizations are far from simple. It is certainly true that in carefully balanced (matched) groups of subjects, rhesus females emit more clear calls than do males; however, a subadult rhesus female (at puberty) paired with a stranger of similar age and sex will almost surely coo less frequently than will a subadult male (at puberty) paired with an infant. Thus, the clear call sex difference is, to a large degree, *situation specific.* If we add to this the fact that early social isolation can also drastically reduce the number of coo vocalizations emitted, we have good evidence for as strong a role for experience and situation as for biological gender in determining vocalizations. (However, we should remember that the differences do exist in the wild.)

Not only are the rhesus vocalizations susceptible to very gross differences in early social experience, they are also susceptible to fairly subtle differences in experience as well. Mitchell and Stevens (1969) have reported that inexperienced mothers (primiparous mothers) bark more frequently than do multiparous mothers, while Scollay (1970) has reported that primiparous rhesus mothers coo more frequently than do multiparous mothers. On the one hand, these findings are not too surprising because primiparous mothers are usually younger than multiparous mothers and *should* vocalize more. However, Mitchell et al. (1966) found more screeching and cooing *in the infants and juveniles* of primiparous mothers, while Stevens and Mitchell (1972) found more girning in primiparously reared infants. In addition, a history of brief repeated separations early in life also increased the number of coo vocalizations emitted later in life (Møller, Harlow, and Mitchell, 1968). Thus, relatively *subtle* differences in early maternal treatment affect rhesus vocalizations (see also Mitchell, 1976).

There is even some evidence that the sights and sounds (and other cues) surrounding brief or prolonged separations can affect rhesus vocalizations. In one study, two infant males became upset *only* when separated from the most disturbed female of their group. In the same study, the mere threat of a separation increased the coo vocalizations of infants (Willott and McDaniel, 1974). Scollay (1970) also reported that animals who merely heard and viewed social separations increased their coo vocalizations. Maple (1974) has shown that rhesus macaques paired with an alien species (baboons) seem to learn the meaning of that species' contact calls. The baboons also responded to the macaque calls. This was true in both interspecific adult dyads and in interspecific juvenile dyads. In separation, the baboons vocalized much more than did the rhesus, indicating that they might be even better than macaques for studies of attachment vocalizations. Erwin (1975) has also emphasized the fact that rhesus vocalizations are under some degree of volitional control.

Zoloth et al. (1977), however, have shown that it is easier for one species of macaque to learn a discrimination based upon its own "coo" vocalizations than it is for an alien species to learn the *same* discrimination. Evidently the coo vocalization is of more relevance to the macaque in question than it is to other macaques or to, for example, vervets.

In short, rhesus monkey sex difference in clear call vocalizations are *real*, but the differences attributable to biological gender can be lessened or increased according to: (1) the specific kind of vocalization in question, (2) the degree of agonism or affiliation, (3) the degree of familiarity of animals, (4) age, (5) rearing experiences, (6) hormonal state, (7) the social situation, (8) social facilitation, (9) anticipation of, a witnessing of, or actual social separation, and (10) discrimination learning of vocalizations.

A gender difference favoring females has also been noted for Japanese macaque vocalizations involving "positive socialness." When grooming, adult male Japanese macaques vocalize far less frequently than do adult females. Females groom more than do males *and* they vocalize more often when they do (Mori, 1975). Separation from social contact also increases adult female Japanese macaque vocalizations more than it does those of adult males and this difference seems to be accentuated if the females are in estrus (Grimm, 1967; Itani, 1963). In addition, mother pigtail (*M. nemestrina*) as well as rhesus (*M. mulatta*) macaques separated from their infants are always more vocal (and active) than are either nonmother females or males (Simons, Bobbitt, and Jensen, 1968; see also Erwin and Flett, 1974). In rhesus preadolescent-infant separations, harsh screeches occur more often in infants separated from and reunited with male preadolescents whereas "girning" (a soft, clear call having positive social significance) occurs more frequently in female dyads (Maple, Brandt, and Mitchell, 1975). A female patas monkey rejoining the troop will "moan" (Struhsaker and Gartlan, 1970). Adult female baboons and to a lesser extent female mangabeys, guenons, langurs, and even female chimpanzees emit "grunts" in female dyads. These grunts are used as close contact calls (Andrew, 1976).

Although the above reports do suggest that adult *male* vocalizations are less often utilized in intragroup positive social behavior than are those of the adult females, the adult males of many species sometimes *respond* to the vocalizations of the conspecifics in their own troop. For example, *Macaca sylvana* adult males go to the aid of an infant if the infant emits a distress call (Lahiri and Southwick, 1966). In addition, adult male stumptail macaques (*Macaca arctoides*) emit soft "grunt-purrs" toward tiny white-furred infants (Trollope and Blurton-Jones, 1975). The adult male spectacled langur (*Presbytis obscurus*) vocalizes with its mouth open while emitting a sound like human laughter and while performing a clasp greeting with an infant (Horwich, 1974). Adult male baboons "grunt" to infants (Andrew, 1976).

On the whole, however, reports on gender-related vocalizations in nonhuman primates suggest that females, in general, emit more of those vocalizations which serve an intragroup function of holding the group together than do males. For example, only female and juvenile mandrills emit contact calls (Jouventin, 1975); and, in gorillas, females tend to answer infant calls more often than do males (Fossey, 1972). In chimpanzees, the female "pant-hoot," which is different from the male pant-hoot in pitch and in the duration of climax, may also be used in minimizing rather than maximizing intragroup tension (Marler and Hobbett, 1975). It is interesting that female chimpanzees are less easily fooled than are males by tape-recorded chimpanzee loud calls played back to them in the wild (Kawanaka and Nishida, 1974).

In the field, there has been too much preoccupation with warning barks, territorial calls, and agonistic vocalizations and too little attention paid to the softer calls and to their function (e.g., the rhesus girn). As in other areas of the biological and life sciences, female behavior has often been ignored while male aggression and dominance behaviors have been the center of interest.

D. CONCLUSION

Marler (1976) has noted that:

> A review of descriptive studies of the vocal ethology of wild and captive non-human primates suggests that although species-specific features abound there are also many points of correspondence between species. Certain sound patterns recur accross species as is illustrated by comparison of the vocalizations of the chimpanzee and the gorilla. (Abstract)

In view of reports of a greater overall frequency of vocalizations in *female* human infants (Bayley, 1968; Lewis, 1969), reports of better development of syntax in four-year-old girls than boys (Koenigsknecht and Friedman, 1976), and of a greater frequency of quasi-agonistic yelling in nursery school boys than in nursery school girls (McGrew, 1972), it behooves us (as primatologists) to examine developmentally the role of gender in the vocalizations of our primate cousins. It is by no means totally impossible that such differences may have some bearing on the frequently reported greater *verbal* abilities of human females (Maccoby and Jacklin, 1974).

In birds, males vocalize to attract females. This is not as true of mammals (Nottebohm, 1972). In primates, intragroup social vocalization is a female trait. During predation, anything as noisy as vocalization would frighten the prey (Tanner and Zihlman, 1976). Among close, same-sexed friends, women are more likely to just sit and talk than are men (Caldwell and Peplau, 1977).

In addition, it is important that we examine all of the factors that affect or produce gender differences in humans, particularly the roles of social situation and early experience. While possible sex differences in socialness and/or vocalization may exist, the fact that familiarity, age, rearing, hormonal state, social facilitation, separation, and learning can all affect the direction of a sex difference in one simple vocalization, the rhesus "coo," suggests that sex differences in human vocalization must indeed be complex.

REFERENCES

Altmann, S. A. Sociobiology of rhesus monkeys. IV: Testing Mason's hypothesis of sex differences in affective behaviour. *Behaviour,* 1968, **32**, 49-69.

Andrew, R. J. Use of formants in the grunts of baboons and other nonhuman primates. *Annals of the New York Academy of Sciences,* 1976, **280**, 673-693.

Apfelbach, R. Electrically elicited vocalizations in the gibbon *Hylobates lar* (*Hylobatidae*), and their behavioral significance. *Zeitschrift für Tierpsychologie,* 1972, **30**, 420-430.

Bayley, N. Behavioral correlates of mental growth: Birth to thirty-six years. *American Psychologist,* 1968, **23** (1), 1-17.

Baysinger, C. M., Brandt, E. M. and Mitchell, G. Development of infant social isolate monkeys (*Macaca mulatta*) in their isolation environments. *Primates,* 1972, **13** (3), 257-270.

Bernstein, I. S. A field study of the male role in *Presbytis cristatus* (*lutong*), *Primates,* 1966, **7** (3), 402-403.

Bishop, N. Vocal behavior of adult male langurs in a high altitude environment. *American Journal of Physical Anthropology,* 1975, **42**, 291 (Abstract).

Bourliére, F., Bertrand, M. and Hunkeler, C. L'ecologie de la mone de Lowe (*Cercopithecus campbelli Lowei*) en cote divoire. *Extrait de la Terre et la Vie,* 1969, **2**, 135-163.

Bramblett, C. A. Social organization as an expression of role behavior among Old World monkeys. *Primates,* 1973, **14** (1), 101-112.

Brandt, E. M., Baysinger, C. and Mitchell, G. Separation from rearing environment in mother-reared and isolation-reared rhesus monkeys (*Macaca mulatta*). *International Journal of Psychobiology,* 1972, **2** (3), 193-204.

Buskirk, W. H., Buskirk, R. E. and Hamilton, W. J., III. Troop mobilizing behavior of adult male chacma baboons. *Folia Primatologica,* 1974, **22**, 9-18.

Caldwell, M. A. and Peplau, L. A. Sex differences in friendship. Paper presented at the *Western Psychological Association* Meeting, Seattle, Washington, April, 1977.

Chivers, D. J., Raemaekers, J. J. and Aldrich-Blake, F. P. G. Long-term observations of siamang behaviour. *Folia Primatologica,* 1975, **23**, 1-49.

Cross, H. A. and Harlow, H. F. Prolonged and progressive effects of partial isolation on the behavior of macaque monkeys. *Journal of Experimental Research in Personality,* 1965, **1**, 39-49.

Dolhinow, P. J. The North Indian langur. In Dolhinow, P. (Ed.) *Primate Patterns.* New York: Holt, Rinehart and Winston, 1972.

Doyle, G. A. The behaviour of the lesser bushbaby. In Martin, R. D., Doyle, G. A. and Walker, A. C. (Eds.) *Prosimian Biology.* London: Duckworth, 1974.

Emlen, J. T. and Schaller, G. B. In the home of the mountain gorilla. In Southwick, C. H. (Ed.) *Primate Social Behavior.* Princeton, N.J.: Van Nostrand, 1963.

Epple, G. The behavior of marmoset monkeys (Callithricidae). In Rosenblum, L. A. (Ed.) *Primate Behavior* (Vol. 4). New York: Academic Press, 1975.

Erwin, J. Rhesus monkey vocal sounds. In Bourne, G. (Ed.) *The Rhesus Monkey* (Vol. 1). New York: Academic Press, 1975.

Erwin, J. and Flett, M. Responses of rhesus monkeys to the separation vocalizations of a conspecific infant. *Perceptual and Motor Skills,* 1974, **39**, 179-185.

Erwin, J. and Mitchell, G. Analysis of rhesus monkey vocalizations: Maturation-related changes in clear call frequency. *American Journal of Physical Anthropology,* 1973, **38** (2), 463-468.

Fossey, D. Vocalizations of the mountain gorilla (*Gorilla gorilla beringei*). *Animal Behaviour,* 1972, **20** (1), 36-53.

Grimm, R. J. Catalogue of sound of the pigtail macaque (*Macaca nemestrina*). *Journal of Zoology* (London), 1967, **152**, 361-373.

Groves, C. P. Notes on the ecology and behaviour of the Angola Colobus (*Colobus angolensis,* P. L. Sclater, 1860) in N. E. Tanzania. *Folia Primatologica,* 1973, **20**, 12-26.

Harrington, J. E. Discrimination between males and females by scent in *Lemur fulvus*. *Animal Behaviour*, 1977, **25**, 147-151.

Hinde, R. A. and Spencer-Booth, Y. Effects of brief separation from mother on rhesus monkeys. *Science*, 1971, **173**, 111-118.

Horr, D. A. The Borneo orang-utan. *Borneo Research Bulletin*, 1972, **4**, 46-50.

Horwich, R. H. Development of behaviors in a male spectacled langur (*Presbytis obscurus*). *Primates*, 1974, **15** (2-3), 151-178.

Hunkeler, C., Bourliére, F. and Bertrand, M. Le comportement social de la Mone de Lowe (*Cercopithecus campbelli lowei*). *Folia Primatologica*, 1972, **17**, 218-236.

Itani, J. Vocal communication of the wild Japanese monkey. *Primates*, 1963, **4** (2), 11-66.

Jouventin, P. Observations sur la socio-ecologie du Mandril. *Extrait de la Terre et la Vie, Revue d'Ecologie Appliquee*, 1975, **29**, 493-532.

Kawabe, M. and Mano, T. Ecology and behavior of the wild proboscis monkey, *Nasalis larvatus* (Wurmb) in Sabah, Malaysia. *Primates*, 1972, **13**, 213-228.

Kawanaka, K. and Nishida, T. Recent advances in the study of inter-unit-group relationships and social structure of wild chimpanzees of the Mahali mountains. *Symposium of the Fifth Congress of the International Primatological Society*, 1974, **5**, 173-186.

King, G. and Fitch, M. Vocal patterns of captive ring-tailed lemurs (*Lemur catta*). Paper presented at the *Western Psychological Association* Meeting, Seattle, Washington, April, 1977.

Koenigsknecht, R. A. and Friedman, P. Syntax development in boys and girls. *Child Development*, 1976, **47**, 1109-1115.

Koyama, N. Observations on mating behavior of wild Siamang gibbons at Fraser's Hill, Malaysia. *Primates*, 1971, **12** (2), 183-189.

Lahiri, R. K. and Southwick, C. H. Parental care in *Macaca sylvana*. *Folia Primatologica*, 1966, **4**, 257-264.

Lewis, M. Infants' responses to facial stimuli during the first year of life. *Developmental Psychology*, 1969, **1** (2), 75-86.

Lindburg, D. G. The rhesus monkey in North India: An ecological and behavioral study. In Rosenblum, L. A. (Ed.) *Primate Behavior: Developments in Field and Laboratory Research* (Vol. II). New York: Academic Press, 1971.

Maccoby, E. E. and Jacklin, C. N. *The Psychology of Sex Differences*. Stanford, California: Stanford University Press, 1974.

Maple, T. L. *Basic studies of interspecies attachment behavior*. Unpublished doctoral dissertation, University of California, Davis, 1974.

Maple, T. L., Brandt, E. M. and Mitchell, G. Separation of preadolescent from infant rhesus monkeys. *Primates*, 1975, **16** (2), 141-153.

Maple, T., Erwin, J. and Mitchell, G. Separation of adult heterosexual pairs of rhesus monkeys: The effect of female cycle phase. *Journal of Behavioral Science*, 1974, **2** (2), 81-86.

Maple, T., Risse, G. and Mitchell, G. Separation of adult males from adult female rhesus monkeys (*Macaca mulatta*) after a short-term attachment. *Journal of Behavioral Science*, 1973, **1** (5), 327-336.

Marler, P. Aggregation and dispersal: Two functions of primate communication. In Jay, P. C. (Ed.) *Primates: Studies in Adaptation and Variability*. New York: Holt, Rinehart and Winston, 1968.

Marler, P. A comparison of vocalizations of red-tailed monkeys and blue monkeys, *Cercopithecus ascanis* and *Cercopithecus mitis,* in Uganda. *Zeitschrift für Tierpsychologie*, 1973, **33**, 223-247.

Marler, P. *The vocal ethology of primates: Implications for psychophysics and psychophysiology.* Paper presented at the *Sixth Congress of the International Primatological Society,* Cambridge, England, August, 1976.

Marler, P. and Hobbett, L. Individuality in a long-range vocalization of wild chimpanzees. *Zeitschrift für Tierpsychologie,* 1975, **38**, 97-109.

Mason, W. A. and Berkson, G. Effects of maternal mobility on the development of rocking and other behaviors in rhesus monkeys: A study with artificial mothers. *Developmental Psychobiology,* 1975, **8** (3), 197-211.

Masure, A. M. and Bourliére, F. Surpeuplement, fecondite, mortalite et agressivite dans une population captive de *Papio papio. Extrait de la Terre et la Vie,* 1971, **25**, 491-505.

McGrew, W. C. Aspects of social development in nursery school children with emphasis on introduction to the group. In Jones, N. G. B. (Ed.) *Ethological Studies of Child Behaviour.* Cambridge, England: Cambridge University Press, 1972.

Mitchell, G. Persistent behavior pathology in rhesus monkeys following early social isolation. *Folia Primatologica,* 1968, **8**, 132-147.

Mitchell, G. Attachment potential in rhesus macaque dyads (*Macaca mulatta*): A sabbatical report. *JSAS Catalogue of Selected Documents in Psychology,* 1976, **6**, 7 MS 1177.

Mitchell, G. and Prassa, S. P. The effects of gender experience and other factors on the vocalizations of nonhuman primates. Paper presented at the *Harlow Symposium.* Tucson, Arizona, December, 1976.

Mitchell, G. and Redican, W. K. Communication in normal and abnormal rhesus monkeys. *Proceedings of the XXth International Congress of Psychology.* Tokyo, Japan: Science Council of Japan, 1972, pp. 171-172.

Mitchell, G., Raymond, E. J., Ruppenthal, G. C. and Harlow, H. F. Long-term effects of total social isolation upon behavior of rhesus monkeys. *Psychological Reports,* 1966a, **18**, 567-580.

Mitchell, G., Ruppenthal, G. C., Raymond, E. J. and Harlow, H. F. Long-term effects of multiparous and primiparous monkey mother rearing. *Child Development*, 1966b, **37**, 781-791.

Mitchell, G. and Stevens, C. W. Primiparous and multiparous monkey mothers in a mildly stressful social situation: First three months. *Developmental Psychobiology*, 1969, **1**, 280-286.

Møller, G. W., Harlow, H. F., and Mitchell, G. Factors affecting agonistic communication in rhesus monkeys (*Macaca mulatta*). *Behaviour*, 1968, **31**, 339-357.

Mori, A. Signals found in the grooming interactions of wild Japanese monkeys of the Koshima troop. *Primates*, 1975, **16** (2), 107-140.

Mukherjee, R. P. and Saha, S. S. The golden langurs (*Presbytis geei*) of Assam. *Primates*, 1974, **15**, 327-340.

Newman, J. D. and Symmes, D. Vocal pathology in socially deprived monkeys. *Developmental Psychobiology*, 1974, **7** (4), 351-358.

Nottebohm, F. The origins of vocal learning. *The American Naturalist*, 1972, **106**, 116-140.

Peters, M. and Ploog, D. Communication among primates. *Annual Review of Physiology*, 1973, **35**, 221-242.

Petter, J. J. and Petter, A. The aye-aye of Madagascar. In Altmann, S. A. (Ed.) *Social Communication Among Primates*. Chicago: University of Chicago Press, 1967.

Pitcairn, T. K. Aggression in natural groups of pongids. In Holloway, R. L. (Ed.) *Primate Aggression, Territoriality and Xenophobia*. New York: Academic Press, 1974.

Poirier, F. E. Nilgiri langur: Territorial behavior. *Proceedings of the Second International Congress of Primatology*, 1968, **1**, 31-35.

Poirier, F. E. Dominance structure of the Nilgiri langur (*Presbytis johnii*) of South India. *Folia Primatologica*, 1970, **12**, 161-186. (a)

Poirier, F. E. The communication matrix of the Nilgiri langur (*Presbytis johnii*) of South India. *Folia Primatologica*, 1970, **13**, 92-136. (b)

Quris, R. Ecologie et organisation sociale de *Cercocebus galeritus* agilis dans le nord-est du gabon. *Terre et la Vie, Revue d'Ecologie Appliquee*, 1975, **29**, 337-398.

Rahaman, H. and Parthasarathy, M. D. The home range, roosting places, and the day ranges of the bonnet macaque (*Macaca radiata*). *Journal of Zoology* (London), 1969, **157**, 267-276.

Reynolds, V. Some behavioral comparisons between the chimpanzee and the mountain gorilla in the wild. *American Anthropologist*, 1965, **67**, 693-706.

Rijksen, H. D. Social structure in a wild orang-utan population in Sumatra. In *Contemporary Primatology*. Basel: Karger, 1975, pp. 373-379.

Rowell, T. E. Agonistic noises of the rhesus monkey. *Symposium of the Zoological Society of London,* 1962, **8**, 91-96.

Rowell, T. E. *The Social Behaviour of Monkeys.* Baltimore: Penguin, 1972.

Rowell, T. E. and Hinde, R. A. Vocal communication by the rhesus monkey (*Macaca mulatta*). *Proceedings of the Zoological Society of London,* 1962, **138**, 279-294.

Rowell, T. E. and Hinde, R. A. Responses of rhesus monkeys to mildly stressful situations. *Animal Behaviour,* 1963, **11**, 235-243.

Saayman, G. S. Baboons' responses to primates. *African Wild Life,* 1971, **25**, 46-49.

Schaller, G. B. *The Mountain Gorilla: Ecology and Behavior.* Chicago: University Press, 1963.

Scollay, P. A. Mother-infant separation in rhesus monkeys (*Macaca mulatta*). Doctoral dissertation, University of California, Davis, 1970.

Simons, R. C., Bobbitt, R. A. and Jensen, G. D. Mother monkeys' (*Macaca nemestrina*) responses to infant vocalizations. *Perceptual and Motor Skills,* 1968, **27**, 3-10.

Sorenson, M. W. A review of aggressive behavior in the tree shrews. In Holloway, R. H. (Ed.) *Primate Aggression, Territoriality, and Xenophobia: A Comparative Perspective.* New York: Academic Press, 1974.

Stevens, C. W. and Mitchell, G. Birth order effects, sex differences, and sex preferences in the peer-directed behavior of rhesus infants. *International Journal of Psychobiology,* 1972, **2**, 117-128.

Stott, K. Sakis: Imps of the rain forest. *Zoonooz,* 1976, **49** (4), 4-8.

Struhsaker, T. T. *Behavior of Vervet Monkeys (Cercopithecus aethiops).* Berkeley, California: University of California Press, 1967.

Struhsaker, T. T. and Gartlan, J. S. Observations on the behaviour and ecology of the patas monkey (*Erythrocebus patas*) in the Waza Reserve, Cameroon. *Journal of Zoology of London,* 1970, **161**, 49-63.

Sugiyama, Y. The ecology of the lion-tailed macaque (*Macaca silenus*) (Linnaeus): A pilot study. *Journal of the Bombay Natural History Society,* 1968, **65** (2), 283-292.

Tanner, N. and Zihlman, A. Discussion paper: The evolution of human communication: What can primates tell us? *Annals of the New York Academy of Science,* 1976, **280**, 467-480.

Tenaza, R. R. I. Monogamy, territory and song among Kloss' gibbons (*Hylobates klossii*) in Siberut Island, Indonesia. II. Kloss' gibbon sleeping trees relative to human predation: Implications for the sociobiology of forest dwelling primates. Doctoral dissertation, University of California, Davis, 1974.

Tenaza, R. R. Songs, choruses and countersinging of Kloss' gibbons (*Hylobates klossii*) in Siberut Island, Indonesia. *Zeitschrift für Tierpsychologie,* 1976, **40**, 37-52.

Trollope, J. and Blurton-Jones, N. G. Aspects of reproduction and reproductive behaviour in *Macaca arctoides. Primates,* 1975, **16** (2), 191–205.

Willott, J. F. and McDaniel, J. Changes in the behavior of laboratory-reared rhesus monkeys following the threat of separation. *Primates,* 1974, **15** (4), 321–326.

Wilson, W. L. and Wilson, C. C. Species-specific vocalizations and the determination of phylogenetic affinities of the *Presbytis aygula-melalophos* group in Sumatra. *Fifth International Congress of Primatology* (Nagoya). Basel: Karger, 1975.

Young, G. H. Paternal influences on the development of behavioral sex differences in infant baboons. Paper presented at the *American Association of Physical Anthropologists* Meeting, Seattle, Washington, April, 1977.

Zoloth, S. R., Peterson, M. R., Beecher, M. D., Moody, D. B. and Stebbins, W. C. Perception of conspecific vocalizations by Japanese macaques (*Macaca fuscata*). Paper presented at the *Animal Behavior Society* Meeting, University Park, Pennsylvania, June, 1977.

23.

Sex Differences in Visual Communications

Primates rely heavily upon the visual mode of communication for several reasons. Primate vision provides excellent acuity and depth perception. Visual communication can be transmitted instantaneously, and several signals can be sent simultaneously. Visual communication is used in a wide variety of contexts. This chapter will attempt to describe some primate visual displays and the contexts in which they occur, with special attention being paid to differential use by males and females.

A. GENERAL COMMENTS ON VISUAL COMMUNICATION

Communication has been described (Rowell, 1972) as occurring when any signal emitted by one animal is used by another to predict the behavior either of the first animal, or of something else in the environment. If this signal involves vision, it is visual communication. The basic types of visual communication are postural, gestural, and physical.

Primate postural communication refers to a subject's overall posture while at rest or in motion. Important aspects of posture are the shape and orientation of the vertebral column, and the orientation of the limbs. Sometimes tail position is also important. This type of visual communication conveys information related to the individual's situational confidence and level of arousal. For example, a confident monkey generally adopts a relaxed, still, posture and/or a smoothly flowing, moving posture. A less confident monkey generally appears quite rigid and the movements are more constrained or jerky. An alert, excited monkey usually adopts a position with raised head and high skeletal muscle tonus.

[1] I thank Steven Coburn for help in the literature search on this topic.

Gestural communication involves movements by a monkey or ape. These movements may be of the animal's entire body, or they may be merely head and facial expressions. Entire body displays are often observed in association with agonistic situations. Some species are known to use a body bounce display as a threat or as a decoy in the presence of a predator. Several species use body displays in which they shake or swing from branches to maintain distance between neighboring groups.

Another full body gesture is the present. In this posture, a monkey or ape displays its posterior towards another primate as a sign of submission or, at other times, to solicit sexual behavior.

Most primate gestures, however, occur as facial expressions. Common facial expressions include the threat, grimace, lip-smack, and yawn displays. These general expressions are observed in a great variety of situations. Facial displays are used in agonistic contexts, sexual contexts, or in association with other ongoing behavior. There are no uniform expressions through the primate order, but many species have similar expressive patterns.

The *lip-smack* expression is used in a wider range of circumstances than any other facial expression. The lower jaw is repeatedly raised and lowered without making tooth contact. At the same time, the tongue repeatedly moves up between the teeth and then back into the mouth. This display is often observed in association with grooming, copulation, submission, and appeasement. Lip-smacking is a general indicator of nonaggression.

The *grimace* can be conceptualized as existing at the end of a continuum opposite the threat. A general grimace display consists of averted eyes, retracted lips, and a backward head-bob. This display is indicative of fright or submission. In various agonistic situations, both forward and backward head-bobs can often be observed, accompanied by the respective threat and grimace expressions. In those "fight or flight" contexts, emphasis on either the forward or backward direction implies information about the monkey's overall propensity to attack or to retreat.

Some general aspects of the *threat* display include a wide open mouth and tension around the eyes. The open mouth component of this expression often allows a primate's canine incisors to be observed. The eyes are directed toward the subject being threatened, usually involving direct eye contact. Prolonged eye contact between primates is generally indicative of threat behavior unless the eye contact is between mating pairs, or between a mother and her infant. The threat expression is often accompanied by a forward head-bob.

Yawn displays can be indicative of fatigue, but are more often indicative of tension and conflict. The primate maximally opens its mouth to expose the entire set of teeth. At the climax of this display, the head may be thrown back and the eyes closed.

Physical visual communication differs from both the postural and gestural forms in that it is not a behavioral signal, but a physical signal. Gender and personal identification can often be made on the basis of a primate's size, color, canine size, or other outstanding characteristics.

> Sexual differences, apart from the primary sexual organs are signalled by dimorphism in body shape, pelage, and dentition. (Peters and Ploog, 1973, P. 228)

Sexual receptivity may be signalled by a color change and swelling of the perineum. Every primate physical feature has the potential of providing another animal with some type of information.

Male and female primates make differential use of these various forms of visual communication in sending information. But in order for a visual signal to be communicable, there must also be a receiver who sees and understands the meaning of the signal. Therefore, what a primate attends to during ongoing visual behavior is also of interest to us. We will now examine the data on sex differences in visual orientation or "looking."

B. VISUAL ORIENTATION

Most studies dealing specifically with visual orientation have been performed with captive monkeys, primarily rhesus macaques (*Macaca mulatta*). One study on rhesus (Baysinger, Brandt, and Mitchell, 1972) found that infant females that had been reared in social isolation visually oriented to observers more than did males. A study by Mitchell (1972) on rhesus infants found several trends: that three- to six-month-old males attend to their mothers longer than do females, but that the females look into their mother's faces and at human observers longer than do males. In addition, six- and twelve-month-old females attend to peers and human observers more than do males, and preadolescent female rhesus look at infants significantly longer per look than do preadolescent males (Mitchell, 1972). Thomsen (1974) also reported that rhesus females hold eye contact with human observers (particularly female humans) longer than do rhesus males. Apparently, rhesus females look at other individuals longer than do rhesus males, at least in captivity.

But, as we know, results on rhesus monkeys do not always generalize to the rest of the primate order. The rhesus is a terrestrial, sexually dimorphic, polygamous primate that lives in multimale groups. What about the visual behavior of a primate that is arboreal, sexually monomorphic, and monogamous? Data on such a primate are provided by Phillips and Mason (1976). In captive titi monkey pairs (*Callicebus moloch*), female titis look at males more than males look at females.

These data are, of course, applicable only to captive rhesus and titi monkeys. What about free-ranging monkeys? One study of free-ranging rhesus macaques (Altmann, 1968) presents data which suggest that adult and juvenile males are more active and receptive than are females in visual communication; however, this does not tell us whether or not males differ from females in long duration *social* looking. A semipopular article on talapoins (*Miopithecus talapoin*), we believe, gives us the best clue regarding possible sex differences in looking in the primate order. Segre (1970) noted that talapoin males spend the greater part of their day sitting and gazing into the distance, whereas the females' gazes are more concerned with events in their immediate proximity.

This distinction between female interest in the proximal and male interest in distal events will help provide a framework for the material we will cover in the rest of this chapter. Let us now discuss visual communication taxonomically.

C. PROSIMIANS

There are very few data on sex differences in visual communication in prosimians. This is to be expected, however. Prosimians, as we know, rely quite heavily upon olfaction (e.g., in scent-marking) in their social interactions with one another. Most prosimians do not even have binocular vision and most display very few facial expressions.

D. NEW WORLD MONKEYS

Relative to the Old World monkeys and, in particular, to the great apes, New World monkeys do not display many facial expressions. As in the prosimians, many New World monkeys rely upon olfaction to a larger extent than do Old World monkeys. However, New World monkeys *do* have stereoscopic vision; and, sex differences in visual communication have been reported for several species.

We know, by now, that all South American monkeys are arboreal. They must, therefore, rely on audition as much as on visual communication. However, as we also know, the South American species can be differentiated from one another on the basis of their social organization. The marmosets, tamarins, titi monkeys, and night monkeys, on the one hand, live in monogamous families; whereas, the squirrel monkeys, howlers, and cebus monkeys live in large troops. We will consider those species which live in large troops first.

In squirrel monkeys (*Saimiri sciureus*), the higher ranked animals show greater variability in agonistic behavior with regard to the choice of the visual signals they employ (Maurus, Hartmann, and Kihlmorgen, 1974). In agonistic contexts, touching of a partner and advancing the mouth towards the partner's neck can be elicited (by brain stimulation) from males more frequently than from females. Straightening of the body and thrusting of the chin toward the partner in agonistic contexts occur only in males (brain stimulated or not); whereas, these two behaviors can be elicited in females only if they are brain stimulated (Maurus et al., 1975). These findings suggest a basic sex difference in agonistic behavior; but, in addition, these findings suggest the existence of a usually unemployed *potential* in some kinds of agonistic behaviors in female squirrel monkeys. In fact, in one study, *dominance* gestures were elicited many more times from brain sites in subordinate females than from brain sites in dominant males (Maurus and Ploog, 1971).

Territorial threats occur more frequently in male howler monkeys than in females (*Alouatta palliata*). This is also true of titi monkeys, *Callicebus moloch* (Marler, 1968).

In a discussion of communication in primates, Peters and Ploog (1973) point out that a supposed lack of sexual dimorphism may be a function of the subtleness of signals rather than of their absence. They use the example of the South American capuchin (*Cebus capuchinus*) whose visual signals evade detection in field observations. Sex differences and visual recognition of gender are aided by vocalizations or patterns of body movements and posture. Visual displays do not occur in a vacuum (by themselves) in New World monkeys (or for that matter in any nonhuman primate). It is the behavior of the whole organism which conveys the social message.

As we have already mentioned, in at least one *monogamous* South American monkey (titi), territorial threats are primarily adult male behaviors (Marler, 1968). Also in the titi (*Callicebus moloch*), as we know, the adult female looks at her partner (a male) much more than the male looks at her (Phillips and Mason, 1976). In the more primitive and primarily (though not exclusively) monogamous New World marmosets and tamarins, the female is more aggressive than is the titi female. In captivity, one dominant female marmoset can inhibit all other females in her presence from reproducing. In *Cebuella pygmaea* (a marmoset), a female usually seems to be the dominant individual; and, in *Saguinus fuscicollis,* dominance interactions are of a more serious nature among *females* than among males. This includes vicious fighting. There is definitely aggressive competition in both sexes (titi monkeys are not as aggressive toward members of their own sex) until the male and female have exclusive rights to "their mate" (Epple, 1975).

Marmosets have twins and the male cares for them more than does the female, although the female *does* nurse and also care for the infants. Sex differences in facial expressions, if they exist at all, are few (Epple, 1975).

Sex differences in posturing have been noted in rufous-naped tamarins (*Saguinous geoffroyi*). Adult females perform an upward tail-coiling when sexually unsatisfied and this is rarely seen in males. In this species of tamarin, although both adult males and females carry the infants, a single female can rear twins successfully in captivity even in the absence of an adult male. Again, sex differences in visual communication other than tail-coiling, *if they exist* at all, are few (Moynihan, 1970).

In summary, monogamous marmosets and tamarins show few, if any, sex differences in visual communication. Monogamous titi monkeys show sex differences as follows: males tend to display more territorial threats while females tend to do more social looking within the family group. Among nonmonogamous New World monkeys, males tend to display more aggressive and territorial visual displays than do females but the females also show at least some potential for performing such displays.

E. OLD WORLD MONKEYS

1. Macaques

a. Rhesus Agonism

According to Altmann (1968), among adult free-ranging rhesus (*Macaca mulatta*), males lip-smack more than do females. The affective behavior of Altmann's males, however, is on the whole more aggressive than is that of adult females which are more submissive. In the laboratory, Harlow and Lauersdorf (1974) reported a similar difference with more threats in males and more passivity in females as early as eighty days of age. Adult females in Altmann's studies are more likely to ignore the social partner, whereas adult males are inclined to lip-smack (a friendly gesture). Juvenile females, however, do not tend to visually ignore partners as much as do adult females. During affective or agonistic displays, adult females tend to "hit-or-withdraw," whereas adult males tend to "come closer-or-scream" (Altmann, 1968).

Dominant male rhesus often attack *directly* without any previous threat. This is not as true of females (Hinde and Rowell, 1962). Branch-shaking is done mostly by males as is chewing and gnashing of teeth (which occur during copulation as well as during threatening bouts) (Cross and Harlow, 1965; Hinde and Rowell, 1962). Female chewing is depressed by human visual stimulation in the laboratory, whereas male chewing is not (Cross and Harlow, 1965). "Territorial" or intertroop threats are also usually made by males (Marler, 1968). However, there is good evidence that the age and/or sex class corresponding to that of introduced monkeys is the one that initiates most of the threat and attack behavior (Southwick, 1967). The "crook tail," or tail held high and forward over the back, signifies dominance primarily in males (Marler, 1968). Most of the facial expressions displayed by rhesus are presented in Rowell (1972) or in Redican (1975).

In the laboratory, Baysinger, Brandt, and Mitchell (1972) reported that infant females looked at human observers more frequently than did male infants, and that the females looked for a slightly longer duration per look. Female infants in this study also *yawned* more than did males, while male infants lipsmacked more.

According to Cross and Harlow (1965), fear grimaces occur more often in females than in males although they decrease with age in both sexes. Threats toward other animals, however, occur more often in males but *increase* with age in both sexes. Many age and sex differences in visual communication are not apparent unless the animals are captive and provoked (Cross and Harlow, 1965). In socially deprived laboratory-reared rhesus, self-directed threats and self-bites (also head-slapping) are more evident in males than in females (if provoked). There is a rapid increase in self-threatening and self-biting in both sexes at three to four years of age. The self-biting actually develops a little earlier than does the self-threat (Cross and Harlow, 1965).

Cross and Harlow (1965) reported that yawning occurred more frequently in males whereas invitations to groom and (of course) presenting occurred more frequently in females. This is evident as early as one year of age (Harlow and Lauersdorf, 1974). Mason (1960), on the other hand, reported that adult rhesus *females* threatened humans *more* than did adult males, even though the adult males would make more actual attacks. Mason (1960) also reported that adult females showed more fear grimaces than did males. If the human observer in the Mason study passively stared at the rhesus adults, the males usually remained placid, averted their gaze, or pressed their chests or sides against the cage to invite grooming (Mason, 1960). Perhaps these males were lab-wise.

According to R. P. Michael (1970), male rhesus aggression directed toward females usually occurs in response to a female's threat or to a female's eating; however, it also sometimes occurs spontaneously. Michael claims that aggression by females to males is less frequent and is usually in response to a male's threat or to a male's attempts to mount her. Redirection of aggression occurs frequently when consort bonds are being established. Threatening away from her consort occurs most frequently in the female rhesus if estrogen levels are high. Threats directed toward the male by the female are highest outside of the period of ovulation. Threats by the male away from his female consort also peak when estrogen is high in the female but decrease when progesterone is high in the female. Progesterone also reduces female threats. (Zumpe and Michael, 1970).

Mitchell (1968) reported that threats were higher in rhesus subadult males than in subadult females, but only when the subadults were paired with animals close to their own age and size. Fear grimaces, on the other hand, occurred more frequently in subadult females if the animals were paired with age-mates or with larger adults (Mitchell, 1968).

b. Rhesus Looking

Developmental data on rhesus "looking" behavior have been collected by Mitchell and collaborators (cf. Mitchell, 1972). During the second three months of life, rhesus male infants look at their mothers more than do female infants; however, the female infants look at their mother's faces more often than do male infants. Female infants, between the ages of three months and one year, tend to look at human observers and at animals other than their mothers more frequently than do male infants (but only if the animals looked at are the same age or older). Between six and twelve months of age, female infants also elicit more looks from other animals than do male rhesus infants.

During rhesus preadolescence, females look at tiny infants longer *during* each look than do males (cf. Mitchell, 1972). The males direct more threats and other forms of hostility toward the infants than do female preadolescents, while the females display significantly more maternal-related behaviors (including lip-smacking) toward the infants than do the males (Chamove et al., 1967). Looks of longest duration are correlated with interest, curiosity, and affection (Mitchell, 1972); hence, the prepubescent females appear to be much more interested in infants than do the males, whereas the males appear to be annoyed by the infants.

Møller, Harlow, and Mitchell (1968) reported that the average duration per occurrence of "crook tail" is longer in subadult male rhesus subjects than in subadult female rhesus subjects (their laboratory rhesus ranged in age from fifteen months to thirty-nine months). They also reported that males display more yawns, social threats, redirected threats, and human-observer-directed threats than do females, while the females display more fear grimaces (Møller, Harlow, and Mitchell, 1968).

Responses of juvenile female rhesus to colored slides of facial expressions have also been monitored (Redican, Kellicutt, and Mitchell, 1971). What was not reported in the Redican study was that the authors were *unable* to train a similar group of males to press a bar in order to look at the pictures. Wilcoxon et al. (1968) also reported a greater tendency on the part of females than males to push buttons for visual self-stimulation.

Suomi, Harlow, and Kimball (1971) reported that, in rhesus reared in social isolation, males show more threats than do females. A similar finding was reported by Mitchell (1968) and by Cross and Harlow (1965). In contrast, as we know, Thomsen (1974) reported that young female rhesus monkeys make longer eye contacts with a human observer than do young male rhesus monkeys.

In summary, most of the data on rhesus visual communication suggest that males use more aggressive expressions and postures whereas females tend to display more fear, appeasement, and social looking.

c. Stumptail Macaques

In the stumptail macaque (*M. arctoides*) there is great variability within and across sex regarding visual communication (cf. Bertrand, 1969). Jones and Trollope (1968), however, have reported that at least one form of facial communication occurs primarily in adult males. They noted that "chomping and teeth grinding" usually occurs in adult *male* stumptails. In regarding others visually, stumptail males have been described as being more status conscious than are females, whereas females, in general, have seemed less predictable, especially with regard to aggressive displays (cf. Nagel and Kummer, 1974).

As in the rhesus, visual communication in the stumptail is affected by estrus. Both male and female *yawning* peaks during the early luteal phase of the female cycle (Slob et al., 1975).

Very little else is known about stumptail sex differences in visual behavior. However, it does appear that males predictably display more aggressively.

d. Japanese Macaques

Physically, the male Japanese macaque (*Macaca fuscata*) is certainly much larger than is the female. However, it has been noted (cf. Mori, 1975) that the frequency of vocal signals is low in adult males and that they (the adult males) tend to use visual signals instead. In food-getting situations, threat responses are not different in males and females, but fear-related responses are elicited primarily from females (Norikoshi, 1971).

It is the adult males who branch-shake and emit warning barks and "territorial" visual threats (Marler, 1968). Thus, for the most part, the data on Japanese macaques corroborate those on the rhesus: males display aggressively and females display fear.

e. Pigtail Macaques

Physically, male adult pigtail macaques (*Macaca nemestrina*) are much larger than the females. Behaviorally, these macaques probably display as many sex differences as do rhesus. Even at less than one year of age there are noticeable sex differences in visual communication. For example, the pouting expression known as the "Flehman face" or "len" (which does *not* occur at all in the rhesus) occurs more frequently in male pigtails than in females (Evans, 1967). In other ways, pigtail macaques have been found to resemble the rhesus monkey. It is probably reasonably safe to assume that sex differences seen in the rhesus also tend to occur in the pigtail macaque.

f. Barbary Macaques

The Barbary or Gibraltar macaque male (*Macaca sylvana*) is much more interested in infants than is the rhesus male. The leader male, for example, carries the neonate from as early as the first day of life (Burton, 1972). In this way the Barbary macaque is very different from rhesus and pigtail macaques. It is the leader *male* that actually encourages sociobiological maturation in the infant's visual displays by reinforcing the infant's mouth sucking movements until they "become" the social chatter gesture. This gesture, in turn, is of primary importance for all future social contact. In one population of Barbary macaques, subadult males, too, played a role in developing the infant's visual communication, whereas subadult females played only a very small role in infant socialization (Burton, 1972). The *adult* females, however, were also important, particularly the infant's mothers. This same population of *Macaca sylvana* had a greater role for young females (particularly older siblings) in infant socialization five or six generations earlier (in 1940). Burton feels that there has been a "tradition drift" in this group of macaques as far as sex differences are concerned. To see Gibraltar macaque behavioral sex differences in visual communication as either genetically fixed or wholly environmentally determined "would be to ignore the basis of evolution." (Burton, 1972, P. 57.)

Lahiri and Southwick (1966) also reported a large role for adult males in infant socialization. Adult male Gibraltar macaques in their study would take the infant from the mother after *lip-smacking* and *teeth-chattering*. Sex differences in the Barbary macaque are (in many ways) apparently much different from those seen in rhesus and pigtail groups.

g. Other Macaques

A relatively ignored species of macaque, the toque macaque (*Macaca sinica*) has recently been closely examined by Dittus (1975). In his groups, most of the visual threat displays were directed toward immature females. He does not report a sex difference in sending signals.

Macaca fascicularis, the crab-eating macaque, is very much like the rhesus (only smaller). There is sexual dimorphism in behavior, but in visual communication only branch-shaking and "territorial" threats have been *reported* as showing sex differences. Males display more branch-shaking and territorial threats than do females (Marler, 1968).

2. Baboons

Closely related to the genus *Macaca*, the Old World monkeys of the genus *Papio*, commonly called baboons, resemble macaques in their visual displays. As in macaques, branch-shaking is performed primarily by male savannah baboons (*P. cynocephalus*). Males also display more prancing (a behavior having a meaning similar to that of branch-shaking). Occasionally, however, branch-shaking is seen in an adult female (Anthoney, 1969). In savannah baboons, adult males protect the troop from predators. While both males and females defend infants, there have been reports of males, after the death of an infant, threatening the mother if she moves away from the dead infant. In these baboons, there appears to be a tendency for young males to imitate threat and attack patterns more than do young females (cf. Hamburg, 1971).

In the hamadryas baboons (*P. hamadryas*), which live in one-male harems, the adult male, as we know, initiates movement of his harem by moving his body with a particular swinging gait (Marler, 1968). Females do not do this.

In *Theropithecus gelada,* a different genus of baboon, males show a greater tendency than females to avoid moving novel objects when they see them (Hughes and Menzel, 1973).

3. Langurs

Langurs of the genus *Presbytis,* are *arboreal* Old World monkeys not closely related to the terrestrial macaques and baboons. Dolhinow and Bishop (1970) have reported that because of the active running and wrestling of male infants, adult langurs (*Presbytis entellus*) threaten and chase male infants more than twice as often as they threaten and chase female infants. Group play among juvenile langurs is unisexual primarily because of a sex difference in roughness. Adult male langurs seldom play but adult female langurs *never* play.

The subadult female langur is subordinate to all adult females; however, the subadult male is not. He may dominate some of the adult females (Dolhinow and Bishop, 1970). These behavioral differences are reflected in visual communication.

Poirier (1970) described large sex differences in langurs (*Presbytis johnii*), primarily in dominance and aggressive displays. Female dominance encounters, when they occurred, involved very loud vocalizations, whereas this was not true of male dominance encounters. Male dominance was primarily silent and *visual,* although more aggressive than female dominance behavior. Poirier did note, however, that a dominance system varies with the species, the troop, and the personalities of the individuals (especially with that of the most dominant male). Suzanne Ripley (1967) has noted that male langurs (but not females) tend to maintain a visual *observational tonus* in which they are more alive to more *distant* conditions and events than are females. This potential sex difference in dominance "vigilance" is in agreement with data presented earlier and with data to be discussed in more detail later in this report.

4. Other Old World Monkeys

In West African vervets (*Cercopithecus spp.*), canine display, body bounce, and crook tail are limited to adult males. The head-bob and erect posture also occur significantly more often in males than in females, while the eye-flash and gape occur only slightly more frequently in males than in females. All or most of these visual displays are related to threat behavior. Females, on the other hand, crouch more than do males and *none* of the threat displays are limited to females alone, whereas three of them are limited to adult males (Durham, 1969). Territorial threats are also usually displayed by male vervets (Marler, 1968).

In the patas monkey (*Erythrocebus patas*), males usually threaten before they attack, whereas females do not. This is apparently also true of Syke's monkeys (*Cercopithecus albogularis*) (Nagel and Kummer, 1974).

In free-ranging Colobus (*Colobus guereza*), territorial aggressive displays are most pronounced in males. In most cases, young subordinate males of this species start the territorial clashes. The females and subordinate males, however, usually stop to watch the dominant males' displays (Schenkel and Schenkel-Hulliger, 1967).

In *Miopithecus talapoin,* the males sit high in branches and visually *survey* activities around them. The males yawn more than do females (particularly the leader male). The *females seem concerned with close details* rather than with extragroup activity. According to Segre "each female has more in common behaviorally with her other female agemates than with males or juveniles." (Segre, 1970, P. 24.)

F. APES

Prior to summarizing the information on sex differences in monkey visual communication, a few words are in order for the apes. First of all, for the monogamous apes (e.g., gibbons and siamangs) few, if any, sex differences in visial displays have been reported. This does not mean, however, that they do not exist.

Chimpanzees (*Pan troglodytes*) have been studied more than any other apes. Young male chimpanzees, like young male baboons, tend to imitate threat and attack patterns more than do young females (Hamburg, 1971). However, in chimpanzees, even the group structure, to say nothing of individual behavior, is extremely variable (Marler, 1968).

Despite the variability in group structure, however, male chimpanzees still show more extragroup and intergroup behavior than do females. Male chimpanzees also seem to be more interested visually in events occurring outside their immediate group. Thus, as in many of the other species of primates discussed in this chapter, males look at *distant* events and females look at close intragroup social events.

It is interesting that, contrary to the usual greater vulnerability of males, the eyesight for distance vision of male chimpanzees appears to be *less* vulnerable than is the eyesight of females. There is apparently a change toward increasing myopia (nearsightedness) with age in chimpanzees, with females showing higher levels of myopia than males. Thus, female chimpanzees are more susceptible to conditions which produce defective vision of this sort (Young, Leary, and Farrer, 1971). In light of the visual vigilance reported for the males of most species of primates, this sex difference would seem to be a very adaptive one for *many* species of primates not just chimpanzees.

With regard to *Gorilla gorilla,* it is the male who visually displays by chest beating and branch-shaking. It is, of course, also the male gorilla who protects the group from extragroup threat (Marler, 1968). The male orangutan (*Pongo pygmaeus*) also tends to orient away from the group more than does the female, but the reasons for this are probably not the same as for the other species of apes (Rijksen, 1975).

G. OVERVIEW

There is an insufficient amount of adequate data on many aspects of sexual differences in primate visual communication. Only a few studies exist on New World monkeys, and for many species of monkeys and apes primatologists know nothing concerning sex differences in visual behavior. In addition to data deficiency, some of the extant information is inconsistent. With these problems in mind, however, certain primate evolutionary trends may be discerned. These trends are of a general character and do not necessarily apply to each individual primate.

This review presents a number of findings which can be accounted for by postulating several basic primate gender-related characteristics. Males are generally considered to exhibit more aggressive behavior than females. Females are typically more passive and submissive. Some evidence for these assertions is provided by the fact that threats not only develop earlier in males, but are more commonly emitted by males than by females. Except for dominance interactions, grimaces are more commonly displayed be females than by males. However, female dominance hierarchies are basically more stable than are those of males and agonistic situations do not arise as frequently for females in contests for status. Yawns are emitted more often by males than by females, which implies that males are generally more tense, a possible result of their more frequent involvement in agonistic behavior.

Male primates are generally less often involved in nonagonistic physical-social contact than are females, as the males occupy much of their time protecting their group. Female primates are generally more occupied with infant care and peaceful social behavior. For example, females groom more than do males. The visual orientation data suggest that females do more observing than do males in a close, social context, while males do more observing than do females over longer distances. We need only survey the studies on extratroop agonistic behavior to determine that these are also predominantly male activities.

The hormonal state of males fluctuates much less than does that of females. It is therefore not surprising to find sexual differences in behavioral cues concerning female receptivity. Females are probably aware of their own receptivity as a result of internal hormonal changes, but males must rely on visual and other cues from the females to be aware of these changes.

It is also interesting to note some special qualities of primate behavior during copulation. When compared with the usual ongoing behavior patterns, copulation appears unusual. This is one situation in which a high degree of social cooperation between the sexes *must* be achieved.

According to the available information, male primates are more inclined to use distant and visual communication than are females. However, females are more active than males in their use of close and vocal communication (see the previous chapter).

Maccoby and Jacklin (1974) say that human females are, on the whole, less aggressive than are males, more competent than males at verbal tasks, and less competent than males at visual-spatial tasks. Some have attributed these sexual differences to early social training. This review finds that, for primates in general, males are more aggressive than are females, more inclined to be visual, and less inclined to be vocal. Therefore, it is possible that these specific human sexual differences are at least partly determined by differences in biology which have evolved over millenia.

One final comment: As we know, chimpanzee males, who are more interested in extragroup events than are chimpanzee females, are not as vulnerable to myopia as are chimpanzee females. The following quote on human myopia may be of some interest to us in light of the material we have just reviewed:

> Girls tend to develop a higher amount of myopia than boys, and more girls tend to develop myopia earlier than boys . . . the boys *never* develop the amount of myopia that girls develop. (Young, 1977, P. 451)

Since human males are more vulnerable than are human females to almost every other kind of physical disorder, this difference is indeed a striking and perhaps important paradox. We will have additional comments concerning general sex differences in aggression and fear in the next three chapters.

REFERENCES

Altmann, S. A. Sociobiology of rhesus monkeys. IV: Testing Mason's hypothesis of sex differences in affective behavior. *Behaviour,* 1968, **32**, 49-69.

Anthoney, T. R. Threat activity in wild and captive groups of savannah baboons. *Proceedings of the Second International Congress of Primatology.* Basel: Karger, 1969, pp. 108-113.

Baysinger, C. M., Brandt, E. M. and Mitchell, G. Development of infant social isolate monkeys (*Macaca mulatta*) in their isolation environments. *Primates,* 1972, **13** (3), 257-270.

Bertrand, M. The behavioral repertoire of the stumptail macaque. *Bibliotheca Primatologica,* 1969, No. 11.

Burton, F. D. The integration of biology and behavior in the socialization of *Macaca sylvana* of Gibraltar. In Poirier, F. E. (Ed.) *Primate Socialization.* New York: Random House, 1972, pp. 29-62.

Chamove, A., Harlow, H. F. and Mitchell, G. Sex differences in the infant-directed behavior of preadolescent rhesus monkeys. *Child Development,* 1967, **38**, 329-335.

Cross, H. A. and Harlow, H. F. Prolonged and progressive effects of partial isolation on the behavior of macaque monkeys. *Journal of Experimental Research in Personality,* 1965, **1**, 39-49.

Dittus, W. P. J. The ecology and behavior of the toque monkey, *Macaca sinica. Dissertation Abstracts,* 1975, B35/09 , 4725.

Dolhinow, P. J. and Bishop, N. The development of motor skills and social relationships among primates through play. In Hill, J. P. (Ed.) *Minnesota Symposia on Child Psychology* (Vol. IV). Minneapolis: University of Minnesota Press, 1970, pp. 141-198.

Durham, N. Sex differences in visual threat displays of West African vervets. *Primates,* 1969, **10**, 91-95.

Epple, G. The behavior of marmoset monkeys (*Callithricidae*). In Rosenblum, L. A. (Ed.) *Primate Behavior* (Vol. 4). New York: Academic Press, 1975, pp. 195-239.

Evans, C. S. Methods of rearing and social interactions in *Macaca nemestrina. Animal Behaviour,* 1967, **15**, 263-266.

Hamburg, D. A. Aggressive behavior of chimpanzees and baboons in natural habitats. *Journal of Psychiatric Research,* 1971, **8**, 385-398.

Harlow, H. F. and Lauersdorf, H. E. Sex differences in passion and play. *Perspectives in Biology and Medicine,* 1974, **17**, 348-360.

Hinde, R. A. and Rowell, T. E. Communication by postures and facial expressions in the rhesus monkey (*Macaca mulatta*). *Proceedings of the Zoological Society of London,* 1962, **138**, 1-21.

Hughes, G. H. and Menzel, E. W., Jr. Use of space and reactions to novel objects in gelada baboons (*Theropithecus gelada*). *Journal of Comparative and Physiological Physiology*, 1973, 83, 1-6.

Jones, N. G. B. and Trollope, J. Social behaviour of stump-tailed macaques in captivity. *Primates*, 1968, 9, 365-394.

Lahiri, R. K. and Southwick, C. H. Parental care in *Macaca sylvana*. *Folia Primatologica*, 1966, 4, 257-264.

Maccoby, E. E. and Jacklin, C. N. *The Psychology of Sex Differences*. Stanford, California: Stanford University Press, 1974.

Marler, P. Aggregation and dispersal: Two functions in primate communication. In Jay, P. C. (Ed.) *Primates: Studies in Adaptation and Variability*. New York: Holt, Rinehart and Winston, 1968, pp. 420–438.

Martenson, J., Sackett, D. P. and Erwin, J. Facial expressions as correlates of overt aggression in pigtail monkeys. *Journal of Behavioral Science* (Durban), in press.

Mason, W. A., Green, P. C. and Posepanko, C. J. Sex differences in affective-social responses of rhesus monkeys. *Behaviour*, 1960, 16, 74-83.

Maurus, M., Hartmann, E. and Kuhlmorgen, B. Invariant quantities in communication processes of squirrel monkeys. *Primates*, 1974, 15 (2-3), 179-192.

Maurus, M., Kuhlmorgen, B., Hartmann-Weisner, E. and Pruscha, H. An approach to the interpretation of the communicative meaning of visual signals in agonistic behavior of squirrel monkeys. *Folia Primatologica*, 1975, 23, 208-226.

Maurus, M. and Ploog, D. Social signals in squirrel monkeys: Analysis by cerebral radio stimulation. *Experimental Brain Research*, 1971, 12, 171-183.

Michael, R. P. Hormonal factors and aggressive behaviour in the rhesus monkey. *Proceedings of the International Society of Psychoneuroendocrinology, Brooklyn, 1970*. Basel: Karger, 1971, pp. 412-423.

Mitchell, G. Persistent behavior pathology in rhesus monkey following early social isolation. *Folia Primatologica*, 1968, 8, 132-147.

Mitchell, G. Looking behavior in the rhesus monkey. *Journal of Phenomenological Psychology*, 1972, 3 (1), 53-67.

Møller, G. W., Harlow, H. F. and Mitchell, G. Factors affecting agonistic communication in rhesus monkeys (*Macaca mulatta*). *Behaviour*, 1968, 31 (3-4), 339-357.

Mori, A. Signals found in the grooming interactions of wild Japanese monkeys of the Koshima troop. *Primates*, 1975, 16 (2), 107-140.

Moynihan, M. Some behavior patterns of platyrrhine monkeys. II: *Saguinus geoffroyi* and some other tamarins. *Smithsonian Contributions to Zoology*, 1970, 28, 1-77.

Nagel, U. and Kummer, H. Variation in cercopithecoid aggressive behavior. In Holloway, R. L. (Ed.) *Primate Aggression, Territoriality, and Xenophobia*. New York: Academic Press, 1974, pp. 159-184.

Norikoshi, K. Tests to determine the responsiveness of free-ranging Japanese monkeys in food-getting situation. *Primates,* 1971, **12** (2), 113-124.

Peters, M. and Ploog, D. Communication among primates. *Annual Review of Physiology,* 1973, **35**, 221-242.

Phillips, M. J. and Mason, W. A. Comparative studies of social behavior in *Callicebus* and *Saimiri;* Social looking in male-female pairs. *Bulletin of the Psychonomic Society,* 1976, **7** (1), 55-56.

Poirier, F. E. Dominance structure of the Nilgiri langur (*Presbytis johnii*) of South India. *Folia Primatologica,* 1970, **12**, 161-186.

Redican, W. K. Facial expressions in nonhuman primates. In Rosenblum, L. A. (Ed.) *Primate Behavior* (Vol. 4). New York: Academic Press, 1975, pp. 103-194.

Redican, W. K., Kellicutt, M. H. and Mitchell, G. Preferences for facial expressions in juvenile rhesus monkeys (*Macaca mulatta*). *Developmental Psychology,* 1971, **5** (3), 539.

Rijksen, H. D. Social structure in a wild orangutan population in Sumatra. In Kondo, S., Kawai, M. and Ehara, A. (Eds.) *Contemporary Primatology.* Basel: Karger, 1975, pp. 373-379.

Ripley, S. Intertroop encounters among Ceylon gray langurs (*Presbytis entellus*). In Altmann, S. (Ed.) *Social Communication Among Primates.* Chicago: University of Chicago Press, 1967, pp. 237-254.

Rowell, T. E. *The Social Behaviour of Monkeys.* Baltimore: Penguin, 1972.

Schenkel, R. and Schenkel-Hulliger, L. On the sociology of free-ranging Colobus (*Colobus guereza catdatus* Thomas 1885). *First Congress of the International Primatological Society,* Frankfurt, 1966. Published in Stark, D., Schneider, R. and Kuhn, H. J. (eds.) *Neue Ergebnisse der Primatologie.* Stuttgart: Gustav Fischer, 1967, pp. 183-194.

Segre, A. Talapoins. *Animal Kingdom,* 1970, **73** (3), 20-25.

Slob, A. K., Goy, R. W., Wilgand, S. J. and Scheffler, G. Gonadal hormones and behaviour in the stumptailed macaque (*Macaca arctoides*) under laboratory conditions: A preliminary report. *Journal of Endocrinology,* 1975, **64**, 38.

Southwick, C. H. An experimental study of intra-group agonistic behaviour in rhesus monkeys (*Macaca mulatta*). *Behaviour,* 1967, **28**, 182-209.

Suomi, S. J., Harlow, H. F. and Kimball, S. D. Behavioral effects of prolonged partial social isolation in the rhesus monkey. *Psychological Reports,* 1971, **29**, 1171-1177.

Thomsen, C. E. Eye contact by non-human primates towards a human observer. *Animal Behaviour,* 1974, **22**, 144-149.

Wilcoxon, H. C., Meier, G. W., Orlando, R. and Paulson, D. G. Visual self-stimulation in socially-living rhesus monkeys. *Proceedings of the Second International Congress of Primatology,* 1968, **1**, 261-266.

Young, F. A. The nature and control of myopia. *Journal of the American Optometric Association,* 1977, **48**, 451–457.

Young, F. A., Leary, G. A. and Farrer, D. N. Four years of annual studies of chimpanzee vision. *American Journal of Optometry,* 1971, **48**, 407–416.

Zumpe, D. and Michael, R. P. Redirected aggression and gonadal hormones in captive rhesus monkeys (*Macaca mulatta*). *Animal Behaviour,* 1970, **18**, 11–19.

24.
Aggression: Prosimians, New World Monkeys, and Macaques

"Males are more aggressive than are females." This statement has been accepted as being true for most primates, including man (Maccoby and Jacklin, 1974). Aggression or aggressiveness in males is presumed to be determined to a large extent by prenatal androgens, and male aggressiveness is supposedly also subtly supported in mother–infant interactions and in interactions between the immature animal and other members of the group.

There is also the belief that aggressive patterns may be learned early in life through observation, imitation, and practice sequences:

> There is a sex difference in the attractiveness of some aggressive patterns from infancy onward, with males tending to be more interested and spending more time in practicing the aggressive patterns. (Hamburg, 1974, P. 219)

But ecological conditions also affect the amount of aggression displayed. As we already know, aggression increases with increased population density and decreases with an increased food supply (Jorde and Spuhler, 1974).

As we know, hormonal correlates of aggression have been reported. Testosterone, for example, while apparently not correlated with dominance per se, *is* correlated with changes in dominance rank and these changes, in turn, typically involve aggression (cf. Mazur, 1976).

The most consistent sex differences we have seen thus far have involved vigilance and defense of the group. Males consistently surpass females in behaviors of these kinds, presumably at least in part because of their greater aggressive potential. We have frequently seen that the adult male of many species attacks an intruder while all other animals retreat (Rowell, 1974).

Situational factors can change the incidence of aggression. Marler (1976) has summarized the factors that are especially liable to provoke aggression. Aggression in an animal is most likely to occur in the proximity of another animal of its own *kind* (whether of gender, age, or species), especially if that animal is a stranger and is behaving aggressively itself, inflicting pain, or otherwise creating frustration. Marler (1976) sees the degree of strangeness or familiarity of the animal as being the most important factor.

In any case, there *are* factors other than gender that play a role in the amount of aggression displayed by primates. In this chapter, however, we will concern ourselves primarily with sex differences, and we will arrange our information in a taxonomic outline as usual.

A. PROSIMIANS

Male–male, male–female, and female–female aggression are seen in most species of tree shrews (*Tupaia spp.*). Males appear to be more aggressive than females. In male aggression, there is more chasing and biting, whereas in female aggression there is more threatening and vocalizing (Sorenson, 1974). It is apparently possible to breed tree shrews for aggressiveness (Schwaier, 1975).

While occasionally agonistic wrestling called "grappling" (Lockard, Heestand, and Begert, 1977) occurs in the slow loris (*Nycticebus coucang*), there is a relative lack of aggression in this species. Threatlike eyebrow movements, for example, are not possible in lorises (or in other prosimians) because they lack the face muscles (Ehrlich and Musicant, 1977; also see Chapter 23).

Among galagos (bushbabies), interactions between animals of the same sex are often agonistic while relations between the sexes are not (Flinn and Nash, 1975). In adult male–adult male fights (*Galago senegalensis*), the male in the home territory usually wins. If the two males are in neutral territory, both become highly aggressive and severe wounds are the result. In adult female–adult female fights, only one of the two females is highly aggressive (in *any* environment). In at least one such fight observed by Flinn and Nash a death resulted. Pregnant females and females with infants are the most aggressive. Strangers of the same age and sex are generally incompatible (Bearder and Doyle, 1974).

Two sequences of behavior which often precede aggression in captive lesser galagos (*Galago senegalensis*) are: (1) leaving the nest box followed by a mutual stare, and (2) leaping behavior followed by a mutual stare. If the females' mates are present, both of them will interfere in the females' fight. *Mutual staring* appears to be an evolutionary precursor to the elaborate threat gestures described in Chapter 23 (Bercovitch, 1977).

Aggressive behaviors are not seen in the lesser bushbaby male until one year of age. There is a male call that occurs only in males over one year of age;

however, silence between two males is a threat or a prelude to a fight. Aggression between females is not as marked as is aggression between males, unless one female is dominant (Doyle, 1974).

In the field, direct confrontations between other species of bushbabies (*G. alleni*) are usually seen between males controlling females or between young males without fixed territories (Charles-Dominique, 1977). From what we have reviewed for prosimians, it appears that most prosimian males are more aggressive than prosimian females.

B. NEW WORLD MONKEYS

In captive black howler monkeys (*Alouatta caraya*), females fight over food and males may often intervene in such fights. Male howlers direct most of their aggression toward outsiders (Benton, 1976). Howlers, of course, show extreme sexual dimorphism in size.

In the more monomorphic spider monkey (*Ateles geoffroyi*), the males express hostile behavior toward intruders (Eisenberg and Kuehn, 1966), but they also attack adult females. In fact, in one zoo colony of spider monkeys (*A. geoffroyi*), 44 percent of the observed fighting involved adult males attacking adult females. In 7.5 percent of the fights, two males were aggressing against one female. Evidently, within the group, there is very little male–male aggression and very little female–male aggression. There is however, *inter*group aggression between males (Klein and Klein, 1971) and more aggressive behavior between *opposite*-sexed subjects than between same-sexed subjects. In this sense, spider monkeys provide us with an interesting exception to Marler's (1976) rules concerning aggression (Rondinelli and Klein, 1976).

Spider monkeys are monomorphic but they are not monogamous. In general, the monogamous New World marmosets show less aggression between opposite-sexed adults than between same-sexed adults. Marmoset males and females show more equal amounts of aggression toward conspecific intruders, but usually toward members of their *own* sex (Epple, 1970). There is, on occasion, female dominance and female aggression toward the male (Kleiman, 1977b).

Apparently, however, there is a substantial amount of variability from group to group and from species to species of marmosets and tamarins with regard to the occurrence of aggression. In some groups (e.g., *Saguinus oedipus*), there is a high incidence of intragroup aggression (Dawson, 1976). In others (*Leontopithecus rosalia*), aggression between males is apparently rare (Kleiman, 1977a). In *Saguinus fuscicollis,* aggression between two females is more serious than is aggression between two males (Epple, 1975). The method of "rejection" from the family group of young adults (*Callithrix jacchus*) includes attacks and threats from the parents, especially from the mother (Ingram, 1976).

In another monogamous New World species *(Callicebus moloch,* the titi monkey), there is not as much aggression toward conspecifics of the same sex as there is in marmosets (Epple, 1975). Titi monkey male aggression is evidently intermediate in frequency between that of the marmoset male and that of the nonmonogamous New World cebids like the spider monkey.

There is an *increase* in female aggression if a strange male is added to an established all-female group of squirrel monkeys *(Saimiri sciureus),* but there is no increase in female aggression when a male is introduced and the females of the group do not know each other (Anschel and Talmage-Riggs, 1976). In stable captive groups of squirrel monkeys, females are less often the object of attack than are males (Strayer, Taylor, and Yanciw, 1975). If two groups of squirrel monkeys are combined, collective aggression appears and *females* play the most active part in the aggressive behavior. However, these aggressive female coalitions usually occur when the dominant male is nearby. If one male is introduced, female collective aggression is directed toward him (Castell, 1969).

In summary, the New World monkey data are not consistent with the overall general statement that males are more aggressive than females. The kinds of sex differences in New World monkey aggression are almost as numerous as are the numbers of species studied.

C. MACAQUES *(MACACA SPP.)*

1. General Statements

As we shall see, among most Old World monkeys, males are indeed more aggressive than females but the nature of this difference seems to be somewhat unpredictable. For example, in the patas monkey *(Erythrocebus patas),* the male threatens before attacking while the female does not (Nagel and Kummer, 1974). However, in the rhesus monkey *(Macaca mulatta),* fully dominant males often attack directly without need of a preliminary threat (Hinde and Rowell, 1962).

Many different factors besides gender affect the way a given group of Old World monkeys reacts aggressively to a stranger. Not only does the sex of a specific stranger make a difference, but the host group's sex composition also makes a difference. Females, in general, respond to an intruder with greater frequencies of noncontact and manual aggression than do males. Males, however, are affected more by an intruder. An introduced male is a greater challenge to most groups and is more likely to respond with aggression himself. Severe wounding is more likely with a male intruder. Males show restraint when fighting with females and often lose to a coalition of several females. Males introduced to all-male groups show *less* submission than do females, even though they are aggressed more. Males introduced to all-female groups, on the other hand, seem quite capable of taking over by threat rather than by direct attack. A vigorous and prolonged attack on an introduced male by a coalition of some Old World monkey females is possible only if there is a resident male to begin with (Bernstein, Gordon, and Rose, 1974).

2. Rhesus Monkeys (*Macaca mulatta*)

In attempts to shape monkey–human contact, the female rhesus is evidently very difficult to tame (even more so than is the pigtail female) (Aarons, 1973). Some have asserted that the adult female rhesus monkey threatens more than does the adult male but that the adult male makes actual attacks whereas the female does not. The same authors have also noted that females display more fear grimaces (Mason, Green, and Posepanko, 1960). Other researchers have been unable to corroborate this finding of greater overall agonistic responsiveness or of mild agonistic responsiveness in rhesus females than in rhesus males (cf. Altmann, 1967). Instead, most researchers find that adult males are most aggressive and adult females submissive (Altmann, 1968). In rhesus, according to Angermeir et al. (1968), only males establish dominance in dyads by physical attack. However, Erwin, Maple, and Welles (1975) found that unfamiliar females in dyads showed very high aggression. Attacked strange females never retaliate in this situation. A report of more aggression in males than in females in rhesus over four years old was published by Drickamer (1974).

In group formation studies, involving four adult males and four adult females, adult male aggression was ten times greater than was female aggression (Bernstein, Gordon, and Rose, 1974). In general, the adult males aggressed one another the most, while the females supported and *instigated* the aggression of the adult males in their groups. Male aggression may be controlled or ritualized, to a certain degree, since most of the wounds were restricted to nonvulnerable areas. (On the other hand, the restriction of the wounds to nonvulnerable areas may be due to the defensive skills of the wounded animal.) The introduction of a single female into an all-male rhesus group produces fighting among males and breaks up male alliances of several years standing (one such alpha male was killed when several females were introduced). Females, however, maintain a stable unit when a male or males are introduced (Bernstein, Gordon, and Rose, 1974). Fairbanks and McGuire (1977), in a different set of group formation studies, noted a correlation between female homosexual behavior and female–female *contact* aggression but not with female-female noncontact aggression.

In the studies involving the introduction of strangers done by Southwick (1967; 1972), intolerance and aggression toward strangers was not often expressed if the stranger was an infant. In addition:

The age and/or sex class corresponding to that of the introduced monkeys was the one which initiated most of the threat and attack behavior. (Southwick, 1967, P. 208)

As we know from our chapters on prenatal and adult hormones, changes in testosterone, estrogen, and progesterone can be correlated with changes in aggression. Prenatal androgen (testosterone) increases "assertive" behavior in young infants so that male infants threaten more and play rougher than do female infants (Harlow and Lauersdorf, 1974). Male infant rhesus *are* more aggressive than are female infant rhesus. At 1½ years of age the sex differences in these behaviors are, if anything, even larger than in early infancy (Hansen, Harlow, and Dodsworth, 1966).

Androgen injected into infant female rhesus monkeys between 6½ and 14½ months of age *increases* their aggressive behavior independent of play and sexual behavior (Joslyn, 1973). It may be that testosterone continues to have an affect on aggression much later than it does on play and sexual posturing. In a way, this makes sense. Sexual posturing in play helps the young rhesus to learn how to copulate as an adult. If the young animal were aggressive at this stage it might interfere with the learning. Aggression between infants is thus much more adverse to sexual outcomes than is aggression between adults (see Goy, Wallen, and Goldfoot, 1974 for support of this idea).

Male rhesus in preadolescence continue to be more aggressive than females in preadolescence (cf. Chamove, Harlow, and Mitchell, 1967; Brandt and Mitchell, 1973; Maple, Brandt, and Mitchell, 1975). In free-ranging rhesus, females are often more active interferers than are males in intragroup fights (Kaplan, 1975) and wounding of females occurs more frequently. However, a wounded male is more likely to have impaired locomotion (Hausfater, 1972).

There is more fighting during the breeding season than outside of the breeding season (Wilson and Boelkins, 1970). Aggression by males to females is usually in response to a female threat, to a female's eating, or, in some cases, it is spontaneous. Aggression by females to males is not very frequent but does occasionally occur in response to a male threat or to his attempt to mount her (Michael, 1970).

Redirected threats occur frequently in consort pairs. A female's threatening away from her male consort occurs most when her estrogen levels are high; threatening toward her male consort occurs away from the time of ovulation. Threatening away by the male consort also increases when estrogen levels in the female are high. It decreases when her progesterone levels are high (Michael, 1970). Ovariectomy increases male–female aggression, particularly in the male. Attacks on females are more frequent around menstruation (Sassenrath, Rowell, and Hendricks, 1973). Progesterone increases aggression in the female by decreasing her tolerance of the male (Michael and Zumpe, 1970).

Testosterone changes are *sometimes* related to aggression in rhesus adulthood. Castration of the male in one study did not eliminate spontaneous aggressive behavior but it did increase the threshold for attack. This effect was reversed by administering testosterone (Perachio, 1976). With a sudden and decisive defeat by a large all-male group, a male's plasma testosterone declined within a week (Rose, Gordon, and Bernstein, 1972). However, in another study, testosterone-treated males displayed high sexual levels but *low* levels of aggressive behavior compared to control males (Vandenbergh and Post, 1976). Also, two castrated males fought and dominated males twice their weight (Wilson and Vessey, 1968). Apparently, aggressiveness in the rhesus cannot be said to be enhanced by postnatal treatment with androgen independent of the individual's natural behavioral predispositions.

3. The Crab-Eating Macaque (*M. fascicularis*)

As in the rhesus monkey, tolerance of a stranger is lowest and aggressiveness most intense in crab-eater individuals of the same age/sex class as the stranger (Angst, 1973). In dyadic encounters between crab-eating macaques there is a high frequency of fighting in male–male pairings, a high frequency of grooming in female–female pairings, and a high frequency of "inspecting" in male–female dyads (Thompson, 1967b)

In dyadic interactions of crab-eaters, males establish dominance by active physical encounter whereas females establish status by a withdrawal of the subordinate animal before actual physical aggression takes place; however, in small groups of *Macaca fascicularis,* females actively engage in physical aggression to establish dominance, particularly when the dominant animal is removed from the group (Hendricks, Seay, and Barnes, 1975).

In larger captive groups, 84 percent of the aggressive bites are received by females. In captive males, in large groups, active aggression against certain individuals seems to facilitate physical assault in the same direction by others (van Hooff and De Waal, 1975). When an old dominant male loses his position through physical aggression from a younger male, there is a transition period during which there is a "joint redirection" of aggression or coalition between the two competing males to maintain the social order (De Waal, 1975). Adult male crab-eaters will attack strangers of all kinds even infants, especially if the mother of the infant is unfamiliar to them (Thompson, 1967a).

4. The Pigtail Macaque (*M. nemestrina*)

According to Bernstein (1972), female pigtail monkeys spend more time in social interaction, including agonistic interaction, than do males. Only in damaging aggression involving biting (as opposed to slapping and pulling hair) do male pigtails surpass females.

When new groups of pigtail monkeys are formed in captivity, the most recently introduced animals are attacked by the residents of approximately the same age/sex classification (Bernstein, 1969). Releasing all animals into an enclosure simultaneously is less likely to result in aggression. There is even less aggression if two entirely different groups are released simultaneously than when one strange animal at a time is released. Pigtail macaques show more behavioral disruption (and aggression) than do rhesus macaques (Bernstein, 1969).

In crowding experiments conducted at the University of Washington Primate Field Station by Joe Erwin and collaborators, increased social density (an increased number of animals) was correlated with increased aggression whereas increased spatial density (decreased space) produced lower levels of aggression (cf. Anderson et al., 1977). Groups of pigtail females containing no adult males exhibited more aggression than did groups containing males. An important social role of the male pigtail is the control of intragroup aggression (Sackett, Oswald, and Erwin, 1975; Oswald and Erwin, 1976). Aggression was higher in two connected rooms than it was in one room, because there was better visual access to others in one room, and because the male was more efficient at performing the "control role" (Anderson et al., 1977). Cover did not prevent aggression in unstable groups that underwent moderate or extreme social change (Erwin et al., 1976). If the males were familiar with some of the females there was more aggression with strangers by the males. Those males with canines showed more aggression than did those without canines (Erwin et al., 1976).

5. The Stumptail Macaque (*M. arctoides*)

In 1964, Orbach and Kling (1964) reported that there were no differences in docility according to sex in the stumptail macaque. However, chomping and teeth grinding occur primarily in males (Blurton-Jones and Trollope, 1968), and there is a sex difference in aggressive boldness vis-à-vis humans if infants are present in the group. In this case adult males become very aggressive toward humans (Rhine and Kronenwetter, 1972).

6. The Japanese Macaque (*M. fuscata*)

In captive groups of Japanese macaques,

> . . . adult dominant males are responsible for virtually all of the attacks on a neonate under the age of 3 months. (Alexander, 1970, P. 284)

But the male is also protective of infants and is responsible for aggressively guarding against extratroop threat (Alexander and Bowers, 1969). Males who are low in overall aggression, however, show the most frequent infant care (Itani, 1963).

Aggressive behavior, particularly among six- to nine-year-old males, increases during the breeding season (Sugawara, 1976). However, testosterone levels do not appear to correlate well with the levels of aggression displayed. The adult level of androgen is less important than are social stimuli (Eaton and Resko, 1974).

Japanese monkeys of both sexes show increased aggression when crowded. Males, however, produce and receive a greater proportion of the total attacks under crowded conditions. In contrast, females attack less often and are less frequently attacked when crowded than are males. Low-ranking males are especially susceptible to receiving attacks from others (Alexander and Roth, 1971). Paradoxically, attacks on humans by Japanese macaques are primarily directed toward females. Female humans are more often bitten by monkeys, and women in their twenties are bitten most often (Itani, 1975). Perhaps women show more interest in infant monkeys and adult monkeys become protective and attack the women.

7. Other Macaques

Barbary macaque males (*M. sylvana*) are probably less aggressive than are rhesus macaques. As we know, the Barbary macaque male is quite interested in infants and a baby may be used in male–male interactions, apparently reducing the likelihood of aggression (Whiten and Rumsey, 1973, P. 421).[1] Females rarely do this.

Bonnet (*M. radiata*), toque (*M. sinica*), and lion-tailed (*M. silenus*) macaques show a greater male–male tolerance than do rhesus macaques (cf. Simonds, 1977).

In summary, male macaques are probably more aggressive than are female macaques. Even in those situations where females show a greater frequency of aggressive acts, the males show more severe aggression than do the females. We will continue our review of sex differences in Old World monkey aggression in Chapter 25.

REFERENCES

Aarons, L. Shaping monkey-human contact. *Perceptual and Motor Skills,* 1973, **36**, 235–243.

Alexander, B. K. Parental behaviour of adult male Japanese monkeys. *Behaviour,* 1970, **36**, (4), 270–285.

Alexander, B. K. and Bowers, J. M. Social organization of a troop of Japanese monkeys in a two-acre enclosure. *Folia Primatologica,* 1969, **10**, 230–242.

[1] Some recent data reported by Taub (personal communication) calls into question the use of such "agonistic buffering" in Barbary macaque males.

Alexander, B. K. and Roth, E. M. The effects of acute crowding on aggressive behavior of Japanese monkeys. *Behaviour,* 1971, **39**, 73-90.

Altmann, S. A. Testing Mason's hypothesis of sex differences in the affective behavior of rhesus monkeys. *American Zoologist,* 1967, **7**, 802.

Altmann, S. A. Sociobiology of rhesus monkeys. IV: Testing Mason's hypothesis of sex differences in affective behavior. *Behaviour,* 1968, **32**, 49-69.

Anderson, B., Erwin, N., Flynn, D., Lewis, L. and Erwin, J. Effects of short-term crowding on aggression in captive groups of pigtail monkeys (*Macaca nemestrina*). *Aggressive Behavior,* 1977, **3**, 33-46.

Angermeir, W. F., Phelps, J. B., Murray, D. and Howanstine, J. Dominance in monkeys: Sex differences. *Psychonomic Science,* 1968, **12**, 344.

Angst, W. Pilot experiments to test group tolerance to a stranger in wild *Macaca fascicularis. American Journal of Physical Anthropology,* 1973, **38**, (2), 625-630.

Anschel, S. and Talmage-Riggs, G. Social structure dynamics in small groups of captive squirrel monkeys. Paper presented at the *International Primatological Society* Meeting, Cambridge, England, August, 1976.

Bearder, S. K. and Doyle, G. A. Field and laboratory studies of social organization in bushbabies (*Galago senegalensis*). *Journal of Human Evolution,* 1974, **3**, 37-50.

Benton, L. The establishment and husbandry of a black howler (*Alouatta caraya*) colony at Columbia Zoo. *International Zoo Yearbook,* 1976, **16**, 149-152.

Bercovitch, F. B. Sequences of aggression in captive female lesser galagos (*G. senegalensis*). Paper presented at the *American Association of Physical Anthropologists* Meeting, Seattle, Washington, April, 1977.

Bernstein, I. S. Introductory techniques in the formation of pigtail monkey troops. *Folia Primatologica,* 1969, **10**, 1-19.

Bernstein, I. S. Daily activity cycles and weather influences on a pigtail monkey group. *Folia Primatologica,* 1972, **18**, 390-415.

Bernstein, I. S., Gordon, T. P. and Rose, R. M. Aggression and social controls in rhesus monkey (*Macaca mulatta*) groups revealed in group formation studies. *Folia Primatologica,* 1974, **21**, 81-107. (a)

Bernstein, I. S., Gordon, T. P. and Rose, R. M. Factors influencing the expression of aggression during introductions to rhesus monkey groups. In Holloway, R. L. (Ed.) *Primate Aggression, Territoriality, and Xenophobia: A Comparative Perspective.* San Francisco: Academic Press, 1974, pp. 211-240. (b)

Blurton-Jones, N. G. and Trollope, J. Social behaviour of stump-tailed macaques in captivity. *Primates,* 1968, **9**, 365-394.

Brandt, E. M. and Mitchell, G. Pairing preadolescents with infants (*Macaca mulatta*). *Developmental Psychology,* 1973, **8**, 222-228.

Castell, R. Communication during initial contact: A comparison of squirrel and rhesus monkeys. *Folia Primatologica,* 1969, **11**, 206-214.

Chamove, A., Harlow, H. F. and Mitchell, G. Sex differences in the infant-directed behavior of preadolescent rhesus monkeys. *Child Development,* 1967, **38** (2), 329-335.

Charles-Dominique, P. Urine marking and territoriality in *Galago alleni* (Waterhouse, 1837 Lorisoidea, Primates) – A field study by radio-telemetry. *Zeitschrift für Tierpsychologie,* 1977, **43**, 113-138.

Dawson, G. A. Behavioral ecology of the Panamanian tamarin *Saguinus oedipus* (Callithricidae, Primates). *Dissertation Abstracts International,* 1976, **B37**, 645-646.

De Waal, F. B. M. The wounded leader: A spontaneous temporary change in the structure of agonistic relations among captive Java monkeys (*Macaca fascicularis*). *Netherlands Journal of Zoology,* 1975, **25** (4), 529-549.

Doyle, G. A. The behaviour of the lesser bushbaby. In Martin, R. D., Doyle, G. A. and Walker, A. C. (Eds.) *Prosimian Biology.* London: Duckworth, 1974, pp. 213-231.

Drickamer, L. C. A ten-year summary of reproductive data for free-ranging *Macaca mulatta. Folia Primatologica,* 1974, **21**, 61-80.

Eaton, G. G. and Resko, J. A. Plasma testosterone and male dominance in a Japanese macaque (*Macaca fuscata*) troop compared with repeated measures of testosterone in laboratory males. *Hormones and Behaviour,* 1974, **5**, 251-259.

Ehrlich, A. and Musicant, A. Social and individual behavior in captive slow lorises. *Behaviour,* 1977, **60**, 195-220.

Eisenberg, J. F. and Kuehn, R. E. The behavior of *Ateles geoffroyi* and related species. *Smithsonian Miscellaneous Collections,* 1966, **151** (8), 1-63.

Epple, G. Quantitative studies on scent marking in the marmoset (*Callithrix jacchus*). *Folia Primatologica,* 1970, **13**, 48-62.

Epple, G. The behavior of marmoset monkeys (*Callithricidae*). In Rosenblum, L. A. (Ed.) *Primate Behavior* (Vol. 4). New York: Academic Press, 1975, pp. 195-239.

Erwin, J., Anderson, B., Erwin, N., Lewis, L. and Flynn, D. Aggression in captive pigtail monkey groups: Effects of provision of cover. *Perceptual and Motor Skills,* 1976, **42**, 319-324.

Erwin, J., Maple, T. and Welles, J. F. Responses of rhesus monkeys to reunion. *Proceedings of the Fifth International Congress of Primatology.* Basel: Karger, 1975, pp. 254-262.

Fairbanks, L. and McGuire, M. Homosexual behavior and female aggression in rhesus macaques. Paper presented at the *Western Psychological Association* Meeting, Seattle, Washington, April, 1977.

Flinn, L. and Nash, L. T. Group formation in recently captured lesser galagos. Paper presented at the Annual Meeting of *American Association of Physical Anthropologists,* Denver, Colorado, 1975.

Goy, R. W., Wallen, K. and Goldfoot, D. A. Social factors affecting the development of mounting behavior in male rhesus monkeys. In Montagna, W. and Sadler, W. A. (Eds.) *Reproductive Behavior.* New York: Plenum, 1974, pp. 223-247.

Hamburg, D. A. Ethological perspectives on human aggressive behaviour. In White, N. F. (Ed.) *Ethology and Psychiatry.* Toronto: University of Toronto Press, 1974, pp. 209-219.

Hansen, E. W., Harlow, H. F. and Dodsworth, R. O. Reactions of rhesus monkeys to familiar and unfamiliar peers. *Journal of Comparative and Physiological Psychology,* 1966, **61**, 274-279.

Harlow, H. F. and Lauersdorf, H. E. Sex differences in passion and play. *Perspectives in Biology and Medicine,* 1974, **17**, 348-360.

Hausfater, G. Intergroup behavior of free-ranging rhesus monkeys (*Macaca mulatta*). *Folia Primatologica,* 1972, **18**, 78-107.

Hendricks, D. E., Seay, B. M. and Barnes, B. The effects of the removal of dominant animals in a small group of *Macaca fascicularis. Journal of General Psychology,* 1975, **92**, 157-168.

Hinde, R. A. and Rowell, T. E. Communication by postures and facial expressions in the rhesus monkey (*Macaca mulatta*). *Proceedings of the Zoological Society of London,* 1962, **138**, 1-21.

Ingram, J. C. Social interactions within marmoset family groups (*C. jacchus*). Paper presented at the *International Primatological Society* Meeting, Cambridge, England, August, 1976.

Itani, J. Paternal care in the wild Japanese monkey, *Macaca fuscata.* In Southwick, C. H. (Ed.) *Primate Social Behavior.* Princeton, N.J.: Van Nostrand, 1963, pp. 91-97.

Itani, J. Twenty years with Mount Takasaki monkeys. In Bermant, G. and Lindburg, P. (Eds.) *Primate Utilization and Conservation.* New York: Wiley, 1975, pp. 101-125.

Jorde, L. B. and Spuhler, J. N. A statistical analysis of selected aspects of primate demography, ecology and social behavior. *Journal of Anthropological Research,* 1974, **30**, 199-224.

Joslyn, W. D. Androgen-induced social dominance in infant female rhesus monkeys. *Journal of Child Psychology and Psychiatry,* 1973, **14**, 137-145.

Kaplan, J. R. Interference in fights and the control of aggression in a group of free-ranging rhesus monkeys. *American Journal of Physical Anthropology,* 1975, **44**, (1), 189.

Kleiman, D. G. The development of pair preferences in the lion tamarin *Leontopithecus rosalia:* Male competition or female choice. Paper presented at the *Animal Behavior Society* Meeting, University Park, Pennsylvania, June, 1977 (a)

Kleiman, D. G. Monogamy in mammals. *The Quarterly Review of Biology,* 1977, **52**, 39-69. (b)

Klein, L. and Klein, D. Aspects of social behaviour in a colony of spider monkeys. *International Zoo Yearbook,* 1971, **11**, 175-181.

Lockard, J. S., Heestand, J. E. and Begert, S. P. Play behavior in slow loris (*Nycticebus coucang*). Paper presented at the *American Society of Primatologists* Meeting, Seattle, Washington, April, 1977.

Maccoby, E. E. and Jacklin, C. N. *The Psychology of Sex Differences,* Stanford, California: Stanford University Press, 1974.

Maple, T., Brandt, E. M. and Mitchell, G. Separation of preadolescents from infants (*Macaca mulatta*). *Primates,* 1975, **16**, 141-153.

Marler, P. On animal aggression: The roles of strangeness and familiarity. *American Psychologist,* 1976, **31**, 239-246.

Mason, W. A., Green, P. C. and Posepanko, C. J. Sex differences in affective social responses of rhesus monkeys. *Behaviour,* 1960, **16**, 74-83.

Mazur, A. Effects of testosterone on status in primate groups. *Folia Primatologica,* 1976, **26**, 214-226.

Michael, R. P. Hormonal factors and aggressive behaviour in the rhesus monkey. Paper presented at the *International Society of Psychoneuroendocrinology,* Brooklyn, 1970.

Michael, R. P. and Zumpe, D. Aggression and gonadal hormones in captive rhesus monkeys (*M. mulatta*). *Animal Behaviour,* 1970, **18**, 1-10.

Nagel, U. and Kummer, H. Variation in cercopithecoid aggressive behavior. In Holloway, R. L. (Ed.) *Primate Aggression, Territoriality and Xenophobia.* New York: Academic Press, 1974, pp. 159-184.

Orbach, J. and K!ing, A. The stump tailed macaque: A docile Asiatic monkey. *Animal Behaviour,* 1964, **12**, 343-347.

Oswald, M. and Erwin, J. Control of intragroup aggression by male pigtail monkeys. *Nature,* 1976, **262**, 686-688.

Perachio, A. A. Hypothalamic regulation of behavioural and hormonal aspects of aggressive and sexual performance. Paper presented at the *International Primatological Society* Meeting, Cambridge, England, August, 1976.

Rhine, R. J. and Kronenwetter, C. Interaction patterns of two newly formed groups of stumptail macaques (*Macaca arctoides*). *Primates,* 1972, **13** (1), 19-33.

Rondinelli, R. and Klein, L. L. An analysis of adult social spacing tendencies and related social interactions in a colony of spider monkeys (*Ateles geoffroyi*) at the San Francisco Zoo. *Folia Primatologica,* 1976, in press.

Rose, R. M., Gordon, T. P. and Bernstein, I. S. Plasma testosterone levels in the male rhesus: Influences of sexual and social stimuli. *Science,* 1972, **178**, 643-645.

Rowell, T. E. Contrasting different adult male roles in different species of nonhuman primates. *Archives of Sexual Behavior,* 1974, **3**, 143-149.

Sackett, D. P., Oswald, M. and Erwin, J. Aggression among captive female pigtail monkeys in all-female and harem groups. *Journal of Biological Psychology,* 1975, **17** (2), 17-20.

Sassenrath, E. N., Rowell, T. E. and Hendricks, A. G. Perimenstrual aggression in groups of female rhesus monkeys. *Journal of Reproduction and Fertility,* 1973, **34**, 509-511.

Schwaier, A. The breeding stock of *Tupaias* at the Batelle Institute. *Laboratory Animal Handbook,* 1975, **6**, 141-149.

Simonds, P. E. Aggression and social bonds in an urban troop of bonnet macaques *(Macaca radiata).* Paper presented at the *American Association of Physical Anthropologists* Meeting, Seattle, Washington, April, 1977.

Sorenson, M. W. A review of aggressive behavior in the tree shrews. In Holloway, R. L. (Ed.) *Primate Aggression, Territoriality and Xenophobia: A Comparative Perspective.* New York: Academic Press, 1974, pp. 13-30.

Southwick, C. H. An experimental study of intragroup agonistic behavior in rhesus monkeys *(Macaca mulatta). Behaviour,* 1967, **28**, 182-209.

Southwick, C. H. and Siddiqi, M. F. Experimental studies on social intolerance in wild rhesus groups. *American Zoologist,* 1972, **12** (4), 651-652.

Strayer, F. F., Taylor, M. and Yanciw, P. Group composition effects on social behaviour of captive squirrel monkeys *(Saimiri sciureus). Primates,* 1975, **16** (3), 253-260.

Sugawara, K. Analysis of the social relations among adolescent males of Japanese monkeys *(Macaca fuscata fuscata)* at Koshima Islet. *Journal of the Anthropological Society of Nippon,* 1976, **83**, 330-354.

Thompson, N. S. Primate infanticide: A note and a request for information. *Laboratory Primate Newsletter,* 1967a, **6** (3), 18-19.

Thompson, N. S. Some variables affecting the behaviour of irus macaques in dyadic encounters. *Animal Behaviour,* 1967b, **15**, 307-311.

Vandenbergh, J. G. and Post, W. Endocrine coordination in rhesus monkeys: Female responses to the male. *Physiology and Behavior,* 1976, **17**, 879-884.

van Hooff, J. A. R. A. M. and de Waal, F. Aspects of an ethological analysis of polyadic agonistic interactions in a captive group of *Macaca fascicularis.* In Kondo, S., Kawai, M. and Ehara, A. (Eds.) *Contemporary Primatology.* Basel: Karger, 1975, pp. 269-274.

Whiten, A. and Rumsey, T. J. Agonistic buffering in the wild Barbary macaque. *Primates,* 1973, **74** (4), 421-425.

Wilson, A. P. and Boelkins, R. C. Evidence for seasonal variation in aggressive behaviour by *Macaca mulatta. Animal Behaviour,* 1970, **18**, 719–724.

Wilson, A. P. and Vessey, S. H. Behavior of free-ranging castrated rhesus monkeys. *Folia Primatologica,* 1968, **9**, 1–14.

Plate I

1

1. **An adult female common chimpanzee** (*Pan troglodytes*). (*Photo by Kingdoms Three Animal Park, Atlanta, Georgia*) Chapter 1

Plate II

2. Facial mask on an adult male orang-utan (*Pongo pygmaeus*). (*Photo by T. Maple*)
Chapter 3

Plate III

3. **Gorilla infants** (*Gorilla gorilla*) **at play.** (*Photo by T. Maple*) Chapter 5

Plate IV

4. An adult female spider monkey (*Ateles geoffroyi*). **Note the large clitoris.** (*Photo by T. Maple*) Chapter 7

Plate V

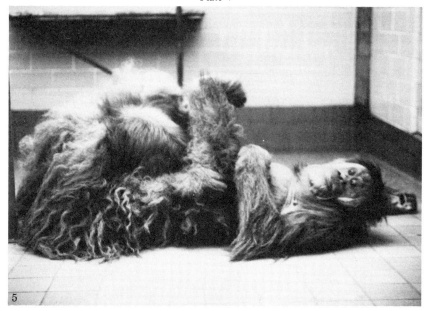

5. Oral-genital sex in the orang-utan (*Pongo pygmaeus*). (*Photo by T. Maple*) Chapter 10
6. Ventro-ventral copulation in the gorilla (*Gorilla gorilla*). (*Photo by J. P. Hess and W. Angst*) Chapter 10

Plate VI

7. A female spider monkey (*Ateles geoffroyi*) carries a juvenile cebus monkey (*Cebus albifrons*). (*Photo by T. Maple, courtesy of the Sacramento Zoo*) Chapter 12

Plate VII

8. An adult male orang-utan with infant (*Pongo pygmaeus*). (*Photo by T. Maple, courtesy of the Atlanta Zoo*) Chapter 13
9. Mother-infant play in lowland gorillas (*Gorilla gorilla gorilla*). (*Photo by T. Maple*) Chapter 30

Plate VIII

10. A gorilla mother and her infant relax at mid-day. (*Photo by T. Maple*) Chapter 31
11. An orang-utan mother *(Pongo pygmaeus)* with her infant. *(Photo by T. Maple)* Chapter 34

25.

Aggression: Other Old World Monkeys, Apes, and Comments on Homo Sapiens

We have seen in Chapter 24 that of the three major groups of primates discussed so far—prosimians, New World monkeys, and macaques—only in macaques were there enough consistent data on the topic to show that males were more aggressive than females. In the present chapter we will continue our review of the primate order to see if the sex difference seen in macaques appears in other non-human primate groups.

A. OTHER OLD WORLD MONKEYS

1. Baboons (*Papio spp.; Theropithecus spp.;* and *Mandrillus spp.*)

Just as ecological variables affect aggression in other species of primates, they also affect aggression in Ugandan olive baboons. Two baboon troops (*P. anubis*) from two different ecological conditions differ in aggressive behavior patterns (Paterson, 1973). As we know, "agonistic buffering" occurs in olive and other baboon males (Gilmore, 1977). Owens (1975) has studied the development of aggressive behavior patterns in olive baboon (*P. anubis*) play. The mean length of contact in play is greater than it is in aggression. Chasing, on the other hand, tends to last longer in adult male aggression than it does in play. Roles are frequently reversed in young male–male play and in adult male–male aggression. In female play and aggression, however, reversal of roles is uncommon. Owens (1975) believes that play may function as practice for aggression, allowing for more prolonged contact to facilitate the learning process. According to Owens (1975), the selection pressures operating upon adult females are likely to be somewhat different from those operating on adult males. Adult females do not come into estrus at the same time and there is less necessity to compete for a mate than there is for males. There is also selection against severe fighting to protect the young. Also, in baboons, young males tend to imitate threat and attack patterns more than do young females (Hamburg, 1971).

In yellow baboons (*P. cynocephalus*), estrus in the adult female increases her aggressive encounters; however, it actually decreases the number of aggressive bouts seen in males. That is, the number of agonistic bouts between males decreases as the number of females in estrus increases. But, when male encounters do occur they are far more likely to lead to wounding when there are females in estrus than when there are no females in estrus (Hausfater, 1975). According to Rhine (1975), in yellow baboons as in olive baboons, nature and experience combine to yield more aggressive behavior from males than from females.

Chacma baboon males (*P. ursinus*) predominate in intertroop conflicts, while adult females rarely participate (Buskirk, Hamilton, and Buskirk, 1975). Males are more hostile in their interactions than are females. The most aggressive male is sometimes the most dominant in individual encounters (Saayman, 1971b). According to Saayman (1971a):

> . . . males were responsible for proportionately more attacks upon other classes of baboons than expected . . . whereas aggression by adult males was largely directed towards the regulation of the social behaviour of the troop, aggression by subadult males often appeared to be unprovoked. (P. 77)

Much of the aggression of subadult chacma baboon males is directed toward females in the late luteal phase of their cycles (Saayman, 1971a)

In baboons living in one-male groups, "agonistic buffering" is used frequently by subordinate males (Gilmore, 1977). In the gelada baboon (*Theropithecus gelada*) one-male group, a male's aggressive behavior occurs only in relation to the maintenance of his harem (Crook, 1966). A male does not tolerate the positioning of another male between himself and any of his females (Bramblett, 1970); but, according to Bernstein (1975), female geladas are more active in agonistic patterns than are males. In hamadryas one-male groups (*P. hamadryas*) the greatest rival of the dominant male is usually one of his own followers with whom the risk of fighting is low (Kummer et al., 1976).

In summary, in most baboon troops, males are more aggressive than are females. Even in one-male groups where females may be involved in more intragroup agonistic encounters than males, the males are more capable of severe wounding.

2. Langurs (*Presbytis spp.*)

Among hanuman langurs (*P. entellus*), multimale bisexual groups and all-male groups are less relaxed and show more aggression among the individuals than do one-male groups (Tiwari and Mukherjee, 1976). As we know, however, when there is a change of leaders in a one-male troop, the new male leader may kill all of the youngest infants (Sugiyama, 1966). There is also intermale agonistic behavior at this time. Ecological conditions like population density and home range size probably contribute extensively to levels of aggression (Boggess, 1977; Dolhinow, 1977).

In the closely related Nilgiri langur (*P. johnii*), most aggressive encounters are between like-sexed individuals and usually involve adult males (Poirier, 1970). According to Poirier (1973):

> Primate social roles are learned social roles. Rather than inheriting their behavioral tendencies, males and females learn them . . . males learn to be assertive, aggressive individuals. (P. 29)

The lutong of Kuala Selangor (*P. cristatus*) shows a similar sex difference. Even when they are juveniles, males show more vigorous play fighting than do females. As we know, only control males show hostility toward other troops (Bernstein, 1968). In short, langur males are more aggressive than langur females.

3. *Cercopithecus* Species

In Struhsaker's (1967) field study of vervets, "juvenile males had more wounds than expected by chance" (P. 60), especially as they neared sexual maturity. Intergroup encounters almost always involved juvenile males. In a captive juvenile group of St. Kitts Island (in the Carribean) vervets (*C. aethiops sabaeus*), however, Raleigh (1977) found no significant sex differences in aggressive behavior.

In direct encounters between adult males and females, males usually win. Lactating female vervets are more aggressive than are other adult female vervets and they are often successful in encounters against adult males. At feeding sites there are more male–male aggressive encounters than there are male–female or female–female encounters and males also have more wounds and scars. Fighting occurs very frequently in vervets between subadults and juveniles as an extension of rough play (Baskin and Krige, 1973).

In *C. sabaeus,* according to Dunbar (1974), agonistic interactions are more frequent in males than in females. One species of the *Cercopithecus* genus (*C. ascanius schmidti*) lives in one-male groups and displays infanticide at the time of group takeover (Struhsaker, 1976). In *C. campbelli loweii,* agonistic interactions are rare and are usually displayed by adult males toward maturing males during the breeding season. Adult females are rarely aggressive but if they do show aggression it is usually toward other females (Hunkeler, Bourliére, and Bertrand, 1972).

4. *Miopithecus* **Talapoin**

In the neotenic-looking talapoin monkey, according to Scruton and Herbert (1970):

> The males were most aggressive towards other females of the group when one female was at midcycle; there were no consistent changes in aggression between the male and the female herself. (P. 49)

When strangers are introduced into a captive talapoin group, residents of the same sex as the introduced stranger attack the newcomer (Scruton and Herbert, 1972). However, females are more aggressive as adults than are males. In fact, according to Wolfheim and Rowell (1972):

> Talapoin courtship seems to involve a high level of fear on the part of the male. Female talapoins are aggressive. (P. 254)

Females rank above males. Estradiol, however, increases female aggression only a little. The effects of gonadal hormones on aggression are dependent upon the preceding social structure (Dixson and Herbert, 1974). A newly impregnated female, however, does show increased aggression (Rowell and Dixson, 1975). Some females in captive groups arc so aggressive during the mating period they *kill* males (Wolfheim, 1977a). It is interesting, however, that although adult female talapoins appear to be more aggressive than adult males, there is no evidence of a similar sex difference among immature talapoins. In fact, immature female talapoins are *less* aggressive than immature male talapoins (Wolfheim, 1977b).

5. *Erythrocebus* **and** *Cercocebus*

An adult male patas monkey (*E. patas*) in the one-male group typical of the species remains aloof much of the time but does drive off extragroup males (Struhsaker and Gartlan, 1970). The onset of puberty of the young males in his harem is followed by attacks on the young males by the adult male (Hall and Mayer, 1967). Between-group aggression, however, often involves adult females (Gartlan, 1976). Aggression is extremely rare, on the other hand, in all-male groups of patas monkeys. Solitary males and those in all-male groups fight males in heterosexual groups over receptive females during the rainy breeding season (Gartlan, 1975).

Chalmers (1968) and Chalmers and Rowell (1971) have reported on free-ranging and captive mangabeys (*Cercocebus albigena*). Adult males in the wild associate with each other less often than what is expected by chance, but when they do associate with each other they show more aggression than do other animals.   In fact, most of the aggression seen is between adult males.   In a captive group where there was apparently only one male, the highest aggression rate was seen in the dominant female.

Bernstein (1971) studied another species of mangabey (*C. atys*) in captivity. In one of his series of studies involving the introduction of strangers to captive groups, Bernstein found that most of the aggression occurred between animals of the same age/sex classes.   Adult male fights were brief but explosive, whereas female fights were prolonged and involved biting and chewing.   Injuries produced by females consisted of abrasions on the skin; those injuries produced by males were punctures or incisions.   Female aggression, although longer, was less intense.   In response to introduced infants, only the juvenile males showed aggression.

B. APES

As we know, gibbon (*Hylobates spp.*) and siamang (*Symphalangus spp.*) groups always contain one adult male and one adult female.   The female as well as the male defends the territory but she does not chase others away as often as does the male (Brockelman, Ross, and Pantuwatana, 1974).   The male and female gibbon are about equally aggressive (Schaller, 1963).

In a captive chimpanzee (*Pan troglodytes*) colony, males attacked females more often than they attacked other males.   Females were just as aggressive as the males.   The females attacked both sexes equally often; in fact, males were attacked more often by females than by males, whereas females were attacked equally often by males and females (Kollar, Beckwith, and Edgerton, 1968).

In early development there is apparently a high level of *potential* aggression in captivity by mothers toward infants not their own.   This is seen in such behaviors as the swagger and shoving away (Savage and Malick, 1977).   Young chimpanzee males, however, tend to imitate threat and attack behaviors more often than do young females (Hamburg, 1971).   In adulthood, the bipedal swagger and sway walk is seen much more frequently in males than in females. The highly aggressive charging display is also more frequent in male chimpanzees than in females; and, in the wild, the adult male chimpanzee initiates many more attacks than does the female.   The female, on the other hand, is more often attacked.   Male chimpanzees sometimes even attack, kill, and eat infants of their own species (Pitcairn, 1974); but, then, so do females.

Reportedly it is impossible to keep two adult male orangutans (*Pongo pygmaeus*) together in a zoo.   This is not true for females.   Orangutan mothers show virtually no aggression toward infants (Pitcairn, 1974). There are some indications from the field that there can be a great deal of male aggression toward the female during sexual interactions (Letson and Ellis, 1977). However, captive studies have indicated that this aggressive behavior is not always the case, although males do often bite and hit females during copulation (Maple and Zucker, 1977). In studies of orangutans in the Sabah, Malaysia rehabilitation project, an older female and a young male about half her size were observed mating. The female, but not the male, turned hostile when observed, disturbed, or provoked (de Silva, 1971).

Bluff charges are made more often by gorilla (*Gorilla gorilla*) males than by females. Chest beating with hooting is also seen more in males than in females (Pitcairn, 1974). Attacks on humans are made primarily by male gorillas (Sabater-Pi, 1966).

C. COMMENTS ON HUMAN AGGRESSION

For some reason, for *non*human primates, little controversy seems to exist regarding whether sex differences in aggression are real. The sex difference of males being more aggressive than females is accepted despite the existence of marked species and individual variability. For *Homo sapiens,* however, social learning theorists frequently see no role for biological differences between the sexes in aggression (Green, 1976).

While heredity probably does play a strong part in species and sex differences in aggression, according to Scott (1970):

> It is equally probable, however, that because of man's unique genetic composition no direct analogies from any other species to man are justified. (P. 568)

As we have seen, in terms of physical sexual dimorphism in size, man is intermediate between the baboons and the gibbons. Those species showing strong sexual dimorphism in size, like the baboon, also show more aggression in males than in females. Those species low in size dimorphism do not show this sex difference. On the basis of these kinds of correlations alone we might expect a sex difference in aggression in people of intermediate degree (Scott, 1974).

On the basis of other physical apparatus needed for aggression, human males seem to have evolved roughly twice as much equipment as have females. In addition, to the extent that testosterone has been implicated in aggression, males also have an edge over females in that regard (Tiger, 1975).

On the other hand, during human evolution, but not during the evolution of any nonhuman primate, males developed small canine teeth. Zihlman (1978), believes that human males become *less* aggressive than male apes, were closely integrated into kin groups, and became "fathers" because they were less aggressive and more nurturing than were their ape cousins.

In terms of direct comparisons of apes and children with regard to aggressive behavior (Braggio et al., 1976), males of all three species show higher frequencies of initiated aggressive contacts than do females. In addition, groups of girls whose mothers received androgen during pregnancy showed heightened aggressiveness as children (Hamburg and Brodie, 1973).

In preschool children, pushing occurs more in boys than in girls, and boys give beatings and chase others more often than do girls (McGrew and McGrew, 1970). Quasi-agonistic yelling also occurs more frequently in nursery school boys than in girls (McGrew, 1972). Whether observed in a neutral environment or in the home of the child, preschool boys in dyads show a generally higher level of aggression than do girls (Jeffers, 1977). Even in communally reared children, boys are more aggressive than girls, although there is overlap and some girls are more aggressive than some boys. Because there are no significant differences between male-male, male-female, female-female, and female-male in communally reared children, Missakian (1976) believes that peer-peer rearing may reduce the sex difference somewhat.

In nine to twelve year olds, females submit more readily when competing against males than when competing against other females (Cronin, Callaghan, and Weisfeld, 1977).

According to Poirier (1974):

> During the transition from adolescence to adulthood we observe the maximum expression of aggression. Among humans, rates of aggression are higher for males than for females. Crime rates are usually two to five times higher for males than for females. Males are more likely to exhibit delinquency in aggressive behavior while females are more often involved in sexual acts. This probably reflects hormonal differences, but also reflects the social tradition of longer family control over females. (P. 149)

As reported by Baron (1977) and Baron and Bell (1977), there is a link between sexual arousal and aggression, and human *females*, but *not* males, respond with increased aggression to arousing erotic stimuli. Highly erotic stimuli inhibits aggression in males. In addition, when a male has been aroused by an attack, he experiences a more rapid return to a normal physiological response level as measured by vascular recovery (change in peripheral blood vessels) than does a female. He evidently accomplishes this, in part, by making a counterattack rather than by making a "peaceful" response (Geen, 1972). There are, those who feel that much behavioral pathology among girls results because of the parents' tendency to discourage overt aggression in them (Lester, 1972).

In terms of self-descriptions, college-age males rate themselves as being significantly more daring than do females (Longstreth, 1970); however, on the College Self-Expression Scale, females are more assertive than males. Each sex is significantly more assertive toward members of the same sex than toward members of the opposite sex (Stebbins et al., 1977). In humor, males appreciate hostile wit more than do females (O'Connell, 1958).

Abnormal behavior is frequently correlated with aggressive behavior. A sex-linked familial neurological disease consisting of cerebral palsy, mental retardation, and compulsive aggressive behavior, for example, is said to be limited to males and to have a genetic locus on the x chromosome (Seegmiller, Rosenbloom, and Kelly, 1967).

Returning to Marler's (1976) review of the causes of aggression, let us recall that he listed the following as being especially liable to provoke aggression:

1. Proximity of an animal.
2. Of one's own kind
3. A stranger
4. Behaving aggressively
5. Inflicting pain
6. Or otherwise creating frustration.

(P. 242)

But what is our conclusion regarding sex differences in human aggression? Maccoby and Jacklin (1974) conclude that there is indeed a well-established sex difference in human aggression. They conclude that males are more aggressive than females and that:

> The sex difference in aggression has been observed in all cultures in which the relevant behavior has been observed. Boys are more aggressive both physically and verbally . . . . The sex difference is found as early as social play begins—at 2 or 2½. Although the aggressiveness of both sexes declines with age, boys and men remain more aggressive through the college years. Little information is available for older adults. The primary victims of male aggression are other males—from early ages, girls are chosen less often as victims. (P. 352)

For an excellent overview of aggression, violence, and even warfare in *Homo sapiens,* the interested reader should consult Maple and Matheson (1973).

REFERENCES

Baron, R. A. Heightened sexual arousal and physical aggression: An extension to females. Unpublished manuscript, 1977.

Baron, R. A. and Bell, P. A. Sexual arousal and aggression by males. Effects of type of erotic stimuli and prior provocation. *Journal of Personality and Social Psychology*, 1977, **35**, 79-87.

Baskin, D. R. and Krige, P. D. Some preliminary observations on the behavior of an urban troop of vervet monkeys (*Cercopithecus aethiops*) during the birth season. *Journal of Behavioural Science*, 1973, **1** (5), 287-296.

Bernstein, I. S. The lutong of Kuala Selangor. *Behaviour*, 1968, **32**, 1-16.

Bernstein, I. S. The influence of introductory techniques on the formation of captive mangabey groups. *Primates*, 1971, **12**, 33-44.

Bernstein, I. S. Activity patterns in a gelada monkey group. *Folia Primatologica*, 1975, **23**, 50-71.

Boggess, J. Social change in a troop of langurs in Nepal. Paper presented at the *American Association of Physical Anthropologists* Meeting, Seattle, Washington, April, 1977.

Braggio, J. T., Nadler, R. D., Lance, J. and Myseko, D. Sex differences in apes and children. Paper presented at the *International Primatological Society* Meeting, Cambridge, England, August, 1976.

Bramblett, C. A. Coalitions among Gelada baboons. *Primates*, 1970, **11**, 327-333.

Brockelman, W. Y., Ross. B. A. and Pantuwatana, S. Social interactions of adult gibbons (*Hylobates lar*) in an experimental colony. In Rumbaugh, D. M. (Ed.) *Gibbon and Siamang*. Basel: Karger, 1974, pp. 137-156.

Buskirk, R. E., Hamilton, W. J., III, and Buskirk, W. H. Defense of space by baboon troops. *American Journal of Physical Anthropology*, 1975, **42**, 293.

Chalmers, N. R. The social behaviour of free living mangabeys in Uganda. *Folia Primatologica*, 1968, **8**, 263-281.

Chalmers, N. R. and Rowell, T. E. Behaviour and female reproductive cycles in a captive group of mangabeys. *Folia Primatologica*, 1971, **14**, 1-14.

Cronin, C. L., Callaghan, J. W. and Weisfeld, G. E. Sex differences in competitive behavior in children. Paper presented at the *Animal Behavior Society* Meeting, University Park, Pennsylvania, June, 1977.

Crook, J. H. Gelada baboon herd structure and movement. *Symposium of Zoological Society of London*, 1966, **18**, 237-258.

de Silva, G. S. Notes on the orangutan rehabilitation project in Sabah. *Malayan Nature Journal*, 1971, **24**, 50-77.

Dixson, A. F. and Herbert, J. Gonadal hormones and aggressive behaviour in captive groups of talapoin monkeys (*Miopithecus talapoin*). *Journal of Endocrinology*, 1974, **61**, 46.

Dolhinow, P. J. Normal monkeys? *American Scientist*, 1977, **65** (letter).

Dunbar, R. I. M. Observations on the ecology and social organization of the green monkey, *Cercopithecus sabaeus*, in Senegal. *Primates*, 1974, **15** (4), 341-350.

Gartlan, J. S. Adaptive aspects of social structure in *Erythrocebus patas*. *Proceedings of the Symposium of the Fifth Congress of the International Primatological Society*. Tokyo: Japan Science Press, 1975, pp. 161–171.

Gartlan, J. S. Ecology and behaviour of the patas monkey. Film presented at the *International Primatological Society* Meeting, Cambridge, England, August, 1976.

Geen, R. G.. *Aggression*. Morristown, N.J.: General Learning Press, 1972, pp. 1–23.

Gilmore, H. A. The evolution of "agonistic buffering" in baboons and macaques. Paper presented at the *American Association of Physical Anthropologists* Meeting, Seattle, Washington, April, 1977.

Green, R. The human primate: Development of sexuality and aggressivity. Paper presented at the *International Primatological Society* Meeting, Cambridge, England, August, 1976.

Hall, K. R. L. and Mayer, B. Social interactions in a group of captive patas monkeys (*Erythrocebus patas*). *Folia Primatologica*, 1967, **5**, 213–236.

Hamburg, D. A. Aggressive behavior of chimpanzees and baboons in natural habitats. *Journal of Psychiatric Research*, 1971, **8**, 385–398.

Hamburg, D. A. and Brodie, H. K. H. Psychological research on human aggressiveness. *Impact of Science on Society*, 1973, **23**, 181–193.

Hausfater, G. Estrous females: Their effects on the social organization of the baboon group. *Proceedings of the Symposium of the Fifth International Congress of Primatology*. Tokyo: Japan Science Press, 1975, pp. 117–127.

Hunkeler, C., Bourliére, F. and Bertrand, M. Le comportement social de la Mone de Lowe (*Cercopithecus campbelli lowei*). *Folia Primatologica*, 1972, **17**, 218–236.

Jeffers, V. Situation specificity of social dominance in preschool children. Paper presented at the *Animal Behavior Society* Meeting, University Park, Pennsylvania, June, 1977.

Kollar, E. J., Beckwith, W. C. and Edgerton, R. B. Sexual behavior of the ARL colony chimpanzees. *Journal of Nervous and Mental Disease*, 1968, **147**, 444–459.

Kummer, H., Abegglen, J. J., Abegglen, C., Falett, J. and Sigg, H. Intimacy and fear of aggression in baboon social relationships. Paper presented at the *International Primatological Society* Meeting, Cambridge, England, August, 1976.

Lester, D. Self-mutilating behavior. *Psychological Bulletin*, 1972, **78**, 119–128.

Letson, G. W. and Ellis, J. E. Analysis of sexual behavior in captive orangutans. Paper presented at the *Animal Behavior Society* Meeting, University Park, Pennsylvania, June, 1977.

Longstreth, L. E. Birth order and avoidance of dangerous activities. *Developmental Psychology*, 1970, **2**, 154.

Maccoby, E. E. and Jacklin, C. N. *The Psychology of Sex Differences.* Stanford, California: Stanford University Press, 1974.

Maple, T. and Matheson, D. W. (Eds.) *Aggression, Hostility, and Violence: Nature or Nurture.* New York: Holt, Rinehart and Winston, 1973.

Maple, T. and Zucker, E. L. Behavioral studies of captive Yerkes orangutans at the Atlanta Zoological Park. *Yerkes Newsletter,* 1977, **14**, (1), 24–26.

Marler, P. On animal aggression: The roles of strangeness and familiarity. *American Psychologist,* 1976, **31** (3), 239–246.

McGrew, W. C. Aspects of social development in nursery school children with emphasis on introduction to the group. In Blurton-Jones, N. (Ed.) *Ethological Studies of Child Behaviour.* Cambridge, England: Cambridge University Press, 1972, pp. 129–156.

McGrew, W. C. and McGrew, P. L. Group formation in preschool children. *Proceedings of the Third International Congress of Primatology,* 1970, **3**, 71–78.

Missakian, E. A. Gender differences in agonistic behavior and dominance relations of communally reared children. Paper presented at the *Animal Behavior Society* Meeting, Durham, North Carolina, 1976.

O'Connell, W. E. A study of the adaptive functions of wit and humor. Unpublished doctoral dissertation, University of Texas, 1958.

Owens, N. W. A comparison of aggressive play and aggression in free-living baboons, *Papio anubis. Animal Behaviour,* 1975, **23**, 757–765.

Paterson, J. D. Ecologically differentiated patterns of aggressive and sexual behavior in two troops of Ugandan baboons, *Papio anubis. American Journal of Physical Anthropology,* 1973, **38** (2), 641–647.

Pitcairn, T. K. Aggression in natural groups of pongids. In Holloway, R. L. (Ed.) *Primate Aggression, Territoriality and Xenophobia.* New York: Academic Press, 1974, pp. 241–272.

Poirier, F. E. Dominance structure of the Nilgiri langur (*Presbytis johnii*) of South India. *Folia Primatologica,* 1970, **12**, 161–186.

Poirier, F. E. Socialization and learning among nonhuman primates. In Kimball, S. T. and Burnett, J. H. (Eds.) *Learning and Culture.* Seattle: University of Washington Press, 1973, pp. 3–41.

Poirier, F. E. Colobinae aggression: A review. In Holloway, R. L. (Ed.) *Primate Aggression, Territoriality and Xenophobia: A Comparative Perspective.* New York: Academic Press, 1974, pp. 123–157.

Raleigh, M. J. Sex differences in the social behavior of juvenile vervet monkeys. Paper presented at the *American Society of Primatologists* Meeting, Seattle, Washington, April, 1977.

Rhine, R. J. The order of movement of yellow baboons (*Papio cynocephalus*). *Folia Primatologica,* 1975, **23**, 72–104.

Rowell, T. E. and Dixson, A. F. Changes in social organization during the breeding season of wild talapoin monkeys. *Journal of Reproduction and Fertility,* 1975, **43**, 419–434.

Saayman, G. S. Aggressive behavior in free-ranging chacma baboons (*Papio ursinus*). *Journal of Behavioural Science,* 1971a, **1** (3), 77–83.

Saayman, G. S. Behaviour of the adult males in a troop of free-ranging chacma baboons. *Folia Primatologica,* 1971b, **15**, 36–57.

Sabater-Pi, J. Gorilla attacks against humans in Rio Muni, West Africa. *Journal of Mammalogy,* 1966, **47**, 123–124.

Savage E. S. and Malick, C. Play and socio-sexual behaviour in a captive chimpanzee (*Pan troglodytes*) group. *Behaviour,* 1977, **60**, 179–194.

Schaller, G. B. Behavioral comparisons of the apes. In DeVore, I. (Ed.) *Primate Behavior.* New York: Holt, Rinehart and Winston, 1963, pp. 474–484.

Scott, J. P. Biology and human aggression. *American Journal of Orthopsychiatry,* 1970, **40** (4), 568–576.

Scott, J. P. Agonistic behavior of primates: A comparative perspective. In Holloway, R. L. (Ed.) *Primate Aggression, Territoriality and Xenophobia: A Comparative Perspective.* New York: Academic Press, 1974, pp. 417–434.

Scruton, D. M. and Herbert, J. The menstrual cycle and its effect on behaviour in the talapoin monkey (*Miopithecus talapoin*). *Journal of Zoology* (London), 1970, **162**, 419–436.

Scruton, D. M.and Herbert, J. The reaction of groups of captive talapoin monkeys to the introduction of male and female strangers of the same species. *Animal Behaviour,* 1972, **20**, 463–473.

Seegmiller, J. E., Rosenbloom, F. M. and Kelly, W. N. Enzyme defect associated with a sex-linked human neurological disorder and excessive purine synthesis. *Science,* 1967, **155**, 1682–1683.

Stebbins, C. A., Kelly, B. R., Tolor, A. and Power, M. E. Sex differences in assertiveness in college students. *The Journal of Psychology,* 1977, **95**, 309–315.

Struhsaker, T. T. *Behavior of Vervet Monkeys* (*Cercopithecus aethiops*) Berkeley: University of California Press, 1967.

Struhsaker, T. T. and Gartlan, J. S. Observations on the behaviour and ecology of the patas monkey (*Erythrocebus patas*) in the Waza Reserve, Cameroon. *Journal of Zoology* (London), 1970, **161**, 49–63.

Struhsaker, T. T. Infanticide in the redtail monkey (*Cercopithecus ascanius schmidti*) of Uganda. Paper presented at the *International Primatological Society* Meeting, Cambridge, England, August, 1976.

Sugiyama, Y. An artificial social change in a Hanuman langur troop (*Presbytis entellus*). *Primates,* 1966, **7**, 41–72.

Tiger, L. Somatic factors and social behaviour. In Fox, R. (Ed.) *Biosocial Anthropology.* New York: Wiley, 1975, pp. 115–132.

Tiwari, K. K. and Mukherjee, R. P. Social structure of Hanuman langur populations around Ramtek, Nagpur District, India. Paper presented at the *International Primatological Society* Meeting, Cambridge, England, August, 1976.

Wolfheim, J. H. A quantitative analysis of the organization of a group of captive talapoin monkeys (*Miopithecus talapoin*). *Folia Primatologica,* 1977a, **27**, 1-27.

Wolfheim, J. H. Sex differences in behavior in a group of captive juvenile talapoin monkeys (*Miopithecus talapoin*). *Behaviour,* 1977b, **63**, 110-128.

Wolfheim, J. H. and Rowell, T. E. Communication among captive talapoin monkeys (*Miopithecus talapoin*). *Folia Primatologica,* 1972, **18**, 244-255.

Zihlman, A. Z. Women in evolution, Part II: Subsistence and social organization among early hominids. *Signs: Journal of Women in Culture and Society,* 1978, **4** (1), 4-20.

26.

Sex Differences in Response to Early Social Deprivation and Separation

Up to this point in our survey of sex differences in the primate order, we have dealt primarily with the normal range of behaviors. In the next two chapters we will deal with the effects of social deprivation and stress. Social deprivation is a catchall term which we will use to refer to such phenomena as rearing in total or partial social isolation, separation from mother or from other attached objects, and the effects of abnormal maternal behavior on male and female infants. Since almost all of the research on social deprivation has been done on the rhesus monkey (*Macaca mulatta*), we will begin our review of sex differences with that species.

A. SOCIAL DEPRIVATION IN RHESUS MACAQUES

Rearing a rhesus monkey infant from birth to six or twelve months of age in social isolation produces marked behavioral abnormalities in the animal. During the first year, the isolate develops self-clinging, repetitive stereotyped rocking, self-mouthing and digit sucking, and crouching. When given the opportunity to interact with others prior to puberty, the isolate-reared rhesus hides its face and crouches and rocks. It withdraws from others in fear (cf. Mitchell, 1970).

At puberty, the behavior of the isolate-reared rhesus changes. Continued pacing replaces the repetitive rocking, self-biting replaces the digit sucking, and abnormal social aggression gradually replaces social fear (Mitchell, 1970).

During adulthood, the isolate-reared rhesus becomes even more aggressive socially and more self-mutilative. In addition, it is unable to display normal sexual posturing (Mitchell, 1970).

With regard to sex differences in response to social deprivation, male rhesus exhibit more disturbance than do females. Self-mouthing and self-clasping are higher in young isolate males than in young isolate females. There is also more rocking in isolate-reared males than in isolate-reared females (Suomi, Harlow, and Kimball, 1971). Older male isolates show more abnormalities in sexual behavior. Female isolates can eventually learn appropriate sexual behavior whereas males cannot. Male isolates also do more self-biting. According to Cross and Harlow (1965), male isolates show higher levels of self-aggression than do female isolates. Gluck and Sackett (1974) and Chamove and Harlow (1970) reported the same finding. In social behavior, male isolates are more disturbed socially than are female isolates (Sackett, 1974).

In group formation studies done by Bernstein, Gordon, and Rose (1974), eighteen socially deprived rhesus monkeys were used as a core group. Initially, the isolates were immobilized. They displayed awkwardness, bizarre movements, stereotyped movements, and low social interaction rates. The isolates seemed to show no coherent group structure, and did not act as a unit against an intruder. When aggression occurred it persisted. When sixteen isolates were placed into a free-ranging environment with a wild troop, only two of six isolate males survived whereas seven of nine females survived (Sackett, 1974). The more enriched the early environment the easier it is for an animal to operate in a team or a coalition (Suomi, 1974).

In terms of sexual behavior, rhesus isolates display strange behavior indeed (See Maple, 1977, for descriptions of the range of unusual sexual behaviors in primates). According to Mason, Davenport, and Menzel (1968):

> The female appears to be less seriously handicapped than the male by social deprivation. (P. 25)

Even as early as two years of age, male isolate-reared rhesus monkeys in particular showed markedly deficient sexual behavior (Mason, 1960). In a group of isolates studied by Missakian (1969), "no socially deprived male, experienced or naive, executed an appropriately oriented mount." (P. 403.)

Female isolates vocalize more than male isolates (Mitchell et al., 1966), particularly when stimulated by a human observer (Cross and Harlow, 1965); they also fear grimace more but threaten less as adults than do isolate males (Mitchell, 1968).

As noted earlier, aggression increases with age in both isolate-reared and socially reared rhesus monkeys (Mitchell, 1975). There is a sex difference in aggression in both rearing groups with males showing more aggression than females. In the case of the isolates this means abnormal aggression. Thus, adult male isolates show more abnormal aggression than do adult female isolates (Mitchell, 1974). This is evident in visual threat behavior as well as in direct attacks (Mitchell and Redican, 1972; Brandt, Stevens, and Mitchell, 1971).

In longitudinal studies of the development of isolate-reared infants from birth to adulthood (cf. Baysinger, Brandt, and Mitchell, 1972), sex differences in behavioral abnormalities are studied as they develop. Self-clasping and rocking, which develop early, occur in males more than in females whereas fear grimaces occur in females more than in males. When brought out of their isolation environments, female isolates emit more distress calls than do males (Brandt, Baysinger, and Mitchell, 1972). When isolate-reared infants are paired with older normal preadolescent animals, male isolates elicit more aggression from the older socially reared animals than do female isolates (Brandt and Mitchell, 1973). Female control animals are better able to establish social contact and an emotional attachment to deprived animals than are males, but there is some evidence as well for a superiority in this regard for heterosexual control-isolate dyads over isosexual control-isolate dyads (Maple, Brandt, and Mitchell, 1975). Normal preadolescents paired with isolate infants of the opposite sex seemed to be more upset (as evidenced by increases in coo vocalizations at separation) when separated from their isolates than did normal preadolescents paired with isolate infants of the same sex (Maple, Brandt, and Mitchell, 1975). At three years of age, isolate-reared males still exhibit more self-clasping behavior than do isolate-reared females, but isolate-reared females exhibit more miscellaneous bizarre behaviors than do the males. Three-year-old isolation-reared males spend more time looking at themselves than do three-year-old isolation-reared females. The average total duration of looking at other animals is also particularly high in three-year-old male isolates (Erwin et al., 1974a).

In the experimental format designed to see whether or not adult male rhesus monkeys can raise infants without the aid of an adult female or any other animal, we have a deprivation experience involving the absence of a mother but with the presence of a conspecific—the adult male (cf. Redican and Mitchell, 1973a; Redican and Mitchell, 1973b). Rhesus infants with a male "parent" are played with more and the play is more intense from their male "parent" than they would obtain from a mother (unless peers are also provided). However, they also receive less ventral contact from a male "parent" than from a female parent (Mitchell, 1977).

If the "parents" are isolate-reared, the infant is exposed to still different kinds of deprivation. There is evidence that adult females in preference tests *prefer* infants to other animals, but not if they had *no* mother themselves when growing up (Sackett, 1970). Isolate-reared mothers ("motherless mothers") are brutal or indifferent to their first offspring (Arling and Harlow, 1967) but not brutal or indifferent to their second and third infants. Apparently, the female isolate mellows somewhat with age and/or experience with infants (Arling, Ruppenthal, and Mitchell, 1969). What is interesting about this abnormal maternal punishment is that the "motherless mothers" are more brutal to male infants than they are to female infants (Sackett, 1974). The outcome of this differential excessive punishment to males is that the young males themselves become hyperaggressive as they mature (Mitchell, Arling, and Møller, 1967).

If the "parent" is an isolate-reared male, there is also a danger of excessive punishment toward the infant, but not in isolate-reared males who are at least ten years old (Gomber and Mitchell, 1974). Infants reared by older isolate-reared males apparently develop much as do normal infants; however, they, like the adopted infants of normally reared males, are played with very frequently and very intensely (see Gomber, 1975). Since all of the infants in the isolate male "parent" study have been female infants, we have no information on infant sex differences for this situation. (The interested reader may also consult Mitchell, 1974; and Mitchell, Redican, and Gomber, 1974, for published accounts of these experiments.)

In an effort to discover the specific kinds of missing input related to various behavioral abnormalities appearing in deprived infants, Mason and Berkson (1975) raised rhesus infants with either mobile or stationary artificial or surrogate mothers. In both groups of animals with surrogate mothers, infant males were more disturbed than were infant females (as seen in more frequent distress calls and barking). At 1½ years of age the contrast between the two sexes in disturbance, as well as in social contact (males *more*), was greatest in the infants raised on stationary surrogates.

In tests utilizing looking behavior as a measure of preference, monkeys reared by real mothers discriminated more sharply between the sexes than did monkeys reared by surrogate mothers; however, those reared on moving surrogate mothers discriminated better than did those reared on immobile surrogates. The first two groups showed a visual preference for a strange adult male over a strange adult female; the group raised with an immobile surrogate mother showed no preference at all (Eastman and Mason, 1975).

It is interesting to note, while we are on the subject of surrogate mothering, that surrogate-reared males will sexually mount surrogate models as they mature, including thrusting to ejaculation. However, as we know from the earlier isolate studies, they will *not* mount real females. Similarly, surrogate-reared females will present only to *surrogate* models (Deutsch and Larsson, 1974).

There are many other kinds of early deprivation that have been examined in rhesus monkeys (see Mitchell, 1970; Mitchell, 1971, for summaries). One of these paradigms involves early single or repeated maternal separation. Agar and Mitchell (1973) have published a bibliography on the effects of deprivation and separation on nonhuman primates.

The most complete and well-designed study of mother–infant separation was done by Scollay (1970). She found a few sex differences in the infants who were separated from their mothers at eight, fourteen and twenty weeks of age. The younger infants and the *male* infants, in particular, seemed to be more sensitive to the separations than did the older and female infants. In addition, when reunited with their mothers after two days, the younger male infants tended to withdraw or run away from their mothers. The female infants tended not to do this. Females, however, emitted more distress vocalizations at separation than did males. Hinde and Spencer-Booth (1971) when separating infants from their mothers at six months of age also found that the mean distress index of male infants was higher than that of female infants; and, in some periods, male infants emitted more distress calls than did females.

Repeated mother–infant separations throughout the first year of life can have long-term effects on rhesus infants. Repeatedly separated infants, and particularly females, continue to emit very high levels of distress calls and fear screeches up to puberty. The effect of separation on the males is the same but it is not as great as it is in females (Møller, Harlow, and Mitchell, 1968).

Interactions with peers can also be extremely important to the behavioral development of a rhesus monkey. The development of the pattern of sexual mount displayed by rhesus males can be interfered with if there are high levels of aggression between peers (Goy, Wallen, and Goldfoot, 1974).

When two peers of the same sex build up a strong 1½ year exclusive emotional attachment for one another, and are then separated at *two years* of age, male-male dyads show more stereotyped movement than do females; however, females spend twice as much time manipulating themselves as do males (Erwin, Mobaldi, and Mitchell, 1971).

In three-year-old heterosexual pairs who are separated after they are allowed one, two, or three weeks to form social attachments, males yawn, cage shake, and self-explore more than do females while females coo, bark, present, and look at their partners and the environment more than do males. The animals having the shortest amount of time together were *most* upset by the separation. This was particularly obvious in the females, however. As the length of pairing increased, the females displayed less and less disturbance at separation; the males did not (Erwin, Brandt, and Mitchell, 1973).

In 4½-year-old like-sexed dyads reunited after being apart for more than two years, there was obvious recognition despite the two-year separation. In male-male reunions there was much less disturbance displayed than was evident in stranger pairings. Familiar females never aggressed one another, even though they had not been in physical contact for two years (unfamiliar females of this age *do* aggress one another). Some of the familiar males, however, did resort to aggression (Erwin et al., 1974b). Despite this tendency toward aggression, the males did recognize each other and, on one occasion there was mutual mounting with anal penetration (Erwin and Maple, 1976).

There is evidence that absence of peers, separation from peers, or rearing in isosexual (same-sexed) groups can have deleterious effects on sexual behavior (cf. Goy, 1976; Erwin and Mitchell, 1975); however, these deficiencies are not as great as are those produced by social isolation. In addition, it is not clear in these cases that males are more severely affected than females. However, in two adult male monkeys who were separated from a "nuclear family" environment:

> Both individually separated males (5 years old) exhibited subsequent sexual deficiencies even though they had been more than adequate prior to separation. (Suomi et al., 1975, P. 578)

Even in short-term adult heterosexual dyads, the male seems to be more disturbed when separated from his consort than is the female. In some males, isolatelike self-biting has been seen in this situation (Maple, Risse, and Mitchell, 1973; see also Maple, Erwin and Mitchell, 1974a; and Maple, Erwin, and Mitchell, 1974b).

Short-term separations from a familiar environment produce different effects in the two sexes. Removing three-year-old rhesus monkeys from their home cages and placing them into new individual cages produced a decrease in visual exploration of the environment and increased jumping and exploration in males, but produced the reverse effect on exploration in females (Mitchell and Gomber, 1976). Sex differences in disturbance due to separation can be affected, however, by social facilitation.

As noted by Willott and McDaniel (1974), the sights, sounds, and other cues surrounding brief or prolonged separations can affect primates. Two males in their study "were usually upset only when separated from the most disturbed female." (P. 326.). A summary of research on attachment and separation in rhesus dyads, including a review of sex differences, has been published by Mitchell (1976).

In summary, male rhesus monkeys appear to be more adversely affected by early deprivation and separation than do female rhesus monkeys. Even in adulthood it appears that social separation is more serious for the male. This finding is surprising in light of the female's greater sociability and the male's tendency to be peripheralized, to change troops, and even to become solitary.

B. SOCIAL DEPRIVATION IN OTHER MONKEYS

Very few other monkeys have been studied at all with regard to the effects which early social deprivation may have on them. Testa and Mack (1977), however, have studied the effects of early social isolation on the crab-eating macaques' (*Macaca fascicularis*) sexual behavior. They found that male isolates showed an abnormal form of mounting and females displayed incomplete sexual presents. In pigtail (*M. nemestrina*) macaques raised in captivity male infants showed more thumb sucking than did females (Jensen, 1966). Jensen (1969) claims that behavioral sex differentiation is severely disturbed by early deprivation and that males are more vulnerable to deprivation than are females. Pigtail males need a richer environment in development than do pigtail females (Jensen, Bobbitt, and Gordon, 1966). Also, in the stumptail macaque (*M. arctoides*), deprivation does not seem to affect males and females equally. In fact, Riesen, Perkins, and Struble (1977) found that isolate males and females tended to be further apart in behavior than did control males and females. Males (apparently) were the most severely affected. (See Goosen and Ribbens' (1973) work where they report significant amounts of female self-biting.)

According to Leonard Rosenblum (1974), who has studied squirrel monkeys (*Saimiri sciureus*), bonnets (*M. radiata*), and pigtail macaques extensively, deprivation devastates male exploration more than it does female exploration. He believes that it is during the period *after* the eleventh week of life that the major effects of different rearing conditions and the sex of the infant can be discerned in the squirrel monkey.

In summary, in the few studies that have been done on deprivation in monkeys other than rhesus macaques, the male seems to be more seriously affected by early deprivation than does the female.

C. SOCIAL DEPRIVATION IN THE APES

In a summary of the effects of early deprivation on the sexual behavior of chimpanzees, Mason, Davenport, and Menzel, (1968) concluded that the sexual roles were not as sharply differentiated for chimpanzees as they were for rhesus monkeys, consequently:

> . . . early social deprivation may have more severe and lasting consequences for masculine sexual development in the monkey than in the chimpanzee. (P. 26)

In the same article, Mason, Davenport, and Menzel (1968) presented some evidence that, for the chimpanzee, *puberty* may be a "critical period for the acquisition or refinement of sex skills." (P. 26.) However, even for chimpanzees, who are more capable of showing recovery from isolation than are rhesus monkeys, males seem to be more adversely affected by restricted rearing than are females (Rogers and Davenport, 1969). Among chimpanzees in captivity:

> The female's behaviour was closer to that of their wild-living counterparts. (Tutin and McGrew, 1973, P. 255)

In the captive gorilla (*Gorilla gorilla*), as in the macaque, there is some evidence for improved mothering with the birth of the second infant (Nadler, 1975b) as a consequence of either learning or age. In addition, there is some evidence that early rearing with peers can compensate somewhat for the lack of a mother (Nadler, 1975a). A mother gorilla who is hand-raised may show restlessness and carelessness with her infant (Faust, 1977, see also Scollay et al., 1975).

In orangutans, deprived mothers show inappropriate maternal responses. An infant orangutan separated from its mother is unable to learn appropriate sexual behavior and, in turn, becomes a poor parent itself (Zucker et al., 1977). Whether such early deprivation conditions are more serious for male than for female gorillas and orangutans is not yet known.

D. COMMENTS ON PEOPLE

Nancy Bayley (1968) has published a well-known study on the behavioral correlates of human mental growth from birth to thirty-six years of age. In this published article she noted that:

> . . . sex differences in patterns of correlations led us to the suggestion that there are genetically determined sex differences in the extent to which the effects of early experiences (such as maternal love and hostility) persist. The girls appeared to be more resilient in returning to their own characteristic inherent response tendencies. Boys on the other hand, were more permanently affected by the emotional climate in infancy whether it was one of warmth and understanding or of punitive rejection. (Pp. 14–15)

In surveys of schizophrenia in childhood, the number of boys with childhood schizophrenia exceeds the number of girls throughout childhood before puberty, and is sometimes eight or ten times that of girls at puberty and beyond (Bender, 1956). In terms of specific abnormal behaviors seen in children's clinics, 22 percent of the boys and only 7 percent of the girls show head-banging (Lester, 1972). Human male suicides outnumber female suicides by about three or four to one (although suicide *attempts* are higher in females) (Jackson, 1954). Despite these differences, when a human male becomes depressed he elicits from others considerably less sympathy and more rejection (often because of the depression) than does a female (Peters and Hammen, 1977). This, of course, makes human males doubly vulnerable to the effects of environmental deprivation.

REFERENCES

Agar, M. E. and Mitchell, G. Bibliography of deprivation and separation with special emphasis on primates. *JSAS: Catalog of Selected Documents in Psychology*, 1973, **3**, 20 pp.

Arling, G. L. and Harlow, H. F. Effects of social deprivation on maternal behavior of rhesus monkeys. *Journal of Comparative and Physiological Psychology*, 1967, **64**, 371-378.

Arling, G. L., Ruppenthal, G. C. and Mitchell, G. Aggressive behavior of the 8 year old nulliparous isolate female monkey. *Animal Behaviour*, 1969, **17**, 190-213.

Bayley, N. Behavioral correlates of mental growth: Birth to thirty-six years. *American Psychologist*, 1968, **23** (1), 1-17.

Baysinger, C. M., Brandt, E. M. and Mitchell, G. Development of infant social isolate monkeys (*Macaca mulatta*) in their isolation environments. *Primates*, 1972, **13** (3), 257-270.

Bender, L. Schizophrenia in childhood: Its recognition, description and treatment. *American Journal of Orthopsychiatry*, 1946, **26**, 499-506.

Bernstein, I. S., Gordon, T. P. and Rose, R. M. Aggression and social controls in rhesus monkey (*Macaca mulatta*) groups revealed in group formation studies. *Folia Primatologica*, 1974, **21**, 81-107.

Brandt, E. M., Baysinger, C. and Mitchell, G. Separation from rearing environment in mother-reared and isolation-reared rhesus monkeys (*Macaca mulatta*). *International Journal of Psychobiology*, 1972, **2** (3), 193-204.

Brandt, E. M. and Mitchell, G. Pairing preadolescents with infants (*Macaca mulatta*). *Developmental Psychology*, 1973, **8**, 222-228.

Brandt, E. M., Stevens, C. W. and Mitchell, G. Visual social communication in adult male isolate-reared monkeys (*Macaca mulatta*). *Primates*, 1971, **12** (2), 105-112.

Chamove, A. S. and Harlow, H. F. Exaggeration of self-aggression following alcohol ingestion in rhesus monkeys. *Journal of Abnormal Psychology*, 1970, **75** (2), 207-209.

Cross, H. A. and Harlow, H. F. Prolonged and progressive effects of partial isolation on the behavior of macaque monkeys. *Journal of Experimental Research in Personality*, 1965, **1**, 39-49.

Deutsch, J. and Larsson, K. Model-oriented sexual behavior in surrogate-reared rhesus monkeys. *Brain, Behavior and Evolution*, 1974, **9**, 157-164.

Eastman, R. F. and Mason, W. A. Looking behavior in monkeys raised with mobile and stationery artificial mothers. *Developmental Psychobiology*, 1975, **8** (3), 213-221.

Erwin, J., Brandt, E. M. and Mitchell, G. Attachment formation and separation in heterosexually naive pre-adult rhesus monkeys (*Macaca mulatta*). *Developmental Psychobiology*, 1973, **6**, 531-538.

Erwin, J. and Maple, T. Ambisexual behavior with male-male anal penetration in male rhesus monkeys. *Archives of Sexual Behavior,* 1976, **5** (1), 9–14.

Erwin, J., Maple, T., Mitchell, G. and Willott, J. A follow-up study of isolation and mother-reared rhesus monkeys which were paired with preadolescent conspecifics in late infancy: Cross-sexed pairings. *Developmental Psychology,* 1974a, **10**, 423–428.

Erwin, J., Maple, T., Willott, J. and Mitchell, G. Persistent peer attachments in rhesus monkeys: Response to reunion after two years of separation. *Psychological Reports,* 1974b, **34**, 1179–1183.

Erwin, J. and Mitchell, G. Initial heterosexual behavior of adolescent rhesus monkeys (*Macaca mulatta*). *Archives of Sexual Behavior,* 1975, **4** (1), 97–104.

Erwin, J., Mobaldi, J. and Mitchell, G. Separation of juvenile rhesus monkeys of the same sex. *Journal of Abnormal Psychology,* 1971, **78** (2), 134–139.

Faust, R. Eighth gorilla born at Frankfurt. *AAZPA Newsletter,* 1977, **18**, 14.

Gluck, J. D. and Sackett, G. P. Frustration and self-aggression in social isolate rhesus monkeys. *Journal of Abnormal Psychology,* 1974, **83** (3), 331–334.

Gomber, J. Caging adult male isolation-reared rhesus monkeys (*Macaca mulatta*) with infant conspecifics. Unpublished doctoral dissertation, Department of Psychology, University of California, Davis, 1975.

Gomber, J. and Mitchell, G. Preliminary report on adult male isolation-reared rhesus monkeys caged with infants. *Developmental Psychology,* 1974, **9**, 419.

Goosen, C. and Ribbens, L. Self-fighting in female stumptail macaques. *Annual Report of the Institute of Experimental Gerontology,* Primate Center TNO, The Netherlands, 1973, pp. 216–217.

Goy, R. W. Hormonal and environmental influences on sexual behaviour in rhesus monkeys. Paper presented at the *International Primatological Society* Meeting, Cambridge, England, August, 1976.

Goy, R. W., Wallen, K. and Goldfoot, D. A. Social factors affecting the development of mounting behavior in male rhesus monkeys. In Montagna, W. and Sadler, W. A. (Eds.) *Reproductive Behavior.* New York: Plenum Press, 1974, pp. 223–247.

Hinde, R. A. and Spencer-Booth, Y. Effects of brief separation from mother on rhesus monkeys. *Science,* 1971, **173**, 111–118.

Jackson, D. D. Suicide. *Scientific American,* 1954 (November), pp. 1–6.

Jensen, G. D. Sex differences in developmental trends of mother-infant monkey behavior (*M. nemestrina*). *Primates,* 1966, **7** (3), 403.

Jensen, G. D. Environmental influences on sexual differentiation: Primate studies. Paper presented at the *Symposium on Environmental Influences on Genetic Expression,* Bethesda, Maryland, April, 1969.

Jensen, G. D., Bobbitt, R. A. and Gordon, B. N. Sex differences in social interaction between infant monkeys and their mothers. *Recent Advances in Biological Psychiatry,* 1966, **9**, 283-293.

Lester, D. Self-mutilating behavior. *Psychological Bulletin,* 1972, **78**, 119-128.

Maple, T. Unusual sexual behavior of nonhuman primates. In Money, J. and Musaph, H. (Eds.) *Handbook of Sexology.* Amsterdam: Elsevier, 1977, pp. 1167-1186.

Maple, T., Brandt, E. M. and Mitchell, G. Separation of preadolescents from infants (*Macaca mulatta*). *Primates,* 1975, **16**, (2), 141-153.

Maple, T., Erwin, J. and Mitchell, G. Separation of adult heterosexual pairs of rhesus monkeys: The effect of female cycle phase. *Journal of Behavioral Science,* 1974, **2**, 81-86 (a)

Maple, T., Erwin, J. and Mitchell, G. Sexually aroused self-aggression in a socialized adult male monkey. *Archives of Sexual Behavior,* 1974, **3**, 471-475 (b)

Maple, T., Risse, G. and Mitchell, G. Separation of adult males from adult female rhesus monkeys (*Macaca mulatta*) after a short term attachment. *Journal of Behavioral Science,* 1973, **1** (3), 327-336.

Mason, W. A. The effects of social restriction on the behavior of rhesus monkeys. I: Free social behavior. *Journal of Comparative and Physiological Psychology,* 1960, **53** (6), 582-589.

Mason, W. A. and Berkson, G. Effects of maternal mobility on the development of rocking and other behaviors in rhesus monkeys: A study with artificial monkeys. *Developmental Psychobiology,* 1975, **8** (3), 197-211.

Mason, W. A., Davenport, R. K., Jr. and Menzel, E. W., Jr. Early experience and the social development of rhesus monkeys and chimpanzees. In Newton, G. and Levine, S. (Eds.) *Early Experience and Behavior.* Springfield, Illinois: Charles C Thomas, 1968, pp. 1-41.

Missakian, E. A. Reproductive behavior of socially deprived male rhesus monkeys (*Macaca mulatta*). *Journal of Comparative and Physiological Psychology,* 1969, **69**, 403-407.

Mitchell, G. Persistent behavior pathology in rhesus monkeys following early social isolation. *Folia Primatologica,* 1968, **8**, 132-147.

Mitchell, G. Abnormal behavior in primates. In Rosenblum, L. A. (Ed.) *Primate Behavior: Developments in Field and Laboratory Research* (Vol. I). New York: Academic Press, 1970, pp. 195-249.

Mitchell, G. Parental and infant behavior. In Hafez, E. S. E. (Ed.) *Comparative Reproduction of Laboratory Primates* (Chapter 14) Springfield, Illinois: Charles C Thomas, 1971, pp. 382-402.

Mitchell, G. Syndromes resulting from social isolation of primates. In Cullen, J. H. (Ed.) *Experimental Behavior: A Basis for the Study of Mental Disturbance.* Dublin: Irish University Press, 1974, pp. 216-223.

Mitchell, G. What monkeys tell us about human violence. *The Futurist,* 1975, **9** (2), 75-80.

Mitchell, G. Attachment potential in rhesus macaque dyads (*Macaca mulatta*): A sabbatical report. *JSAS: Catalogue of Selected Documents in Psychology,* 1976, **6**, 7, MS 1177.

Mitchell, G. Parental behavior in nonhuman primates. In Money, J. and Musaph, H. (Eds.) *Handbook of Sexology.* Amsterdam: Elsevier/North Holland Biomedical Press, 1977, pp. 749-759.

Mitchell, G., Arling, G. L. and Møller, G. W. Long-term effects of maternal punishment on the behavior of monkeys. *Psychonomic Science,* 1967, **8**, 209-210.

Mitchell, G. and Gomber, J. Moving laboratory rhesus monkeys (*Macaca mulatta*) to unfamiliar home cages. *Primates,* 1976, **17** (4), 543-546.

Mitchell, G., Raymond, E. J., Ruppenthal, G. C. and Harlow, H. F. Long-term effects of total social isolation upon behavior of rhesus monkeys. *Psychological Reports,* 1966, **18**, 567-580.

Mitchell, G. and Redican, W. K. Communication in normal and abnormal rhesus monkeys. *Proceeding of the XXth International Congress of Psychology.* Tokyo, Japan: Science Council of Japan, 1972, pp. 171-172.

Mitchell, G., Redican, W. K. and Gomber, J. Lesson from a primate: Males can raise babies. *Psychology Today,* 1974, **7** (11), 63-68.

Møller, G. W., Harlow, H. F. and Mitchell, G. Factors affecting agonistic communication in rhesus monkeys (*Macaca mulatta*). *Behaviour,* 1968, **31**, 339-357.

Nadler, R. D. Determinants of variability in maternal behavior of captive female gorillas. In Kondo, S., Kawai, M., Ehara, A. and Kawamura, S. (Eds.) *Proceedings of the Symposium of the Fifth Congress of the International Primatological Society.* Tokyo: Japan Science Press, 1975, pp. 207-216 (a)

Nadler, R. D. Second gorilla birth at the Yerkes Regional Primate Research Center. *International Zoo Yearbook,* 1975, **15**, 134-137 (b)

Peters, S. D. and Hammen, C. Differential rejection of male and female depression. Paper presented at the *Western Psychological Association* Meeting, Seattle, Washington, April, 1977.

Redican, W. K. and Mitchell, G. A longitudinal study of paternal behavior in adult male rhesus monkeys: I. Observations on the first dyad. *Developmental Psychology,* 1973, **8** (1), 135-136. (a)

Redican, W. K. and Mitchell, G. The social behavior of adult male infant pairs of rhesus monkeys in a laboratory environment. *American Journal of Physical Anthropology,* 1973, **38** (2), 523-526. (b)

Riesen, A. H., Perkins, M. and Struble, R. G. Open-field behavior in socially deprived stumptail monkeys. Paper presented at the *American Society of Primatologists* Meeting, Seattle, Washington, April, 1977.

Rogers, C. M. and Davenport, R. K. Effects of restricted rearing on sexual behavior of chimpanzees. *Developmental Psychology,* 1969, **1**, 200-204.

Rosenblum, L. A. Sex differences in mother-infant attachment in monkeys. In Friedman, R. C., Richart, R. M. and Vande Wiele, R. L. (Eds.) *Sex Differences in Behavior.* New York: Wiley, 1974, pp. 123-145.

Sackett, G. P. Unlearned responses, differential rearing experiences, and the development of social attachments by rhesus monkeys. In Rosenblum, L. A. (Ed.) *Primate Behavior: Developments in Field and Laboratory Research* (Vol. 1). New York: Academic Press, 1970, pp. 111-140.

Sackett, G. P. Sex differences in rhesus monkeys following varied rearing experiences. In Friedman, R. C., Richert, R. M. and Vande Wiele, R. L. (Eds.) *Sex Differences in Behavior.* New York: Wiley, 1974, pp. 99-122.

Scollay, P. A. Mother-infant separation in rhesus monkeys (*Macaca mulatta*). Doctoral dissertation, Department of Anthropology and Psychology, University of California, Davis, 1970.

Scollay, P. A., Joines, S., Baldridge, C. and Cuzzone, A. Learning to be a mother. *Zoonooz,* 1975, **48** (4), 4-9.

Suomi, S. J. Social interactions of monkeys reared with mothers and peers. *Primates,* 1974, **15** (4), 311-320.

Suomi, S. J., Eisele, C. D., Grady, S. A. and Harlow, H. F. Depressive behavior in adult monkeys following separation from family environment. *Journal of Abnormal Psychology,* 1975, **84** (5), 576-578.

Suomi, S. J., Harlow, H. F. and Kimball, S. D. Behavioral effects of prolonged partial social isolation in the rhesus monkey. *Psychological Reports,* 1971, **29**, 1171-1177.

Testa, T. J. and Mack, D. The effects of social isolation on sexual behavior in *Macaca fascicularis.* In Chevalier-Skolnikoff, S. and Poirier, F. E. (Eds.) *Primate Biosocial Development.* New York: Garland, 1977, pp. 407-438.

Tutin, C. E. G. and McGrew, W. C. Chimpanzee copulatory behavior. *Folia Primatologica,* 1973, **19**, 237-256.

Willott, J. F. and McDaniel, J. Changes in the behavior of laboratory-reared rhesus monkeys following the threat of separation. *Primates,* 1974, **15** (4), 321-326.

Zucker, E. L., Wilson, M. E., Wilson, S. F. and Maple, T. The development of sexual behavior in infant and juvenile male orang-utans (*Pongo pygmaeus*). Paper presented at the *American Society of Primatologists* Meeting, Seattle, Washington, April, 1977.

27.
Fear, Stress, and Crowding

As we know from Chapters 22 and 23, a nonhuman primate indicates its fear by posture, movement, facial expressions, and vocalizations. A fearful animal crouches, moves rigidly and awkwardly, looks quickly, frequently, and furtively at more dominant animals, bares its teeth in a grimace, and often emits squeak or screech vocalizations (cf. Jolly, 1972). In addition to these displays of fear, the body of the highly fearful and stressed animal responds physiologically with increased adrenalin or epinephrine in the blood, pupillary dilation, and with piloerection (hair on back, shoulders, and arms standing up). If the stress or fear lasts over a relatively long period of time there is also an increase in adrenal corticosteroids which are driven by ACTH (adrenocorticotrophic hormone). Increased ACTH in turn is a result of stimulation of the hypothalamo-hypophyseal axis.

We have also seen in our chapters on sex differences in vocalizations, facial expressions, aggression, and early experience that, in many species of nonhuman primates, females display more fear than do males. Despite the tendency for males to *display* less fear, they are nevertheless *more* susceptible to the effects of early deprivation and to the *stress* and depression of early social separation than are females (see Chapter 26).

A. STRESS, ACTH, AND SEX DIFFERENCES

As Hinde (1974) has pointed out, *stress* is a vague term referring to the physical and psychological changes accompanying the physiological functioning that occurs during sustained fear, anxiety, or aggression. For example:

> If an animal is subjected to frequent and prolonged stress, the adrenal glands may enlarge and produce more steroids in response to a given amount of ACTH in the blood reaching them. (Hinde, 1974, P. 327)

415

By injecting ACTH into rhesus monkeys and measuring the adrenal steroids produced in the urine as a result of the injection, Sassenrath (1970) was able to measure the degree to which the adrenals had enlarged in each individual. The steroid level produced was *lowest* in the most dominant animals and *highest* in the lower ranked rhesus monkeys. If low-ranked rhesus monkeys were removed from the group, their steroid levels (produced by ACTH injections) fell. The steroid levels of rhesus females fell when they were in consort with a male (Sassenrath, 1970).

Sassenrath's (1970) data:

> . . . suggest a sex difference in responsiveness to high levels of stress . . . such a sex difference appears reasonable in view of the evidence that estrogens can exert sensitizing or stimulating effects on the pituitary-adrenal axis, while testosterone appears to exert inhibitory effects . . . . In the present study, sex differences in ACTH-responses become more pronounced as the initially subadult cagemates approach sexual maturity. (P. 296)

There is also evidence that increased stress in adult female rhesus interferes with reproduction. For example, Goo and Sassenrath (1977) reported that rhesus females in the top one-third of the status hierarchy have a greater number of pregnancies than do those in the bottom one-third. High levels of glucocorticoids (as a result of the effects of ACTH on the adrenals) interfere with reproduction.

When captive female squirrel monkeys (*Saimiri sciureus*) are first brought into captivity, they undergo a lengthy adjustment process such that there are very few viable infants produced until their third year in captivity. Males, on the other hand, require only a short period of adjustment to captivity before they adequately reproduce (Lorenz, Anderson, and Mason, 1973). In the quarantine period immediately following capture, low conception rates and amenorrhea are common in rhesus (*Macaca mulatta*), baboon (*Papio spp.*), and chimpanzee (*Pan troglodytes*) females (Nieman, 1970). In Japanese macaques, pubescent and adult females are affected differently by a change in environment (Wolfe, 1977). Malelike homosexual copulatory behavior appears in the female of many species when they are in a new environment or undergoing stress. This malelike homosexual copulatory behavior is true of tree shrews (Autrum and von Holst, 1968), squirrel monkeys (Talmage-Riggs and Anschel, 1973), pigtail macaques (Bernstein, 1969), and rhesus macaques (Goy, 1976).

It is interesting, in light of the above findings, that there is some disagreement concerning which sex of rhesus monkey displays more agonistic behavior (fear and aggression). It is probably important that this disagreement is related to the degree of adjustment to captivity. For example, Aarons (1973) found that, in shaping monkey–human contact, the female rhesus showed less progress than did the male rhesus. Females evidently do not respond to stress as well as do males. Mason, Green, and Posepanko (1960) also reported that the *female* rhesus in captivity displayed more fear grimaces and threats than did the male rhesus; and Rowell and Hinde (1963), in testing the responses of captive rhesus monkeys to mildly stressful situations, reported there was more vocalizing in females. Altamnn (1967), on the other hand, studied free-ranging rhesus monkeys and reported that:

> These data gave no indication that females showed a higher incidence of affective responses as a whole nor that they were more inclined to exhibit the milder forms of agonistic behavior. (P. 802)

It is possible that females removed from their stable kinship groups experience more stress than do males who are removed from stable groups, but that males *within* the stable groups experience more stress than do females. This differential stress might then be reflected in the numbers of agonistic displays.

How can these sex differences be explained in light of what we have found regarding the effects of adverse early experience (reported in Chapter 26)? We may recall that males seemed to be more vulnerable to early social separation than did females and that this finding was remarkably consistent from study to study. If females show more stress in captivity than do males, how can it be true that males are more susceptible to the adverse affects of early social isolation?

There is no readily available clear answer to this question. More observations of different kinds of deprivation and stress at various ages and in both field and laboratory settings will be required to clarify the role of gender in response to deprivation and stress. We can speculate, however.

It is possible that there are different critical periods for males and females with regard to physiological and psychological vulnerability. The data *are* fairly consistent regarding early social isolation. Males *are* more vulnerable to *early* isolation than are females (cf. Mitchell, 1970; also see Chapter 26). Even in studies of early mother–infant separation (without social isolation), male infants seem to be more vulnerable (cf. Hinde and Spencer-Booth, 1971; Scollay, 1970).

The data on the response to stress in the late juvenile, adolescent, subadult, and adult periods of life are not as consistent. It appears, however, that estrogen levels affect a female's response to stress. Subadult females do not respond to stress as much as do adult females (Wolfe, 1977), and the ACTH response in captivity becomes more pronounced as subadults approach maturity (Sassenrath, 1970). Perhaps this later sensitivity to stress is adaptive for the adult female and helps keep her in her natal group as she approaches maturity. Since it is adaptive for subadult males to leave their natal groups, to travel alone, or to join all-male groups and new heterosexual groups, a high sensitivity to stress at this age would be maladaptive for them. It is interesting that high testosterone levels inhibit the pituitary-adrenal axis resulting in the production of less ACTH (Sassenrath, 1970).

There is evidence, in species which are more complex and more intelligent than the rhesus, that puberty and adolescence may be critical for the acquisition of some social behaviors. For the chimpanzee, for example, puberty may be a critical period for the acquisition and refinement of sexual behavior (Mason, Davenport, and Menzel, 1968).

Even in people, puberty and adolescence has been referred to as a period of "storm and stress." Females, at these ages, are treated much differently than are males. Even Nancy Bayley (1968), in her studies of people ranging from birth to thirty-six years of age has said:

> It may be that the emotional turmoil of adolescence is more disruptive of the girls' than of the boys' cognitive processes. (P. 9)

The human female's reproductive system responds to stress in ways similar to the nonhuman primate female's reproductive system. Constant social or environmental change and stress can cause amenorrhea. There is, for example, psychogenic amenorrhea in airline stewardesses (Greep, 1968). Apparently, prolonged stress can even affect the sex of a child at conception. Schizophrenic women who develop their psychoses within one month of conception tend to deliver only *female* infants. Apparently, at this *early* age of development the male *embryo* is destroyed by the physiological stress mechanisms (Taylor, 1969). This is in keeping with the notion that males are more vulnerable at earlier ages than are females.

B. CROWDING

As we may remember from Chapters 24 and 25, one of Marler's (1976) external conditions most likely to provoke aggression was proximity of another animal. As Hinde (1974) has stated:

> Individuals of a social species seek the proximity of others—but only up to a point, and not necessarily all the time . . . this may result in the maintenance of "individual distance." (P. 326)

Crowding in many species of mammals, and allegedly in man, seems to have the affect of increasing aggression. However, recent research on nonhuman primates suggests that the relationship between crowding and aggression is not a simple one. Jorde and Spuhler (1974) concluded, after multivariate analyses done on twenty-one primate species, that population density seems to be higher overall in arboreal than in terrestrial species and that when density is high, aggression is high. However, these overgeneralized analyses do not apply to specific species like the orangutan, which is arboreal yet has low density, or to the rhesus monkey, which is terrestrial yet very aggressive.

The truth is that very few primate species have been studied carefully enough to determine the effects of crowding per se. And, in those studies that have been done, only a few of the researchers have carefully defined what they meant by crowding (cf. Erwin, 1976 as an example of good definitions). The three species of primates that have been studied most thoroughly with regard to the effects of crowding are *Macaca fuscata, Macaca nemestrina,* and *Papio anubis.*

In studies of crowding done on the Japanese macaque (*M. fuscata*), Alexander and Roth (1971) reported that both sexes showed increased aggressiveness under crowded conditions relative to uncrowded conditions, but that *male* Japanese macaques attacked more than did females and were attacked more than were females under crowded conditions:

> Low-ranking adult, subadult, and juvenile males showed the greatest increase in attacks received. In contrast, females were proportionately less aggressive and less frequently attacked when crowded. (Alexander and Roth, 1971, P. 88)

In the pigtail macaque (*M. nemestrina*), several different variables have been manipulated independently. Erwin and his associates (Erwin, 1976; Erwin et al., 1976; Anderson et al., 1977) have manipulated *social* density and *spatial* density separately. Crowding by increased *spatial* density involves maintaining the same number of animals but reducing the space. Crowding by increased *social* density, on the other hand, consists of maintaining the same amount of space but increasing the number of animals. The two types of crowding have different effects on the animals. Increased social density increases aggression. Increased spatial density can actually *decrease* aggression in some situations. The introduction of new animals, or the removal of familiar animals, complicates the results regarding change in aggression with changes in social density. However, with changes in spatial density, we frequently get changes in control male visual access to group members. The overall conclusion that increased social density increases aggression more than does increased spatial density still applies in short-term crowding, however (cf. Anderson et al., 1977).

Relatively long-term crowding, of the increased *spatial* density variety, has been studied by Elton and Anderson (1977) using olive baboons (*Papio anubis*).

A captive baboon group containing one adult male was crowded by "periodically moving one wall of their cage until available space had been reduced to 50 percent." (P. 225.) The result, according to Elton and Anderson (1977), was "social disintegration and individual pathology" (P. 225):

> The result of crowding was an increase in intratroop agonistic and sexual behaviors. . . . Approximately one and a half weeks after each of these wall movements, there was a noticeable increase in tension and general activity which lasted from seven to ten days. (Pp. 232–233)

Elton and Anderson (1977) go on to say that, under extreme crowding, individual abnormalities, such as chewing on their own hands and feet and pulling hair out of other animals by the handful and eating it, occurred regularly. Under these conditions of spatial crowding, social disintegration began first with the infants and juveniles and second with the *females lowest in dominance.* Chronic crowding apparently increased ACTH response levels and therefore induced stress as we have defined it earlier (Sassenrath, 1970).

Robert Sommer (1969); of the University of California at Davis, has examined the effects of social proximity on people in his book entitled *Personal Space.* It is clear from his investigations that people, particularly strangers, have limits beyond which they do not tolerate another person to come closer. Increased arousal is apparently caused by intrusion into one's personal space and such intrusions tend to increase self-directed stereotypical and nonstereotypical motor activity, particularly if the intruding stranger is of the opposite sex (Coss, Jacobs, and Allerton, 1975).

According to Hamburg (1971), in people:

> . . . overcrowding can have dramatic consequences, including an enhanced propensity to aggression. (Cited in Hinde, 1974, P. 327)

Thus, it is possible that too much social stimulation over prolonged periods of time (particularly in adulthood) can increase pathological aggression just as too little social stimulation over prolonged periods during critical periods in early childhood can increase aggression. Human sex differences in response to social and spatial crowding have not been carefully examined; however, it is likely (based on what we know so far about sex differences in people) that men would show a greater increase in aggressiveness as a result of crowding than would women.

REFERENCES

Aarons, L. Shaping monkey-human contact. *Perceptual and Motor Skills,* 1973, **36**, 235-243.

Alexander, B. K. and Roth, E. M. The effects of acute crowding on aggressive behavior of Japanese monkeys. *Behaviour,* 1971, **39**, 73–90.

Altmann, S. A. Testing Mason's hypothesis of sex differences in the affective behavior of rhesus monkeys. *American Zoologist,* 1967, **7** (4), 802.

Anderson, B., Erwin, N., Flynn, D., Lewis, L. and Erwin, J. Effects of short-term crowding on aggression in captive groups of pigtail monkeys (*Macaca nemestrina*). *Aggressive Behavior,* 1977, **3**, 33–46.

Autrum, H. and von Holst, D. Socialer "stress" bei Tupajas (*Tupaia glis*) und seine Wirkung auf Wachstum, Korpergewicht und Fortpflanzung. *Zeitschrift für vergleichende Physiologie,* 1968, **58**, 347–355.

Bayley, N. Behavioral correlates of mental growth: Birth to thirty-six years. *American Psychologist,* 1968, **23** (1), 1–17.

Bernstein, I. S. Introductory techniques in the formation of pigtail monkey troops. *Folia Primatologica,* 1969, **10**, 1–19.

Coss, R. G., Jacobs, L. S. and Allerton, M. W. Changes of stereotypical and non-stereotypical motor activity of subjects in free-field settings during intrusions of personal space by unfamiliar people. Paper presented at the *Human Ethology Meeting of the Animal Behavior Society,* University of North Carolina, Durham, North Carolina, May, 1975.

Elton, R. H. and Anderson, B. V. The social behavior of a group of baboons (*Papio anubis*) under artificial crowding. *Primates,* 1977, **18**, 225–234.

Erwin, J. Aggressive behavior of captive pigtail macaques: Spatial conditions and social controls. *Laboratory Primate Newsletter,* 1976, **15** (2), 1–10.

Erwin, J., Anderson, B., Erwin, N., Lewis, L. and Flynn, D. Aggression in captive pigtail monkey groups: Effects of provision of cover. *Perceptual and Motor Skills,* 1976, **42**, 319–324.

Goo, G. P. and Sassenrath, E. N. Dominance rank and reproduction in captive rhesus monkeys. Paper presented at the *American Society of Primatologists* Meeting, Seattle, Washington, April, 1977.

Goy. R. W. Hormonal and environmental influences on sexual behaviour in rhesus monkeys. Paper presented at the *International Primatological Society* Meeting, Cambridge, England, August, 1976.

Greep, R. O. Hypothalamic-pituitary-ovarian relationships. In Hoffman, M. and Kleiman, R. L. (Eds.) *Advanced Concepts in Contraception.* Amsterdam: *Excerpta Medica Foundation,* 1968, pp. 108–110.

Hamburg, D. A. Crowding, stranger contact and aggressive behaviour. In Levi, L. (Ed.) *Society, Stress and Disease* (Vol. 1). Cambridge, England: Oxford University Press, 1971, pp. 209–218.

Hinde, R. A. *Biological Bases of Human Social Behavior.* New York: McGraw-Hill, 1974.

Hinde, R. A. and Spencer-Booth, Y. Effects of brief separation from mother on rhesus monkeys. *Science,* 1971, **173**, 111-118.

Jolly, A. *The Evolution of Primate Behavior.* New York: Macmillan, 1972.

Jorde, L. B. and Spuhler, J. N. A statistical analysis of selected aspects of primate demography, ecology, and social behavior. *Journal of Anthropological Research,* 1974, **30**, 199-224.

Marler, P. On animal aggression: The roles of strangeness and familiarity. *American Psychologist,* 1976, **31** (3), 239-246.

Mason, W. A., Davenport, R. K., Jr. and Menzel, E. W., Jr. Early experience and the social development of rhesus monkeys and chimpanzees. In Newton, G. and Levine, S. (Eds.) *Early Experience and Behavior.* Springfield, Illinois: Charles C Thomas, 1968, pp. 1-41.

Mason, W. A., Green, P. C. and Posepanko, C. J. Sex differences in affective-social responses of rhesus monkeys. *Behaviour,* 1960, **16**, 74-83.

Mitchell, G. Abnormal behavior in primates. In Rosenblum, L. A. (Ed.) *Primate Behavior: Developments in Field and Laboratory Research* (Vol. I, Chapter 5). New York: Academic Press, 1970, pp. 195-249.

Nieman, W. H. Comparative aspects of reproduction in primates. *Proceedings of the Third International Congress of Primatology,* 1970, **1**, 234-237.

Rowell, T. E. and Hinde, R. A. Response of rhesus monkeys to mildly stressful situations. *Animal Behaviour,* 1963, **11**, 235-243.

Sassenrath, E. N. Increased adrenal responsiveness related to social stress in rhesus monkeys. *Hormones and Behavior,* 1970, **1**, 283-298.

Scollay, P. A. Mother-infant separation in rhesus monkeys (*Macaca mulatta*). Doctoral dissertation, Department of Anthropology and Psychology, University of California, Davis, 1970.

Sommer, R. *Personal Space.* Englewood Cliffs, N.J.: Prentice-Hall, 1969.

Talmage-Riggs, G. and Anschel, S. Homosexual behavior and dominance hierarchy in a group of captive female squirrel monkeys (*Saimiri sciureus*). *Folia Primatologica,* 1973, **19**, 61-72.

Taylor, M. A. Sex ratios of Newborns: Associated with prepartum and postpartum schizophrenia. *Science,* 1969, **164**, 723-724.

Wolfe, L. Behavior patterns of estrous females of the Arashiyama West troop of Japanese macaques (*Macaca fuscata*). Paper presented at the *American Association of Physical Anthropologists* Meeting, Seattle, Washington, April, 1977.

28.

Vulnerability and Mortality

We have seen in Chapter 26 that the two sexes differ in vulnerability to early social deprivation, with males being more vulnerable than females. In Chapter 27, however, we found that females were particularly susceptible to stress in adulthood, particularly if they were low-ranking females kept in captivity. In the present chapter, we will review the question of vulnerabiltiy in general, whether it be to infectious diseases, to wounding, or to congenital disorders. We will employ our usual taxonomic format.

A. GENERAL COMMENTS

There is not always an obvious reason for differential vulnerability to various diseases in the two sexes. According to Freeland (1976), however:

> Social interactions involving close association with conspecific individuals not belonging to the group, or travel in areas not frequented by the group, can lead to the acquisition of diseases. (P. 12)

Males, of course, interact more than do females with conspecifics outside the home troop. Males also tend to travel more. These factors should make males more susceptible to pathogens. In addition, if there is competition for females within the home troop, the stress of the competition should also make males more vulnerable to disease and wounding. Freeland (1976) also claims that males protect females with whom they mate partly to prevent the pregnant females from being exposed to diseases carried by other animals. That is, he believes that disease helps select for such behavior. Even without the primate male's impregnating the female, Freeland (1976) believes that:

> Sexual fidelity of individual primates to other members of their sexual group is the result of selection for the avoidance of new diseases. (Freeland, 1976, P. 19)

Differential vulnerability to disease is not the only kind of differential vulnerability seen in nonhuman primates. The two sexes are also differentially vulnerable to injury and, in some cases, the differential vulnerability itself is different depending upon the age of the individuals being considered.

For example, in object discrimination reversal skills, *normal* infant male rhesus monkeys perform better than do females at forty-six days of age. The difference at forty-six days of age appears to be testosterone dependent, since postnatal injections of testosterone proprionate enhance object discrimination reversal skills in infant females. Orbital prefrontal brain lesions lead to decreased performance for both males and androgenized females, but not for normal females. The earlier the operation is performed, however, the greater the recovery there is, and the less the sex difference (Goldman, 1976).

We mention these specific differences above so that we may remind ourselves that sex differences are never static but change with age. Sex differences result from developmental processes. Sex differences in vulnerability to disease or to injury are not exceptions to this general rule.

B. NEW WORLD MONKEYS

As we have seen in Chapter 26, deprivation rearing is more devastating to male squirrel monkey (*Saimiri sciureus*) exploratory reactions than it is to these same reactions in females (Rosenblum, 1974). Male squirrel monkeys also seem to be more vulnerable to color blindness. Adult males cannot see the color red as well as can females (Jacobs, 1977).

Among howler monkeys (*Alouatta palliata*), a universal preponderance of females in howler troops is a function of greater postnatal male mortality and the tendency for males to become peripheralized and isolated from troops (Scott, Malmgren, and Glander, 1976). In New World species with less sexual dimorphism there is little evidence of differential vulnerability to wounding and disease.

C. RHESUS MACAQUES (*MACACA MULATTA*)

Even prior to birth, the rhesus male embryo and fetus are more vulnerable to abnormal development during pregnancy than are rhesus monkey females (Sackett, Holm, and Landesman-Dwyer, 1975). As we already know, male rhesus monkeys are also more vulnerable to social deprivation in infancy than are females:

Females reared in deprived situations are less detrimentally affected than males. The source of this difference must be in neural, sensory, or biochemical differences between males and females. (Sackett and Ruppenthal, 1973, P. 84)

This susceptibility to the effects of early deprivation is seen in higher thumb-sucking scores (Benjamin, 1961), more self-clasping, more disturbance (Suomi, Harlow, and Kimball, 1971), and more self-directed aggression (Gluck and Sackett, 1974) in male isolate-reared rhesus monkeys than in female isolate-reared rhesus monkeys. If isolate-reared rhesus are placed into a free-ranging environment, only 33 percent of the males survive whereas over 77 percent of the females survive (Sackett, 1974).

But this difference in survival among isolate-reared animals should not surprise us. There have been reports of such sex differences in survival even among rhesus monkeys that have grown up in the free-ranging situation. If free-ranging rhesus monkeys are removed from their troops (for up to 103 days) and then are subsequently reintroduced to their troops, over 44 percent of the males fail to rejoin their original groups (some of them become solitary) whereas only 11 percent of reintroduced females fail to rejoin their original groups (Vessey, 1971).

Even among unmolested free-ranging rhesus monkeys, bone fractures (particularly of the clavicle) are found more frequently in males than in females (Buikstra, 1975). At four years of age, free-ranging rhesus males have a higher mortality rate than do females. By the age of eight years, only 26 percent of the original group of males is still alive whereas 60 percent of the original females are still living. Out of 183 females observed by Drickamer (1974), none died between the ages of four and eight. Out of 179 males, sixty-three died between the ages of four and eight. "Mortality among the reproductively mature males is greater than among mature females." (Wilson and Boelkins, 1970, P. 723.)

In some studies, observers report that free-ranging female rhesus receive more wounds than do males but that the males' wounds are more likely to impair locomotion and hence threaten survival (Hausfater, 1972).

As far as rhesus monkeys are concerned, the only anomaly to which females seem to be more susceptible than males is myopia. Both myopia and eye enlargement are more likely to develop in female rhesus than in male rhesus monkeys (Wiesel and Raviola, 1977). (Perhaps this is because female rhesus groom at close contact more than do males.)

D. OTHER OLD WORLD MONKEYS

In a study of high risk infants (prematurity and low birth weight) in *Macaca nemestrina,* males were more vulnerable than females, especially those males rejected by their mothers or those receiving inadequate mothering.

According to Dazey and Erwin (1976), in captivity, the mortality rate between birth and 30 days of age is higher in male pigtail macaques (*Macaca nemestrina*) than in females, *but* for the rate between 30 and 180 days the reverse is true. This study, based upon eight years of records at the University of Washington Regional Primate Research Center Field Station (on 1,174 animals), does not corroborate most studies of this sort, which show that there are more male deaths at all ages. As we will see, however, there are other studies showing early female vulnerability.

Male pigtail macaques (like rhesus) are more vulnerable to early deprivation than female pigtails. For example, as a result of deprivation, males show more thumb sucking (Jensen, 1966), and masturbation. Males need a richer early environment for normal development than do females (Jensen, Bobbitt, and Gordon, 1966). Sexual behavioral differentiation is severely disturbed by early deprivation and males are especially vulnerable to deprivation in this regard (Jensen, 1969). This is also true of stumptail macaques, *Macaca arctoides* (Riesen, Perkins, and Struble, 1977).

In large captive groups of pigtail macaques, males exceed females in biting and other damaging aggression (Bernstein, 1972). In zoo colonies of crab-eating macaques (*M. fascicularis*), male mortality during the first year of life is higher than is female mortality (Angst, 1976).

Surprisingly, in free-ranging toque macaques (*M. sinica*), mortality is actually higher in infant and juvenile *females* than it is in infant and juvenile males, but by four to five years of age female mortality rates become low and male rates surpass female rates. The high mortality in immature females is a direct result of exclusion from food. The later male deaths result from the rigors of peripheralization and migration (Dittus, 1975a). Senile females are more numerous than senile males. In provisional colonies, where young females do not have to compete for food, more females reach breeding age than is true in free-ranging groups. In free-ranging troops, however, after maturity, females live an average of 17 years whereas mature toque males live only 10½ years (Dittus, 1975b).

Among Japanese macaques (*M. fuscata*) congenital malformations of the limbs occur more frequently in males than in females (Iwamoto, 1967).

Among baboons (*Papio papio*), there are more male infants than female infants at birth but many deaths occur at birth and during the first year of life. Also, more males succumb to wounding so that by adulthood, more females than males are still alive (Masure and Bourliere, 1971). In free-ranging *Papio anubis* (olive baboons), male mortality is low until the male reaches three years of age. Mortality in males then increases until five years of age. For females, on the other hand, there is only a slight increase in mortality just before two years of age and a slight increase between four and 4½ years of age. There is a slight preponderance of males at birth and in infancy, but females outnumber males as adults (Berger, 1972).

Mangabey males (*Cercocebus atys*) produce and receive more severe wounding in their fights than do females (Bernstein, 1971). It is likely that the mortality rates of this species resemble those of macaques and baboons. Among patas monkeys (*Erythrocebus patas*), heterosexual groups have priority over all-male groups (and over solitary males) when they arrive at water holes. Because of this priority there is a differential mortality in the dry season with all-male groups having a higher mortality rate than do heterosexual groups (Gartlan, 1975).

Juvenile male vervets (*Cercopithecus aethiops*) receive more wounds than expected by chance, especially as they near sexual maturity (Struhsaker, 1967). Among langurs (*Presbytis entellus*), adults threaten and chase young males, even male infants, more than twice as often as they chase immature females (Dolhinow and Bishop, 1970).

In summary, for Old World monkeys, males appear to be more vulnerable to many kinds of injury. Congenital abnormalities, abnormalities resulting from early social deprivation, wounds from others of their own species, and vulnerability to adverse ecological conditions are all higher in males than in females.

E. APES

With what little is known concerning differential vulnerability in the great apes it is fairly clear that male chimpanzees (*Pan troglodytes*) seem to be more adversely affected by early social restriction than do females (Rogers and Davenport, 1969). Chimpanzee males, however, seem to be more capable of recovering from the devastating effects of early isolation than macaque males. Still, in adulthood, captive female sexual behaviors are closer to the behaviors of their wild-living counterparts than are male sexual behaviors (Tutin and McGrew, 1973).

In studies of "alcoholism" in chimpanzees and orangutans (*Pongo pygmaeus*), male chimpanzees drank more alcohol than did female chimpanzees. Of the chimpanzee males, 54 percent consumed enough vodka to become intoxicated at least once, compared to only 25 percent of the females. None of the orangutans (males or females) showed intoxication (Fitz-Gerald, Barfield, and Warrington, 1968).

As in monkeys, female apes are more susceptible to visual myopia than are male apes. There is a natural increase in myopia with age and with prolonged caging, but female chimpanzees show higher levels of myopia than do males. Thus, there is more defective vision for distant objects in females than there is in males (Young, Leary, and Farrer, 1971).

F. COMMENTS ON *HOMO SAPIENS*

In *most* studies involving monkeys and apes, cultural studies are eliminated. However, in most studies of human vulnerability, sex differences can be hopelessly confounded by sex-role stereotypes having a cultural basis (Sackett and Ruppenthal, 1973). As we have seen in the previous two chapters, however, girls appear to be more resilient to and more resistant to the adverse effects of early experience than do boys. The months thirteen to thirty-six may be critical for boys. Emotional factors operating at these times may depress or enhance the cognitive abilities of boys more than they do those of girls. Adolescence, on the other hand, may be more disruptive for girls (cf. Bayley, 1968).

Prior to birth, male embryos are more vulnerable to teratogens than are female embryos (Taylor, 1969). There is more schizophrenia among boys than among girls (Bender, 1956) and there are three or four boys with infantile autism for every girl with the same disorder (Rimland, 1964; Wing, 1976). Prior to puberty, male human beings show a much higher average vulnerability to illness than do females. There is higher perinatal, neonatal, and early childhood disorder in boys than in girls (Luce and Wand, 1977; Hamburg and Lunde, 1966).

In sexual behavior, human males again display greater fragility and vulnerability than do human females (Gadpaille, 1972). Broken marriages through death or divorce are more serious for males than for females. Men also show more alcoholism and suicide than do women (Luce and Wand, 1977; Jackson, 1954).

However, there is some evidence that human females, once past puberty, show *higher* average referral rates for mental health than do human males (Luce and Wand, 1977); and, depression may be higher in the human female (Price, 1967). It is also possible that some forms of mental illness (e.g., catatonia), although occurring less frequently in women than in men, may be more serious physically for women, and may shorten the lives of women more than they shorten the lives of men (Niswander, Haslerud, and Mitchell, 1963a; Niswander, Haselrud, and Mitchell, 1963b).

On the other hand, for other psychoses, males show the greater symptom severity and less benefit from treatment than do females (Bingham and Kunin, 1977).

A problem with many of these findings is that clinical judgments about healthy and ill people very often parallel stereotypic sex-role differences for humans. For example, healthy characteristics resemble behaviors *considered to be* healthy for men but not for women. If a woman shows stereotyped male-role behavior, she may be judged unhealthy (Broverman et al., 1970). There is even stereotypic bias and discrimination which appears regarding causes of death and the details concerning death (Kastenbaum, Peyton, and Kastenbaum, 1977).

Two final points should be made at this time regarding differential vulnerability and mortality in nonhuman and human primates. All three major groups of primates that have been studied—monkeys, apes, and people—show a sex difference in *myopia* (Young, 1977). Female primates (including humans) show more defective distance vision than do male primates. In light of the consistent finding of a greater role for primate males than for females in group protection and vigilance, this sex difference makes a great deal of evolutionary sense.

Finally, in all species of primates studied (including humans), females live longer than do males. The commonly heard speculation that, with changing sex roles in modern society, women will begin to feel the pressures that men have felt and will begin to die off from heart attacks, ulcers, and other disorders of "stress" is probably not correct. There are evidently some basic biological primate tendencies in *Homo sapiens* which make women inherently more resilient and stronger than men, physically and/or psychologically.

REFERENCES

Angst, W. Breeding statistics of *Macaca fascicularis* in Basel Zoo. Paper presented at the *International Primatological Society* Meeting, Cambridge, England, August, 1976.

Bayley, N. Behavioral correlates of mental growth: Birth to thirty-six years. *American Psychologist,* 1968, **23** (1), 1-17.

Bender, L. Schizophrenia in childhood: Its recognition, description and treatment. *American Journal of Orthopsychiatry,* 1956, **26**, 499-506.

Benjamin, L. S. The effect of bottle and cup feeding on the nonnutritive sucking of the infant rhesus monkey. *Journal of Comparative and Physiological Psychology,* 1961, **54** (3), 230-237.

Berger, M. E. Population structure of olive baboons (*Papio anubis*) in the Laikipia district of Kenya. *East African Wildlife Journal,* 1972, **10**, 159-164.

Bernstein, I. S. The influence of introductory techniques on the formation of captive mangabey groups. *Primates,* 1971, **12**, 33-44.

Bernstein, I. S. Daily activity cycles and weather influences on a pigtail monkey group. *Folia Primatologica,* 1972, **18**, 390-415.

Bingham, L. R. and Kunin, R. A. Sex differences in patient self-evaluation one year post-treatment: A follow-up study. Paper presented at the *Western Psychological Association* Meeting, Seattle, Washington, April, 1977.

Broverman, I. K., Broverman, D. M., Clarkson, F. E., Rosenkrantz, P. S. and Vogel, S. R. Sex role stereotypes and clinical judgments of mental health. *Journal of Counseling and Clinical Psychology,* 1970, **34** (1), 1-7.

Buikstra, J. E. Healed fractures in *Macaca mulatta:* Age, sex, symmetry. *Folia Primatologica,* 1975, **23**, 140-148.

Dazey, J. and Erwin, J. Infant mortality in *Macaca nemestrina:* Neonatal and post-neonatal mortality at the Regional Primate Research Center Field Station, University of Washington, 1967-1974. *Theriogenology,* 1976, **5**, 267-279.

Dittus, W. P. J. The ecology and behavior of the toque monkey, *Macaca sinica. Dissertation Abstracts,* 1975, B35/09, 4725. (a)

Dittus, W. P. J. Population dynamics of the toque monkey, *Macaca sinica.* In Tuttle, R. H. (Ed.) *Socioecology and Psychology of Primates.* The Hague: Mouton Publishers, 1975, pp. 125-151. (b)

Dolhinow, P. J. and Bishop, N. The development of motor skills and social relationships among primates through play. In Hill, J. P. (Ed.) *Minnesota Symposia on Child Psychology* (Vol. IV). Minneapolis: University of Minnesota Press, 1970, pp. 141-198.

Drickamer, L. C. A ten-year summary of reproductive data for free-ranging *Macaca mulatta. Folia Primatologica,* 1974, **21**, 61-80.

Fitz-Gerald, F. L., Barfield, M. A. and Warrington, R. J. Voluntary alcohol consumption in chimpanzees and orangutans. *Quarterly Journal of Studies on Alcohol,* 1968, **29** (2), 330-336.

Freeland, W. J. Pathogens and the evolution of primate sociality. *Biotropica,* 1976, **8** (1), 12-24.

Gadpaille, W. J. Research into the physiology of maleness and femaleness: Its contributions to the etiology and psychodynamics of homosexuality. *Archives of General Psychiatry,* 1972, **26**, 193-206.

Gartlan, J. S. Adaptive aspects of social structure in *Erythrocebus patas. Symposia of the Fifth Congress of the International Primatological Society.* Tokyo: Japan Science Press, 1975, pp. 161-171.

Gluck, J. D. and Sackett, G. P. Frustration and self-aggression in social isolate rhesus monkeys. *Journal of Abnormal Psychology,* 1974, **83** (3), 331-334.

Goldman, P. S. Maturation of the mammalian nervous system and the ontogeny of behavior. *Advances in the Study of Behavior,* 1976, **7**, 1-90.

Hamburg, D. A. and Lunde, D. T. Sex hormones in the development of sex differences in human behavior. In Maccoby, E. E. (Ed.) *The Development of Sex Differences.* Stanford, California: Stanford University Press, 1966, pp. 1-24.

Hausfater, G. Intergroup behavior of free-ranging rhesus monkeys (*Macaca mulatta*). *Folia Primatologica,* 1972, **18**, 78-107.

Iwamoto, M. Morphological observations on the congenital malformation of limbs in the Japanese monkey. *Primates,* 1967, **8**, 247-270.

Jackson, D. D. Suicide. *Scientific American,* 1954 (November), 1-6.

Jacobs, G. H. Visual sensitivity: Significant within-species variations in a nonhuman primate. *Science,* 1977, **197**, 499-500.

Jensen, G. D. Sex differences in developmental trends of mother-infant monkey behavior (*M. nemestrina*). *Primates,* 1966, **7** (3), 403.

Jensen, G. D. Environmental influences on sexual differentiation: Primate studies. Paper presented at the *Symposium on Environmental Influences on Genetic Expression,* Bethesda, Maryland, April, 1969.

Jensen, G. D., Bobbitt, R. A. and Gordon, B. N. Sex differences in social interaction between infant monkeys and their mothers. *Recent Advances in Biological Psychiatry,* 1966, **9**, 283-293.

Kastenbaum, R., Peyton, S. and Kastenbaum, B. Sex discrimination after death. *Omega,* 1977, **7** (4), 351-359.

Lorenz, R., Anderson, C. O. and Mason, W. A. Notes on reproduction in captive squirrel monkeys *(Saimiri sciureus).* *Folia Primatologica,* 1973, **19**, 286-292.

Luce, S. R. and Wand, B. Sex differences in health and illness. *Canadian Psychological Review,* 1977, **18**, 79-91.

Masure, A. M. and Bourliere, F. Surpluplement, fecondite, mortalite et agressivite dans une population captive de *Papio papio.* *Extrait de la Terre et la Vie,* 1971, **25**, 491-505.

Niswander, G. D., Haslerud, G. M. and Mitchell, G. Differences in longevity of released and retained schizophrenic patients. *Diseases of the Nervous System,* 1963, **24** (6), 348-352. (a)

Niswander, G. D., Haslerud, G. M. and Mitchell, G. Effects of catatonia on schizophrenic mortality. *Archives of General Psychiatry,* 1963, **9**, 548-551. (b)

Price, J. The dominance hierarchy and the evolution of mental illness. *The Lancet,* 1967 (July), 243-246.

Riesen, A. H., Perkins, M. and Struble, R. G. Open-field behavior in socially deprived stumptail monkeys. Paper presented at the *American Society of Primatologists* Meeting, Seattle, Washington, April, 1977.

Rimland, B. *Infantile Autism.* New York: Appleton-Century-Crofts, 1964.

Rogers, C. M. and Davenport, R. K. Effects of restricted rearing on sexual behavior of chimpanzees. *Developmental Psychology,* 1969, **1**, 200-204.

Rosenblum, L. A. Sex differences, environmental complexity, and mother-infant relations. *Archives of Sexual Behavior,* 1974, **3**, 117-128.

Sackett, G. P. Sex differences in rhesus monkeys following varied rearing experiences. In Friedman, R. C., Richert, R. M. and Vande Wiele, R. L. (Eds.) *Sex Differences in Behavior.* New York: Wiley, 1974, pp. 99-122.

Sackett, G. P., Holm, R. A., Davis, A. E. and Fahrenbruch, C. E. Prematurity and low birth weight in pigtail macaques: Incidence, prediction, and effects on infant development. *Symposia of the Fifth Congress of the International Primatological Society.* Nagoya, Japan, 1974, pp. 189-205.

Sackett, G., Holm, R. and Landesman-Dwyer, S. Vulnerability for abnormal development: Pregnancy, outcomes and sex differences in macaque monkeys. In Ellis, N. R. (Ed.) *Aberrant Development in Infancy: Human and Animal Studies.* Hillsdale, N.J.: Lawrence Erlbaum Associates, 1975, pp. 59-76.

Sackett, G. P. and Ruppenthal, G. C. Development of monkeys after varied experiences during infancy. In Barnett, S. A. (Ed.) *Ethology and Development.* London: Heinemann Medical Books Ltd., 1973, pp. 52-87.

Scott, N. J., Malmgren, L. A. and Glander, K. E. Grouping behaviour and sex ratio in mantled howling monkeys (*Alouatta palliata*). Paper presented at the *International Primatological Society* Meeting, Cambridge, England, August, 1976.

Struhsaker, T. T. *Behavior of Vervet Monkeys (Cercopithecus aethiops)* Berkeley: University of California Press, 1967.

Suomi, S. J., Harlow, H. F. and Kimball, S. D. Behavioral effects of prolonged partial social isolation in the rhesus monkey. *Psychological Reports,* 1971, **29**, 1171-1177.

Taylor, M. A. Sex ratios of newborns: Associated with prepartum and postpartum schizophrenia. *Science,* 1969, **164**, 723-724.

Tutin, C. E. G. and McGrew, W. C. Chimpanzee copulatory behavior. *Folia Primatologica,* 1973, **19**, 237-256.

Vessey, S. H. Free-ranging rhesus monkeys: Behavioral effects of removal, separation and reintroduction of group members. *Behaviour,* 1971, **1**, 216-227.

Wiesel, T. N. and Raviola, E. Myopia and eye enlargement after neonatal lid fusion in monkeys. *Nature,* 1977, **266**, 66-68.

Wilson, A. P. and Boelkins, R. C. Evidence for seasonal variation in aggressive behaviour by *Macaca mulatta. Animal Behaviour,* 1970, **18**, 719-724.

Wing, L. (Ed.) *Early Childhood Autism.* New York: Pergamon, 1976, pp. 67-71.

Young, F. A. The nature and control of myopia. *Journal of the American Optometric Association,* 1977, **48**, 451-457.

Young, F. A., Leary, G. A. and Farrer, D. N. Four years of annual studies of chimpanzee vision. *American Journal of Optometry,* 1971, **48**, 407-416.

29.
Learning and Performance[1]

A. GENERAL COMMENTS

In our chapter on early deprivation we saw that male nonhuman primates are more seriously affected by adverse experiences in early life than are female nonhuman primates (see also Sackett, 1974; Mitchell, 1970). In our chapter on vulnerability we found that there is greater male vulnerability in other areas of biological development besides behavioral development. For example, males show a greater incidence of prematurity (Sackett et al., 1974), more infantile autism (Rimland, 1964), and more prenatal abnormalities (Maccoby and Jacklin, 1974).

This differential vulnerability in biology and behavior in males has been used as rather indirect evidence for greater plasticity in males than in females. The argument is that if males are more susceptible to the environment in early life, they must be more susceptible to positive as well as negative influences. If they are more susceptible to positive influences *behaviorally,* we may be tempted to assume that they have less prewired nervous systems or that they are more plastic than are females.

However, there is little or no evidence that male nonhuman primates have a greater learning capacity than do female nonhuman primates (cf. Mitchell, 1977). Greater susceptibility to disease or to the effects of early experience does *not* necessarily mean greater learning capacity. A plausible alternative explanation to a greater flexibility and learning capacity for males may be that females simply adjust faster or recover more quickly from the environmental effects that are so devastating for males. Another possible reason for differences in vulnerability

[1]This chapter is based on the following previously published article: Mitchell, G. A note on sex differences in learning or motivation in nonhuman primates. *Laboratory Primate Newsletter,* 1977, **16**, 1–5.

might be the differences in rates of maturation or in levels of maturity in the two sexes. However, to argue that males learn faster than females *or* that females learn faster than males on the basis of deprivation or vulnerability data alone will shed very little light on the topic of sex differences in learning. What we need to do is to examine male and female performance in the learning situation itself. In the present short chapter we will do just that.

The recent primate literature includes very little on sex differences in learning and performance. Nearly everything that has been published on this topic has involved Old World monkeys (primarily macaques) and apes (primarily chimpanzees). We will therefore discuss only these two major groups of nonhuman primates in the first sections of this chapter.

B. MONKEYS, APES, AND PEOPLE

1. Old World Monkeys

Higher order learning abilities such as delayed response and reversal learning are especially well developed in primates. After studies of the bushbaby (*Galago senegalensis*) and macaque nervous systems, Skeen and Masterson (1976) concluded:

> . . . despite their remote common ancestry, it can be concluded that the differentiation of the medial dorsal nucleus of the thalamus *prefrontal* system into distinct divisions and the involvement of this system in delayed alternation and spatial reversal are features probably as old as the order Primates itself. (P. 179)

In fact, there is evidence that such a differentiated system exists even in tree shrews (Skeen and Masterson, 1976). In rhesus macaques, Goldman (1975; 1976) has learned that, in object discrimination reversal, normal infant males (forty-six days old) perform better than do normal infant females and that neonatal injections of androgen enhance the early performances of the females. Lesions of the orbital prefrontal system lead to decreased performance in males and in androgenized females but not in normal females (Goldman, 1976). Apparently, males have a slight advantage over females in this ability (object discrimination reversal) when they are very young, and this difference appears to be hormonally dependent.

Interestingly enough, behavioral data accumulated at later ages show that females are more adept at delayed response than are males. Blomquist (1960), for example, found that female rhesus monkeys were superior to males in delayed response performance. McDowell, Brown, and McTee (1960) also found that female rhesus were better than male rhesus at spatial delayed response.

In a task related to delayed response (sand-digging task) observed in free-ranging animals, female Japanese macaques (*M. fuscata*) performed better than did males. During the sand-digging tests, in which a reward was buried at different depths (hence with different delays), *all* (except extremely unstable individuals) adult females showed more enduring attention to the test situation than did males. This was true even when the females failed the test (Tsumori, 1966; 1967). Tsumori (1967), however, feels that the sex difference is not as much in the cue-producing or cue-retaining processes as it is in the motivational processes. The female superiority, however, can be generalized to more than one troop.

It is possible that prolonged attentiveness to close detail evolved more readily in females than in males because the male roles of troop protector, guardian, and, in particular, lookout demand attention to distant extratroop and inter-troop activities (cf. Ripley, 1967). A delayed-response task involving more distant cues and incentives might produce the opposite sex difference.

Most studies of tool using and the development of new skills in Old World monkeys find either no difference between males and females or report that females are superior to males in acquisition or performance. For example, Wilcoxon et al. (1968) reported that only five of twenty-two rhesus monkeys worked consistently at pressing buttons to visually self-stimulate. Of these five, four were:

. . . young, mature, middle-status females without infants, and not in estrus. (P. 266)

According to French and Candland (1977), however, in a captive colony of Japanese macaques, a metal rod that had been formed into a swing by a juvenile male was used primarily by two young males, both offspring of the most dominant female. The invention and spread of two new locomotor techniques—"chimney climbing" and "the swinging technique"—in a captive group of stumptail macaques (*M. arctoides*), however, were each developed by a female juvenile (Bertrand, Desprez, and Laban, 1976). According to Wechkin (1970), social facilitation in object manipulation in macaques is not affected by dominance, age, or sex.

In social learning, however, there seem to be grounds for believing that female Old World monkeys have an advantage over males. Kawamura (1963) has reported that males (particularly subordinate peripheral males) are the last to pick up new food eating habits during the process of subculture propagation. Moreover, if infant-care behavior is as dependent upon experience as some recent researchers have suggested (cf. Seay, 1966; Lancaster, 1971; Poirier, 1973; Scollay et al., 1975; Arling and Harlow, 1967), then the so-called "maternal instinct" may not be as much instinct as very facile learning or at least a behavior easier to learn for females than for males.

In a case study of one female pigtail macaque's ability to acquire behavior patterns beyond the natural behavioral repertoire of the species, Bertrand (1976) reported that her female pigtail (*M. nemestrina*) was toilet trained by eight months of age and learned to control both urination and defecation. This particular female pigtail also spontaneously learned to give food to another but gave only what was of no value to herself. She also cheated when giving and discovered bartering and symbolic giving (Bertrand, 1976). There is no information on similar behaviors for male pigtail macaques.

In the Koshima troop of Japanese macaques (*M. fuscata*), Kawai (1967) threw sweet potatoes into the troop and reported that many more females than males learned to catch them before they hit the ground (predominantly left-handed). Kawai explains the sex difference by saying that the dominant males did not *have* to learn to catch and that the peripheral males were kept away by the leaders. However, Rose (1976) reported that female olive baboons (*P. anubis*) also use bipedalism more frequently in feeding than do males, and his study did not involve catching behavior. Thus, it may be that something more than dominance or social inhibition of the males is involved in this sex difference.

2. Apes

In captive chimpanzees (*Pan troglodytes*), it is frequently noted that the female's behavior is closer to that of their wild-living counterparts than is the male's behavior (Tutin and McGrew, 1973). We may wonder whether the chimpanzee female is less affected by environmental change or, alternatively, quicker to recover from this environmental change.

As in the macaque delayed-response studies, female apes also surpass males in delayed response. In an examination of sex differences in delayed matching-to-sample performance in chimpanzees, Grilly (1975) found a superiority of females over males in matching accuracy. This sex difference was exhibited over an extended period of time and was found with a sample size of seven females and ten males. The finding was also generalizable to two different retention intervals.

In tool-using and "tool-making," Menzel (1972) reported that the spontaneous invention of ladders occurred in both sexes and possibly more in males than in females. McGrew (1974), however, noted that an old female was most effective in tool use by wild chimpanzees in feeding upon driver ants. In tool use for dental grooming, McGrew and Tutin (1973) also believe that female chimpanzees use tools to groom the teeth of other chimpanzees more than do males. Ellis (1975) found no sex differences in tool use in a small group of orangutans at the Oklahoma City Zoo, but Nadler and Braggio (1974) noted that both female orangutans *and* female chimpanzees interacted more with inanimate objects than did conspecific males.

3. Homo sapiens

In extant anthropological writings, the evolution of human tool use has been classically related to decreased canine size in *Homo sapiens* and to the consequent greater need for weapons or tools in defense (cf. Pilbeam, 1968). However, Tanner and Zihlman (1976) have focused upon the role of females in the evolution of human tool use. They emphasize the mother's innovative economic role, *gathering* with tools and *sharing* food with her offspring. Because of the female's social centrality she is ideal as a primary socializer and teacher of new skills, including tool-using. The smaller canine size in males is believed by Tanner and Zihlman (1976) to be secondary to female centrality and comes about because females choose less aggressive males to help them with child rearing.

According to Etkin (1963), the integration of the male into a permanent family, the reduction of intermale competition, the elimination of estrus in females, along with the substitution of diffuse sexual receptivity and the development of cultural transmission, *all* require a nervous system involving well-developed and refined social-emotional systems needed for interpersonal relations as much as a well-evolved cerebral cortex for intelligence. The use of symbols, plans, and tools therefore evolved in a social context that was already quite sophisticated. The female primate was almost always a central figure in that social context.

Tanner and Zihlman (1976) carry their ideas concerning the importance of women in evolution further by suggesting that language first developed in females:

> While gathering plants, speech would add little, although do no harm; during predation or hunting, anything as noisy as speech would likely frighten away the prey. (P. 475)

As the species *Homo sapiens* became less sexually dimorphic physically, it came to resemble superficially the more monomorphic and monogamous species of primates like the gibbon. In monogamous birds, there are correlations between vocal learning and long-lasting *pair bonds*. In birds, where vocal learning has been identified, *females* are the selective agents and males use vocal signals to attract females. This is not true of most mammals and probably not true of primates. For primates, *female* vocalizations may have been more important (see Nottebohm, 1972).

It is certainly true that human females are superior to human males in verbal learning (cf. Maccoby and Jacklin, 1974). At two, three, four, five, and six years of age, girls have higher scores on syntax maturity, sentence length, and specific grammatical categories, especially starting at four years of age (Koenigsknecht and Friedman, 1976). Boys, on the other hand, seem to be more capable at visual-spatial tasks, discrimination of forms, and mathematical ability (Klopfer et al., 1977; Etaugh and Turton, 1977; and Maccoby and Jacklin, 1974).

Overall, however, Maccoby and Jacklin (1974) conclude that:

. . . there is no difference in *how* the two sexes learn. Whether there is a difference in *what* they find easier to learn is a different question. (P. 62)

REFERENCES

Arling, G. L. and Harlow, G. F. Effects of social deprivation on maternal behavior of rhesus monkeys. *Journal of Comparative and Physiological Psychology,* 1967, **64**, 371-378.

Bertrand, M. Acquisition by a pigtail macaque of behavior patterns beyond the natural repertoire of the species. *Zeitschrift für Tierpsychologie,* 1976, **42**, 139-169.

Bertrand, M., Desprez, M. and Laban, M. The invention and spread of two locomotor techniques in a captive group of *Macaca arctoides.* Paper presented at the *International Primatological Society* Meeting, Cambridge, England, August, 1976.

Blomquist, A. J. Variables influencing delayed response performance by rhesus monkeys. *Dissertation Abstracts,* 1960, **21**, 1634-1635.

Ellis, J. Orangutan tool use at Oklahoma City Zoo. *The Keeper,* 1975, **1**, 5-6.

Etaugh, C. and Turton, W. J. Sex differences in discrimination of forms by elementary school children. *The Journal of Genetic Psychology,* 1977, **130**, 49-55.

Etkin, W. Social behavioral factors in the emergence of man. *Human Biology,* 1963, **35** (3), 299-310.

French, J. A. and Candland, D. K. The genesis of object play in *Macaca fuscata.* Paper presented at the *Animal Behavior Society* Meeting, University Park, Pennsylvania, June, 1977.

Goldman, P. S. Age, sex, and experience as related to the neural basis of cognitive development. *UCLA Forum on Medical Science,* 1975, **18**, 379-392.

Goldman, P. S. Maturation of the mammalian nervous system and the ontogeny of behavior. *Advances in the Study of Behavior,* 1976, **7**, 1-90.

Grilly, D. M. Sex differences in delayed matching-to-sample performance of chimpanzees. *Psychological Reports,* 1975, **37**, 203-207.

Kawai, M. Catching behavior observed in the Koshima troop. A case of newly acquired behavior. *Primates,* 1967, **8**, 181-186.

Kawamura, S. The process of sub-culture propagation among Japanese macaques. In Southwick, C. H. (Ed.) *Primate Social Behavior.* Princeton, N.J.: Van Nostrand, 1963, pp. 82-89.

Klopfer, F. J., Jackson, T. T., Wolfe, G. and Jeffrey, G. More than meets the eye: Spatial visualization and sex differences. Paper presented at the *Western Psychological Association* Meeting, Seattle, Washington, April, 1977.

Koenigsknecht, R. A. and Friedman, P. Syntax development in boys and girls. *Child Development,* 1976, **47**, 1109–1115.

Lancaster, J. B. Play-mothering: The relations between juvenile females and young infants among free-ranging vervet monkeys (*Cercopithecus aethiops*). *Folia Primatologica,* 1971, **15**, 161–182.

Maccoby, E. E. and Jacklin, C. N. *The Psychology of Sex Differences.* Stanford, California: Stanford University Press, 1974.

McDowell, A. A., Brown, W. L. and McTee, A. C. Sex as a factor in spatial delayed-response performance by rhesus monkeys. *Journal of Comparative and Physiological Psychology,* 1960, **53**, 429–432.

McGrew, W. C. Tool use by wild chimpanzees in feeding upon driver ants. *Journal of Human Evolution,* 1974, **3**, 501–508.

McGrew, W. C. and Tutin, C. E. G. Chimpanzee tool use in dental grooming. *Nature,* 1973, **241**, 477–478.

Menzel, E. W., Jr. Spontaneous invention of ladders in a group of young chimpanzees. *Folia Primatologica,* 1972, **17**, 87–106.

Mitchell, G. Abnormal behavior in primates. In Rosenblum, L. A. (Ed.) *Primate Behavior* (Vol. 1). New York: Academic Press, 1970, pp. 195–249.

Mitchell, G. A note on sex differences in learning or motivation in nonhuman primates. *Laboratory Primate Newsletter,* 1977, **16**, 1–5.

Nadler, R. D. and Braggio, J. T. Sex and species differences in captive-reared juvenile chimpanzees and orang-utans. *Journal of Human Evolution,* 1974, **3**, 541–550.

Nottebohm, F. The origins of vocal learning. *The American Naturalist,* 1972, **106**, 116–140.

Pilbeam, D. The earliest hominids. *Nature,* 1968, **219**, 1335–1338.

Poirier, F. E. Socialization and learning among nonhuman primates. In Kimball, S. T. and Burnett, J. H. (Eds.) *Learning and Culture.* Seattle: University of Washington Press, 1973, pp. 3–41.

Rimland, B. *Infantile Autism.* New York: Appleton-Century-Crofts, 1964.

Ripley, S. Intertroop encounters among Ceylon gray langurs (*Presbytis entellus*). In Altmann, S. (Ed.) *Social Communication Among Primates.* Chicago: University of Chicago Press, 1967, pp. 237–254.

Rose, M. D. Bipedal behavior of olive baboons (*Papio anubis*) and its relevance to an understanding of the evolution of human bipedalism. *American Journal of Physical Anthropology,* 1976, **44**, 247–261.

Sackett, G. P. Sex differences in rhesus monkeys following varied rearing experiences. In Friedman, R. C., Richert, R. M. and Vande Wiele, R. L. (Eds.) *Sex Differences in Behavior.* New York: Wiley, 1974, pp. 99–122.

Sackett, G. P., Holm, R. A., Davis, A. E. and Fahrenbruch, C. E. Prematurity and low birth weight in pigtail macaques: Incidence, prediction, and effects on infant development. *Symposia of the Fifth Congress of the International Primatological Society.* Nagoya, Japan, 1974, pp. 189-205.

Scollay, P. A., Joines, S., Baldridge, C. and Cuzzone, A. Learning to be a mother. *Zoonooz,* 1975, **48** (4), 4-9.

Seay, B. M. Maternal behavior in primiparous and multiparous rhesus monkeys. *Folia Primatologica,* 1966, **4**, 146-168.

Skeen, L. C. and Masterson, R. B. Origins of anthropoid intelligence. III. Role of prefrontal system in delayed-alternation and spatial-reversal learning in a prosimian (*Galago senegalensis*). *Brain, Behavior and Evolution,* 1976, **13**, 179-195.

Tanner, N. and Zihlman, A. Discussion paper: The evolution of human communication: What can primates tell us? *Annals of the New York Academy of Sciences,* 1976, **280**, 467-480.

Tanner, N. and Zihlman, A. Women in evolution. Part I: Innovation and selection in human origins. *Signs: Journal of Women in Culture and Society,* 1976, **1** (3), 585-608.

Tsumori, A. Delayed response of wild Japanese monkeys by the sand-digging method. II: Cases of the Takasakiyama troops and the Ohirayama troop. *Primates,* 1966, **7**, 363-380.

Tsumori, A. Newly acquired behavior and social interactions of Japanese monkeys. In Altmann, S. A. (Ed.) *Social Communication Among Primates.* Chicago: University of Chicago Press, 1967, pp. 207-219.

Tutin, C. E. G. and McGrew, W. C. Chimpanzee copulatory behavior. *Folia Primatologica,* 1973, **19**, 237-256.

Wechkin, S. Social relationships and social facilitation of object manipulation in *Macaca mulatta. Journal of Comparative and Physiological Psychology,* 1970, **73**, 456-460.

Wilcoxon, H. C., Meier, G. W., Orlando, R. and Paulson, D. G. Visual self-stimulation in socially-living rhesus monkeys. *Proceedings of the Second International Congress of Primatology,* 1968, **1**, 261-266.

30.
Sex Differences in the Brain

As a way of bracketing our basic review on behavioral sex differences with information on underlying structural differences and, as a reminder or review of the hormonal and neurological data presented in the opening chapters of this book, we will now return to a discussion of the central nervous systems of male and female nonhuman primates. With as many *behavioral* sex differences as are probable or at least possible in nonhuman primates, we should expect to see at least *some* sex differences in their central nervous systems.

A. GENERAL COMMENTS

As we have seen in past chapters, prenatal androgens determine, to a large extent, the sexual differentiation of the external genitalia and of sexually dimorphic behaviors such as play, aggressiveness, and sexual mounting. Prenatal genetic females are masculinized physically and behaviorally by the administration of testosterone proprionate during the first trimester of the pregnancy of the mother.

Underlying the behavioral sex differences are presumed sex differences in the central nervous system, most likely in the hypothalamus. But, as Goldman (1976) has suggested, it is probable that other brain structures besides the hypothalamus are affected by prenatal androgen, even central nervous system structures outside of the limbic system.

We have seen in the preceding chapter that the development of the prefrontal cortex may be sensitive to the effects of testosterone proprionate. We may recall that, in object discrimination reversal, normal infant male rhesus (forty-six days old) perform better than do females and that neonatal injections of androgen improve female performance. Orbital prefrontal lesions,[1] on the other hand, decrease performance on object discrimination reversal in males and in androgenized females but not in normal females (Goldman, 1976). This is good evidence for a role for sex hormones in differential development of the central nervous system outside of the hypothalamus:

> . . . the prefrontal cortex, among other structures, may form part of the neural basis of sex-dependent differences in a wide variety of behaviors. (Goldman, 1976, P. 61)

The anterior association (temporal) cortex also plays a role in a wide variety of social behaviors. Lesions in this area reduce facial expressions, vocalizations, aggression, grooming, physical proximity, infant care, and sexual receptivity. Lesions, in effect, lead to the loss of social contact and effectively isolate the individual from others socially. Goldman (1976) believes that both the prefrontal and the anterior temporal areas may be important for sexually dimorphic play behavior.

It is clear from data such as those presented above that, while primates may have become somewhat more free from hormonal control of behavior than have some other mammals, this freedom applies primarily to freedom from hormones in adult life and not to freedom from the organizational effects of prenatal and other early hormonal effects (Phoenix, 1974). There is no doubt, for example, that early administration of testosterone changes the nervous system both in the hypothalamus and outside the hypothalamus.

Besides the hypothalamus, the rest of the limbic system and, more than likely, higher cerebral mechanisms, spinal systems, neurovascular (circulatory) systems, and neurotransmitters are also affected by gonadal hormones (Karczmar, 1975). Most obvious in this regard are those structures that deal with sex differences in sexual behavior. But, as we now know, there are many sex differences in other behavioral patterns more or less closely correlated with reproduction and, of course, ultimately dependent upon it for their existence. The *primary* role of testosterone in the central nervous system is likely to be "the suppression of those neural systems which mediate feminine sexual behavior" (Neuman, Steinbeck, and Hahn, 1970, P. 21), but testosterone also suppresses behaviors which are not directly *sex*-related but which are nevertheless, associated with the behavior of females.

---

[1] There is evidence that the differentiation of the medial dorsal nucleus of the thalamus from the prefrontal system occurred sometime *before* the separation of monkeys from prosimians. The involvement of this system in delayed response and reversal learning is therefore as old as the primate order itself (Skeen and Masterson, 1976) which is to say that the sex differences in the central nervous system discussed here may be somewhat more generally applicable than is at first apparent.

As for the hypothalamus, it *is* organized differently (especially with regard to its cyclicity), depending upon the relative concentrations of prenatal steroids:

> . . . tonic gonadotrophic release can be maintained by the MBH (medial basal hypothalamus) alone, but connections with the anterior hypothalamus are necessary for cyclicity . . . one of the functions of early exposure to androgen is to disrupt this connection between the anterior hypothalamus and the medial basal hypothalamus. (Valenstein, 1968, P. 30)

Let us now explore sex differences in the brains of nonhuman primates at the various taxonomic levels.

B. PROSIMIANS AND NEW WORLD MONKEYS

Pfaff (1976) has published an article showing that there are receptors for sex hormones in vertebrate brains. For example, there are target sites for progesterone in the hypothalamus and there is a direct action of this hormone on certain hypothalamic structures (Sar and Stumpf, 1973).

In the tree shrew, and in the squirrel monkey, estrogen is concentrated and retained in certain neurons of the limbic and brainstem areas (preoptic area, basal hypothalamus, and amygdala) (Keefer and Stumpf, 1975a; Stumpf, Sar, and Keefer, 1974). These areas are interconnected so as to permit an interaction between hormonal and *sensory* stimuli (Keefer and Stumpf, 1975b).

Let us look at an example of how sensory stimulation affects this estrogen-concentrating nerve network in the squirrel monkey. There are units responding to vaginal stimulation in the midbrain, pons, tectum, central gray, raphe nuclei, and reticular formation. Estrogen facilitates the responses of these single units to genital and somatosensory stimuli. Units responding to tactile stimuli have larger receptive fields (and more units responding) when the female squirrel monkey is estrogenized (Rose, 1977).

Also in squirrel monkeys, *morphologic* sex differences have been found in brain areas involved in the regulation of reproduction (Bubenik and Brown, 1973):

> A marked sex difference was found in the nuclear diameter of neurons in the nucleus medialis amygdalae (NMA). (P. 619) . . . Some constitutional differences could exist which could be further modified by the level of circulating hormones . . . no sex differences were found in . . . the cerebral cortex. (P. 620)

The neurons in the amygdala area mentioned above were smaller in females than they were in males. It is interesting that aggressive calls, as well as sexual behaviors, are elicited from the squirrel monkey's amygdala and hypothalamus by brain stimulation (Jurgens et al., 1967) and that sex differences in aggressive, dominance, and sexual behaviors have been elicited by brain stimulation in this species (Maurus et al., 1975; Maurus and Ploog, 1971; McLean, 1975).

C. OLD WORLD MONKEYS

Since the early work of Zuckerman and Fisher (1938) on growth of the brain in rhesus monkeys, much has been learned about the development of sex differences in the brains of Old World monkeys. Zuckerman and Fisher (1938) did point out, however, that: (1) there is no significant correlation between brain weight and body weight, and (2) the rhesus brain is 90 percent grown by six months (postnatal) of age and 100 percent grown by two years of age. Based upon 92 adult male rhesus and 102 adult female rhesus monkeys, the average weight of the adult male brain (89.45 grams) is slightly larger than the average weight of the adult female brain (82.03 grams). This gross weight measure, however, means very little as far as behavior is concerned.

The more important sex differences in the rhesus monkey central nervous system are attributable to differences in prenatal testosterone. The source of the extra testosterone present for the male fetus can be traced to the fetal testes (Resko, 1974):

> . . . differentiation is not an all-or-none process . . . different amounts of testosterone produce different amounts of androgenization in the female (P. 216)

It is highly probable that the hypothalamus is the primary focal point of male hormone action, although the action undoubtedly extends to extrahypothalamic levels and, according to some, perhaps to the entire nervous system (Livrea, 1972). The hypothalamic feedback system which secretes LH (leutenizing hormone), in response to estrogens, differs in males and females. Males are more sensitive to estrogen, i.e., there is more suppression of LH with estradiol in males (Resko, 1975).

The sexual receptivity of female rhesus monkeys can be increased by basal hypothalamic implants of estrogen (Michael, 1968). Bilateral lesions made in the ventral preoptic and anterior hypothalamic area in female rhesus monkeys block spontaneous ovulation and interfere with the release of LH in response to rising titers of estrogen (Norman, Resko, and Spies, 1976). Actually, however, estrogen has a dual action on the central nervous system: it inhibits the pituitary and stimulates the brain to release LH (see Spies and Norman, 1975, for elaboration).

Whatever the specific effects of estrogen, the hypothalamus and pituitary glands of male and female rhesus monkeys respond differently to estrogen and this difference is a function of prenatal hormones (Steiner et al., 1974). At sexual maturity, the hypothalamic-pituitary axis of male rhesus monkeys and baboons does not change (Tabei and Heinrichs, 1974); in females it probably does.

In adult female baboons, a frontal cut in the medial preoptic hypothalamus initially produces anovulation and amenorrhea. With time, however, the female shows a temporary recovery of LH release, of ovulation, and of regular menstruation. Apparently, there is another level (besides the medial preoptic area) in the hypothalamus of the baboon which can regulate gonadotrophin secretion and influence periodicity. There is also evidence that mating cycles in the female Japanese monkey also have a central nervous system origin (Hayashi et al., 1975), probably in the hypothalamus.

But what of hormone-concentrating cells in the central nervous system of Old World monkeys? Like fish, amphibians, birds, various mammals, tree shrews, and squirrel monkeys, rhesus monkeys have sex-hormone-concentrating cells in preoptic, hypothalamic, and other limbic structures (Pfaff et al., 1976). Some of these cells are undoubtedly in the amygdala because lesions in the amygdala have different effects on male and female macaques. Lesions in males decrease aggression whereas lesions in females *increase* aggression (Kling, 1974; Mass and Kling, 1975; Kling, 1975).

There is evidence that bilateral frontal lobectomies in rhesus monkeys produce reversals in sex-typical preferences and, perhaps, in sex-typical behaviors (Suomi, Harlow, and Lewis, 1970). Frontal females behave inappropriately in interactions with normal males. While nonoperated control females are submissive, frontal females direct threats and sham attacks at males (Deets et al., 1970). The frontal cortex area develops earlier in male macaques (by forty-six days of age) than it does in female macaques (Goldman, 1975) and, as we know, it is also used in solving delayed response and delayed alternation problems. Thus:

> . . . the gonadal hormones may play an inductive role in the postnatal differentiation of cortical mechanisms similar to that played in the differentiation of hypothalamic mechanisms. (Goldman, 1975, P. 62)

However, since the action in this case is postnatal, the frontal area evidently has a different "critical period" for sexual differentiation than does the hypothalamic area. Male rhesus monkeys with orbital prefrontal lesions are impaired at 2½ months of age, females are not impaired until fifteen to eighteen months of age (Goldman et al., 1974).

In the male rhesus monkey, as well as in the female, the hypothalamus and pituitary gland are important for sexual behavior. When testosterone is radio-actively labelled and infused into the brain's ventricular system, the:

> Uptake of labelled testosterone appeared to be greater in the pituitary and hypothalamus than in other areas of the brain. (Sholitan, Taylor, and Lewis, 1974, P. 537)

With regard to macaque penile erection, evoked by brain stimulation, the highest response probability is found in the preoptic area of the hypothalamus (Robinson and Mishkin, 1968).

It is likely, however, that the sex hormones do not act directly on the brain to change behavior but rather act through chemical mediators such as neuro-transmitters. For example, sex hormones probably regulate reproductive behavior by modifying the activity of serotonin-containing neural systems. Serotonin is but one of a group of neurophysiologically important chemicals called biogenic amines (cf. Gradwell, Everitt, and Herbert, 1975; Vestergaard, 1977):

> One of the major growing points in this general area concerns the problem of *how* hormones act within the CNS to modulate neural activity and hence induce alterations in behaviour. From a large number of experiments on sub-human species . . . it has become clear that the biogenic amines noradrenaline, dopamine, and 5-hydroxytryptamine may play a fundamentally important role in mediating the effects of hormones on behaviour. (Everitt, 1976, abstract)

Drugs which specifically affect biogenic-amine-mediated neurotransmission, also change the action of the sex hormones. The neurotoxic destruction of biogenic-amine-mediated neurons affects sex-hormone-mediated behavior. Finally, sex hormones have a marked effect on biogenic amine metabolism in the cerebral cortex (Everitt, 1976).

D. THOUGHTS ON *HOMO SAPIENS*

Despite all of the basic biological information on the importance of hormones, other chemicals, and specific neural structures on sex differences in nonhuman primates, the same brain can *still* contain mechanisms for both male and female behavior and the most important determinants of sex differences in people almost certainly remain as "cognitive templates in the brain built up through experience." (Beach, 1976, P. 473.) We are not denying this when we suggest that the neurophysiological research done on nonhuman primate sex differences should, in some way, be compared to clinical data on *Homo sapiens*. There are, indeed, some sex differences in emotional and cognitive behavior in humans. We need to examine their adaptive and neural bases (Gray, 1971). It is still possible that, as June Reinisch (1974) suggested, prenatal and later hormonal effects:

> . . . must be seen as limited in scope and of a diffuse quality rather than directly related to specific behaviors . . . as setting a *bias* on the neural substratum, which in turn predisposes the individual to the acquisition and expression of sexually dimorphic patterns of response and behavior. (P. 51)

We also do not want to leave this chapter by giving the impression that all nonhuman primate brains are alike. They are not. While it is true that the hormone-concentrating cells, frontal areas, and hypothalamic-pituitary axis go back to a most remote common ancestry, prosimians, monkeys, apes, and man definitely differ in other aspects of central neurology and neurophysiology. Here is but one example:

> . . . the human Sylvian fissure is longer on the left than on the right. The chimpanzee brains had a similar asymmetry but to a lesser degree than the human brains. The rhesus brains, however, showed no significant differences between left and right fissure lengths. (Yeni-Komshian and Benson, 1976, P. 387)

We can add to this such large differences between apes and people as the lack of motor and sensory speech areas in apes, and a smaller overall cerebral cortex in apes than in *Homo sapiens*.

REFERENCES

Beach, F. A. Cross-species comparisons and the human heritage. *Archives of Sexual Behavior,* 1976, **5**, 469–485.

Bubenik, G. A. and Brown, G. M. Morphologic sex differences in the primate brain areas involved in regulation of reproductive activity. *Experientia,* 1973, **15** (5), 619–621.

Deets, A. C., Harlow, H. F., Singh, S. D. and Blomquist, A. J. Effects of bilateral lesions of the frontal granular cortex on the social behavior of rhesus monkeys. *Journal of Comparative and Physiological Psychology,* 1970, **72** (3), 452–461.

Everitt, B. J. Growing points in research on sexual and aggressive behavior. Paper presented at the *International Primatological Society* Meeting, Cambridge, England, August, 1976.

Goldman, P. S. Age, sex, and experience as related to the neural basis of cognitive development. In Buchwald, N. A. and Brazier, M. A. B. (Eds.) *Brain Mechanisms in Mental Retardation.* New York: Academic Press, 1975, pp. 379-392.

Goldman, P. S. Maturation of the mammalian nervous system and the ontogeny of behavior. *Advances in the Study of Behavior,* 1976, **7**, 1-90.

Goldman, P. S., Crawford, H. T., Stokes, L. P., Galkin, T. W. and Rosvold, H. E. Sex-dependent behavioral effects of cerebral cortical lesions in the developing rhesus monkey. *Science,* 1974, **186**, 540-542.

Gradwell, P. B., Everitt, B. J. and Herbert, J. 5-hydroxytryptamine in the central nervous system and sexual receptivity of female rhesus monkeys. *Brain Research,* 1975, **88**, 281-293.

Gray, J. A. Sex differences in emotional and cognitive behaviour in mammals including man: Adaptive and neural bases. *Acta Psychologica,* 1971, **35**, 89-111.

Hagino, N. Hypothalamic regulation of ovulation and menstruation in the baboon (nonhuman primate). *Federation Proceedings,* 1976, **35**, 701.

Hayashi, M., Oshima, K., Yamaji, T. and Shimamoto, K. LH levels during various reproductive states in the Japanese monkey (*Macaca fuscata fuscata*). In Kondo, S., Kawai, M. and Ehara, A. (Eds.) *Contemporary Primatology.* Basel: Karger, 1975, pp. 152-157.

Jurgens, U., Maurus, M., Ploog, D. and Winter, P. Vocalization in the squirrel monkey (*S. sciureus*) elicited by brain stimulation. *Experimental Brain Research,* 1967, **4**, 114-117.

Karczmar, A. G. Neurotransmitters in the modulation of sexual behavior. *Psychopharmacology Bulletin,* 1975, **11**, 40-42.

Keefer, D. A. and Stumpf, W. E. Atlas of estrogen-concentrating cells in the central nervous system of the squirrel monkey. *Journal of Comparative Neurology,* 1975, **160** (4), 419-441. (a)

Keefer, D. A. and Stumpf, W. E. Estrogen-concentrating neuron systems in the brain of the tree shrew. *General and Comparative Endocrinology,* 1975, **26** (4), 504-516. (b)

Kling, A. Differential effects of amygdalectomy in male and female nonhuman primates. *Archives of Sexual Behavior,* 1974, **3**, 129-134.

Kling, A. Brain lesions and aggressive behavior of monkeys in free living groups. In Fields, W. S. and Sweet, W. H. (Eds.) *Neural Bases of Violence and Aggression.* St. Louis: Warren H. Green, Inc., 1975, pp. 146-160.

Livrea, G. La base neuroendocrine del comportamento sessuale. *Archives of Science and Biology,* 1972, **55**, 61-102.

MacLean, P. Role of pallidal projections in species-typical display behavior of squirrel monkey. *Transactions of the American Neurological Association,* 1975, **100**, 25-28.

Mass, R. and Kling, A. Social behavior in stumptailed macaques (*Macaca speciosa*) after lesions of the dorsolateral frontal cortex. *Primates,* 1975, **16** (3), 239-252.

Maurus, M., Kuhlmorgen, B., Hartmann-Wiesner, E. and Pruscha, H. An approach to the interpretation of the communicative meaning of visual signals in agonistic behavior of squirrel monkeys. *Folia Primatologica,* 1975, **23**, 208-226.

Maurus, M. and Ploog, D. Social signals in squirrel monkeys: Analysis by cerebral radio stimulation. *Experimental Brain Research,* 1971, **12**, 171-183.

Michael, R. P. Neural and non-neural mechanisms in the reproductive behavior of primates. *Progress in Endocrinology: Proceedings of the Third International Congress of Endocrinology.* Mexico, D.F., June-July 1968, pp. 302-309.

Neuman, F., Steinbeck, H. and Hahn, J. D. Hormones and brain differentiation. In Martini, L., Motta, M. and Fraschini, F. (Eds.) *The Hypothalamus.* New York: Academic Press, 1970, pp. 1-35.

Norman, R. L., Resko, J. A. and Spies, H. G. The anterior hypothalamus: How it affects gonadotropin secretion in the rhesus monkey. *Endocrinology,* 1976, **99** (1), 59-71.

Pfaff, D. W. The neuroanatomy of sex hormone receptors in the vertebrate brain. In Kumar, T. C. A. (Ed.) *Neuroendocrine Regulation of Fertility.* Basel: Karger, 1976, pp. 30-45.

Pfaff, D. W., Gerlach, J. L., McEwen, B. S., Ferin, M., Carmel, P. and Zimmerman, E. A. Autoradiographic localization of hormone concentrating cells in the brain of the female rhesus monkey. *Journal of Comparative Neurology,* 1976, **170**, 279-293.

Phoenix, C. H. Prenatal testosterone in the nonhuman primate and its consequences for behavior. In Friedman, R. C., Richart, R. M. and Vande Wiele, R. L. (Eds.) *Sex Differences in Behavior.* New York: Wiley, 1974, pp. 19-32.

Reinisch, J. M. Fetal hormones, the brain, and human sex differences: A heuristic, integrative review of the recent literature. *Archives of Sexual Behavior,* 1974, **3**, 51-90.

Resko, J. A. The relationship between fetal hormones and the differentiation of the central nervous system in primates. In Montagna, W. and Sadler, W. A. (Eds.) *Reproductive Behavior.* New York: Plenum, 1974, pp. 211-222.

Resko, J. A. Fetal hormones and their effect on the differentiation of the central nervous system. *Federation Proceedings,* 1975, **34** (8), 1650-1655.

Robinson, B. W. and Mishkin, M. Penile erection evoked from forebrain structures in *Macaca mulatta. Archives of Neurology,* 1968, **19**, 184-198.

Rose, J. D. Facilitation by estradiol of single unit responses to genital and somatosensory stimuli in the midbrain and pons of the female squirrel monkey. *Anatomical Record,* 1977, **187**, 697-698.

Sar, M. and Stumpf, W. E. Neurons of the hypothalamus concentrate (^{3}H) progesterone or its metabolites. *Science,* 1973, **183**, 1266-1268.

Sholiton, L. J., Taylor, B. B. and Lewis, H. P. The uptake and metabolism of labelled testosterone by the brain and pituitary of the male rhesus monkey (*Macaca mulatta*). *Steroids,* 1974, **24** (4), 537-547.

Skeen, L. C. and Masterson, R. B. Origins of anthropoid intelligence. III: Role of prefrontal system in delayed-alternation and spatial-reversal learning in a prosimian (*Galago senegalensis*). *Brain, Behavior and Evolution,* 1976, **13**, 179-195.

Spies, H. G. and Norman, R. L. Interaction of estradiol and LHRH on LH release in rhesus females: Evidence for a neural site of action. *Endocrinology,* 1975, **97** (3), 685-692.

Steiner, R. A., Clifton, D. K., Spies, H. G. and Resko, J. A. Feedback control of LH by estradiol in female, male, and female pseudo-hermaphroditic rhesus monkeys. *Endocrinology,* 1974, **94** (Supp.), A-195, 280.

Stumpf, W. E., Sar, M. and Keefer, D. A. Anatomical distribution of estrogen in the central nervous system of mouse, rat, tree shrew, and squirrel monkey. *Advances in Bioscience,* 1974, **15**, 77-88.

Suomi, S. J., Harlow, H. F. and Lewis, J. K. Effect of bilateral frontal lobectomy on social preferences of rhesus monkeys. *Journal of Comparative and Physiological Psychology,* 1970, **70** (3), 448-453.

Tabei, T. and Heinrichs, W. L. Metabolism of progesterone by the brain and pituitary gland in subhuman primates. *Neuroendocrinology,* 1974, **15**, 281-289.

Valenstein, E. S. Steroid hormones and the neuropsychology of development. In Isaacson, R. L. (Ed.) *The Neuropsychology of Development.* New York: Wiley, 1968, pp. 1-39.

Vestergaard, P. Sexual mechanisms in the brain. Neurophysiological and neurochemical aspects. *Ugeskr. Laeg.,* 1977, **139**, 10-13.

Yeni-Komshian, G. H. and Benson, D. A. Anatomical study of cerebral asymmetry in the temporal lobe of humans, chimpanzees and rhesus monkeys. *Science,* 1976, **192**, 387-389.

Zuckerman, S. and Fisher, R. B. Growth of the brain in the rhesus monkey. *Proceedings of the Zoological Society of London,* 1938, **107**, 529-538.

31.

Cyclical Variation

Up to this point in our review of sex differences, we have discussed the degree of variability shown by male and female primates primarily in regard to age differences. Variability in sex differences is also apparent in regard to cyclical and noncyclical factors. In the present chapter, we will discuss cyclical variability and in Chapter 32, we will examine noncyclical variability.

A. GENERAL COMMENTS

As pointed out by Neuman, Steinbeck, and Hahn (1970), "most patterns of behavior are more or less closely correlated with the reproductive cycle." (P. 15.) Sex differences in behavior, as we know, are also correlated with reproductive cycles. Many differences in behavior associated with gender and with reproductive cycles are in large part determined by the levels of fetal testicular androgens present in the first trimester of pregnancy. For example, the hypothalamus is organized differently with regard to differences in cyclicity. According to Valenstein (1968):

> . . . tonic gonadotrophic release can be maintained by the MBH (medial basal hypothalamus) alone, but connections with the anterior HTH (hypothalamus) are necessary for cyclicity . . . one of the functions of early exposure to androgen is to disrupt this connection between the anterior HTH and the MBH. (P. 30)

Thus, we might expect female primates, particularly those of species showing strong physical sexual dimorphism, to show more behavioral cyclicity than we might expect of males, especially in those behaviors which vary with the estrous cycle. Many if not most primates *do* have estrous cycles, and they are behaviorally different not only within their monthly cycles but also at different seasons of the year (Butler, 1974).

However, there are other kinds of behavioral cyclicity besides those directly associated with the estrous cycle. There are diurnal cycles, for example, that are less obviously associated with reproduction than are the monthly cycles.

The present chapter will remind us of much of the material covered in our chapters on sexual behavior and hormones, but this chapter has a different emphasis. Instead of concentrating on sexual behavior or on any specific behavior per se, we will focus on sex differences in *cyclicity,* regardless of the specific behavior in question.

B. PROSIMIANS AND NEW WORLD MONKEYS

1. Prosimians

Sex differences in behavior are sometimes revealed in activity cycles. Among tree shrews (*Tupaia glis*) males are more active and rest less often than do females (Sprankel and Richarz, 1976). However, most of the behavioral cyclicity seen in prosimians is related directly to reproduction. For example, in the ring-tailed lemur (*Lemur catta*), mating is restricted to a brief period and coincides with vaginal estrus. There is little or no emancipation of sexual behavior from hormonal influences among prosimians (Evans and Goy, 1968). For example, among *Microcebus murinus,* sexual activation of the male occurs during January. There is a loss of body weight, an increase in oxygen consumption, and an increase in testes size in the male. But the control of prosimian mating is not completely *internal.* Mating in *Microcebus,* for example, occurs only when the female's body weight is less than that of the male's (Perret, 1977). In *Lemur catta,* the cyclicity of the breeding season may differ by as much as six months when lemurs kept in Portland, Oregon are compared to those remaining in their native Madagascar. Lemurs are evidently sensitive to photoperiod changes (Van Horn, 1975). There is also seasonal variation in the size of the testes in lemurs (Bogart, Cooper, and Benirschke, 1977). In studies of ring-tailed lemur vocalizations, daily or diurnal rhythms also have been found (King and Fitch, 1977).

One other very important reproductive cycle should be discussed at this point—reproductive turnover. Prosimians and, to a lesser extent New World monkeys, show high reproductive turnover relative to other primates. This means that they have short gestation periods, multiple births, and short interbirth intervals (Leutenegger, 1977).

2. New World Monkeys

Among some of the more primitive New World marmosets and tamarins (e.g., *Saguinus oedipus*) there are fewer sex differences overall than exist among the cebids. For example, there are no seasonal differences in the reproductive tracts of males and females in *Saguinus oedipus* (Dawson, 1976).

Activity cycles in the cebids depend upon whether the species in question is nocturnal or diurnal. In the nocturnal owl monkey (*Aotus trivirgatus*), for example, most of the activity occurs during the animal's dark cycle but activity also increases at feeding times (Leibrecht and Kelley, 1977).

Squirrel monkeys in the laboratory (*Saimiri sciureus*) have a mating season during which there is a dramatic increase in the amount of sniffing of females by males (Hennessy et al., 1977). Males exhibit a six to twenty week season of copulation and ejaculation, the onset of which coincides with an annual peak in male body weight (the "fatted male" phenomenon, Wilson, 1977). The male squirrel monkey also undergoes an annual testes cycle with increased spermatogenesis evident during the "fatted" condition (DuMond and Hutchinson, 1967). However, in recent years it has become evident that, for the squirrel monkey:

> Animals of *both* sexes show significant physiological and behavioral changes during the yearly cycle. Whether the environmental timing factors affect both the males and females equally, or affect only one sex directly and the other indirectly through social communication is not clear. (Baldwin, 1970, P. 317)

It *is* clear, however, that androgen and castration affect body weight in both the male and the female. The "fatted" seasonal condition in the field is regulated by androgen and females share the responsiveness to androgen with males (Nadler and Rosenblum, 1972).

In addition to their seasonal sensitivity to androgens, squirrel monkeys also show cyclicity in response to estrogens and progesterone. If females receive estrogen they show more affiliative behavior and more grasping of males. The males, in turn, show more following of the female. If the female receives progesterone, on the other hand, grasping by the female and following by the male both decline. There are seasonal changes in these behaviors which are hormonally mediated (Anderson and Mason, 1977).

During the breeding season of *Saimiri sciureus* (the squirrel monkey), there are also smaller cyclic changes in behavior associated with normal cyclic changes in estrogens and progesterone. Because squirrel monkeys (and many other New World monkeys) do not exhibit a conspicuous menstruation, it is difficult to estimate the length of their reproductive cycles. However, Wolf, O'Connor, and Robinson (1977) have estimated that the "monthly" cycle of the squirrel monkey lasts only nine days!

C. THE RHESUS MONKEY (*MACACA MULATTA*)

The rhesus monkey, unlike the squirrel monkey, has a relatively obvious menses. The reproductive cycle for the rhesus is twenty-eight days, much longer than the cycle of the squirrel monkey.

Bilateral lesions of the ventral-preoptic-anterior hypothalamic area in rhesus females block spontaneous ovulation and interfere with cyclicity (Norman, Resko, and Spies, 1976). In the nonlesioned female rhesus, however, sexual behavior is dependent upon, and can be induced by the administration of ovarian hormones (Michael and Bonsall, 1977). Estrogen increases lever pressing for access to males and progesterone decreases lever pressing for males (Michael and Keverne, 1972). Grooming of the male by the female reaches a minimum at midcycle when male grooming of the female reaches a maximum (Michael and Herbert, 1963).

At ovulation, the time spent near the male by the female increases, male ejaculations increase, female sex-skin coloration increases, and female sexual solicitations increase (Czaja and Bielert, 1975).

There is also a *premenstrual* increase in some of the above activities including an increase in vaginal responsiveness (contractility and vasocongestion of the vagina). Hence, there are apparently two cyclic peaks of female receptivity in the rhesus—one at ovulation and one during the premenstrual period (Erikson, 1967). Ovarian hormones apparently exert a greater cyclic influence over rhesus sexual behavior in a social context than they do in a standard laboratory pair-test paradigm (Gordon, 1977).

Sexual receptivity in reasonably stable social groups is cyclical in each female and confined to relatively discrete periods of from four to seventeen days per cycle. These cycles terminate abruptly when the females' progesterone concentrations rise (Gordon, Rose, and Bernstein, 1977). In the summer months, however, cycles are irregular and menstruation is infrequent (Keverne and Michael, 1970). Also, when preadolescent females are brought into captivity from the wild they sometimes become retarded in developing adult cyclicity (Mahoney, 1975). Females are more distressed when separated from males during estrus than when separated from them outside of estrus (Maple, Erwin, and Mitchell, 1974). Attacks on females by males are more frequent at menstruation than at ovulation (Sassenrath, Rowell, and Hendricks, 1973). Cyclicity in the monthly behaviors of female rhesus monkeys continue until the females reach menopause at twenty-seven to twenty-eight years of age (van Wagenen, 1970).

There are seasonal as well as monthly cycles in rhesus behavior. In free-ranging rhesus the number of grooming pairs varies with the time of year. There is a peak in the number of grooming pairs during the fall and the lowest number of grooming pairs occurs in the spring (Teas et al., 1977). Vandergrift (1977) has examined seasonal fluctuations in behavior in captivity.

There are also *physical* changes which occur seasonally in rhesus. Moltings, beginning on the tail and crown and continuing to the legs, hips, arms, back, and terminating on the flanks, occur annually. Adult males and nonpregnant females are the first to molt. Molting begins at the end of the mating season and it is probably affected by changes in hormones (Vessey and Morrison, 1970).

An annual rhythm in plasma testosterone levels, which peak in the fall, occurs in rhesus monkeys maintained in a laboratory (Plant et al., 1974). In free-ranging rhesus males, there are seasonal changes in the production of semen with increases in the fall breeding season (Zamboni, Conaway, and van Pelt, 1974). Maximum ejaculations in English laboratories are said to occur in December whereas in free-ranging rhesus in India they occur between November and January. Testicular regression starts in February (Michael and Keverne, 1971).

Apparently, these seasonal variations do not appear in young adult males, even though they are hormonally and sexually mature (Resko, 1967). Moreover, mating seasonality *does* occur in castrated adult males. According to Michael and Wilson (1975):

> . . . the castrated males demonstrated a clear-cut annual behavioural rhythm that could not have been mediated through the gonads. These behavioural changes may depend on an alteration in the threshold of a brain mechanism that is not mediated through the gonad. (P. 328)

There is definitely a breeding season in *Macaca mulatta* (the rhesus). On Cayo Santiago (an island off Puerto Rico) the mating season lasts from August to March (Vandenbergh and Vessey, 1966). In India the mating season starts somewhat later and ends a little earlier. However, rhesus males in India show a marked change in the color of the sexual skin nearly two months before the onset of mating (Lindburg, 1971). During the breeding season, there are increases in group changing (Lindburg, 1969), leadership change (Neville, 1968), the formation of all-male groups (Boelkins and Wilson, 1972), troop fission (Missakian, 1973), and wounding (Hausfater, 1972; Wilson and Boelkins, 1970).

There are some who feel that most of the seasonal changes seen in males are indirect changes communicated to males socially by females after hormonal changes in females. Gordon and Bernstein (1973) believe that seasonal influences act *only* on females but females communicate their endocrine status to males even without direct physical contact. With access to receptive females, plasma testosterone levels in males increase two or threefold within a week (Rose, Gordon, and Bernstein, 1972).

In captivity, *Macaca mulatta* sometimes breeds throughout the year and, although seasonal peaks sometimes occur, a season based on exposure to light is not seen in some laboratories (Birkner, 1970). In other captive groups, however, the timing of the annual increase in mating (the fall) *is* similar to that seen in free-ranging groups in India; and, if photoperiod is controlled for two years, the seasonal cycle disappears (Michael and Zumpe, 1976). After a change from one environment to another, two years are required for a full shift in seasonal breeding cycles (see Varley and Vessey, 1977).

There are also *diurnal* rhythms in hormones in rhesus monkeys (Bernstein, 1974). Testosterone reaches a high point in rhesus males at 10 P.M. and a low point between 4 and 10 A.M. (just the reverse of the human cycle) (Southren and Gordon, 1975; Plant and Michael, 1971). Since castration removes the cyclicity, the nocturnal rise is dependent on the testes. Corticosteroid levels decline when testosterone rises (Michael, Setchell, and Plant, 1974). Rhesus also shows diurnal sleep rhythms aside from sleeping at night. Approximately 6 percent of the day is spent in sleeping. Female rhesus monkeys sleep more than do males, but females do not sleep in the early afternoon when males sleep the most (Swett, 1969).

There are also diurnal variations in spontaneous uterine activity in nonpregnant females. Uterine activity is highest during periods of light and lowest in darkness. Norepinephrine has a causal role in the occurrence of this uterine activity (Harbert and Zuspan, 1977).

D. OTHER OLD WORLD MONKEYS

1. Other Macaques

The menstrual cycles of other macaques are similar to those of the rhesus. As in rhesus monkeys, there are no menstrual cycles (or there are irregular ones) outside of the mating season in the Japanese macaque (Nigi, 1975). Japanese monkeys have a definite breeding season. The factors involved in controlling the breeding season in Japanese macaques is, as in rhesus, not of pituitary or gonadal origin but apparently lies in the higher parts of the brain (Hayashi et al., 1975).

Adult Japanese macaque males leave troops and change troops more often during the breeding season than they do outside the breeding season (cf. Kawamura, 1965). At about seven years of age they leave a solitary existence and approach, mate, and attempt to join a strange troop during the *breeding season* (Sugiyama and Ohsawa, 1974). Agonistic behavior among six- to nine-year-old males increases during the breeding season (Sugawara, 1976).

The birth season among Japanese macaques is from March to July (Nigi, 1976). In some troops, male care of infants increases at this time (Itani, 1963).

Among stumptailed macaques (*Macaca arctoides*), the menstrual cycle resembles that of the rhesus and behavior changes with the stage of estrus. Male and female yawning increases during the early luteal phase. Male grooming is lowest before menstruation and female grooming declines sharply just after ovulation (Slob et al., 1975).

With regard to annual cycles, the stumptail macaque differs from the rhesus and Japanese macaques. Stumptails show an unusual pattern of three birth peaks throughout the year (Estrada and Estrada, 1976).

With regard to diurnal cycles, the pigtail macaque (*M. nemestrina*) male's testosterone cycle might be unique. Unlike the rhesus, but *like* the human male, the pigtail male has sexual behavior levels that are highest in the early morning and lowest in the evening (Martenson et al., 1977). Diurnal variation in pigtail testosterone levels has not been studied, however.

2. Baboons and Other Old World Monkeys

As in macaques, there is hypothalamic regulation of ovulation, menstruation, and cyclical behavior associated with estrus in baboons (Hagino, 1976). However, there are some behaviors that are not affected by estrus in baboons. Adult female dominance, for example, does not change during estrus in the yellow baboon (*Papio cynocephalus*), although dominance changes among males are related to the presence of estrous females (Hausfater, 1975). There are increases in male grooming of swollen females and increases in grooming of males by estrous females in olive baboons (Rowell, 1968).

With regard to the estrous cycle itself, tactile isolation leads to shorter cycles but to greater swelling (Rowell, 1970). But estrus itself causes only slight changes in intrasexual social organization in *Papio anubis* (Rowell, 1969). Estrus also has little effect on the behavior of one-male groups of gelada baboons (*Theropithecus gelada*) (Dunbar, 1976). In chacma baboons, however, females in the luteal phase of their cycles are more often victims of male attacks (Saayman, 1971).

Among mangabeys (*Cercocebus galeritus*), some calls of adult males depend upon the sexual activity cycles of the males (Quris, 1975). Vervet males (*Cercopithecus aethiops*) increase the number of their red, white, and blue dominance displays during the breeding season (Struhsaker, 1967). The departure of subadult males from *C. campbelli* groups occurs during the birth season (Hunkeler, Bourliere, and Bertrand, 1972).

In the neotenous talapoin monkey (*Miopithecus talapoin*), females join male subgroups at the height of the mating season. (There is no consort behavior, however.) (Rowell and Dixson, 1975).

E. APES

1. Chimpanzees (*Pan spp.*)

In data comparing the hormonal patterns during pregnancy of rhesus monkeys, chimpanzees (*Pan troglodytes*), and humans it is apparent that chimpanzees and humans, but not rhesus, show a separate fetoplacental hormone contribution (estriol). Thus, the hormonal cyclicity of apes is more like that of humans than is the hormonal cyclicity of monkeys (Hobson, 1976).

Female hormonal cyclicity in chimpanzees and humans is present in early infancy. Serum FSH levels are higher in males than in females during infancy in both species. Moreover, in chimpanzee infant females, but not in males, the FSH levels show periodic variation (Faiman, Winter, and Grotts, 1973). The periodic variation disappears during childhood and returns at puberty when menstruation starts. Menstrual cycles in the chimpanzee continue until the late forties. Chimpanzee females apparently die before the age of fifty without cessation of the menstrual cycle (even though they are infertile) (Graham and McClure, 1977).

Despite these many years of reproductive cycling, there are very few studies on chimpanzees which show behavioral changes (other than sex) that correlate with the stage of the cycle. Apparently, however, estrogen *does* increase dominance in the female chimpanzee (see Chapter 17). There is, of course, periodic variation in sexual behavior in female common chimpanzees that is associated with the cycle stage.

Data collected throughout the females' sexual cycle for the species *Pan paniscus* (the pigmy chimpanzee), show that this chimpanzee differs from the common chimpanzee (*Pan troglodytes*). The common chimpanzee female engages in sexual behavior only when in estrus, whereas *Pan paniscus* females engage in copulatory behavior during all phases of the sexual cycle. The genitalia of female pigmy chimpanzees remain turgid throughout the cycle (Savage and Bakeman, 1976).

2. Gorillas (*Gorilla spp.*)

The gorilla estrous cycle is twenty-eight days long and the female displays estrous behavior for only two to three days. There is some copulation during pregnancy but the intervals between pregnancy copulations are extremely variable (Harcourt and Stewart, 1976). In subadult gorillas, there is increased intersexual play when the female is in estrus (Keiter, 1977). Gorillas, unlike pigmy chimpanzees, mate in a cyclic manner. Matings are closely related to the degree of female genital swelling (Nadler, 1975). Sexual swelling, although much less noticeable than in chimpanzees, can be used as a useful predictor of ovulation and sexual behavior (Nadler et al., 1977). Very few other data on behavioral cyclicity have been published on gorillas.

3. Orangutans

Maple, Zucker, and Dennon (in press) have reported on a female orangutan which exhibited proceptive sexual behavior for four to six days every twenty-six to thirty days. They believe that these peaks coincide with ovulation (but also see Maple and Zucker, 1977; Nadler, 1976). There is no swelling with estrus in the orangutan female (cf. Rijksen, 1975). The adult male's long call varies in frequency according to the reproductive cycle of the female (Rijksen, 1975).

F. COMMENTS ON HUMANS

As we know, in humans there is continual rather than periodic sexual activity (Jensen, 1976). This one difference between human and nonhuman primates has been discussed by Butler (1974). He suggests that the term menstrual cycle should be confined to the sex cycle of women because there is no evidence that proceptivity or receptivity is affected by ovulation in women. However, the reproductive pattern of *Pan paniscus* is very much like that of humans in this regard. In addition, the question of whether or not ovulation affects female proceptivity and receptivity is still an open one. Apparently, humans are capable of using olfaction to identify the sex of a conspecific in their environment and the stimulus value of women for men varies with their ovulatory cycle (Spencer, 1977). Furthermore, women are sensitive to musklike odors at menstruation and at ovulation. The quality of these odors for women is altered by progesterone. Women living together and seeing males fewer than three times per week experience a synchrony of their menstrual cycles (Mai, 1972).

As in nonhuman female primates, the menstrual cycles of human female primates can be disrupted by environmental change. As we have noted psychogenic amenorrhea occurs frequently in airline stewardesses (Greep, 1968). There is indeed apparently "some interpersonal physiological process which affects the menstrual cycle." (McClintock, 1971, P. 245.) Moreover, at the time of ovulation, some women do indeed feel a strong sex drive (Jones, Shainberg, and Byer, 1977).

Human females show large increases in estrogen at around the age of eleven, but the estrogen levels do not become cyclical until eighteen months prior to menarche (at the time of the growth spurt). Removal of the ovaries in human females has no effect on sexual desire. Removal of the adrenals, on the other hand, decreases sexual desire. Apparently androgens, not estrogens, are responsible for desire in both sexes (Hamburg and Lunde, 1966).

In *Homo sapiens*, as we already know, androgens peak in the early morning hours in men, whereas estrogens and FSH are higher in both sexes early in the morning. There is no good evidence for the male daily cycle in testosterone appearing in women unless the women are ovariectomized or on the pill (Curtis, 1972).

The material presented in the present chapter should serve to remind us that cyclical variability as well as age differences overlay many of the sex differences in behavior that we have referred to in earlier chapters of our book. In the next chapter we will discuss noncyclical variability in male and female behavior.

REFERENCES

Anderson, C. O. and Mason, W. A. Hormones and social behavior of squirrel monkeys (*Saimiri sciureus*). I. Effects of endocrine status of females on behavior within heterosexual pairs. *Hormones and Behavior,* 1977, **8**, 100-106.

Baldwin, J. D. Reproductive synchrony in squirrel monkeys (*Saimiri*). *Primates,* 1970, **11**, 317-326.

Bernstein, I. S. Behavioral and environmental events influencing primate testosterone levels. *Journal of Human Evolution,* 1974, **3**, 517-525.

Birkner, F. E. Photic influences on primate (*Macaca mulatta*) reproduction. *Laboratory Animal Care,* 1970, **20** (2), 181-185.

Boelkins, R. C. and Wilson, A. P. Intergroup social dynamics of the Cayo Santiago rhesus (*Macaca mulatta*) with special reference to changes in group membership by males. *Primates,* 1972, **13** (2), 125-140.

Bogart, M. H., Cooper, R. W. and Benirschke, K. Reproductive studies of black and ruffed lemurs. *International Zoo Yearbook,* 1977, **17**, 177-182.

Butler, H. Evolutionary trends in primate sex cycles. *Contributions to Primatology,* 1974, **3**, 2-35.

Curtis, G. C. Psychosomatics and chronobiology: Possible implications of neuroendocrine rhythms: A review. *Psychosomatic Medicine,* 1972, **34** (3), 235-256.

Czaja, J. A. and Bielert, C. Female rhesus sexual behavior and distance to a male partner: Relation to stage of menstrual cycle. *Archives of Sexual Behavior,* 1975, **4**, 583-597.

Dawson, G. A. Behavioral ecology of the Panamanian tamarin *Saguinus oedipus* (*Callitrichidae–Primates*). *Dissertation Abstracts International,* 1976, **B37**, 645-646.

DuMond, F. V. and Hutchinson, T. C. Squirrel monkey reproduction: The "fatted" male phenomenon and seasonal spermatogenesis. *Science,* 1967, **158**, 1467-1470.

Dunbar, R. I. M. Oestrus behaviour and social relations among gelada baboons. Paper presented at the *International Primatological Society* Meeting, Cambridge, England, August, 1976.

Erikson, L. B. Relationship of sexual receptivity to menstrual cycles in adult rhesus monkeys. *Nature,* 1967, **216** (5112), 299-301.

Estrada, A. and Estrada, R. Birth and breeding cyclicity in an outdoor living stumptail macaque (*Macaca arctoides*) group. *Primates,* 1976, **17** (2), 225-231.

Evans, C. S. and Goy, R. W. Social behaviour and reproductive cycles in captive ring-tailed lemurs (*Lemur catta*). *Journal of Zoology of London,* 1968, **156**, 181-197.

Faiman, C., Winter, J. S. D. and Grotts, D. Gonadotropins in the infant chimpanzee: A sex difference. *Proceedings of the Society for Experimental Biology and Medicine,* 1973, **144**, 952-955.

Gordon, T. P. The influence of ovarian hormones on male sexual behavior in a social group of rhesus monkeys. Paper presented at the *Animal Behavior Society* Meeting, University Park, Pennsylvania, June, 1977.

Gordon, T. P. and Bernstein, I. S. Seasonal variation in sexual behavior of all-male rhesus troops. *American Journal of Physical Anthropology,* 1973, **38**, 221-225.

Gordon, T. P., Rose, R. M. and Bernstein, I. S. Social and hormonal influences on sexual behavior in the rhesus monkey. Paper presented at the *American Society of Primatologists* Meeting, Seattle, Washington, April, 1977.

Graham, C. E. and McClure, H. M. Ovarian tumors and related lesions in aged chimpanzees. *Veterinary Pathology,* 1977, **14**, 380-386.

Greep, R. O. Hypothalamic-pituitary-ovarian relationships. In Hoffman, M. and Kleinman, R. L. (Eds.) *Advanced Concepts in Contraception.* Amsterdam: Excerpta Medica Foundation, 1968, pp. 108-110.

Hagino, N. Hypothalamic regulation of ovulation and menstruation in the baboon (non-human primate). *Federation Proceedings,* 1976, **35**, 701.

Hamburg, D. A. and Lunde, D. T. Sex hormones in the development of sex differences in human behavior. In Maccoby, E. E. (Ed.) *The Development of Sex Differences.* Stanford, California: Stanford University Press, 1966, pp. 1-24.

Harbert, G. M. and Zuspan, F. P. Relationship between catecholamines and the periodicity of spontaneous uterine activity in a nonpregnant primate (*Macaca mulatta*). *American Journal of Obstetrics and Gynecology,* 1977, **129**, 51-58.

Harcourt, A. H. and Stewart, K. J. Sexual behaviour of wild mountain gorilla. Paper presented at the *International Primatological Society* Meeting, Cambridge, England, August, 1976.

Hausfater, G. Intergroup behavior of free-ranging rhesus monkeys (*Macaca mulatta*). *Folia Primatologica,* 1972, **18**, 78-107.

Hausfater, G. Estrous females: Their effects on the social organization of the baboon group. *Proceedings of the Symposia of the Fifth International Congress of Primatology.* Tokyo: Japan Science Press, 1975, pp. 117-127.

Hayashi, M., Oshima, K., Yamaji, T. and Shimamoto, K. LH levels during various reproductive states in the Japanese monkey (*Macaca fuscata fuscata*). In Kondo, S., Kawai, M. and Ehara, A. (Eds.) *Contemporary Primatology.* Basel: Karger, 1975, pp. 152-157.

Hennessy, M. B., Coe, C. C., Mendoza, S. P., Lowe, E. L. and Levine, S. Seasonal fluctuations and sex differences in olfactory behavior of the squirrel monkey (*Saimiri sciureus*). Paper presented at the *American Society of Primatologists* Meeting, Seattle, Washington, April, 1977.

Hobson, W. Reproductive endocrinology of female chimpanzees: A suitable model of humans. *Journal of Toxicology and Environmental Health,* 1976, **1**, 657-668.

Hunkeler, C., Bourliere, F. and Bertrand, M. Le comportement social de la Mone de Lowe (*Cercopithecus campbelli lowei*). *Folia Primatologica,* 1972, **17**, 218-236.

Itani, J. Paternal care in the wild Japanese monkey, *Macaca fuscata.* In Southwick, C. H. (Ed.) *Primate Social Behavior.* Princeton, N.J.: Van Nostrand, 1963, pp. 91-97.

Jensen, G. D. Comparisons of sexuality of chimpanzees and humans. Paper presented at the *International Primatological Society* Meeting, Cambridge, England, August, 1976.

Jones, K. L., Shainberg, L. W. and Byer, C. O. *Sex and People.* New York: Harper & Row, 1977.

Kawamura, S. Matriarchal social ranks in the Minoo-B troop. In Imanishi, I. and Altmann, S. (Eds.) *Japanese Monkeys.* Atlanta, Georgia: Altmann, 1965, pp. 105-112.

Keiter, M. D. Reproductive behavior in subadult captive lowland gorillas (*Gorilla gorilla gorilla*). Paper presented at the inagural meeting of the *American Society of Primatologists,* Seattle, April, 1977.

Keverne, E. B. and Michael, R. P. Annual changes in the menstruation of rhesus monkeys. *Journal of Endocrinology,* 1970, **48**, 669-670.

King, G. and Fitch, M. Vocal patterns in captive ring-tailed lemurs (*Lemur catta*). Paper presented at the *Western Psychological Society* Meeting, Seattle, Washington, April, 1977.

Leibrecht, B. C. and Kelley, S. T. Some observations of behavior in breeding pairs of owl monkeys. Paper presented at the *American Society of Primatologists* Meeting, Seattle, Washington, April, 1977.

Leutenegger, W. Evolution of litter size in primates. Paper presented at the *American Society of Primatologists* Meeting, Seattle, Washington, April, 1977.

Lindburg, D. G. Rhesus monkeys: Mating season mobility of adult males. *Science,* 1969, **166**, 1176-1178.

Lindburg, D. G. A field study of the reproductive behavior of the rhesus monkey *(Macaca mulatta).* In Rosenblum, L. A. (Ed.) *Primate Behavior* (Vol. 4). New York: Academic Press, 1971, pp. 1-106.

Mahoney, C. J. Aberrant menstrual cycles in *Macaca mulatta* and *Macaca fascicularis. Laboratory Animal Handbook,* 1975, **6**, 243-255.

Mai, L. Chemical communication in primates. *Anthropology UCLA,* 1972, **4** (2), 57-83.

Maple, T., Erwin, J. and Mitchell, G. Separation of adult heterosexual pairs of rhesus monkeys: The effect of female cycle phase. *Journal of Behavioral Science,* 1974, **2** (2), 81-86.

Maple, T. and Zucker, E. L. Behavioral studies of captive Yerkes orangutans at the Atlanta Zoological Park. *Yerkes Newsletter,* 1977, **14** (1), 24-26.

Maple, T., Zucker, E. L. and Dennon, M. B. Cyclic proceptivity in a captive female orang-utan *(Pongo pygmaeus abelii). Behavioural Processes,* in press.

Martenson, J., Jr., Oswald, M., Sackett, D. and Erwin, J. Diurnal variation of common behaviors of pigtail monkeys *(Macaca nemestrina). Primates,* unpublished manuscript, 1977.

McClintock, M. K. Menstrual synchrony and suppression. *Nature,* 1971, **229**, 244-245.

Michael, R. P. and Bonsall, R. W. Periovulatory synchronisation of behaviour in male and female rhesus monkeys. *Nature,* 1977, **265**, 463-465.

Michael, R. P. and Herbert, J. Menstrual cycle influences grooming behavior and sexual activity in the rhesus monkey. *Science,* 1963, **140**, 500-501.

Michael, R. P. and Keverne, E. B. An annual rhythm in the sexual activity of the male rhesus monkey, *Macaca mulatta,* in the laboratory. *Journal of Reproduction and Fertility,* 1971, **25**, 95-98.

Michael, R. P. and Keverne, E. B. Differences in the effects of oestrogen and androgen on the sexual motivation of female rhesus monkeys. *Journal of Endocrinology,* 1972, **55**, 40.

Michael, R. P., Setchell, K. D. R. and Plant, J. M. Diurnal changes in plasma testosterone and studies on plasma corticosteroids in non-anesthetized male rhesus monkeys *(M. mulatta). Journal of Endocrinology,* 1974, **63**, 325-335.

Michael, R. P. and Wilson, M. I. Mating seasonality in castrated male rhesus monkeys. *Journal of Reproduction and Fertility,* 1975, **43**, 325-328.

Michael, R. P. and Zumpe, D. Environmental and endocrine factors influencing annual changes in sexual potency in primates. *Psychoneuroendocrinology,* 1976, **1**, 303-313.

Missakian, E. A. The timing of fission among free-ranging rhesus monkeys. *American Journal of Physical Anthropology,* 1973, **38** (2), 621-624.

Nadler, R. D. Sexual cyclicity in captive lowland gorillas. *Science,* 1976, **189,** 812-814.

Nadler, R. D. Sexual behavior of captive orang-utans. Paper presented at the *International Primatological Society* Meeting, Cambridge, England, August, 1972.

Nadler, R. D., Graham, C. E., Neill, J. D. and Collins, D. C. Genital swelling, pituitary gonadotropins and progesterone in the menstrual cycle of the lowland gorilla. Paper presented at the *American Society of Primatologists* Meeting, Seattle, Washington, April, 1977.

Nadler, R. D. and Rosenblum, L. A. Hormonal regulation of the "fatted" phenomenon in squirrel monkeys. *Anatomical Record,* 1972, **173,** 181-187.

Neuman, F., Steinbeck, H. and Hahn, J. D. Hormones and brain differentiation. In Martini, L., Motta, M. and Fraschini, F. (Eds.) *The Hypothalamus.* New York: Academic Press, 1970, pp. 1-35.

Neville, M. K. Male leadership change in a free-ranging troop of Indian rhesus monkeys (*Macaca mulatta*). *Primates,* 1968, **9,** 13-27.

Nigi, H. Menstrual cycle and some other related aspects of Japanese monkeys (*Macaca fuscata*). *Primates,* 1975, **16** (2), 207-216.

Nigi, H. Some aspects related to conception of the Japanese monkey (*Macaca fuscata*). *Primates,* 1976, **17** (1), 81-87.

Norman, R. L., Resko, J. A. and Spies, H. G. The anterior hypothalamus: How it affects gonadotropin secretion in the rhesus monkey. *Endocrinology,* 1976, **99** (1), 59-71.

Perret, M. Influence du groupement social sur l'activation sexuelle saisonniere chez le ♂ de *Microcebus murinus* (Miller, 1777). *Zeitschrift für Tierpsychologie,* 1977, **43,** 159-179.

Plant, T. M. and Michael, R. P. Diurnal variations in plasma testosterone levels of adult male rhesus monkeys. *Acta Endocrinologica Congress,* 1971, Abstract No. 69.

Plant, T. M., Zumpe, D., Sauls, M. and Michael, R. P. An annual rhythm in the plasma testosterone of adult male rhesus monkeys maintained in the laboratory. *Journal of Endocrinology,* 1974, **62,** 403-404.

Quris, R. Ecologie et organisation sociale de *Cercocebus galeritus agilis* dans le nord-est du gabon. *Terre et la Vie, Revue d'Ecologie Appliquee,* 1975, **29,** 337-398.

Resko, J. A. Plasma androgen levels of the rhesus monkey: Effects of age and season. *Endocrinology,* 1967, **81,** 1203-1225.

Rijksen, H. D. Social structure in a wild orang-utan population in Sumatra. In Kondo, S., Kawai, M. and Ehara, A. (Eds.) *Contemporary Primatology.* Basel: Karger, 1975, pp. 373-379.

Rose, R. M., Gordon, T. P. and Bernstein, I. S. Plasma testosterone levels in the male rhesus: Influences of sexual and social stimuli. *Science*, 1972, **178**, 643-645.

Rowell, T. E. Grooming by adult baboons in relation to reproductive cycles. *Animal Behaviour*, 1968, **16**, 585-588.

Rowell, T. E. Intra-sexual behaviour and female reproductive cycles of baboons (*Papio anubis*). *Animal Behaviour*, 1969, **17**, 159-167.

Rowell, T. E. Baboon menstrual cycles affected by social environment. *Journal of Reproduction and Fertility*, 1970, **21**, 133-141.

Rowell, T. E. and Dixson, A. F. Changes in social organization during the breeding season of wild Talapoin monkeys. *Journal of Reproduction and Fertility*, 1975, **43**, 419-434.

Saayman, G. S. Aggressive behavior in free-ranging chacma baboons (*Papio ursinus*). *Journal of Behavioural Science*, 1971, **1**, 77-83.

Sassenrath, E. N., Rowell, T. E. and Hendricks, A. G. Perimenstrual aggression in groups of female rhesus monkeys. *Journal of Reproduction and Fertility*, 1973, **34**, 509-511.

Savage, E. S. and Bakeman, R. Comparative observations on sexual behaviour in *Pan paniscus* and *Pan troglodytes*. Paper presented at the *International Primatological Society* Meeting, Cambridge, England, August, 1976.

Slob, A. K., Goy, R. W., Wiegand, S. J. and Scheffler, G. Gonadal hormones and behaviour in the stumptail macaque (*Macaca arctoides*) under laboratory conditions: A preliminary report. *Journal of Endocrinology*, 1975, **64**, 38.

Southren, A. L. and Gordon, G. G. Rhythms and testosterone metabolism. *Journal of Steroid Metabolism*, 1975, **6**, 809-813.

Spencer, J. O. Olfactory discrimination of sex by humans. Paper presented at the *Western Psychological Association* Meeting, Seattle, Washington, April, 1977.

Sprankel, H. and Richarz, K. Nicht reproduktires Verhalten von *Tupaia glis* Diard, 1820 im raum-zeitlichen Bezug: Eine quantitative Analyse. *Zeitschrift für Saugetierkunde Bd.*, 1976, **41**, 77-101.

Struhsaker, T. T. *Behavior of Vervet Monkeys* (*Cercopithecus aethiops*). Berkeley: University of California Press, 1967.

Sugawara, K. Analysis of the social relations among adolescent males of Japanese monkeys (*Macaca fuscata fuscata*) at Koshima Islet. *Journal of Anthropology Society of Nippon*, 1976, **83** (4), 330-354.

Sugiyama, Y. and Ohsawa, H. Life history of male Japanese macaques at Ryozenyama. *Fifth International Congress of Primatology*, Nagoya, Japan, 1974, pp. 407-410.

Swett, C. Daytime sleep patterns in free-ranging rhesus monkeys. *Psychophysiology*, 1969, **6**, 227-228.

Teas, J., Taylor, H. G., Richie, T. L. and Southwick, C. H. Seasonal influences on grooming pair composition in *Macaca mulatta.* Paper presented at the *American Society of Primatologists* Meeting, Seattle, Washington, April, 1977.

Valenstein, E. S. Steroid hormones and the neuropsychology of development. In Isaacson, R. L. (Ed.) *The Neuropsychology of Development.* New York: Wiley, 1968, pp. 1-39.

Vandegrift, P. Perimenstrual behavior in the rhesus monkey (*Macaca mulatta*). Paper presented at the *American Association of Physical Anthropologists* Meeting, Seattle, Washington, April, 1977.

Vandenbergh, J. G. and Vessey, S. Seasonal breeding in free-ranging rhesus monkeys and related ecological factors. *American Zoologist,* 1966, **6** (3), 207.

Van Horn, R. N. Primate breeding season: Photoperiodic regulation in captive *Lemur catta. Folia Primatologica,* 1975, **24**, 203-220.

van Wagenen, G. Menopause in a subhuman primate. *Anatomical Record,* 1970, **166**, 392.

Varley, M. A. and Vessey, S. H. Effects of geographic transfer and age on the timing of seasonal breeding of rhesus monkeys. *Folia Primatologica,* 1977, **28**, 52-59.

Vessey, S. H. and Morrison, J. A. Molt in free-ranging rhesus monkeys, *Macaca mulatta. Journal of Mammalogy,* 1970, **51**, 89-93.

Wilson, A. P. and Boelkins, R. C. Evidence for seasonal variation in aggressive behaviour by *Macaca mulatta. Animal Behaviour,* 1970, **18**, 719-724.

Wilson, M. I. Characterization of the oestrous cycle and mating season of squirrel monkeys from copulatory behaviour. *Journal of Reproduction and Fertility,* 1977, **51**, 57-63.

Wolf, R. L., O'Connor, R. F. and Robinson, J. A. Cyclic changes in plasma progestins and estrogens in squirrel monkeys. *Biology of Reproduction,* 1977, **17**, 228-231.

Zamboni, L., Conaway, C. H. and van Pelt, L. Seasonal changes in production of semen in free-ranging rhesus monkeys. *Biology of Reproduction,* 1974, **11**, 251-267.

32.
Noncyclical Variability

A. GENERAL COMMENTS

As we have seen in previous chapters, sex differences in behavior are not static but change with age. In Chapter 31, we saw that sex differences in behavior also sometimes vary cyclically, especially if the behaviors involved are closely related to reproductive behavior. Thus, as Mason (1976) has reminded us, sex differences in behavior should be viewed as processes not patterns. They develop and change with the situation.

In the present chapter, we will concern ourselves with noncyclical variability and with variability that is not explicitly age-related. After having read the present chapter we should gain an appreciation for the complexity and relative unpredictability of primate behavior and, in particular, for the unpredictability of sex differences in primate behavior. We have seen, for example, that while sex differences in dominance are frequently reported, no one kind of social behavior is correlated with dominance and not all groups of primates even have dominance hierarchies (cf. Kolata, 1976). As in the case of dominance, many of the sex differences in behavior we have covered do not appear as consistently and predictably as we may have suggested heretofore.

Let us look at an example other than dominance. A major sex difference appearing in nonhuman primates is a sex difference in parental behavior. Males, in most species but not in all species, show less interest in the young than do females. However, males *do* show interest in the young. Their interest is more variable from species to species as well as from individual to individual than is the interest of females. Even the interest in infants from one situation to another for a given individual is more variable for males than it is for females. Female interest, on the other hand, is usually strong and persistent (Hamburg, 1968).

As Burton (1977) has emphasized, regardless of sex differences, the outstanding behavioral characteristic of primates is *variability*. Behaviors are individual characteristics not just age, sex, or regular rhythmic characteristics. Let us now examine some of this individual variability in a taxonomic framework.

B. PROSIMIANS AND NEW WORLD MONKEY EXAMPLES

Among lorises, some species of the genus *Galago* do not seem to live in permanent social groups while others do (Flinn and Nash, 1975). Some galagos (e.g., *Galago alleni*) have a social organization in which adult males have large territories containing several females which, in turn, have smaller territories within their male's territory. There is even a sex difference in variability with regard to home range size. Females have home ranges of equivalent area but the territories of males are larger and show greater individual variation than do those of females (Charles-Dominique, 1977).

Among prosimians in general, there is tremendous variability in behavior. Some prosimians are solitary and others form temporary sleeping groups. *Propithecus* and *Indrii* appear to form family groups. Even within one genus, *Lemur,* there is tremendous variability in mother–infant care (cf. Klopfer, 1972; Klopfer, 1976). In the amount of contact between adults of the opposite sex, *Lemur fulvus* shows friendly and frequent intersexual interaction, whereas *Lemur variegatus* shows an almost complete mutual avoidance by adult males and adult females (Kress, 1975). In addition, Tattersall (1976) has shown that *Lemur mongoz* is very flexible behaviorally and that this variability correlates with environmental factors. The prosimians are not inflexible just because we label them more primitive. Based upon these kinds of variations alone, it may be that we have not been cautious *enough* throughout this book in our generalizations from one prosimian species to another.

With regard to visual communication we have already seen that prosimians differ substantially from one another and especially from monkeys and apes in regard to the presence or absence of binocular vision. Some prosimians (e.g., tree shrews) have no overlap of their two visual fields and hence no stereoscopic vision. Others, like lemurs, have some overlap while still others, like the tarsier, have substantial overlap. With regard to color vision, only the tree shrew among prosimians has color vision (cf. Noback et al., 1969). Differences such as these produce a great deal of variability within the suborder.

Although all of the "primitive" New World marmosets and tamarins have color vision (Noback et al., 1969), there are other sources of substantial variability. As Epple (1975a) has noted, the size and composition of groups of marmosets in the wild are extremely variable, even within a single species. To be sure, most marmosets are monogamous and live in family groups, but there are also larger permanent and temporary aggregations in the wild (Epple, 1975a). In addition, while the adult male usually carries the infants, in *some* groups the female carries the infants more than the male (Epple, 1975b).

Among the New World cebids there is also a great potential for variability and for unusual individual behaviors. Four female squirrel monkeys (*Saimiri sciureus*) deprived of adult male contact for two years came to respond to one another as heterosexual mating couples do (Talmage-Riggs and Anschel, 1973). Also in squirrel monkeys, variability in behavior, particularly in agonistic behavior, seems to be greater in animals of high rank than in animals of low rank particularly with regard to the choice of signals used (Maurus, Hartmann, and Kuhlmorgen, 1974). There are also sex differences in sensory processes in cebids that undoubtedly lead to variability in behavior. Male cebus and squirrel monkeys, for example, cannot see the red end of the spectrum as well as females (Jacobs, 1977a). Sex differences in sensory abilities have been, as we know, largely ignored in our coverage of other aspects of primate sex differences in the present book. Nevertheless, they could be a source of great variability.

As we know, all living prosimians (except *Tupaia*), whether diurnal or nocturnal, are either color-blind or almost color-blind. All diurnal New World monkeys, on the other hand, can see color. *Cebus* and *Saimiri* show some deficiencies in the red end of the spectrum, however. The one nocturnal genus of New World monkeys, *Aotus,* is color-blind (Noback, et al., 1969), relative to the diurnal species.[1] These sensory differences are undoubtedly sources of variability not yet considered in our discussions of behavioral sex differences.

C. RHESUS MONKEYS (*MACACA MULATTA*)

After an in-depth study of the degree of biological variability in *Macaca mulatta,* Hromada (1968) concluded that man does *not* show the greatest amount of variability among the primates. There is evidently marked anatomical variability in *Macaca mulatta* which equals or exceeds that of *Homo sapiens.*

We will not, of course, discuss every kind of biological variability that could possibly affect behavioral sex differences in the rhesus monkey. This would be an endless task. Instead, we will concentrate on that variability which is most intimately tied to reproductive success, variability in the reproductive organs and in reproductive behavior.

On Cayo Santiago, Puerto Rico, DeRousseau noted substantial variability in the accessory sex structures of rhesus monkeys. For example, in studies of 219 males, there was marked variability in penis size. Moreover, social status seemed to be associated with baculum and glans length in rhesus male adults. Since these organs are, to a large extent, determined by the amount of prenatal testosterone available in early development, these findings suggest that the range of prenatal testosterone levels in males must be substantial. As Resko (1974) has written:

[1] Jacobs (1977b) concluded that *Aotus* was in fact capable of color vision but that its color vision capacity was both weak and aberrant relative to normal primate trichromacy.

> There is evidence to suggest that differentiation is not an all-or-none process but that different amounts of testosterone produce different levels of androgenization . . . (P. 216)

The fact that prenatal testosterone levels vary along a continuum can be employed here to remind us that the ideal condition of complete sexual dimorphism does not exist. There is evidently a continuum within and across sex (Goy and Goldfoot, 1975). Even in the most prenatally masculinized of the rhesus males, feminine sexual behaviors are still displayed. Female neural mechanisms are evidently protected from the effects of androgen in both males and females (Goy and Goldfoot, 1975). Similarly, malelike mounting is part of the normal repertoire in rhesus females (Michael, Wilson, and Zumpe, 1974).

Early androgens, however, *do* produce variability. Using large doses of androgen early in life, van Wagenen (1966) was able to produce a small, adult male rhesus monkey which reached sexual maturity as early as two years of age. Of course, other factors can also affect age of sexual maturity. Animals raised in captivity reach sexual maturity at earlier ages than do animals in the wild. Maple, Erwin, and Mitchell (1973), for example, reported sexual maturity in nontreated captive male rhesus at as early as three years of age. Early maturation, however, is also correlated with other changes which sometimes affect the degree of variability seen in rhesus monkeys. For example, the early maturing rhesus female shows *less* variability in dentition than does the later maturing female or the later maturing male. The dental development of the male is less predictable than is the dental development of the female. On these grounds alone, Hurme and van Wagenen (1956) suggest that male and female rhesus should be treated as if they represent two different species in studies of biological age appraisal.

In rhesus sexual behavior, hormones alone cannot explain performance. Individual preferences play a very strong role in mating (Phoenix, 1974). According to Michael and Saayman (1967) overall levels of sex are determined by the male, and the female's cyclical variability does not influence differences between males. These researchers noted that different males with the same female showed a large range of different behaviors (i.e., great variability), whereas different rhesus females with the same male showed very little variability. However, when Goy and Goldfoot (1975) studied different pairs of rhesus, they found that males do not differ in maximal sexual expression as much as pairs do. They also found that *no* females show consistent rejection and/or frigidity. In the rhesus monkey, at least, impotence is really a result of partner incompatibility and/or a bad bond. There is no doubt however that individuals do differ in finickiness (Goy and Goldfoot, 1975).

We have already seen that age and cyclical variation also contribute somewhat to the variability frequently seen in sex difference research. No one would question the statement that a rhesus female's mood is affected by her stage of estrus. However, a male's mood can also be affected by factors which are sometimes cyclical and sometimes not. For example, when six young rhesus males were able to control how long they sat in different colors of light, they spent less time with red light than with blue light. The researchers involved in this project (Humphrey and Keeble, 1977) are convinced that their finding does not mean that male rhesus monkeys have a preference for blue light over red light. On the contrary, they believe that *time* passes faster (twice as fast) for the males when they are in red light than when they are in blue light. The monkeys do things faster in red light than in blue light leading the reasearchers to believe that color has an influence on the passage of subjective time. Is blue calming? Is red arousing? Although squirrel and cebus male monkeys cannot see red as well as females, the same is *not* true for rhesus males (cf. Noback et al., 1969).

In any case, the above research findings should suffice as examples of noncyclical sources of variability in male and female rhesus. Sex differences in behavior, even in strongly dimorphic species, are not invariably present or absent.

One final comment on variability in rhesus macaques is in order. Apparently rhesus monkeys reared under conditions of social isolation show *more* behavioral variability from animal to animal and more behavioral variability within an individual than do monkeys raised by very different mothers. We might expect the reverse to be true. After all, each socially reared monkey has experienced a very unique kind of social experience while each social isolate has experienced the same isolation environment as any other social isolate. However, in this case, it is obviously the individuality from within each animal (genetic variability) and not differential environmental input that is producing most of the variability (see Fittinghoff et al., 1974, for examples). Changes in within-group variability with early deprivation can undoubtedly change sex differences in behavior.

D. OTHER OLD WORLD MONKEYS

All Old World monkeys have color vision, so this characteristic does not contribute substantially to variability in communication from species to species. However, there is variability in other optical characteristics. For example, in most species, males have larger eyes than do females (but with lower power) (cf. Young and Leary, 1977).

In macaques other than rhesus, there is clearly tremendous variability on many behavioral characteristics. Yamada (1971), for example, studied five different natural troops of Japanese macaques (*M. fuscata*) and found large differences from troop to troop in numbers of males and females per troop and in the age distributions of the males. In addition, Kawamura (1963) reported that each wild Japanese macaque troop has a food list all its own. He also found great variability in culturally propagated male care of infants. Itani (1975) found that, even in some kinds of cyclical variation, troops differed from each other. The mating seasons, for example, differ slightly from troop to troop.

In stumptail macaques (*M. arctoides*) sexual posturing is extremely variable and idiosyncratic. Individual males differ greatly in their responses to females, and ovarian hormones have inconstant activating effects on male–female interactions (Slob, 1975). Among free-ranging bonnet macaques (*M. radiata*), the sex ratio varies. Rahaman and Parthasarathy (1967), for example, found a sex ratio of one-to-one in seven groups, more males than females in one group, and more females than males in six groups.

Not only do macaques display great variability, they are also tolerant of it. Berkson (1977) has released blind and defective individuals into wild groups of macaques and found that these primates even tolerate totally blind individuals.

With regard to sex differences in infant mortality in macaques, most researchers have found that infant males show higher mortality rates than do infant females. However, in at least one species (*M. fascicularis*) apparently the reverse is true (Erwin, 1977).

Such differences appearing in behavioral comparisons of the many species of the genus *Macaca* have forced primatologists to admit that there is no single species of macaque that can adequately represent the genus. This is true even (and we might say *especially*) in the case of sexual behavior (Nadler and Rosenblum, 1973).

Despite the great variability seen in macaques, different species of macaques will breed with one another if given the opportunity. While they do not typically hybridize in nature, they do so on occasion. In captivity, Avise and Duvall (1977) have been able to produce viable hybrids among six different species of macaques. Attempts to hybridize macaques with more distantly related monkeys [e.g., baboons (*Papio*)] have usually failed because of fetal abortion or early infant death but not because the two different genera will not copulate with one another (cf. Maple, 1974).

Baboons, of course, will also hybridize with one another when permitted to do so in captivity. The behavior of the hybrids often appears to include elements of both parents' behavioral repertoires (Nagel, 1970). Hybridization, of course, can add considerable amounts of behavioral variability to a troop; and, in some cases, can even reverse typical sex differences in behavior.

Aside from genetic sources of variability, baboons also show variability related to ecological variables. Baboon troops (*Papio anubis*) from different ecological conditions show different aggressive and sexual behaviors. Thus, there is genetic, ecological, and even protocultural malleability in this genus (Paterson, 1973). With regard to baboon structural variability there is wide variability across species with respect to intraspecific (intersexual) scaling of tooth size (Post, 1977).

According to Lancaster (1971), sex differences in play behavior in vervets (*Cercopithecus aethiops*) can be characterized as much by variability as by overall frequencies. Male vervets show more variability in play than do females.

We have seen that the langur (*Presbytis entellus*) male's propensity toward killing infants during group takeovers can be extremely variable. Bogess (1977) has suggested that the variation reflects ecological conditions, e.g., high population density leading to more infanticides. Curtin (1977), for example, never observed the typical pattern of male takeovers. Also varying in hanuman langurs are such factors as birth season, size of all-male groups, sex ratios, and number of all-male groups (Vogel, 1970).

In *Presbytis johnii*, adult males are said to be less conservative, more flexible, and more variable in their behavior patterns than are adult females (Poirier, 1969).

With regard to other species of langurs, the species of the *Presbytis aygula-melalophos* group display variability in pelage color, color pattern, lie of the hair on the head, and in male loud calls (Wilson and Wilson, 1975). There are also variations in the coat color patterns of the so-called "white" *Colobus* (Hull, 1977). These structural differences can possibly affect generalizations concerning sex differences.

E. APES

In the genus *Hylobates,* different species have very distinct pelage coloration and vocalizations. Despite this, hybridization occurs in nature. Mixed species groups are found and so are hybrids. The hybrids emit atypical territorial vocalizations and often have a different pelage color. This suggests that both of these characteristics—coloration and vocalization—are under a high degree of genetic control (Brockelman, 1976). [There is a siamang by gibbon hybrid at the Atlanta Zoo (Maple, personal communication).]

Of all the primates besides *Homo sapiens,* it is commonly assumed that the great apes demonstrate the greatest behavioral diversity. Parker (1974) compared ten different genera of primates and found this to be true. Individual differences can be striking on such dimensions as mother's parity, effects of experience, age, and even "personality" (Clark, 1977). Whereas, in chimpanzees, for example, there is more hunting in males than in females, females *do* hunt (van Lawick-Goodall, 1973). There are marked voice differences (individual differences in voice) and differences in vocalizations from group to group in gorillas (Fossey, 1972). This variability has resulted in some confusion concerning sex differences in vocal repertoire size in groups of gorillas. The size and sex composition of gorilla groups also vary (Marler, 1968).

The sexual behavior patterns of the three genera of great apes could hardly differ more than they do. In fact, even within one genus (*Pan*), the two existing species (*P. troglodytes* and *P. paniscus*) are extremely different from each other in this regard. Aside from genus and species differences there is remarkable variability from group to group and from individual to individual in great ape sexual behavior. It is extremely difficult to get reliable behavioral sex differences in these remarkable animals (cf. Zucker et al., 1977; Maple, 1977; Nadler, 1977).

As we shall see in the next chapter, the term *individual* variability in the great apes means something very different from the same term as applied to the lesser apes, monkeys, and prosimians.

F. COMMENTS ON HUMAN VARIABILITY

There is no question that heredity plays a strong role in determining individual differences in people as well as in nonhuman primates. However, it is also possible that man's unique genetic composition produces individual differences for which nonhuman analogies would be unjustified (see Scott, 1970). After all, it should be remembered that:

> The nonhuman primates are not our grandfathers but our distant cousins. (Scott, 1974, P. 428)

It is certainly true that in human beings there is often so much individual variation *within* the sexes that there is a great deal of overlap between them (Scott, 1974).

On several dimensions of behavior and/or structure (e.g., dentition) human females are much less variable between groups than are males (Harris, 1977). There is also some evidence that males vary more than females on more complex behavioral constructs (e.g., intelligence as measured by IQ tests) (cf. Maccoby and Jacklin, 1974). We will not go into the possible reasons for such differences except to say that differential vulnerability, differential maturation rates and differential treatment by the cultures in which male and female humans are reared must surely have something to do with the differences in variability reported, (to say nothing about variability in the measuring instruments).

REFERENCES

Avise, J. C. and Duvall, S. W. Allelic expression and genetic distance in hybrid macaque monkeys. *The Journal of Heredity,* 1977, **68**, 23–30.

Berkson, G. The social ecology of defects in primates. In Chevalier-Skolnikoff, S. and Poirier, F. E. (Eds.) *Primate Bio-Social Development: Biological, Social and Ecological Determinants.* New York: Garland, 1977, pp. 189–204.

Boggess, J. Social change in a troop of langurs in Nepal. Paper presented at the *American Association of Physical Anthropologists* Meeting, Seattle, Washington, April, 1977.

Brockelman, W. Y. Preliminary report on relations between *Hylobates lar* and *Hylobates pileatus.* Paper presented at the *International Primatological Society* Meeting, Cambridge, England, August, 1976.

Burton, F. D. Ethology and the development of sex and gender identity in nonhuman primates. *Acta Biotheoretica,* 1977, **26**, 1-18.

Charles-Dominique, P. Urine marking and territoriality in *Galago alleni* (Waterhouse, 1837 Lorisoidea, Primates) — A field study by radio-telemetry. *Zeitschrift für Tierpsychologie,* 1977, **43**, 113-138.

Clark, C. B. A preliminary report on weaning among chimpanzees of the Gombe National Park, Tanzania.   Paper presented at the *American Association of Physical Anthropologists* Meeting, Seattle, Washington, April, 1977.

Curtin, R. A. Socioecology of langurs in the Nepal Himalaya. Paper presented at the *American Association of Physical Anthropologists* Meeting, Seattle, Washington, April, 1977.

DeRousseau, C. J. Variability of accessory sex structures in *Macaca mulatta. American Journal of Physical Anthropology,* 1974, **41**, 475.

Epple, G. The behavior of marmoset monkeys (*Callithricidae*). In Rosenblum, L. A. (Ed.) *Primate Behavior* (Vol. 4). New York: Academic Press, 1975, pp. 195-239. (a)

Epple, G. Parental behavior in *Saguinus fuscicolliss ssp.* (Callithricidae). *Folia Primatologica,* 1975, **24**, 221-238. (b)

Erwin, J. Infant mortality in *Macaca fascicularis:* Neonatal and postnatal mortality at the Regional Primate Research Center Field Station, University of Washington, 1967-1976. *Theriogenology,* 1977, **7**, 357-366.

Fittinghoff, N. A., Lindburg, D. G., Gomber, J. and Mitchell, G. Consistency and variability in the behavior of mature, isolation-reared, male rhesus macaques. *Primates,* 1974, **15**, 111-139.

Flinn, L. and Nash, L. T. Group formation in recently captured lesser galagos. Paper presented at the Annual Meeting of the *American Association of Physical Anthropologists,* Denver, Colorado, 1975.

Fossey, D. Vocalizations of the mountain gorilla (*Gorilla gorilla beringei*). *Animal Behaviour,* 1972, **20** (1), 36-53.

Goy, R. W. and Goldfoot, D. A. Neuroendocrinology: Animal models and problems of human sexuality. *Archives of Sexual Behavior,* 1975, **4** (4), 405-420.

Hamburg, D. A. Evolution of emotional responses: Evidence from recent research on nonhuman primates. *Science and Psychoanalysis,* 1968, **12**, 39-54.

Harris, E. F. Biologic implications of sexually dimorphic qualitative traits in the human dentition. Paper presented at the *American Association of Physical Anthropologists* Meeting, Seattle, Washington, April, 1977.

Hromada, J. Contribution to the question of variability in *Macaca mulatta. Sbornik vedeskych praci Lekarski fakulty KU v Hradci Kralove,* 1968, **11** (1), 169-175.

Hull, D. B. The "white" Colobus of Mt. Kenya. Paper presented at the *American Society of Primatologists* Meeting, Seattle, Washington, April, 1977.

Humphrey, N. K. and Keeble, G. R. Do monkeys' subjective clocks run faster in red light than in blue? *Perception,* 1977, **6**, 7-14.

Hurme, V. O. and van Wagenen, G. Emergence of permanent first molars in the monkey (*M. mulatta*): Association with other growth phenomena. *The Yale Journal of Biology and Medicine,* 1956, **28**, 538-566.

Itani, J. Twenty years with Mount Takasaki monkeys. In Bermant, G. and Lindburg, P. (Eds.) *Primate Utilization and Conservation.* New York: Wiley, 1975, pp. 101-125.

Jacobs, G. H. Visual sensitivity: Significant within-species variation in a nonhuman primate. *Science,* 1977, **197**, 499-500. (a)

Jacobs, G. H. Visual capacities of the owl monkey (*Aotus trivirgatus*). I. Spectral sensitivity and color vision. *Vision Research,* 1977, **17**, 811-820. (b)

Kawamura, S. The process of sub-culture propagation among Japanese macaques. In Southwick, C. H. (Ed.) *Primate Social Behavior.* Princeton, N.J.: D. Van Nostrand, 1963, pp. 82-89.

Klopfer, P. H. Patterns of maternal care in lemurs: II. Effects of group size and early separation. *Zeitschrift für Tierpsychologie,* 1972, **30**, 277-296.

Klopfer, P. H. and Dugard, J. Patterns of maternal care in lemurs. III. *Lemur variegatus. Zeitschrift für Tierpsychologie,* 1976, **40**, 210-220.

Kolata, G. B. Primate behavior: Sex and the dominant male. *Science,* 1976, **191**, 55-56.

Kress, J. H. Socialization patterns of *Lemur variegatus.* Paper presented at the *American Association of Physical Anthropologists* Meeting, Denver, Colorado, April, 1975.

Lancaster, J. B. Play-mothering: The relations between juvenile females and young infants among free-ranging vervet monkeys (*Cercopithecus aethiops*). *Folia Primatologica,* 1971, **15**, 161-182.

Maccoby, E. and Jacklin, C. *The Psychology of Sex Differences.* Stanford, California: Stanford University Press, 1974.

Maple, T. L. Basic studies of interspecies attachment behavior. Doctoral dissertation, Department of Psychology, University of California, Davis, 1974.

Maple, T. L. Unusual sexual behavior of nonhuman primates. In Money, J. and Musaph, H. (Eds.) *Handbook of Sexology.* New York: Elsevier, 1977, pp. 1167-1188.

Maple, T., Erwin, J. and Mitchell, G. Age of sexual maturity in laboratory-born pairs of rhesus monkeys (*Macaca mulatta*). *Primates,* 1973, **14** (4), 427–428.

Marler, P. Aggregation and dispersal: Two functions in primate communication. In Jay, P. C. (Ed.) *Primates: Studies in Adaptation and Variability.* New York: Holt, Rinehart and Winston, 1968, pp. 420–438.

Mason, W. A. Primate social behavior: Pattern and Process. In Masterson, R. B. et al. (Eds.) *Evolution of Brain and Behavior in Vertebrates.* Hillsdale, N.J.: Lawrence Erlbaum Associates, 1976, pp. 425–455.

Maurus, M., Hartmann, E. and Kuhlmorgen, B. Invariant quantities in communication processes of squirrel monkeys. *Primates,* 1974, **15** (2–3), 179–192.

Michael, R. P. and Saayman, G. S. Individual differences in the sexual behaviour of male rhesus monkeys (*Macaca mulatta*) under laboratory conditions. *Animal Behaviour,* 1967, **15**, 460–466.

Michael, R. P., Wilson, M. I. and Zumpe, D. The bisexual behavior of female rhesus monkeys. In Friedman, R. C., Richart, R. M. and Vande Wiele, R. L. (Eds.) *Sex Differences in Behavior.* New York: Wiley, 1974, pp. 399–412.

Nadler, R. D. Sexual behavior of the chimpanzee in relation to the gorilla and orang-utan. In Bourne, G. (Ed.) *Progress in Ape Research.* New York: Academic Press, 1977, pp. 191–206.

Nadler, R. D. and Rosenblum, L. A. Sexual behavior of male pigtail macaques in the laboratory. *Brain, Behavior and Evolution,* 1973, **7**, 18–33.

Nagel, U. Social organization in a baboon hybrid zone. *Proceedings of the Third International Congress of Primatology,* 1970, **3**, 48–57.

Noback, C. R., Berger, M., Laemle, L. K. and Shriver, J. E. Phylogenetic aspects of the visual systems in primates and *Tupaia. Proceedings of the Second International Congress of Primatology* (Vol. 3). New York: Karger, 1969, pp. 49–54.

Parker, C. E. Behavioral diversity in ten species of nonhuman primates. *Journal of Comparative and Physiological Psychology,* 1974, **87**, 930–937.

Paterson, J. D. Ecologically differentiated patterns of aggressive and sexual behavior in two troops of Ugandan baboons, *Papio anubis. American Journal of Physical Anthropology,* 1973, **38** (2), 641–647.

Phoenix, C. H. The role of androgens in the sexual behavior of adult male rhesus monkeys. In Montagna, W. and Sadler, W. A. (Eds.) *Reproductive Behavior.* New York: Plenum, 1974, pp. 249–258.

Poirier, F. E. Behavioral flexibility and intertroop variation among Nilgiri langurs (*Presbytis johnii*) of South India. *Folia Primatologica,* 1969, **11**, 119–133.

Post, D. G. Baboon feeding behavior and the evolution of sexual dimorphism. Paper presented at the *American Association of Physical Anthropologists* Meeting, Seattle, Washington, April, 1977.

Rahaman, H. and Parthasarathy, M. D. A population survey of the bonnet monkey, *Macaca radiata* (Geoffroy) in Bangalore, South India. *Journal of the Bombay Natural History Society,* 1967, **64** (2), 251–255.

Resko, J. A. The relationship between fetal hormones and the differentiation of the central nervous system in primates. In Montagna, W. and Sadler, W. A. (Eds.) *Reproductive Behavior.* New York: Plenum, 1974, pp. 211–222.

Scott, J. P. Biology and human aggression. *American Journal of Orthopsychiatry,* 1970, **40** (4), 568–576.

Scott, J. P. Agonistic behavior of primates: A comparative perspective. In Holloway, R. L. (Ed.) *Primate Aggression, Territoriality and Xenophobia: A Comparative Perspective.* New York: Academic Press, 1974, pp. 417–434.

Slob, A. K. Effects of ovariectomy on male-female interactions in the stumptail macaque (*M. arctoides*). *Acta Endocrinologica,* 1975, **10**, 145.

Talmage-Riggs, G. and Anschel, S. Homosexual behavior and dominance hierarchy in a group of captive female squirrel monkeys (*Saimiri sciureus*). *Folia Primatologica,* 1973, **19**, 61–72.

Tattersall, I. Behavioural variability in *Lemur mongoz.* Paper presented at the *International Primatological Society* Meeting, Cambridge, England, August, 1976.

van Lawick-Goodall, J. Cultural elements in a chimpanzee community. In Menzel, E. W. (Ed.) *Symposia of the Fourth International Congress of Primatology* (Vol. 1): *Precultural Primate Behavior.* Basel: Karger, 1973, pp. 144–184.

van Wagenen, G. Studies in reproduction (*Macaca mulatta*). In Miller, C. O. (Ed.) *Proceedings, Conference on Nonhuman Primate Toxicology.* Warrenton, Virginia: Department of Health, Education and Welfare, 1966, pp. 103–113.

Vogel, C. Behavioral differences of *Presbytis entellus* in two different habitats. *Proceedings of the Third International Congress of Primatology,* 1970, **3**, 41–47.

Wilson, W. L. and Wilson, C. C. Species-specific vocalizations and the determination of phylogenetic affinities of the *Presbytis aygula-melalophos* group in Sumatra. *Proceedings of the Fifth International Congress of Primatology.* Basel: Karger, 1975, pp. 459–463.

Yamada, M. Five natural troops of Japanese monkeys on Shodoshima Island: II. A comparison of social structure. *Primates,* 1971, **12** (2), 215–250.

Young, F. A. and Leary, G. A. Sexual differences in the optical characteristics of major subhuman primates. Paper presented at the *American Society of Primatologists* Meeting, Seattle, Washington, April, 1977.

Zucker, E. L., Wilson, M. E., Wilson, S. F. and Maple, T. The development of sexual behavior in infant and juvenile male orang-utans (*Pongo pygmaeus*). Paper presented at the Inaugural Meeting of the *American Society of Primatologists,* Seattle, Washington, April, 1977.

33.
Self-Awareness

A. GENERAL COMMENTS

In the first chapter of this book, we defined some of the terms we would use. We admitted that in using the term "behavior" we were not assuming that the organism "behaving" was conscious of its behavior. We were assuming, for the most part, that the typical primate did not intend to perform a particular act before the act was performed. We accepted the belief that the average primate had no conceptual representation of its own "actions" (cf. Reynolds, 1976). However, from a long line of research extending over many decades it has become obvious that at least some nonhuman primates *are* conscious of who they are, *do* intend to perform given actions, *do* have a conceptual representation of their actions and, in fact, *are* aware of their own existence. In this chapter we will first review the evidence for the latter assertion and then show its importance to an examination of behavioral sex differences.

B. INDIVIDUALITY

As we have seen in the last two chapters, there is a substantial amount of individual variation in nonhuman primate behavior. There is so much variability from genus to genus, from species to species, from troop to troop, from individual to individual, and from situation to situation within an individual that it has become wise for primatologists to be very careful about making statements concerning what a given primate will *not* do. The words "always" and "never" have become rare words in the behavioral primatologist's vocabulary. Classifying certain behaviors according to age and sex can also be risky (cf. Burton, 1977) because of individual variability.

C. SPECIES IDENTIFICATION OR SPECIES IDENTITY

In August of 1977, M. Aaron Roy (1977) of the Delta Regional Primate Research Center in Covington, Louisiana organized a symposium at the *American Psychological Association* meetings in San Francisco, California. The topic of the symposium was: Early experience and the development of an adequate species identity.

This symposium grew out of the growing body of evidence that atypical or deprived rearing conditions in early life can produce a failure to develop a normal orientation toward one's own species. Evidently, complex organisms like the nonhuman primates require physical social contact with conspecifics in order to develop adequate species affiliation. Species-typical behavioral patterns depend upon an awareness by the organism that it belongs to the same species as does its partner (Roy, 1977).

A female gorilla, Toto, was raised in a human home for nine years. Socially, Toto behaved much like a human in interactions with humans even to the extent of "falling in love" with one or the other of the male caretakers whenever she was in estrus (cf. Roy, in press). Apparently, no "innate" mechanism for "knowing" to which species one belongs exists for primates. Early experiences with conspecifics are necessary for the development of this knowledge.

However, primates *are* born with *some* knowledge about their own species; they are not *completely* dependent upon early experience. Infant rhesus monkeys show selective attention to monkey pictures in preference over nonmonkey pictures. Even human infants show preferences for schematic human faces (Roy, in press). This partial innate recognition cannot account for complete species identity, however, since as we already know, rearing in social isolation interferes with species identity.

Species identity is not always completely exclusive or specific. Cross-species affinities are common in the laboratory and even occur in the wild (Roy, in press; Maple, 1974). There is therefore some amount of generalizability of attachment processes within the primate order (see Mitchell, 1976).

D. CONSEQUENCE OF EARLY ISOLATION FOR SELF-DIRECTED BEHAVIOR

Abnormal behavior is invariably a result of early social isolation. Social behaviors directed toward conspecifics become abnormal. In the case of isolate rhesus monkeys, the social abnormality gradually changes from one in which the social isolate infant withdraws from others in fear, to one in which the social isolate adult frequently, inappropriately, and, often, in an arbitrary way, aggresses his or her conspecific (Mitchell, 1975).

Thus, the monkey reared in social isolation changes as it matures from an animal showing abnormal social fear to an animal evincing abnormal social aggression. As these changes occur, so do changes in self-directed behaviors. The rhesus monkey reared in social isolation shows exaggerated self-clinging, self-sucking, self-rocking, and other idiosyncratic self-directed movements and postures. As the isolate-reared monkey gets older, its self-rocking changes to more whole-bodied swaying, somersaulting, repetitive pacing, or circling; and its self-sucking behavior gradually changes to self-chewing, self-hitting, and self-biting (Mitchell, 1975).

It is interesting that the changes in the isolates' self-directed behaviors seem to parallel those seen in the development of their abnormal social behaviors. Early in development, the isolate-reared monkey hides from its own arms and legs as well as from other monkeys. When it looks at its bodily parts, it often fear grimaces as though afraid of them. As the animal grows older, however, it begins to respond to its own limbs with threats and bites. But perhaps more importantly, for this particular chapter, the isolate-reared rhesus appears to respond to its own bodily parts as if they were not part of himself (or herself). It looks as though the animal does not seem to be able to discriminate self from nonself (Mitchell, 1975). But does a monkey have a self it can recognize?

E. SELF-RECOGNITION

After a series of experiments (e.g., Gallup, 1977a, b), Gordon G. Gallup, Jr. concluded that only the great apes and humans are capable of recognizing the self. Gallup went so far as to provide a preadolescent crab-eating macaque with 2,400 hours of mirror exposure to the self (to help the animal get to know itself). The monkey failed to show signs of self-recognition.

Self-recognition is demonstrated when an animal or human displays self-directed behavior toward some object or mark on its own body that it cannot feel or see without a mirror. In a vigorous test of self-recognition, in which animals are anesthetized and marked with red dye, it has been shown that both chimpanzees and orangutans take advantage of a mirror to touch or inspect marked areas of the skin (Gallup, 1970; Gallup et al., 1971; Lethmate and Ducker, 1973). None of the following species show self-recognition in mirrors: rhesus monkeys, crab-eater monkeys, stumptail monkeys, spider monkeys, capuchins, mandrills, hamadryas baboons, or gibbons (cf. Gallup, 1977b). In all of these species, the tested individuals respond to the mirror image as though they were in the presence of another conspecific individual. Does this mean that there is a prewired "innate" difference between the great apes and other nonhuman primates in self-recognition ability?

The available evidence suggests that the difference is *not* completely dependent upon innate differences. Even in normal chimpanzees there is an initial tendency to treat the reflection of the self as though it were another animal. This tendency wanes with time, however, so that after about three days with a mirror it is essentially replaced by self-inspection. During self-inspection, the chimpanzee uses the mirror to look at and touch parts of the body it cannot see without the mirror. When anesthetized, marked with a red dye, and returned to the mirror, self-inspection, particularly toward the red dye marks, is exaggerated. Gallup (1977b) believes that monkeys and lesser apes do not recognize themselves in mirrors because they fail to develop a "sufficiently well integrated self-concept." (P. 283.)

But chimpanzees are not born with this ability. In a study of chimpanzees reared in social isolation, it was found that isolation rearing did away with the self-recognition ability. Social isolates cannot recognize themselves in mirrors (Gallup et al., 1971). Remedial social experience (social "therapy"), however, increases chimpanzee self-recognition to some extent.

Thus, the development of an individual identity in a chimpanzee requires social interaction with others. Surprisingly, this interaction need not be with a conspecific. Home-reared (human-reared) chimpanzees recognize themselves in mirrors even though they prefer human company and show disdain for conspecifics (Gallup, in press). Thus, an accurate species identity is not necessary for an accurate self-recognition or for self-awareness. Vicki, a chimpanzee who was human-raised by K. J. Hayes, recognized herself and declared herself *human*! One of the things Vicki learned in her human home was to sort snapshots into a human versus animal pile. She placed her own picture in the human pile (Gallup et al., 1977).

With the exception of the other great apes and man, attempts to demonstrate self-recognition in all other species have failed (Gallup, 1977b). It is interesting that only the great apes, among nonhuman primates, are capable of learning sign language. There are some researchers (e.g., Terrace and Bever, 1976) who believe that self-recognition may be necessary for the development of sign language:

> All the ingredients for human language are present in other species—they do not become language until an animal learns that it can refer to itself symbolically. (Terrace and Bever, 1976, P. 580)

Because there are sex differences in human verbal ability (Maccoby and Jacklin, 1974), with females showing superiority over males, it might be interesting to know whether there are also sex differences in sign language ability and self-recognition in chimpanzees. Terrace, Petitto, and Bever (1977) taught signs to a male chimpanzee (Nim) and found that the rates at which he acquired signs (and the content of his vocabulary) did not differ from those of a female chimpanzee (Washoe). Both Nim and Washoe made their first signs at the beginning of their fourth month.

Another interesting question concerning self-recognition and signing is the extent to which the pigmy chimpanzee is capable of performing these two complex skills. Since *Pan paniscus* (the pigmy chimpanzee) in some respects appears to be more humanlike than does *Pan troglodytes* (the common chimpanzee) (Savage and Bakeman, 1976), it might not be surprising to find that this species could develop more complex language skills and more integrated and complex self-concepts.

F. COMMENTS ON HUMANS

Amsterdam (1972) has traced the development of self-recognition in human infants employing methods similar to those which Gallup (1970) used on monkeys and apes (see also Papousek and Papousek, 1974; and Schulman and Kaplowitz, 1977).

Using videotape, Amsterdam and Greenberg (1977) demonstrated that twenty-month-old human infants were different from ten- and fifteen-month-old infants in their display of self-consciousness (when seeing their own image). Self-consciousness appears some time after ten months of age and occurs with greatest frequency in response to a simultaneous self-image. On the subject of sex differences, self-consciousness in a young child is less likely to occur in response to a female than to a male (Amsterdam and Greenberg, 1977).

Self-observation in a mirror during the first two years of life has been studied by Schulman and Kaplowitz (1977). Self-observation does not vary between one and twenty-four months of age on a videotape. However, somewhere between nineteen and twenty-four months of age, self-recognition is evident. Before this age (at thirteen to eighteen months) an avoidance reaction (perhaps due to self-consciourness) is apparent. No sex differences were reported.

The self-concept of *Homo sapiens* proceeds far beyond what we have seen in the great apes. Language and other cognitive skills make the human being particularly able to recognize, describe, fantasize, and define the self. For example, between seven and twelve years of age there is an increase in altruistic behavior which has been assumed to reflect increasing *awareness* of social behaviors by both boys and girls. At this time, girls seem to show more altruism than do boys (but before this age they do not surpass boys). Apparently, in humans, knowledge of one's self includes knowledge of one's sex and of what one's sex is *supposed* to do (cf. Skarin and Moely, 1976). [According to Shapiro (1978), chimpanzees can conceptualize *people* into male and female categories.]

As in chimpanzees, atypical rearing affects the development of the self-concept in humans. Becoming human requires early social contact. Institutionalized children often display the abnormal self-directed behaviors seen in socially deprived monkeys and apes. These symptoms include self-rocking, self-mouthing, self-biting, stereotyped movements, etc. Also among the abnormalities displayed by the emotionally disturbed are those associated with the concept of self. Many disturbed and retarded children show a minimal capacity for "self-object" differentiation. The process of becoming human probably requires interaction with humans during the child's first and second years of life (cf. Roy and Roy, in press). In those individuals suffering the earliest and most severe forms of social and emotional disturbance (autistic children) there is often a complete absence of the use of the pronouns "I" or "me" and, frequently, there is complete mutism (cf. Wing, 1976).

It is interesting that, in human adults, mirrors frequently increase feelings of self-consciousness and/or tend to produce positive feedback to the self-viewer of his or her own affective state. When angered, for example, a highly self-conscious person will agress more than will a low self-conscious person. A mirror increases anger in an already angered self-conscious person:

> Increased awareness of one's affective state enhances the tendency to respond to that state. (Scheier, 1976, P. 643)

Perhaps the horrendous aggressive acts of human beings cannot be attributed to inherent aggressiveness as much as to inherent self-awareness.

In recent years, self-recognition, self-awareness, and consciousness have become respectable subjects in comparative psychology. Powers (1973), for example, has developed a model for consciousness based upon feedback principles. Shafton (1976) has written an extensive monograph on subjective factors in the social adaptations of man and other primates in which he outlines the "Conditions of Awareness." The recent emphasis on awareness is affecting behavioral primatology and running headlong into another major theoretical framework for our discipline, that of sociobiology.

As Tobach (1976) has pointed out, the comparative method is useful in studies of behavior only when:

> . . . the questions to be answered are based on stated assumptions which are testable and when the levels of the phenomena being compared are equivalent. (P. 185)

Like Tobach (1976), Porges (1976) has warned that, in *ontogenetic* as well as in phylogenetic comparisons, there are problems associated with response equivalence. Often, the way in which a two-year old responds to a given situation cannot be directly compared to the way in which an eight-year old responds. (We suggest here that the same problems can occur in sex and/or gender comparisons.)

In phylogenetic comparisons this principle applies to both the "objective" concepts of sociobiology *and* to the more "subjective" concepts of the self-awareness advocates of behavioral primatology. Those behavioral primatologists most interested in self-recognition or self-awareness are quick to point out the weaknesses of the sociobiological approach to comparative behavior in this regard. Those aware of self-recognition believe that primates and particularly great apes and man are *not* driven by genes, *can* control their own behaviors, and *do* act with intention and with purposiveness.

Many respond to sociobiology as a new form of social Darwinism. Klopfer (1977), however, argues that an unfortunate alliance of some sociobiologists with politicians is a poor basis for discrediting a field or theory. He argues *for* a science of sociobiology even if man is ultimately shown to be "so infinitely flexible as not to be biologically restricted to any particular social forms." (P. 83.) We agree with Klopfer (1977) that we need to know "*the* nature of" our social heritage; however, we do *not* feel, (knowing the intellectual capacities of nonhuman primates as we do) that the behavior of goats and lemurs will reveal human behavior as much as will the behavior of humans themselves and their *nearest* relatives. In short, the proof of self-awareness in chimpanzees demands that we *also* develop sophisticated alternatives to sociobiology.

Even the sociobiologists themselves recognize a need for something to account for individuality, the concept of self, and *cultural* evolution. Dawkins (1976) believes that genes cannot account for all kinds of replication:

> What, after all, is so special about genes? The answer is that they are replicators. The *meme* is a new kind of replicator. (P. 207)

The meme, according to Dawkins (1976), is an idea or concept replicated from generation to generation by cultural evolution. The idea of a God is a meme. As Dawkins (1976) suggests:

> God is a meme with a high survival value. (P. 207) (and) . . . A meme for celibacy can be successful in the meme pool. (P. 213)

Dawkins admits that memes are advantageous only to themselves but that people have *conscious foresight*:

> We have the power to *defy* the selfish genes of our birth and, if necessary, the selfish memes of our indoctrination. (P. 215)

The "meme" is a clever concept but perhaps too cute to deal with that awful and awesome body of data that has been the subject matter of psychology for generations. These data cannot be brushed aside by appeals to the gene. There is no reason, therefore, to believe they will be defined away by a "meme." There is an area of research in psychology called cognition toward which the sociobiologists should at least occasionally nod even if they do not always have direct interest in the area.

G. SELF-AWARENESS AND GENDER DIFFERENCES

What does all of the above have to do with gender differences? Awareness of self suggests that behavior is conscious, intentional, and purposive. Awareness of self suggests that an individual is aware of his/her gender and can intentionally, purposefully, and consciously *act* as he/she feels he/she should or wants to. While the existence of self-awareness does *not* mean that phylogeny, genetics, physiology, indeed, biology have nothing to do with behavioral sex differences, it does presuppose that the individual is *capable* of a wider range of behaviors than would be the case without self-awareness. It is no wonder that groups of male and female humans show such extensive overlap on so many characteristics. The possible range of behaviors within each sex is very large indeed. So large, in fact, that some researchers have stated that natural science approaches to human sexuality and sex differences are inappropriate (Whitsett, 1977).

But it is not our intention to make the present chapter a disclaimer for our entire book. Sex differences in nonhuman primate behavior (without intent or purpose) certainly *do* exist. So do differences in *human* primate behavior (e.g., sex differences are seen in infancy before self-recognition has developed). In fact, it is probable that there are even sex differences in human *action* (in behavior where purpose and intent play an obvious role). An area of research of interest to us would be one involving research on sex differences in self-awareness itself. Since there appears to be a relationship between verbal skills and self-awareness in nonhuman and retarded humans, it may be that female humans develop self-recognition earlier than do males. We know of no studies on this question.

REFERENCES

Amsterdam, B. Mirror self-image reactions before age two. *Developmental Psychobiology,* 1972, **5**, 297-305.

Amsterdam, B. and Greenberg, L. M. Self-conscious behavior of infants: A videotape study. *Developmental Psychobiology,* 1977, **19** (1), 1-6.

Burton, F. D. Ethology and the development of sex and gender identity in nonhuman primates. *Acta Biotheoretica,* 1977, **26** (1), 1-18.

Dawkins, R. *The Selfish Gene.* New York: Oxford University Press, 1976.

Gallup, G. G. Chimpanzees: Self-recognition. *Science,* 1970, **167**, 86-87.

Gallup, G. G. Absence of self-recognition in a monkey (*Macaca fascicularis*) following prolonged exposure to a mirror. *Developmental Psychobiology,* 1977a, **10**, 281-284.

Gallup, G. G. Self-recognition in primates: A comparative approach to the bi-directional properties of consciousness. *American Psychologist,* 1977b, **32**, 329-338.

Gallup, G. G. Consequences of atypical rearing in chimpanzees. In Roy, M. A. (Ed.) *Species Identification: A phylogenetic Evaluation,* in press.

Gallup, G. G., Boren, J. L., Gagliardi, G. J. and Wallnau, L. B. A mirror for the mind of man or will the chimpanzee create an identify crisis for *Homo sapiens. Journal of Human Evolution,* 1977, **6**, 303-313.

Gallup, G. G., McClure, M. K., Hill, S. D. and Bundy, R. A. Capacity for self-recognition in differentially reared chimpanzees. *Psychological Record,* 1971, **21**, 69-74.

Klopfer, P. H. Social Darwinism lives! (Should it?) *The Yale Journal of Biology and Medicine,* 1977, **50**, 77-84.

Lethmate, J. and Dücker, G. Untersuchungen zum Selbsterkennen im Soigel bei Orang-utans und linigen anderen Affenarten. *Zeitschrift für Tierpsychologie,* 1973, **33**, 248-269.

Maccoby, E. E. and Jacklin, C. N. *The Psychology of Sex Differences.* Stanford, California: Stanford University Press, 1974.

Maple, T. L. Basic studies of interspecies attachment behavior. Doctoral dissertation, Department of Psychology, University of California, Davis, 1974.

Mitchell, G. What monkeys can tell us about human violence. *The Futurist,* 1975, **9** (2), 75-80.

Mitchell, G. Attachment potential in rhesus macaque dyads (*Macaca mulatta*): A sabbatical report. *Catalogue of Selected Documents in Psychology,* 1976, **6**, M.S. No. 1177.

Papousek, H. and Papousek, M. Mirror image and self-recognition in young human infants: I. A new method of experimental analysis. *Developmental Psychobiology,* 1974, **7**, 149-157.

Porges, S. W. Ontogenetic comparisons. *International Journal of Psychology,* 1976, **11**, 203-214.

Powers, W. *Behavior: The control of Perception.* Chicago: Aldine, 1973.

Reynolds, V. *The Biology of Human Action.* San Francisco: W. H. Freeman, 1976.

Roy, M. A. Early experience and the development of an adequate species identity. Symposium presented at the *American Psychological Association* Meeting, San Francisco, California, August, 1977.

Roy, M. A. Introduction. In Roy, M. A. (Ed.) *Species Identification: A Phylogenetic Evaluation,* in press.

Roy, R. L. and Roy, M. A. Consequences of atypical rearing experiences in humans. In Roy, M. A. (Ed.) *Species Identification: A Phylogenetic Evaluation,* in press.

Savage, E. S. and Bakeman, R. Comparative observations on sexual behaviour in *Pan paniscus* and *Pan troglodytes.* Paper presented at the *International Primatological Society* Meeting, Cambridge, England, August, 1976.

Scheier, M. F. Self-awareness, self-consciousness and angry aggression. *Journal of Personality,* 1976, **44**, 627-644.

Schulman, A. H. and Kaplowitz, C. Mirror-image responses during the first two years of life. *Developmental Psychobiology,* 1977, **10**, 133-142.

Shafton, A. *Conditions of Awareness.* Portland, Oregon: Riverstone Press, 1976.

Shapiro, G. Is an ecological understanding of sign language in apes necessary? Lecture given to the Ecology Group, University of California, Davis, California, February, 9, 1978.

Skarin, K. and Moely, B. E. Altruistic behavior: An analysis of age and sex differences. *Child Development,* 1976, **47**, 1159-1165.

Terrace, H. S. and Bever, T. G. What might be learned from studying language in the chimpanzee? The importance of symbolizing oneself. *Annals of the New York Academy of Sciences,* 1976, **280**, 579-588.

Terrace, H. S., Petitto, L. and Bever, T. G. Project Nim: Progress report. I. Unpublished manuscript, 1977.

Tobach, E. Evolution of behavior and the comparative method. *International Journal of Psychology,* 1976, **11** (3), 185-201.

Whitsett, G. A critique of the natural science approach to human sexuality. Paper presented at the *Animal Behavior Society* Meeting, University Park, Pennsylvania, June, 1977.

Wing. L. (Ed.) *Early Childhood Autism.* New York: Pergamon, 1976, pp. 67-71.

34.

Summary and Epilogue

Our review of behavioral sex differences in nonhuman primates began with two definitions. One definition distinguished behavior from action and the other definition differentiated sex differences from gender differences. We decided that when we used the term *behavior* we would assume that the primates in question had no conception of their own behavior and had no intent. We reserved the word *action* for those behaviors in which there was conscious choice, purposiveness, or intent. Thus, many great ape and human behaviors are intentional or purposive actions. The title of the book uses the word *behavior* because we assumed that most sex differences were not intentional, purposive, or conscious.

In Chapter 2, we outlined the primate order. There are eight major groups of primates on which we concentrated throughout the book. The first group was the suborder of prosimians, a primitive group of primates showing extreme variability from species to species and even within the species. The second major group included the New World marmosets and tamarins, monomorphic and monogamous, yet relatively primitive primates, that so often seemed to show few if any sex differences in behavior. Our third major group was the cebids, New World monkeys showing great variability in social structure, in physical size, and in sex differences. Our fourth group was the very well-known terrestrial Old World monkeys, including macaques and baboons. The arboreal Old World monkeys, including langurs and other leaf-eaters, made up our fifth major group. The sixth and seventh groups were the monogamous lesser apes (the gibbons and siamangs) and the three great apes, chimpanzee, gorilla, and orangutan), respectively. Our eighth group was *Homo sapiens*. While we did not go into detail in our examination of behavioral sex differences in people we *did* comment on them at the conclusion of each chapter.

Our next chapter, Chapter 3, dealt with physical sexual dimorphism in primates. In this chapter we examined physical dimorphism in size, in coloration, in dentition, and in other physical characteristics.

In Chapter 4, we introduced the importance of prenatal androgens and other prenatal hormones. We learned that prenatal androgens, at least in macaques, predispose an individual toward rough play, aggressive behavior, and foot-clasp mounting.

The development of play was discussed in Chapter 5. Male infants and male juveniles play more often and more intensely than do females. Chapter 6 began with a reminder that sex differences are actually dynamic developmental processes. Puberty and adolescence were discussed at various taxonomic levels. We noted that sex differences may disappear or may even be reversed at puberty in some species of primates. We noted the role of sex hormones in adulthood in Chapters 6, 7, and 8 and to a lesser extent later in Chapter 31.

Our discussions of sexual behavior in Chapters 9 and 10 revealed that, in general, sexual posturing varies greatly from species to species and becomes more variable as we go from prosimians to monkeys to apes. We learned that males and females of the most complex nonhuman primates frequently display the normal sexual posturings of the opposite sex. Homosexual behavior is seen in both male and female nonhuman primates but probably does not involve as exclusive a preference for a same-sexed partner as is true for human homosexuality. We also learned that nonhuman as well as human primates practice incest avoidance.

In Chapter 11 and in Chapters 12 and 13, on birth and on infant care respectively, we saw how the complexity of parental behavior increases through the primate order. The mouth-to-mouth resuscitation given to newborn infants by chimpanzee mothers and the "obstetrical" behaviors displayed by marmoset and orangutan males were particularly impressive. In some species of primates (e.g., marmosets), males actually display more infant care than do females.

In Chapters 14 and 15 we learned that in the social spacing of most primates, males are peripheralized and females remain in the core of the group. Males are usually (but not always) dominant over females (Chapter 17) but leadership (Chapter 18) depends upon the situation. Females are frequently leaders. Females can also overcome male dominance by forming coalitions with other females.

In Chapter 19, we discussed what is perhaps the most consistent sex difference in primate behavior, that involving vigilance, protection, and intertroop behavior. Males surpass females in these behaviors in virtually all species of primates.

In feeding situations, males often dominate females. In addition, male primates eat more meat than do female primates. Males also do most of the hunting for meat and most of the killing (see Chapter 20).

Females generally surpass males in friendly social contacts, especially in grooming (Chapter 21) and as well as in friendly, social, and soft vocaizations (Chapter 22). Males, on the other hand, in keeping with their role as sentinels and group protectors, tend to emit more alarm and territorial calls.

In visual communication, females emit more expressions of fear, males more aggressive expressions (Chapter 23). A consistent but not invariable sex difference is that males are more aggressive than females (Chapters 24 and 25).

Males appear to be more adversely affected by early social deprivation than are females (Chapter 26), but females are apparently more disturbed by a new environment in adulthood, by stress in adulthood, and often, by crowding. The female has apparently evolved to survive trauma in a reasonably stable group, but not trauma without kin and/or a stable group (Chapter 27).

We saw that male primates, in general, are more vulnerable to infection and to congenital abnormalities. Almost *all* defects occur more frequently in males than in females, with the exception of visual myopia. It is interesting that males *need* distance vision for their vigilance behavior (see Chapters 19 and 28).

In Chapter 29, we saw that female primates are more adept at delayed response and delayed alternation that are males. There are many sex differences in the brain, but few of them, with the exception of those dealing with sexual behavior and cyclicity, have been explained (see Chapters 30 and 31).

Despite there being a large number of fairly consistent sex differences among nonhuman primates, there are a great many changes in these sex differences from age to age, from troop to troop, and from individual to individual. Always affecting sex differences are cyclical changes such as monthly fluctuations in hormones associated with estrus, diurnal rhythms, and seasonal (annual) rhythms (see Chapter 31). Over and above age-related and cyclical variation, there are idiosyncratic behaviors involving individual or noncyclical variation from troop to troop, from generation to generation, and even from time to time within an individual animal (Chapter 32).

A significant source of variability in the higher nonhuman primates is purposiveness or intentionality. Gallup and others have shown that only the great apes (chimpanzees and orangutans) can recognize themselves in mirrors. This means that only orangutans, chimpanzees, and people are self-aware. They can behave with intention, and with purposiveness. The ability to recognize the self is correlated with language ability. These two abilities (self-awareness and language) can make behavioral sex differences (in those species which have them) even more variable than they are in those species which lack self-awareness and languagelike skills. Thus, our final three chapters (Chapters 31, 32 and 33) serve as partial disclaimers, or at least serve to remind us, that behavioral sex differences in primates, and particularly in the most intelligent primates—great apes and man—are certainly not simple, always predictable, never changing differences but are, rather, very complex, unpredictable, changing processes under volitional control.

Our final comments dealt with two major theoretical forces concerned with human evolution. One of these, sociobiology (which emphasizes the role of genetics in determining sex differences), tends to ignore a great share of what is known concerning cognitive skills, degree of choice, and self-awareness in the higher primates, especially in people. The other force has actually risen as a counterforce to sociobiology and has no readily agreed upon title. It notes that *genetic* constraints evolved through prehistory and no longer apply because cultural evolution now overlays (to a large extent) much of genetic evolution. Sociobiology, according to the "backlash" group, underrates the human brain, consciousness, and culture. Sociobiologists stress the differences between the sexes—between their physiques, their behaviors, and their attitudes. Those in the other camp emphasize plasticity, choice, awareness; and, in some subgroups, *androgyny*. Some, for example, say that both men and women should have the freedom to act as both the nurturant and the aggressive individual. The resulting androgyny (*andro* = male and gyne = female) should promote better mental health because the individual would not be restricted by sex-role stereotypes. The problem with this approach is that *all* gender differences are not *sex-role stereotypes* simply because there *is* great variability and a measure of choice in whether or not they are displayed. It seems to us that, if there are some evolutionary foundations for gender differences in people, it would also be of help to the mental health of people to know what these differences are. Our position on this matter then is an adamant middle-of-the-road position. We are not being wishy-washy in taking this stand. We *insist* that those promoting sociobiology do more than nod toward self-awareness in man and apes. We also *insist,* however, that there are indeed some *biological* sex and gender differences in primates, including some well-substantiated sex differences in people. It was with this spirit that the present book was written.

One final point. Because of ever-increasing populations and a possible need for new roles for people of both genders, genes *cannot* have priority. In the crowded world we now know, rearing as large a family as possible would be sociobiologically sane only in the vaguest theoretical sense. Thus, sex must truly exist for reasons other than procreation. Genes cannot have priority any more than one gender can have priority over another. Many human gender differences, however, may ultimately relate to courtship and mating, others to stages of reproduction somewhat further removed from the sexual act itself though ultimately guided by it. However, since overpopulation is today's *major* problem, *sex* first and *then* gender (only in so far as it relates to sex) must have priority in our scientific investigations. As Freud knew, sex *can* and *does* have priority. But in our trying times it must be sex *without* reproductive success.

Appendix

Genera, species and common names of primates (based on Napier, J. R. and Napier, P. H. *A Handbook of Living Primates.* New York: Academic Press, 1967).

Genus	*Species*	*Common name*
A. Prosimians		
Arctocebus	*calabarensis*	angwantibos
Avahi	*laniger*	Avahi
Cheirogaleus	*major*	greater dwarf lemur
Daubentonia	*madagascariensis*	aye-aye
Galago	*crassicaudatus*	thick-tailed bush baby
	senegalensis	bushbaby
	alleni	Allen's galago
Hapalemur	*griseus*	grey gentle lemur
	simus	broad-nosed gentle lemur
Indri	*indri*	Indri
Lemur	*catta*	ring-tailed lemur
	variegatus	ruffed lemur
	macaco (fulvus)	black lemur
	mongoz	mongoose lemur
Lepilemur	*mustelinus*	sportive lemur
Loris	*tardigradus*	slender loris
Microcebus	*murinus*	lesser mouse lemur
Nycticebus	*coucang*	slow loris
Perodicticus	*potto*	potto
Phaner	*furcifer*	fork-marked dwarf lemur
Propithecus	*verreauxi*	Verreaux's sifaka
Tarsius	*spectrum*	spectral tarsier
Tupaia	*glis*	common treeshrew
	montana	mountain treeshrew
	gracilis	slender treeshrew

B.   New World Monkeys

Genus	*Species*	*Common name*
Alouatta	*palliata (villosa)*	mantled howler
	caraya	black howler
	seniculus	red howler
Aotus	*trivirgatus*	owl or night monkey
Ateles	*geoffroyi*	black handed spider monkey
	paniscus	black spider monkey
	belzebuth	long-haired spider monkey
Brachyteles	*arachnoides*	woolly spider monkey
Cacajao	*rubicundus*	red uakari
Callicebus	*moloch*	dusky titi
	torquatus	widow titi
Callimico	*goeldii*	Goeldi's marmoset
Callithrix	*jacchus*	common marmoset
Cebuella	*pygmaea*	pygmy marmoset
Cebus	*apella*	black-capped capuchin
	albifrons	white-fronted capuchin
	capuchinus	white-throated capuchin
Chiropotes	*satanus*	black saki
Lagothrix	*lagothrica*	woolly monkey
Leontideus	*rosalia*	golden lion marmoset
Pithecia	*pithecia*	saki
Saguinus	*fuscicollis*	brown-headed tamarin
	pileatus	red-capped tamarin
	oedipus	cotton-top tamarin
Saimiri	*sciureus*	squirrel monkey
	oerstedii	red-backed squirrel monkey

C.   Old World Monkeys

Cercocebus	*atys*	sooty mangabey
	albigena	grey-cheeked mangabey
	galeritus	agile mangabey
	torquatus	white-collared mangabey
Cercopithecus	*aethiops*	grivet, vervet
	sabaeus	green monkey
	mitis	blue monkey
	albogularis	Syke's monkey
	campbelli	Campbell's monkey
Colobus	*polykomos*	King colobus
	guereza	Abyssinian colobus
Cynopithecus	*niger*	Celebes black "ape"
Erythrocebus	*patas*	patas monkey

Genus	Species	Common name
Macaca	mulatta	rhesus macaque
	arctoides	stumptail macaque
	nemestrina	pigtail macaque
	radiata	bonnet macaque
	fuscata	Japanese macaque
	silenus	lion-tailed macaque
	sinica	Toque macaque
	sylvana	Barbary macaque
	fascicularis	crab-eater macaque
	assamensis	Assamese macaque
	cyclopis	Formosan macaque
	maurus	Celebes macaque
Mandrillus (Papio)	sphinx	mandrill
Miopithecus	talapoin	talapoin
Nasalis	larvatus	proboscis monkey
Papio	anubis	olive baboon
	papio	guinea baboon
	cynocephalus	yellow baboon
	hamadryas	hamadryas baboon
	ursinus	chacma baboon
Presbytis	entellus	Hanuman langur
	johnii	John's langur
	cristatus	silvered leaf monkey
	obscurus	dusky leaf monkey
	pileatus	capped langur
	melalophos	banded leaf monkey
	aygula	Sunda Island langur
	geei	golden langur
Pygathrix	nemaeus	Douc langur
Rhinopithecus	roxellanae	snub-nosed langur
Simias	concolor	Pagai Island langur
Theropithecus	gelada	gelada baboon

D. Lesser Apes

Hylobates	lar	white-handed gibbon
	klossii	Kloss's gibbon
	hoolock	hoolock gibbon
Symphalangus	syndactylus	siamang

E. Great Apes

Gorilla	gorilla gorilla	lowland gorilla
	gorilla beringei	mountain gorilla
Pan	troglodytes	common chimpanzee
	paniscus	Pygmy chimpanzee
Pongo	pygmaeus pygmaeus	Borneo orangutan
	pygmaeus abelii	Sumatra orangutan

F. People

Homo	sapiens	people

Author Index

Aarons, L., 308, 309, 310, 316, 378, 382, 416, 420
Abbott, D.H., 50, 63, 74, 83, 92, 102, 128, 136
Abegglen, H., 160, 245, 274, 398
Abegglen, J.J., 160, 398
Adachi, K., 110, 124
Adams, R.M., 83, 86, 243, 245
Adey, W.R., 120, 124
Agar, M.E., 132, 133, 134, 136, 323, 328, 405, 410
Alcock, J., 59, 65, 153, 159
Aldrich-Blake, F.P.G., 148, 158, 243, 350
Alexander, B.K., 56, 64, 96, 106, 189, 193, 255, 256, 280, 287, 297, 301, 381, 382, 383, 419, 421
Alexander, M., 252, 256, 260
Allerton, M.W., 420, 421
Altman, I., 306, 333
Altmann, S.A., 224, 229, 336, 344, 349, 359, 361, 370, 378, 383, 417, 421
Alvarez, F., 221, 229, 250, 256, 283, 287
Amsterdam, B., 483, 486
Anderson, B., 192, 194, 381, 383, 384, 419, 420, 421
Anderson, C.O., 54, 64, 92, 103, 222, 229, 285, 287, 416, 431, 453, 460
Ando, S., 231, 238, 245
Andrew, R.J., 348, 349
Angermeier, W.F., 253, 256, 378, 383
Angst, W., 228, 229, 380, 383, 426, 429

Anschel, S., 131, 136, 250, 251, 256, 261, 377, 383, 416, 422, 469, 478
Anthoney, T.R., 265, 271, 366, 370
Apfelbach, R., 148, 157, 342, 349
Arling, G.L., 404, 405, 410, 413, 435, 438
Atkinson, L.E., 43, 48, 115, 124
Atwood, R.J., 165, 175
Autrum, H., 127, 136, 416, 421
Avise, J.C., 472, 474
Ayats, H., 288, 302
Azuma, S., 141, 157, 231

Bachman, C., 160, 245, 274
Badham, M., 204, 212
Bailey, R.C., 286, 287
Bailey, S.M., 19, 24
Bakeman, R., 17, 28, 150, 162, 458, 465, 483, 488
Bakwin, H., 154, 157
Baldridge, C., 17, 29, 414, 440
Baldwin, J.D., 50, 61, 64, 131, 136, 221, 222, 229, 283, 287, 453, 460
Baldwin, J.I., 50, 61, 64, 222, 229, 283, 287
Balkin, J., 271, 273
Barfield, M.A., 427, 430
Barnes, B., 264, 273, 380, 385
Baron, R.A., 154, 155, 157, 395, 396, 397
Bartlett, L., 282, 287
Bartol, K.M., 282, 287
Basckin, D.R., 391, 397

Bauer, H.R., 150, 157, 241, 243
Baum, M.J., 109, 119
Bayley, N., 349, 409, 410, 418, 421, 428, 429
Baysinger, C.M., 336, 344, 349, 350, 358, 362, 370, 404, 410
Baxter, M.J., 141, 157, 325, 328
Beach, F.A., 115, 119, 125, 126, 131, 134, 136, 446, 447
Bearder, S.K., 250, 256, 375, 383
Becker, H., 158
Beckwith, W.C., 206, 216, 393, 398
Beecher, M.D., 355
Beg, M.A., 53, 70
Begeman, M.L., 73, 88
Begert, S.P., 49, 67, 128, 138, 375, 386
Bekker, T., 127, 137
Bekoff, M., 179, 193
Bell, P.A., 154, 155, 157, 395, 397
Bender, L., 409, 410, 428, 429
Benirschke, K., 452, 460
Benjamin, H., 62, 64
Benjamin, L., 425, 429
Benson, D.A., 447, 450
Benton, L., 75, 83, 251, 256, 295, 301, 308, 315, 376, 383
Bercovitch, F.B., 375, 383
Berger, M.E., 426, 429, 477
Berkson, G., 56, 64, 190, 193, 205, 212, 285, 287, 306, 333, 336, 337, 352, 405, 412, 472, 474

Bernstein, I.S., 15, 16, 24, 37, 44, 52, 53, 55, 64, 94, 99, 103, 104, 106, 110, 119, 141, 142, 157, 201, 204, 212, 228, 229, 249, 252, 253, 256, 258, 260, 262, 265, 272, 278, 279, 280, 284, 287, 288, 296, 298, 301, 302, 321, 322, 324, 325, 326, 329, 338, 349, 378, 380, 381, 383, 387, 390, 391, 393, 397, 403, 410, 416, 421, 426, 427, 429, 454, 455, 456, 460, 461, 465
Bert, J., 281, 288, 299, 302
Bertrand, B., 330
Bertrand, M., 51, 56, 64, 66, 75, 86, 202, 212, 215, 239, 243, 244, 266, 274, 280, 288, 297, 302, 303, 309, 315, 321, 324, 329, 330, 339, 350, 351, 364, 370, 391, 398, 435, 436, 438, 457, 462
Bever, T.G., 482, 488
Bielert, C., 35, 44, 76, 83, 97, 99, 103, 119, 134, 136, 454, 460
Biller, H., 210, 212
Bingham, H.C., 206, 212
Bingham, L.R., 428, 429
Birch, H.G., 23, 24, 113, 119, 267, 272
Birdwhistell, R.L., 18, 24
Birkner, F.E., 456, 460
Bishop, N., 14, 25, 52, 65, 203, 213, 267, 272, 298, 302, 321, 326, 329, 349, 366, 370, 427, 430
Blackwell, K., 313, 315
Blander, R., 124
Blomquist, A.J., 434, 438, 447
Blurton-Jones, N.G., 13, 24, 79, 88, 142, 143, 157, 163, 190, 193, 198, 348, 355, 381, 383
Bobbitt, R.A., 192, 197, 228, 231, 347, 354, 408, 412, 426, 431
Boelkins, R.C., 225, 229, 248, 256, 284, 288, 379, 388, 425, 432, 455, 466
Bogart, M.H., 452, 460
Boggess, J., 203, 212, 391, 397, 473, 475
Bolwig, N., 111, 120, 200, 212, 265, 272, 299, 302, 326, 329
Bonsall, R.W., 97, 103, 123, 454, 463

Boren, J.L., 487
Bourlière, F., 51, 64, 66, 75, 86, 202, 212, 215, 238, 239, 244, 245, 266, 274, 280, 288, 297, 302, 303, 338, 339, 350, 351, 352, 391, 398, 426, 431, 457, 462
Bowden, D.M., 110, 121, 171, 176
Bowers, J.M., 280, 287, 297, 301, 382
Boyd, T.C., 310, 315, 326, 329
Braggio, J.T., 58, 59, 62, 64, 66, 153, 161, 207, 216, 268, 272 , 275, 395, 397, 436, 438
Bramblett, C. A., 5, 11, 13, 14, 24, 51, 52, 56, 64, 71, 126, 136, 206, 213, 237, 243, 281, 288, 297, 302, 326, 329, 337, 340, 350, 390, 397
Brandt, E. M., 54, 64, 76, 84, 165, 166, 167, 168, 170, 171, 172, 173, 175, 177, 180, 181, 190, 193, 196, 243, 336, 343, 344, 347, 349, 350, 358, 362, 370, 379, 383, 386, 403, 404, 406, 410, 412
Brenner, M., 269, 275
Breuggeman, J. A., 53, 64, 76, 84, 97, 103, 188, 193
Brockelman, W. Y., 300, 302, 393, 397, 473, 475
Brodie, H. K. H., 40, 45, 117, 121, 158, 395, 398
Bromley, L. J., 55, 69, 79, 88, 111, 123
Brotherton, J., 35, 44
Broverman, D. M., 119, 120, 428, 429
Broverman, I. K., 429
Brown, B. B., 306, 333
Brown, G. M., 38, 44, 93, 103, 443, 447
Brown, W. L., 434, 439
Bubenik, G. A., 38, 44, 93, 103, 443, 447
Buckley, J. S., 321, 329
Buergel-Goodwin, U., 74, 89, 92, 107, 294, 306
Buettner-Janusch, J., 181, 193, 312, 315
Buikstra, J. E., 425, 429
Bundy, R. A., 487
Bunger, 192, 194
Burton, F. D., 55, 65, 80, 84, 191, 193, 210, 213, 365, 370, 467, 475, 479, 486

Buskirk, R. E., 238, 243, 281, 288, 299, 302, 303, 338, 350, 390, 397
Butler, F. D., 82, 84
Butler, H., 115, 119, 451, 459, 460
Butler, R. J., 19, 24
Byer, C. O., 32, 46, 83, 86, 104, 114, 115, 122, 156, 160, 459, 462
Bygott, J. D., 312, 315

Caine, N. G., 53, 165, 167, 168, 169, 170, 172, 173, 174, 175
Caldwell, M. A., 62, 65, 328, 329, 349, 350
Callaghan, J. W., 269, 272, 395, 397
Cameron, P., 328, 329
Candland, D. K., 56, 65, 190, 195, 222, 229, 250, 255, 256, 257, 258, 435, 438
Carducci, B., 243
Carmel, P., 39, 47, 106, 449
Carpenter, C. R., 133, 134, 136, 251, 257, 278, 279, 288
Castell, R., 222, 229, 377, 384
Chalmers, N. R., 112, 119, 145, 157, 201, 202, 213, 238, 243, 285, 288, 322, 329, 392, 397
Chamove, A., 53, 65, 363, 370, 379, 384, 403, 410
Chance, M. R. A., 284, 288
Chapman, M., 147, 157, 203, 213
Charles-Dominique, P., 50, 127, 136, 220, 229, 294, 302, 376, 384, 468, 475
Chepko-Sade, B. D., 135, 139, 254, 260
Chevalier-Skolnikoff, S., 15, 22, 25, 142, 143, 144, 146, 157, 158, 296, 297, 302, 324, 329
Chinn, R., 311, 314
Chivers, D. J., 148, 158, 205, 213, 241, 243, 288, 342, 350
Chivers, S. T., 205, 213
Clark, C. B., 206, 213, 473, 475
Clark, D. L., 251, 257
Clark, G., 23, 24, 113, 119, 267, 272
Clark, W. E. LeGros., 4
Clarkson, F. E., 429
Clifton, D. K., 48, 450

Clutton-Brock, T. H., 13, 25, 219, 229
Coburn, S., 356
Cochran, C. G., 94, 99, 101, 103, 252, 257
Cockett, A. T. K., 110, 120, 124
Coe, C. C., 137, 462
Coe, C. L., 75, 85, 93, 103
Coelho, A. M., 12, 25
Cohen, J. E., 242, 243
Collins, D. C., 464
Collumb, H., 288, 302
Comfort, A., 117, 120, 153, 158
Conaway, C. H., 67, 86, 95, 105, 107, 181, 197, 455, 456
Cooper, R. W., 452, 460
Coss, R. G., 420, 421
Cramer, D. L., 21, 25
Crawford, H. T., 448
Cronin, C. L., 269, 272, 277, 395, 397
Crook, J. H., 55, 65, 237, 243, 266, 272, 286, 288, 390, 397
Cross, H. A., 76, 84, 322, 329, 343, 344, 350, 361, 362, 363, 370, 403, 410
Cubicciotti, D. D., III, 130, 131, 136, 222, 229
Curtin, R. A., 147, 158, 239, 244, 473, 475
Curtis, G. C., 114, 120, 460
Cuzzone, A., 17, 29, 414, 440
Czaja, J. A., 99, 103, 136, 454, 460

Dang, D. C., 111, 120, 171, 175
Dalgard, D. W., 15, 16, 29
Dare, R., 252, 257, 308, 315
Davenport, R. K., Jr., 161, 403, 408, 412, 414, 418, 422, 427, 431
Davis, A. E., 177, 317, 431, 440
Dawkins, R., 485, 486
Dawson, G. A., 128, 136, 257, 376, 384, 453, 460
Dazey, J., 55, 67, 171, 175, 192, 195, 426, 429
Deag, J. M., 55, 65
Deets, A. C., 445, 447
Demarest, W. J., 135, 137, 156, 157, 158
Deming, M., 270, 276
Dennon, M. B., 459, 460

DeRousseau, C. J., 44, 94, 103, 469, 475
de Silva, G. S., 17, 25, 152, 158, 394, 397
Desprez, M., 435, 438
Deutsch, J., 405, 410
DeVore, I., 12, 25, 80, 85, 200, 213, 298, 303
DeWaal, F. B. M., 264, 272, 280, 288, 380, 384, 387
DeGiacomo, R. F., 15, 25, 172, 175, 286, 287
Dillon, J. E., 251, 257
Din, N. A., 16, 28, 200, 217
Dittus, W. P. J., 15, 25, 79, 84, 144, 158, 228, 260, 263, 272, 365, 370, 426, 430
Dixson, A. F., 112, 120, 146, 158, 239, 246, 264, 265, 272, 276, 321, 332, 392, 397, 400, 457, 465
Dodsworth, R. O., 54, 66, 322, 330, 379, 385
Doering, C. H., 154, 158
Dolan, K. J., 51, 65, 202, 213
Dolhinow, P. J., 14, 20, 25, 52, 65, 86, 170, 175, 177, 203, 213, 267, 272, 276, 321, 326, 329, 338, 350, 366, 370, 391, 397, 427, 430
Dolnick, E. H., 23, 25
Doyle, G. A., 14, 25, 74, 84, 127, 137, 181, 194, 220, 230, 250, 256, 257, 294, 302, 341, 350, 375, 376, 383, 384
Draper, W. A., 53, 64, 284, 288
Drickamer, L. C., 225, 230, 378, 384, 425, 430
Drickamer, L. D., 225, 230, 253, 257
Driscoll, J. W., 210, 214
Drobeck, H. P., 16, 29, 36, 48, 124
Ducker, G., 481, 487
Dugard, J., 182, 195
Duggleby, C. R. 135, 137
DuMond, F., 22, 25, 93, 103, 453, 460
Dunbar, P., 201, 213
Dunbar, R.I.M., 112, 120, 145, 158, 201, 213, 237, 238, 244, 265, 266, 273, 280, 288, 297, 299, 302, 321, 329, 391, 397, 457, 461
Dunne, K., 39, 46, 102, 105, 110, 122, 143, 160, 180, 195, 263, 274

Durham, N. M., 20, 26, 223, 230, 266, 273, 367, 370
Duvall, S. W., 472, 474

Eastman, R. F., 405, 410
Eaton, G. G., 111, 120, 141, 161, 254, 257, 259, 382, 384
Edelman, M., 62, 68
Edgerton, R. B., 206, 215, 393, 398
Ehrhardt, A. A., 42, 44, 63, 65
Ehrlich, A., 50, 65, 220, 230, 375, 384
Eichorn, D. H., 42, 44, 154, 158
Eisele, C. D., 69, 332, 414
Eisele, S., 103, 136
Eisenberg, J. F., 14, 22, 26, 132, 137, 223, 230, 252, 257, 295, 302, 320, 330, 376, 384
Elbadawl, A., 120
Elias, M. E., 210, 213
Ellefson, J. O., 205, 214, 240, 244, 300, 302
Ellis, J. E., 160, 394, 398, 436, 438
Elton, R. H., 419, 420, 421
Emlen, J. T., 208, 214, 282, 288, 342, 350
Emory, G., 237, 244
Epple, G., 74, 84, 92, 103, 128, 137, 168, 176, 183, 194, 221, 232, 251, 257, 295, 302, 341, 350, 360, 370, 376, 377, 384, 468, 475
Erickson, L. B., 132, 137, 153, 158, 454, 461
Erwin, J., 55, 67, 77, 79, 84, 86, 122, 134, 135, 137, 138, 161, 171, 175, 192, 194, 195, 225, 230, 262, 275, 279, 289, 290, 291, 323, 331, 335, 343, 344, 347, 350, 352, 371, 378, 381, 383, 384, 386, 387, 404, 406, 407, 410, 411, 412, 419, 421, 426, 429, 454, 463, 470, 472, 475, 477
Erwin, N., 79, 84, 383, 384, 421
Esser, A. H., 240, 246
Estrada, A., 79, 84, 142, 159, 171, 176, 190, 194, 263, 273, 325, 330, 457, 461
Estrada, R., 142, 159, 171, 176, 457, 461

Etaugh, C., 438
Etkin, W., 210, 214, 437, 438
Evans, C. S., 92, 103, 364, 370, 452, 461
Everitt, B. J., 97, 101, 103, 104, 118, 120, 125, 134, 137, 446, 447, 448
Exline, R. V., 154, 159, 286, 289

Fady, J. C., 56, 65
Fahrenbruch, C. E., 177, 317, 431, 440
Faiman, C., 43, 44, 80, 83, 84, 113, 115, 120, 458, 461
Fairbanks, L. A., 134, 137, 146, 159, 205, 214, 280, 283, 289, 378, 384
Falett, J., 160, 245, 274, 398
Farrer, D. N., 368, 373, 427, 432
Faust, R., 82, 84, 174, 176, 409, 411
Fedigan, L., 189, 194, 202, 214, 297, 303
Fedigan, L. M., 51, 65, 141, 157, 189, 194, 297, 303, 325, 328
Ferin, M., 39, 47, 106, 449
Fisher, R. B., 23, 30, 41, 48, 237, 244, 444, 450
Fitch, M., 341, 351, 452, 462
Fittinghoff, N. A., 471, 475
Fitz-Gerald, F. L., 427, 430
Fitzgerald, H. E., 87
Flett, M., 347, 350
Flinn, L., 50, 65, 319, 338, 375, 385, 468, 475
Flynn, D., 383, 384, 421
Fontenelle, G., 242, 247, 286, 292
Fooden, J., 23, 26, 80, 84
Forest, M. G., 43, 48, 93, 94, 107
Fossey, D., 208, 214, 268, 273, 342, 348, 350, 473, 475
Foster, D. B., 46
Foster, D. L., 36, 44, 72, 85
Fox, G. J., 80, 85, 240, 244
Fox, R., 156, 159, 285, 289
Fragaszy, D. M., 130, 131, 137, 222, 230, 250, 251, 257, 279, 289, 308, 309, 315
Fragga, A. M., 10, 11
Fraser, M. D., 73, 88
Freeland, W.J., 423, 424, 430

Freeman, H. E., 59, 65, 153, 159
Freeman, S. K., 104
French, J. A., 56, 65, 435, 438
Freudenberg, R., 210, 214
Friedman, P., 349, 351, 437, 439
Fujii, H., 227, 230, 255, 257

Gabow, S. L., 225, 230, 253, 257, 296, 303
Gadpaille, W. J., 44, 119, 120, 428, 430
Gagliardi, G. J., 487
Galdikas-Brindamour, B., 242, 244
Gale, C. C., 124
Galkin, T. W., 448
Gallup, G. G., 1, 4, 481, 482, 483, 486, 487, 491
Gandolfo, R., 56, 70
Gartlan, J. S., 15, 22, 26, 52, 65, 76, 85, 146, 159, 202, 214, 238, 244, 247, 281, 285, 289, 291, 298, 306, 310, 315, 348, 354, 392, 398, 400, 427, 430
Gautier, J. P., 75, 85, 146, 159
Gautier-Hion, A., 15, 26, 75, 85, 146, 159
Gavan, J. A., 18, 26, 82, 85
Geen, R. G., 395, 398
Gerber, G. L., 271, 273
Gerlach, J. L., 39, 47, 106, 449
Gerwitz, H. B., 211, 214, 309, 315
Gewirtz, J. L., 211, 214, 309, 315
Gijzen, A., 18, 26, 81, 85
Gil, D. G., 212, 214
Gilmore, H. A., 200, 214, 266, 273, 389, 390, 398
Gingerich, P. D., 23, 26, 74, 85
Glander, K. E., 168, 176, 184, 194, 223, 234, 251, 258, 424, 432
Gloor, P., 119, 121
Gluck, J. D., 403, 411, 425, 430
Goldberg, S., 62, 65
Goldfoot, D. A., 36, 37, 45, 54, 65, 95, 98, 99, 104, 110, 117, 118, 121, 134, 137, 141, 159, 263, 273, 379, 385, 406, 411, 470, 475
Goldin, P. G., 269, 273

Goldman, P. S., 60, 66, 180, 194, 424, 430, 434, 438, 441, 442, 445, 448
Gomber, J., 405, 407, 411, 413, 475
Gondos, B., 124
Goo, G. P., 254, 258, 416, 421
Goodall, J., 66, 81, 85, 149, 162, 206, 214, 241, 246
Goosen, C., 408, 411
Gordon, B. N., 228, 231, 408, 412, 426, 431
Gordon, G. G., 114, 124, 456, 465
Gordon, T. P., 93, 99, 103, 104, 106, 119, 157, 252, 256, 258, 260, 296, 302, 322, 329, 378, 380, 383, 387, 403, 410, 454, 455, 456, 461, 465
Gouzoules, H., 79, 85, 142, 159, 189, 190, 194, 263, 273, 297, 303, 324, 330
Goy, R. W., 12, 28, 33, 34, 35, 36, 37, 45, 47, 49, 66, 92, 95, 98, 100, 101, 103, 104, 107, 118, 121, 124, 134, 136, 137, 163, 171, 177, 263, 276, 333, 372, 379, 385, 406, 407, 411, 416, 421, 451, 461, 465, 470, 475
Gradwell, P. B., 101, 104, 446, 448
Grady, S. A., 414
Graham, C. E., 81, 85, 113, 116, 121, 148, 159, 458, 461, 464
Gray, J. A., 447, 448
Gray, J. L., 210, 211, 217
Green, P. C., 371, 378, 386, 417, 422
Green, R., 42, 45, 119, 121, 394, 398
Green, S., 208, 216, 322, 331
Greenberg, L. M., 483, 486
Greep, R. O., 118, 121, 418, 421, 459, 461
Grilly, D. M., 436, 438
Grimm, R. J., 344, 347, 350
Gross, R. J., 110, 121, 171, 176
Grotts, D., 43, 44, 80, 83, 84, 113, 115, 120, 458, 461
Groves, C. P., 240, 244, 340, 350
Gucwinski, A., 181, 194, 294, 303
Gucwinska, H., 181, 194, 294, 303

Hagemenas, F. C., 34, 45
Hagino, N., 38, 45, 112, 121, 448, 457, 461
Hahn, J. D., 33, 37, 47, 442, 449, 451, 464
Hale, P. A., 15, 28
Hall, K. R. L., 76, 80, 85, 203, 214, 238, 244, 298, 303, 392, 398
Hamburg, D.A., 38, 42, 44, 45, 63, 66, 83, 85, 116, 117, 121, 158, 180, 194, 298, 303, 326, 330, 366, 367, 370, 374, 385, 390, 393, 395, 398, 420, 421, 428, 430, 459, 461, 467, 475
Hamilton, M. E., 19, 26, 33
Hamilton, W. J., 23, 26, 238, 243, 281, 288, 299, 302, 303, 338, 350, 390, 397
Hammen, C., 409, 413
Hammer, S., 44, 45
Hansen, E. W., 49, 53, 54, 66, 322, 330, 379, 385
Harbert, G. M., 456, 461
Harcourt, A.H., 150, 159, 208, 214, 241, 244, 458, 461
Harding, R. S. O., 238, 244, 311, 315
Harlow, H. F., 33, 34, 49, 53, 54, 62, 65, 66, 69, 70, 73, 76, 84, 88, 224, 226, 234, 253, 259, 279, 284, 289, 291, 296, 303, 322, 329, 330, 332, 333, 343, 344, 346, 350, 352, 353, 361, 362, 363, 370, 371, 372, 379, 384, 385, 403, 404, 406, 410, 413, 414, 425, 432, 435, 438, 445, 447, 450
Harlow, M. K., 69, 332
Harper, L. V., 42, 46, 62, 69, 242, 246, 286, 289
Harrington, J. E., 127, 137, 294, 303, 341, 351
Harris, E. F., 19, 21, 26, 32, 474, 476
Harrison, V., 32
Hartmann, E., 250, 259, 359, 371, 469, 477
Hartmann-Wiesner, E., 138, 259, 371, 449
Harvey, P. A., 13, 25, 219, 229
Haslerud, G. M., 428, 431
Hausfater, G., 77, 80, 85, 112, 121, 144, 147, 157, 159, 162, 203, 206, 213, 265, 273, 295, 303, 312, 315, 379, 385, 390, 398, 425, 430, 455, 457, 461, 462

Hayashi, M., 111, 121, 445, 448, 456, 462
Heestand, J. E., 49, 67, 128, 138, 375, 386
Heinrichs, W. L., 48, 73, 88, 102, 107, 445, 450
Heisler, P. S., 195, 231, 289, 316
Hendrichs, A. G., 99, 106, 379, 387, 454, 465
Hendricks, D. E., 264, 273, 380, 385
Hennessey, M. B., 130, 137, 453, 462
Herbert, J., 35, 46, 97, 101, 103, 104, 107, 112, 120, 122, 134, 137, 146, 158, 162, 264, 265, 272, 322, 323, 330, 331, 392, 397, 400, 446, 448, 454, 463
Herbst, A. L., 23, 26, 33, 46
Hess, J. P., 268, 273
Hill, C. A., 205, 214
Hill, C. W., 16, 28
Hill, S. D., 487
Hinde, R. A., 53, 54, 66, 70, 133, 137, 224, 230, 248, 253, 259, 335, 344, 351, 354, 361, 370, 377, 385, 406, 411, 415, 417, 418, 420, 421, 422
Hobbett, L., 300, 304, 343, 348, 351, 352
Hobson, W., 42, 46, 113, 122, 458, 462
Hoff, M. P., 59, 60, 66, 70, 161, 176, 208, 215, 218, 334
Holm, R. A., 171, 177, 317, 424, 431, 440
Hopf, S., 51, 67, 131, 138
Horr, D. A., 152, 159, 173, 176, 209, 210, 215, 242, 244, 300, 303, 343, 351
Horwich, R. H., 75, 85, 147, 159, 172, 176, 204, 205, 215, 299, 303, 321, 330, 348, 351
Howanstine, J., 256, 383
Howard, S., 270, 274
Hrdy, D. B., 170, 176, 204, 215, 267, 273
Hrdy, S. B., 147, 159, 170, 176, 203, 204, 215, 239, 244, 267, 273, 285, 289, 294, 303
Hromada, J., 16, 26, 469, 476
Hughes, G. H., 366, 371
Hughes, J., 255, 256
Huhtaniemi, I. T., 34, 43, 46
Hull, D. B., 473, 476
Humphrey, N. K., 471, 476

Hunkeler, C., 51, 64, 66, 75, 86, 202, 212, 215, 239, 244, 266, 274, 280, 288, 297, 302, 303, 321, 339, 350, 351, 391, 398, 457, 462
Hunter, J. L., 80, 86, 201, 215
Hurme, V. O., 20, 21, 26, 77, 86, 470, 476
Hutchinson, T. C., 22, 25, 93, 103, 453, 460
Hutt, C., 83, 86

Illner, P., 124
Imanishi, K., 56, 66, 189, 194, 210, 215, 227, 230, 255, 259, 303
Ingram, J. C., 50, 66, 183, 194, 320, 330, 376, 385
Irons, R., 190, 193
Itani, J. A., 227, 231, 240, 244, 325, 330, 344, 347, 351, 382, 385, 457, 462, 472, 476
Itoigawa, N., 227, 231, 255, 259
Iwamoto, M., 426, 430
Izawa, K., 224, 233

Jacklin, C. N., 2, 4, 62, 63, 67, 155, 160, 271, 274, 282, 286, 287, 290, 301, 304, 327, 331, 349, 351, 369, 371, 374, 386, 396, 399, 433, 437, 438, 439, 474, 476, 482
Jackson, D. D., 409, 411, 428, 430
Jackson, T. T., 439
Jacobs, G. H., 424, 430, 469, 476
Jacobs, L. S., 420, 421
Jacobsen, J., 62, 66
Jaffe, R. B., 46
Jay, P., 147, 159, 203, 215, 267, 274
Jeffers, V., 268, 274, 395, 398
Jeffrey, G., 439
Jensen, G. D., 113, 122, 153, 154, 160, 192, 194, 197, 228, 231, 269, 274, 347, 354, 408, 411, 426, 430, 431, 459, 462
Johnson, D. F., 98, 104
Joines, S., 17, 29, 414, 440
Jolly, A., 9, 11, 14, 26, 181, 183, 195, 415, 422
Jones, E. C., 78, 86, 116, 122
Jones, P. A. (See Preface)

Jones, K. L., 32, 46, 83, 86, 104, 114, 115, 122, 156, 160, 459, 462
Jones, N. G. B., 325, 330, 364, 371
Jorde, L. B., 13, 27, 219, 231, 307, 316, 374, 385, 419, 422
Joslyn, W. D., 37, 46, 122, 252, 259, 379, 385
Jouventin, P., 16, 22, 27, 237, 244, 266, 274, 281, 289, 338, 343, 351
Judge, D. S., 171, 176, 228, 231, 297, 303
Juno, C. J., 267, 274
Jurgens, U., 444, 448

Kaplan, J., 185, 195, 222, 231, 279, 289, 379, 385
Kaplowitz, C., 483, 488
Karczmar, A. G., 41, 46, 118, 119, 122, 442, 448
Kastenbaum, B., 428, 431
Kastenbaum, R., 428, 431
Katchadourian, H., 83, 86, 156, 160
Kaufman, I. C., 55, 67, 69, 263, 276, 324, 332
Kaufman, J. H., 53, 67
Kavanagh, M., 310, 316
Kawabe, M., 245, 289, 299, 303, 340, 351
Kawai, M., 145, 161, 227, 231, 237, 238, 245, 281, 290, 309, 316, 436, 438
Kawamura, S., 78, 86, 190, 195, 227, 231, 280, 289, 309, 316, 435, 438, 456, 462, 471, 476
Kawanaka, K., 348, 351
Keeble, G. R., 471, 476
Keefer, D. A., 38, 46, 48, 91, 93, 104, 107, 180, 195, 443, 448, 450
Keifer, G., 67, 86, 105
Keiter, M. D., 60, 67, 82, 86, 151, 160, 268, 274, 458, 462
Kelley, S. T., 130, 138, 185, 195, 283, 290, 320, 331, 453, 462
Kellicutt, M. H., 363, 372
Kelly, B. R., 276, 400
Kelly, J. T., 20, 27
Kelly, W. N., 396, 400
Kemper, T. D., 212, 215
Kern, J. A., 15, 22, 27
Keverne, E. B., 98, 99, 100, 101, 104, 105, 146, 160, 265, 274, 323, 330, 454, 455, 462, 463

Kimball, S. D., 322, 333, 363, 372, 403, 414, 425, 432
Kimble, D. P., 35, 46
King, G., 341, 351, 452, 462
Kinzey, W. G., 21, 27, 185, 195, 223, 231, 279, 289, 309, 316
Kirk, J. H., 15, 16, 27, 77, 86
Kittinger, G. W., 34, 45
Klaiber, E. L., 120
Kleiman, D. G., 50, 67, 83, 86, 129, 138, 148, 160, 168, 169, 170, 176, 185, 195, 219, 231, 251, 259, 295, 303, 309, 316, 320, 326, 330, 376, 386
Klein, D., 132, 138, 223, 231, 283, 289, 295, 304, 376, 386
Klein, L. L., 132, 138, 223, 231, 234, 283, 389, 295, 304, 376, 386
Kling, A., 39, 46, 102, 104, 105, 110, 122, 143, 160, 180, 195, 263, 274, 381, 386, 445, 448, 449
Klopfer, F. J., 438
Klopfer, P. H., 182, 195, 220, 231, 463, 476, 485
Kobayaski, Y. 120
Koenigsknecht, R. A., 349, 351, 437, 439
Koford, C. B., 141, 157
Kolata, G., 248, 265, 274, 467, 476
Kollar, E. J., 206, 216, 393, 398
Korenbrot, C. C., 46
Kortlandt, A., 206, 216
Kortmulder, K., 135, 138, 156, 160, 226, 231, 248, 258
Kotera, S., 171, 177
Koyama, N., 78, 86, 148, 160, 190, 195, 226, 227, 231, 232, 233, 255, 258, 259, 263, 274, 285, 289, 324, 330, 342, 351
Kravetz, M. A., 104
Kress, J. H., 127, 138, 182, 195, 468, 476
Kressler, P. L., 16, 30, 309, 317
Krige, P. D., 203, 216, 391, 397
Kronenwetter, C., 296, 305, 325, 332, 381, 386
Kuehn, R. E., 14, 22, 26, 132, 137, 223, 230, 252, 257, 295, 302, 320, 330, 376, 384

Kühlmorgen, B., 138, 250, 259, 359, 371, 449, 469, 477
Kummer, H., 53, 68, 145, 160, 220, 232, 236, 237, 245, 265, 266, 274, 275, 364, 367, 371, 377, 386, 390, 398
Kunin, R. A., 428, 429
Kurland, J. A., 56, 67
Kurokawa, T., 226, 232, 255, 258, 279, 289
Kurt, F., 160, 236, 245
Kuyk, K., 55, 67, 192, 195

Laban, M., 435, 438
Laemle, L. K., 477
Lahiri, R. K., 264, 274, 324, 330, 348, 351, 365, 371
Lamb, M. E., 62, 67, 210, 211, 216
Lancaster, J. B., 51, 52, 67, 201, 202, 216, 239, 245, 266, 274, 280, 290, 435, 439, 473, 476
Lance, J., 64, 272, 397
Landesman-Dwyer, S., 171, 177, 424, 431
Langford, J. B., 183, 195
Lanman, J. T., 43, 46, 92, 105, 113, 122, 165, 168, 176
Larsson, K., 405, 410
Latta, J., 51, 67, 131, 138
Lauersdorf, H. E., 49, 53, 66, 279, 289, 296, 303, 361, 362, 370, 379, 385
Laus, J., 86
Leary, G. A., 368, 373, 427, 432, 471, 478
Lee, M. Q., 209, 216
Leibrecht, B. C., 130, 138, 185, 195, 283, 290, 320, 331, 453, 462
Lester, D., 395, 398, 409, 412
Lethmate, J., 481, 487
Letson, G. W., 160, 394, 398
Leutenegger, W., 13, 20, 27, 167, 176, 452, 462
Levine, S., 32, 46, 137, 462
Lewis, H. P., 48, 96, 107, 446, 450
Lewis, J. K., 445, 450
Lewis, L., 383, 384, 421
Lewis, M., 62, 65, 349, 351
Li, Su-Chen., 16, 28
Lichstein, L., 53, 67
Lieff, J. D., 240, 246
Lin, P. M., 19, 27

Lindburg, D. G., 22, 27, 53, 54, 67, 132, 133, 138, 254, 258, 279, 290, 296, 304, 335, 351, 455, 463, 475
Livrea, G., 39, 47, 95, 105, 444, 448
Lockard, J. S., 49, 50, 67, 83, 86, 128, 138, 243, 245, 375, 386
Longstreth, L. E., 396, 398
Lorenz, R., 312, 317, 416, 431
Lorinc, G. A., 190, 195, 255, 258
Lowe, A., 51, 69
Lowe, E. L., 137, 462
Loy, J., 22, 27, 53, 54, 67, 78, 86, 95, 105, 112, 122
Loy, D., 53, 67, 86, 105
Lucas, J. W., 202, 216
Luce, S. R., 428, 431
Lunde, D. T., 38, 45, 63, 66, 83, 85, 116, 121, 326, 330, 428, 430, 459, 461
Lynn, D. B., 210, 216

Maccoby, E. E., 2, 4, 62, 63, 67, 155, 160, 271, 274, 282, 286, 287, 290, 301, 304, 327, 331, 349, 351, 366, 371, 374, 386, 396, 399, 433, 437, 438, 439, 474, 476, 482, 487
Mack, D., 407, 414
Mackey, W. C., 211, 216
MacLean, P., 39, 47, 444, 448
MacRoberts, M. H., 55, 67
Mahoney, C. J., 171, 176, 225, 228, 232, 454, 463
Mai, L., 117, 122, 459, 463
Malick, C., 58, 63, 69, 148, 149, 156, 162, 207, 217, 310, 317, 327, 332, 393, 400
Malinowski, B., 210, 216
Malmgren, L. A., 223, 234, 424, 432
Mano, T., 245, 289, 299, 303, 340, 351
Manski, D., 75, 85, 205, 215, 299, 303, 321, 330
Maple, T., 56, 59, 60, 66, 67, 68, 70, 77, 86, 113, 122, 126, 134, 135, 137, 138, 149, 151, 152, 153, 156, 160, 161, 164, 173, 174, 176, 208, 209, 215, 218, 270, 274, 310, 312, 323, 327, 331, 334, 336, 343, 344, 347, 351, 352, 378, 379, 384, 386, 394, 396,

Maple (Con't.)
399, 403, 404, 406, 407, 411, 412, 414, 454, 459, 463, 470, 472, 474, 476, 477, 478, 480, 487
Marler, P. A., 23, 27, 219, 232, 239, 245, 253, 259, 278, 290, 300, 304, 337, 338, 339, 340, 342, 343, 348, 352, 360, 361, 364, 365, 366, 367, 368, 371, 375, 376, 386, 396, 399, 418, 422, 473, 477
Marr, L. D., 96, 106, 260
Marsden, H. M., 187, 198, 224, 225, 232, 253, 259
Martenson, J., Jr., 110, 114, 122, 142, 161, 371, 457, 463
Martin, D., 150, 161
Martin, R. D., 181, 196
Martino, A., 288, 302
Mason, W. A., 2, 4, 54, 58, 64, 68, 72, 87, 92, 103, 130, 131, 136, 149, 161, 185, 196, 208, 216, 221, 222, 229, 230, 232, 241, 245, 267, 275, 285, 287, 322, 326, 331, 336, 337, 352, 358, 360, 362, 371, 372, 378, 386, 403, 405, 408, 410, 412, 416, 417, 418, 422, 431, 453, 460, 467, 477
Mass, R., 445, 449
Masserman, J. H., 253, 259, 284, 290, 308, 309, 310, 316
Masterson, R. B., 434, 440, 442, 450
Masui, K., 15, 27, 79, 87, 226, 232
Masure, A. M., 238, 245, 338, 352, 426, 432
Matheson, D. W., 396, 399
Maurus, M., 39, 47, 131, 138, 250, 259, 359, 360, 371, 444, 448, 449, 469, 477
Mayer, B., 76, 85, 203, 214, 238, 244, 392, 398
Mazur, A., 93, 95, 105, 113, 116, 122, 249, 252, 259, 271, 275, 374, 386
McCary, J. L., 156, 161
McClintock, M. K., 117, 122, 459, 463
McClure, H. M., 81, 85, 113, 116, 121, 148, 159, 458, 461
McClure, M. K., 487
McDaniel, J., 347, 355, 407, 414

McDowell, A. A., 434, 439
McEwen, B. S., 39, 47, 106, 449
McGrew, P. L., 62, 68, 395, 399
McGrew, W. C., 62, 68, 349, 352, 395, 399, 408, 414, 427, 432, 436, 439, 440
McGuire, M. T., 134, 137, 146, 159, 280, 289, 378, 384
McKenna, J. J., 86, 170, 177, 267, 276
McKinney, J. P., 82, 87
McMahon, C. A., 16, 27, 169, 177
McTee, A. C., 434, 439
Meier, G. W., 372, 440
Mendoza, S. P., 137, 462
Menzel, E. W., Jr., 161, 249, 259, 267, 268, 275, 282, 290, 304, 308, 312, 316, 366, 371, 403, 408, 412, 418, 422, 436, 439
Merz, E., 191, 196
Michael, R. P., 41, 47, 77, 87, 95, 96, 97, 98, 99, 100, 101, 103, 104, 105, 106, 108, 116, 117, 122, 123, 126, 127, 128, 129, 130, 131, 133, 135, 139, 140, 141, 142, 143, 144, 145, 146, 147, 148, 151, 161, 323, 330, 331, 362, 371, 373, 379, 386, 444, 449, 454, 455, 456, 462, 463, 464, 470, 477
Migeon, C. J., 43, 48, 93, 94, 107
Milch, K. H., 62, 68, 268, 275
Miller, L. C., 267, 274
Milton, K., 223, 232, 295, 304
Mishkin, M., 40, 48, 446, 449
Missakian, E. A., 135, 139, 253, 259, 268, 275, 323, 325, 328, 331, 395, 399, 403, 412, 455, 463
Missakian-Quinn, E. A., 62, 68, 77, 87, 135, 139, 172, 177, 225, 232, 254, 259, 268, 275
Mitchell, G., 53, 54, 60, 64, 65, 68, 69, 76, 77, 84, 86, 132, 133, 134, 135, 136, 138, 165, 166, 167, 168, 169, 170, 171, 172, 173, 174, 175, 177, 180, 181, 182, 183, 184, 185, 186, 187, 188, 189, 190, 191,

Mitchell, G. (Con't.)
192, 193, 196, 199, 200,
201, 207, 208, 209, 216,
218, 225, 230, 253, 259,
284, 290, 296, 304, 314,
319, 323, 328, 331, 335,
336, 343, 344, 345, 346,
347, 349, 350, 352, 353,
354, 358, 362, 363, 370,
371, 372, 379, 383, 384,
386, 402, 403, 404, 405,
406, 407, 410, 411, 412,
413, 417, 422, 428, 431,
433, 439, 454, 463, 470,
475, 477, 480, 481, 487
Mizuno, A., 238, 245
Mobaldi, J., 225, 230, 406,
411
Modahl, K. B., 141, 161, 254,
259
Moely, B. E., 286, 291, 483,
488
Møller, G. W., 253, 259, 343,
344, 346, 353, 363, 371,
405, 406, 413
Montagna, W., 110, 124
Moody, D. B., 355
Moore, G. T., 16, 27, 169,
177
Mori, A., 226, 232, 325,
331, 347, 353, 364, 371
Mori, U., 145, 161, 237, 245,
− 281, 290
Mörike, D., 239, 245
Morrison, J. A., 96, 107, 455,
466
Mortenson, B. K., 221, 235
Moynihan, M., 129, 139, 183,
196, 361, 371
Muckenhirn, N. A., 221, 235
Mukherjee, R. P., 239, 240,
245, 247, 298, 304, 306,
338, 353, 391, 401
Murakami, M., 210, 211, 217
Murray, S., 256, 383
Murrell, S., 221, 232
Musicant, A., 50, 65, 220,
230, 375, 384
Myers, S. A., 212, 216
Myseko, D., 64, 272, 397

Nadler, R. D., 23, 27, 58, 59,
62, 64, 66, 68, 70, 93,
106, 113, 123, 131, 139,
141, 150, 151, 152, 153,
161, 164, 174, 177, 207,
208, 215, 216, 218, 267,
268, 272, 274, 275, 334,
397, 409, 413, 436, 439,
453, 458, 459, 464, 472,
474, 477

Nagel, U., 53, 68, 236, 245,
265, 275, 331, 364, 367,
371, 377, 386, 472, 477
Napier, J. R., 3, 4, 9, 10, 11,
493
Napier, P. H., 3, 4, 9, 10, 11,
493
Nash, L. T., 50, 65, 265, 275,
319, 330, 375, 385, 468,
475
Nass, G. G., 20, 27
Nathan, J. F., 238, 244, 265,
273, 299, 302
Neill, J. D., 464
Neuman, F., 33, 37, 47, 442,
449, 451, 464
Neville, M. K., 184, 196, 223,
225, 233, 253, 259, 279,
290, 320, 331, 455, 464
Newkirk, J. B., 313, 316
Newman, J. D., 344, 353
Newton, N., 119, 123, 153,
161
Nicolson, N. A., 207, 216
Nieman, W. H., 118, 123,
416, 422
Nigi, H., 79, 87, 111, 123,
141, 161, 456, 457, 464
Nishida, T., 78, 87, 226, 233,
348, 351
Nishima, A., 87
Nishimura, A., 27, 224, 232,
233
Nissen, H. W., 113, 122
Niswander, G. D., 428, 431
Noback, C. R., 468, 469,
471, 477
Nolte, A., 55, 68
Normura, T., 43, 47, 111,
123, 140, 162
Norikoshi, K., 226, 233, 255,
259, 309, 316, 364, 372
Norman, R. L., 40, 48, 101,
106, 107, 444, 449, 450,
454, 464
Nottebohm, F., 249, 353,
437, 439

Oates, J. F., 321, 333
Oatley, T. B., 313, 316
O'Connell, W. E., 62, 68, 396,
399
O'Connor, R. F., 453, 466
Ohsawa, H., 27, 78, 87, 88,
140, 163, 226, 232, 234,
456, 465
Ohsawa, N., 43, 47, 111, 123,
140, 162
Okano, J. A., 59, 68, 152,
162, 209, 217, 242, 245,
268, 275

O'Leary, S., 210, 211, 212,
217
Omar, A., 16, 28, 200, 217
Omark, D. R., 62, 68, 269,
275, 277
Oppenheimer, E. C., 223, 233,
251, 260, 320, 331
Oppenheimer, J. R., 75, 87,
170, 177, 223, 233, 251,
260, 295, 304, 320, 331
Orbach, J., 381, 386
Orkin, M., 23, 27
Orlando, R., 372, 440
Orlosky, F. J., 20, 29
Oshima, K., 121, 448, 462
Oswald, M., 122, 161, 262,
275, 279, 290, 291, 381,
386, 387, 463
Owens, N. W., 56, 68, 200,
217, 281, 290, 299, 305,
389, 399

Pantuwatana, S., 300, 302,
393, 397
Page, C., 310, 315, 326, 329
Palthe, T. V. W., 206, 217
Paluck, R. J., 240, 246
Papousek, H., 483, 487
Papousek, M., 483, 487
Parer, J. T., 46
Parke, R. D., 210, 211, 212,
217
Parker, C. E., 473, 477
Parthasarathy, M. D., 79, 87,
227, 228, 233, 280, 290,
297, 304, 309, 316, 337,
338, 353, 472, 478
Paterson, J. D., 145, 162,
389, 399, 472, 477
Patterson, D., 67, 86, 105
Paulson, D. G., 372, 440
Pechtel, C., 308, 309, 310,
316
Peffer, P. G., 191, 196
Pelletier, A., 127, 137
Peplau, L. A., 62, 65, 328,
329, 349, 350
Perachio, A. A., 39, 47, 94,
96, 101, 103, 106, 252,
256, 257, 260, 380, 386
Perkins, M., 408, 413, 426,
431
Perret, M., 127, 139, 221,
233, 250, 260, 451, 464
Peters, M., 251, 260, 308, 316
341, 353, 358, 360, 372
Peters, S. D., 409, 413
Peterson, M., 355
Peterson, M. R., 119, 157
Petitto, L., 482, 488
Petter, A., 220, 221, 233,
341, 353

Petter, J. J., 220, 221, 233, 341, 353
Pettet, A., 299, 304
Peyton, S., 428, 431
Pfaff, D. W., 39, 41, 47, 102, 106, 118, 123, 443, 445, 449
Pfeiffer, C. A., 31, 47
Phelps, J. B., 256, 383
Phillips, M. J., 358, 360, 372
Phoenix, C. H., 12, 28, 33, 34, 35, 36, 45, 47, 96, 97, 98, 104, 106, 442, 449, 470, 477
Pilbeam, D., 21, 28, 437, 439
Pitcairn, T. K., 153, 162, 242, 246, 300, 304, 342, 343, 353, 393, 394, 399
Plant, T. M., 95, 96, 105, 106, 123, 455, 456, 463, 464
Ploog, D., 39, 47, 51, 67, 131, 138, 250, 259, 341, 353, 358, 360, 371, 372, 444, 448, 449
Poglayen-Neuwall, I., 173, 177
Poirier, F. E., 43, 47, 53, 69, 75, 82, 87, 144, 158, 204, 217, 219, 233, 240, 246, 267, 275, 294, 298, 304, 321, 323, 331, 339, 353, 366, 372, 391, 395, 399, 435, 439, 473, 477
Pope, B. L., 14, 28, 223, 233
Pope, C., 61, 70, 286, 292
Porges, S. W., 484, 487
Posepanko, C. J., 322, 331, 371, 378, 386, 417, 422
Post, D. G., 20, 28, 310, 316, 472, 477
Post, W., 380, 387
Power, M. E., 276, 400
Powers, W., 484, 487
Prassa, S. P., 335, 352
Price, J., 270, 275, 428, 431
Prowse, D. L., 195, 231, 289, 316
Pruscha, H., 138, 259, 371, 449
Pusey, A. E., 81, 87, 207, 217, 241, 246, 285, 290, 327, 332

Quiatt, D., 186, 196
Quick, L. B., 226, 233, 255, 260
Quris, R., 80, 87, 238, 246, 290, 299, 304, 340, 353, 457, 464

Raemaekers, J. J., 243, 350
Rahaman, H., 227, 233, 280, 290, 297, 304, 309, 316, 337, 338, 353, 472, 478
Raleigh, M. J., 52, 69, 121, 201, 217, 239, 246, 310, 316, 321, 332, 391, 399
Ralls, K., 13, 14, 17, 18, 28, 50, 69
Raphael, D., 180, 196
Raviola, E., 425, 432
Raymond, E.J., 352, 353, 413
Redican, W. K., 54, 69, 179, 180, 188, 196, 345, 352, 361, 363, 372, 403, 404, 405, 413
Reichler, M. L., 212, 215
Reinisch, J. M., 41, 47, 447, 449
Reisen, S. M., 263, 276
Resko, J. A., 12, 28, 34, 36, 45, 47, 48, 73, 77, 87, 96, 101, 106, 111, 120, 172, 177, 254, 257, 382, 384, 444, 449, 450, 454, 455, 464, 469, 478
Reynolds, F., 17, 23, 28, 58, 69, 81, 87, 300, 305
Reynolds, V., 1, 2, 4, 17, 23, 28, 58, 69, 81, 87, 241, 246, 300, 304, 305, 326, 332, 343, 353, 479, 487
Rhine, R. J., 200, 217, 237, 238, 246, 281, 290, 296, 298, 299, 305, 310, 317, 325, 332, 381, 386, 390, 399
Ribbens, L., 408, 411
Richarz, K., 452, 465
Riesen, A. H., 408, 413, 426, 431
Rifkin, I., 43, 48, 115, 124
Rijksen, H. D., 17, 23, 28, 151, 162, 242, 246, 268, 275, 343, 353, 368, 372, 459, 464
Rimland, B., 428, 431, 433, 439
Riopelle, A. J., 10, 11, 15, 16, 28
Ripley, S., 298, 305, 366, 372, 435, 439
Riss, D., 149, 162, 241, 246
Risse, G., 135, 138, 336, 344, 352, 407, 412
Ritchie, T. L., 466
Robboy, S. J., 23, 26, 33
Roberts, P., 319, 332
Robinson, B. W., 40, 48, 446, 449
Robinson, J. A., 103, 136, 454, 466

Robinson, J. M., 20, 29
Rodman, P. S., 17, 171, 176, 228, 231, 242, 246, 297, 304
Rogers, C. M., 408, 414, 427, 431
Rondinelli, R., 223, 234, 376, 386
Rose, J. D., 443, 449
Rose, M. D., 310, 317, 436, 439
Rose, R. M., 93, 94, 99, 103, 104, 106, 119, 157, 252, 256, 258, 260, 296, 302, 322, 329, 378, 380, 383, 387, 403, 410, 454, 456, 461, 465
Rosenberger, A. L., 195, 231, 289, 316
Rosenbloom, F. M., 396, 400
Rosenblum, L. A., 51, 55, 67, 69, 75, 79, 84, 88, 93, 103, 106, 111, 123, 131, 139, 141, 161, 185, 192, 196, 197, 228, 234, 324, 332, 408, 414, 424, 431, 453, 464, 472, 477
Rosenkrantz, P. S., 429
Rosenson, L. M., 181, 197, 319, 332
Rosvold, H. E., 448
Ross, B. A., 300, 302, 306, 333, 393, 397
Roth, E. M., 382, 383, 419, 421
Rothe, H., 168, 177, 184, 197, 312, 319, 320, 332
Rowell, R. M., 374, 387
Rowell, T. E., 16, 28, 37, 48, 53, 66, 82, 88, 99, 106, 112, 115, 123, 132, 133, 137, 139, 145, 146, 153, 157, 162, 164, 179, 180, 197, 200, 217, 238, 239, 243, 246, 253, 259, 264, 265, 275, 276, 278, 281, 285, 288, 291, 294, 305, 311, 315, 317, 321, 326, 332, 335, 338, 344, 354, 356, 361, 370, 372, 377, 379, 385, 387, 392, 397, 400, 401, 417, 422, 454, 457, 465
Roy, M. A., 480, 484, 487, 488
Roy, R. L., 484, 488
Rudd, B. T., 112, 120, 146, 158, 264, 272
Rudran, R., 204, 217, 240, 246
Rumsey, T. J., 191, 198, 382, 387

Ruppenthal, G. C., 54, 69, 322, 332, 352, 353, 404, 410, 413, 425, 428, 431
Russell, C., 13, 28, 269, 276
Russell, W. M. S., 13, 28, 269, 276

Saayman, G. S., 135, 139, 145, 162, 266, 276, 281, 291, 298, 305, 338, 354, 390, 400, 457, 465, 470, 477
Sabater-Pi, J., 300, 305, 394, 400
Sackett, G. P., 70, 73, 88, 122, 161, 171, 177, 186, 197, 224, 226, 234, 279, 284, 291, 309, 317, 371, 381, 387, 403, 404, 405, 411, 414, 424, 425, 428, 430, 431, 433, 439, 440, 463
Sade, D. S., 77, 88, 135, 139, 224, 234, 253, 254, 260, 284, 291, 323, 332
Saha, S. S., 240, 245, 298, 304, 338, 353
Sanders, K. M., 42, 46, 62, 69, 242, 246, 286, 291
Sar, M., 40, 48, 91, 93, 107, 118, 123, 443, 450
Sassenrath, E. N., 78, 88, 99, 106, 123, 254, 258, 379, 387, 416, 418, 420, 421, 422, 454, 465
Satnick, R., 325, 332
Sauer, E. G. F., 28, 50, 69, 197
Sauer, E. M., 28, 50, 69, 197
Sauls, M., 106, 464
Saunders, C. D., 144, 162, 265, 276
Savage, E. S., 17, 28, 58, 63, 69, 148, 149, 150, 156, 162, 207, 217, 310, 317, 327, 332, 393, 400, 458, 465, 483, 488
Schaller, G. B., 56, 59, 69, 150, 162, 208, 214, 217, 241, 246, 281, 282, 288, 291, 299, 305, 327, 332, 342, 350, 354, 393, 400
Scheffler, G., 103, 124, 136, 163, 333, 372, 465
Scheier, M. F., 484, 488
Schenkel, R., 75, 88, 299, 305, 367, 372
Schenkel-Hulliger, L., 75, 88, 299, 305, 367, 372
Schiller, H. S., 110, 121, 124, 171, 176
Schlottman, R. S., 56, 70

Schneider, J., 135, 139, 254, 260
Schreiber, G. R., 182, 197
Schroers, L., 172, 177
Schulman, A. H., 483, 488
Schusterman, R. J., 310, 315, 326, 329
Schwaier, A., 28, 73, 88, 375, 387
Scollay, P. A., 17, 29, 344, 346, 354, 405, 409, 414, 417, 422, 435, 440
Scott, J. P., 29, 269, 276, 394, 400, 474, 478
Scott, N.J., 223, 234, 424, 432
Scruton, D. M., 112, 123, 146, 158, 162, 265, 272, 392, 400
Seay, B. M., 53, 69, 70, 264, 273, 380, 385, 435, 440
Seegmiller, J. E., 396, 400
Segre, A., 15, 20, 29, 52, 70, 203, 217, 281, 291, 298, 305, 321, 333, 359, 367, 372
Seilz, M. E., 17, 29
Serón-Ferré, M., 46
Setchell, K. D. R., 95, 105, 456, 463
Shafton, A., 1, 4, 484, 488
Shainberg, L. W., 32, 46, 83, 86, 104, 114, 115, 122, 156, 160, 459, 462
Shapiro, G., 483, 488
Sharpe, L. G., 253, 256, 284, 288
Shaughnessey, P. W., 15, 25, 172, 175
Sheposh, J. P., 270, 276
Shimamoto, K., 121, 448, 462
Sholiton, L. J., 48, 96, 107, 446, 450
Shotland, R. L., 301, 305
Shriver, J. E., 477
Siddigi, M. R., 53, 70, 220, 234, 296, 305, 387
Siegel, M. E., 19, 29
Sigg, H., 160, 245, 274, 398
Simonds, P. E., 55, 70, 143, 162, 171, 177, 192, 193, 197, 227, 228, 234, 263, 276, 285, 291, 324, 333, 382, 287
Simons, R. C., 192, 197, 347, 354
Singh, S. D., 447
Sirianni, J. E., 20, 29
Skarin, K., 286, 291, 483, 488
Skeen, L. C., 434, 440, 442, 450

Slimp, J., 34, 35, 45
Slob, A. K., 111, 124, 143, 163, 325, 333, 364, 372, 457, 465, 472, 478
Smith, E. O., 69, 73, 88, 191, 196, 276, 279, 291, 296, 305, 323, 331
Snipes, C. A., 43, 48, 93, 94, 107
Snyder, P. A., 183, 197, 251, 260
Sommer, R., 242, 246, 301, 305, 420, 422
Sorenson, M. W., 181, 197, 220, 234, 250, 260, 294, 305, 340, 354, 375, 387
Southren, A. L., 114, 124, 456, 465
Southwick, C. H., 53, 54, 70, 220, 234, 264, 274, 296, 305, 324, 330, 333, 348, 351, 365, 371, 378, 379, 387, 466
Spencer, J. O., 459, 465
Spencer-Booth, Y., 53, 54, 66, 70, 179, 197, 224, 230, 344, 351, 406, 411, 417, 422
Spies, H. G., 40, 48, 101, 106, 107, 444, 449, 450, 454, 464
Sprankel, H., 452, 465
Spuhler, J. N., 13, 27, 219, 231, 307, 316, 374, 385, 419, 422
Stebbins, C. A., 270, 276, 396, 400
Stebbins, W. C., 355
Steinbeck, H., 33, 37, 47, 442, 449, 451, 464
Steiner, R. A., 40, 48, 110, 115, 124, 445, 450
Stephenson, G. R., 141, 163, 255, 260
Sternglanz, S. H., 210, 217
Stevens, C. W., 55, 56, 70, 172, 177, 346, 353, 354, 403, 410
Stevenson, M. F., 184, 198, 202, 217
Stewart, K. J., 150, 159, 208, 214, 241, 244, 458, 461
Stokes, L. P., 448
Stott, K., 14, 22, 29, 234, 306, 341, 354
Straw, M. K., 301, 305
Strayer, F.F., 185, 198, 222, 234, 283, 291, 377, 387
Strobel, D. A., 309, 318
Strommen, E. A., 87
Struble, R. G., 408, 413, 426, 431

Struhsaker, T. T., 22, 29, 75, 88, 121, 201, 202, 218, 238, 247, 285, 291, 298, 306, 321, 333, 339, 348, 354, 391, 392, 400, 427, 432, 457, 465

Strum, S.C., 238, 244, 311, 317

Stumpf, W. E., 38, 40, 46, 48, 91, 93, 104, 107, 118, 123, 180, 195, 443, 448, 450

Stynes, A. J., 69, 263, 276, 324, 332

Sugawara, K., 140, 163, 226, 234, 382, 387, 456, 465

Sugiyama, Y., 27, 51, 55, 70, 78, 87, 88, 140, 144, 163, 203, 218, 226, 232, 234, 263, 276, 280, 291, 297, 306, 310, 317, 324, 333, 337, 354, 391, 400, 456, 465

Sullivan, D. J., 16, 29, 36, 134

Sunderland, 75

Suomi, S. J., 54, 69, 70, 73, 88, 188, 198, 224, 226, 234, 253, 260, 284, 285, 291, 322, 332, 333, 363, 372, 403, 407, 414, 425, 432, 445, 450

Sussman, R. W., 221, 234, 295, 306

Swett, C., 456, 465

Swindler, D. R., 18, 20, 26, 82, 85

Symmes, D., 253, 261, 284, 292, 344, 353

Syme, G. J., 249, 260

Symons, D., 60, 61, 70, 249, 260

Tabei, T., 48, 73, 88, 102, 107, 445, 450

Taft, P. D., 23, 26, 33

Takashima , I., 110, 124

Talmage-Riggs, G., 131, 136, 250, 251, 256, 261, 377, 383, 416, 422, 469, 478

Tanaka, T., 171, 177

Tanner, N., 211, 218, 249, 354, 437, 440

Tanticharvenyos, P., 306, 333

Tanzer, D., 153, 163, 210, 218

Tarrant, L. H., 15, 16, 23, 29, 79, 88, 171, 178

Tasch, R. J., 62, 70

Tattersall, I., 221, 234, 295, 306, 468, 478

Tavris, C., 163

Taylor, B. B., 48, 96, 107, 446, 450

Taylor, H. G., 466

Taylor, M., 185, 198, 222, 234, 283, 291, 377, 387

Taylor, M. A., 418, 422, 428, 432

Teas, J., 455, 466

Teleki, G., 313, 314, 317

Tenaza, R. R., 240, 247, 291, 294, 300, 306, 313, 317, 319, 333, 342, 354

Terrace, H. S., 482, 488

Terry, R. L., 325, 328, 333

Testa, T. J., 407, 414

Thomsen, C. E., 286, 291, 328, 333, 358, 363, 372

Thompson, N. S., 144, 163, 324, 333, 380, 387

Tiger, L., 116, 117, 124, 210, 218, 282, 291, 394, 400

Tijskens, J., 18, 26, 81, 85

Tiwar, K. K., 239, 247, 298, 306, 391, 401

Tobach, E., 484, 488

Tokuda, K., 171, 177

Tokunaga, D. H., 319, 331

Tolor, A., 276, 400

Torii, M., 254, 261, 309, 310, 317

Travis, C. B., 242, 247, 286, 292

Trilling, J. S., 195, 231, 289, 316

Trimble, M. R., 101, 107

Trollope, J., 79, 88, 142, 143, 157, 163, 171, 178, 190, 193, 198, 325, 330, 348, 355, 364, 371, 381, 383

Truelove, J. K., 16, 30, 309, 317

Tsumori, A., 435, 440

Turton, W. J., 438

Tutin, C. E. G., 149, 163, 207, 218, 310, 317, 333, 408, 414, 427, 432, 436, 439, 440

Tyrrell, D. S., 229, 257

Ullrich, W., 173, 178, 218

Valenstein, E. S., 40, 48, 102, 107, 443, 450, 451, 466

Valerio, D. A., 15, 16, 29

Vandenbergh, J. G., 224, 234, 253, 276, 279, 292, 380, 387, 455, 466

van den Berghe, P. L., 17, 18, 29

Vandergrift, P., 455, 466

van Hooff, J. A. R. A. M., 149, 163, 206, 217, 326, 333, 380, 387

Van Horn, R. N., 452, 466

van Lawick-Goodall, J., 58, 70, 81, 88, 148, 149, 163, 207, 208, 218, 241, 247, 312, 317, 473, 478

van Pelt, L., 95, 96, 107, 455, 466

van Wagenen, G., 16, 20, 21, 26, 29, 33, 77, 78, 86, 88, 454, 466, 470, 476, 478

Varley, M. A., 77, 87, 135, 139, 172, 177, 225, 232, 253, 254, 259, 261, 284, 292, 322, 333, 456, 466

Vencl, F., 313, 317

Vessey, S. H., 96, 97, 107, 187, 198, 225, 235, 252, 261, 380, 388, 425, 432, 455, 456, 466

Vestergaard, P., 93, 107, 118, 124, 446, 450

Vinsel, A., 301, 306, 328, 333

Vogel, C., 147, 163, 240, 247, 473, 478

Vogel, S. R., 429

Vogel, W., 120

vonder Haar Laws, J., 170, 177, 267, 276

von Holst, D., 74, 89, 91, 92, 107, 127, 136, 294, 306, 416, 421

Wade, T. D., 143, 163, 193, 198, 225, 235, 284, 292, 324, 333

Wagner, D. S., 229, 257

Wagner, J., 269, 276

Wagner, N. M., 229, 257

Wallen, K., 34, 35, 37, 45, 100, 107, 379, 385, 406, 411

Wallnau, L. B., 487

Wand, B., 428, 431

Warner, P., 123

Warrington, R. J., 427, 430

Waser, P. M., 240, 247

Washburn, S. L., 12, 25

Watts, E. S., 15, 28, 29

Webber, A. W., 243

Weber, I., 147, 163, 240, 247

Wechkin, S., 253, 259, 284, 290, 435, 440

Weigand, S. J., 124, 163, 333, 372, 465

Weisband, C., 171, 177, 263, 276

Weisfeld, G. E., 269, 272, 277, 395, 397

Weiss, G., 43, 48, 115, 124

Weitz, S., 242, 247

Welles, J. F., 378, 384

Wells, 33
Westlund, B. J., 310, 317
Whiten, A., 191, 198, 382, 387
Whiting, B., 61, 70, 286, 292
Whitsett, G., 155, 164, 486, 488
Wickler, N., 21, 23, 29, 154, 164, 249, 261
Wiesel, T. N., 425, 432
Wigodsky, H. S., 16, 27, 169, 177
Wilcoxon, H. C., 372, 435, 440
Willes, R. R., 16, 30, 309, 317
Williams, L., 184, 198
Willott, J. F., 347, 355, 407, 411, 414
Wilson, A. P. 225, 229, 252, 261, 284, 288, 379, 380, 388, 425, 432, 455, 466
Wilson, C. C., 339, 355, 473, 478
Wilson, J., 306, 333
Wilson, M. E., 59, 70, 77, 87, 153, 164, 176, 208, 209, 218, 334, 414, 478
Wilson, M. I., 95, 97, 105, 107, 123, 135, 139, 453, 455, 463, 466, 470, 477
Wilson, S. F., 164, 414, 478
Wilson, W. L., 339, 355, 473, 478
Wing, L., 428, 432, 484, 488

Winter, J. S. D., 43, 44, 80, 83, 84, 113, 115, 120, 458, 461
Winter, P., 448
Wise, L. A., 309, 318
Wolf, R. L., 453, 466
Wolfe, G., 439
Wolfe, L., 111, 124, 141, 164, 416, 417, 422
Wolfheim, J. H., 15, 30, 52, 63, 70, 72, 78, 89, 146, 164, 239, 247, 264, 277, 285, 292, 298, 321, 334, 392, 401
Wood, B. A., 18, 19, 23, 24, 30
Wolf, M., 253, 259, 284, 290
Wortman, M. S., 282, 287
Wright, F., 270, 277, 282, 292
Wurman, C., 172, 176, 205, 215

Yamada, M., 226, 235, 471, 478
Yamaji, T., 121, 448, 462
Yanciw, P., 185, 198, 222, 234, 283, 291, 377, 387
Yeni-Komshian, G. H., 447, 450
Yerkes, R. W., 112, 123
Yoshiba, K., 231
Young, D., 221, 232

Young, F. A., 368, 369, 373, 427, 428, 432, 471, 478
Young, G. H., 56, 71, 200, 218, 338, 355
Young, L. E., 270, 276

Zamboni, L., 95, 96, 107, 455, 466
Zemjanis, R., 110, 120, 124
Zihlman, A. L., 21, 25, 30, 211, 218, 249, 354, 394, 401, 437, 440
Zimmerman, E. A., 39, 47, 106, 449
Zimmerman, R. R., 309, 318
Zinser, O., 286, 287
Zoloth, S. R., 347, 355
Zucker, E. L., 59, 60, 67, 68, 70, 71, 113, 122, 151, 152, 153, 160, 161, 164, 176, 209, 218, 327, 334, 394, 399, 409, 414, 459, 463, 474, 478
Zuckerman, S., 23, 30, 41, 48, 444, 450
Zumpe, D., 96, 99, 100, 105, 106, 108, 126, 127, 128, 129, 130, 131, 133, 135, 139, 140, 141, 142, 143, 144, 145, 146, 147, 148, 151, 161, 362, 373, 379, 386, 456, 463, 464, 470, 477
Zuspan, F. P., 456, 461

Subject Index

ACTH, 74, 78, 415, 416, 417, 418, 420
action, 1, 479, 489
activity cycles, 452. *See also* cyclicity
adolescence, 72ff, 418, 428
adrenalin, 415
adrenal steroids. *See* stress
adrenocortical trophic hormones. *See* ACTH
adrenogenital syndrome, 42
adult hormones, 37, 90ff, 109ff
affiliation, 347
afterbirth, 172. *See also* placenta
age of social partner, 345
aggregation, 241
aggression, 133, 155, 287, 296, 374–401, 403, 406, 418
aggressive display, 366
agonistic buffering, 191, 200, 266, 389, 390
AGS, 42
alarm vocalizations, 294, 297, 337. *See also* vocalizations
alcoholism, 427
alliances, 175, 253, 278–292, 378
all-male groups, 240, 280, 310
allomothers, 204
alpha male, 239, 262, 263, 264
altruism, 286, 483
amenorrhea, 418, 459. *See also* captivity
amygdala, 38, 180, 443
androgens, 31, 34, 35, 36, 37, 42, 90
androgyny, 492
anthropoidea, 5

anxiety, 415
apes, 17, 80
appendix, 493
arboreal, 13, 14, 219
arousal, 61
artificial respiration, 173
assertiveness, 379
attachment, 323, 347, 406, 407, 480
attacks, 374ff, 377
attractivity, 125
aunting, 184, 201, 202, 209. *See* infant care
authority, 270
autistic children, 484
autonomy, 185
awareness, 479–488

bachelor males, 201
baboons, 53, 56, 80, 144, 199–201
baby-sitting, 83, 184, 208, 211
back-riding, 151
baldness, 110
Barbary macaque, 55
bark, 336. *See also* vocalizations
behavior, 1, 2, 479, 489
behavioral pathology, 395, 404
biogenic amines, 446
bipedalism, 310
bipedal swagger, 393
birth, 165ff, 205, 320, 409, 426, 428, 490
birth season, 457. *See also* cyclicity
bisexuality, 126
biting, 381. *See also* aggression

bizarre movements, 403
black infants, 200
bluff charges, 300, 394
body bounce, 367
body movements, 360
bonnet monkey, 69. *See also* macaques
brachiation, 152
brain, 38, 41, 102, 441–447
branch shaking, 299, 365, 366, 368
breeding season, 321, 455

Caesarean delivery, 171
call for contact, 344. *See also* coo;
    vocalizations
canines, 20, 21, 255, 266, 367, 394
cannibalism, 168, 312
captivity, 208, 297, 308, 312, 322, 327,
    358, 381, 393, 408, 416, 417, 426, 455
castration, 380
Catarrhini, 6
catching behavior, 436
Cayo Santiago, 77, 133, 254, 469
Cebidae, 184. *See also* New World monkeys
Cebus, 75. *See also* New World monkeys
central hierarchy, 193, 199
central nervous system, 441–447
cerebral cortex, 447
chacma baboon, 111. *See also* baboon; Old
    World monkeys
chest-beating display, 300, 394
chewing, 361
child abuse, 212
children, 62
chimpanzees, 58. *See also* great apes
chinning, 73, 91
clear calls. 335, *See also* vocalizations
coalitions, 221, 237, 253, 266, 278–292,
    326, 378, 380
cognition, 486
collective aggression, 377. *See also* alliances
College Self-Expression Scale, 270, 396
Colobus, 172. *See also* Old World monkeys
coloration, 22, 473
color blindness, 424, 469
color-phases, 22, 473
color vision, 468, 471
common names, 9, 10, 493ff
communal displays, 241, 343. *See also*
    vocalizations

communally reared children, 395
communication, 319, 401
competition, 270, 308
congenital malformations, 426, 427
conscious foresight, 485
consciousness, 479
consort, 132, 135, 225, 254, 325, 362,
    379, 416
conspecifics, 480
control role, 248, 262, 263, 265, 278, 279,
    293
coo, 336. *See also* vocalizations
coo-screech, 336. *See also* vocalizations
copulation, 126, 133
copulins, 98
courtship, 127. *See also* sexual behavior
crab-eating macaque, 55. *See also* Old
    World monkeys
crepuscular, 221
critical period, 408, 445
crook tail, 361, 363, 367
crowding, 203, 381, 418–420, 490
cultural evolution, 485
cyclicity, 321, 451–460, 491

defense, 179
defensive stoning, 299
delayed response, 434, 435, 436, 445, 491
dentition, 19, 20, 21
dependent rank, 226, 255
depression 270, 409
deprivation. 404, *See also* social deprivation
derived dominance, 248
DHTP, 35, 94, 100, 101, 109, 252
digit sucking, 402
dihydrotestosterone. *See* DHTP
distance vision, 367, 429, 490
distress vocalizations, 406. *See also*
    vocalizations
diurnal, 456, 457
diversionary display, 294
dominance, 94, 101, 141, 149, 168, 171,
    183, 200, 225, 227, 248–278, 280, 282,
    283, 307, 321, 325, 327, 360, 368, 378,
    467, 490
dominance displays, 248, 250
dominance hierarchies, 249, 251
dorsoventral mounts, 148. *See also* sexual
    behavior

dyads, 324, 344ff, 380, 404, 407

early deprivation, 405, 426, 433, 480, 490
early experience, 36, 37, 402ff
ecological variables, 472
egoistic, 286
embrace, 285
estradiol, 392. *See also* sex hormones
estrogens, 31, 34, 37, 38, 100. *See also* sex
    hormones
estrous cycle, 112, 457. *See also* sex hor-
    mones; cyclicity
estrus, 132, 325, 344, 390
exploration, 408
extra troop behavior, 293–318, 382
eye contact, 153, 154, 286, 328, 363
eye size, 471

facial expressions, 357. *See also* visual
    communication
familiarity, 347, 375
families of primates, 7
    family group, 283, 299, 341
    See also monogamy
fathers, 62, 174, 184, 188, 210, 211, 212,
    395
fatted condition, 93, 130, 453. *See also*
    cyclicity
fear, 338, 402, 415–420
fear grimace, 362, 363. *See also* fear;
    grimace
feeding, 307–318, 490
female choice, 279
feminine, 271
fighting, 360, 389. *See also* play
finickiness, 470
Flehman face, 364
focus of attention, 263
follicle stimulating hormone. *See* FSH
food-eating habits, 309
food sharing, 251, 308, 309, 310, 437
free-ranging, 323, 359, 379, 425, 426,
    472
friend, 286
frontal lobectomies, 445
frugivorous, 242
frustration, 396
FSH, 43, 113. *See also* sex hormones
function of play, 60

gathering, 437
gaze, 362. *See also* eye contact; looking
gelada, 111. *See also* baboon; Old World
    monkey
gender difference, 1, 2, 212, 489
gender preference, 224
genera, 9, 10, 11, 493ff
genetic variability, 471
genital display, 39, 250
genital swelling, 458, 459. *See also* swelling
genus, 493ff
gestural communication, 357. *See also*
    visual communication
gibbons, 57. *See also* lesser apes
girning, 347. *See also* vocalizations
gonadotrophins, 31
gorilla, 59. *See also* great apes
grappling, 375
great apes, 57, 80, 113, 148, 173, 206,
    241, 267, 299, 310, 326, 367, 393,
    408, 427, 434, 458, 473, 481
grimace, 357. *See also* visual communication
grooming, 111, 133, 207, 267, 319–328,
    362, 490
group change, 220
group defense, 264, 293, 311
group focus role, 226, 255, 279
group formation, 262
group protection, 293, 311
group size, 13, 219
group structure, 219ff, 236ff
grunt, 348. *See also* vocalizations
grunt-purr, 190. *See also* vocalizations,
    348
guenons, 146. *See also* Old World monkeys

Handbook of Living Primates, 493ff
harassment, 142, 206, 266
harems, 199, 200, 236, 238, 266, 326
harsh noises, 335
head-bob, 357. *See also* visual
    communication
herding, 237
HMG, 115
home range, 223
Homo sapiens, 24. *See also* human
homosexual, 134, 142, 143, 416
honks of warning, 340. *See also*
    vocalizations

hooting, 342, 394
hormonal information, 12ff. *See also* sex
　hormones; ACTH
hormone-concentrating cells, 445
hostile wit, 396
howler, 75. *See also* New World monkeys
human, 18, 41, 61, 82, 114, 153, 156, 174,
　210, 242, 268, 285, 301, 322, 327, 348,
　368, 382, 394, 396, 409, 420, 427, 434,
　437, 446, 459, 474, 483
human gonadotrophins. *See also* HMG
hunting, 19, 311, 312, 313, 490
hybrid, 236, 326, 472
hypothalamus, 35, 38, 39, 40, 41, 72, 82,
　96, 102, 118, 441–447, 451
hypothesis of neural bisexuality, 126, 131

ICSH 40, 97. *See also* sex hormones
incest avoidance, 129, 135, 156, 157, 210
independence, 208, 209
individuality, 467–478, 479-488
infant care, 179ff, 199ff, 490
infanticide, 203, 204, 239, 473
infant sharing, 204, 205
infra orders. *See* Contents
inhibition to steal, 309
intent, 486, 491
interstitial cell stimulating hormone. *See*
　ICSH
intertroop behavior, 293–306, 390, 490
intragroup aggression, 376. *See also*
　aggression
intragroup organization, 219–292
intratroop cohesion, 295
intruders, 300, 377
invitations to groom, 322
isolation, 179. *See also* social deprivation
isosexual groups, 221, 222, 236, 239

Japanese macaque, 56. *See also* Old World
　monkeys
jealousy, 130, 222
jumping-roaring display, 340. *See also*
　vocalizations

Kibbutzim, 156
kidnapping, 186
kin selection, 175
kinship, 190, 224, 225, 227, 327, 417

Koshima, 436

langurs, 147, 203
leadership, 189, 248, 271, 278–292, 490
learning, 61, 433–438
lemuriformes, 6, 167. *See also* prosimians
len, 364
lesser apes, 57, 80, 148, 173, 205, 240, 267,
　299, 310, 326, 367, 393, 427, 434, 458,
　473
LH, 40, 43, 101. *See also* sex hormones
lion-tailed macaque, 56. *See also* Old World
　monkeys
lipsmack, 191, 357, 363, 365
long call, 268, 300, 343. *See also*
　vocalizations
looking, 130, 358, 363
lookout behavior, 297
lordosis, 127. *See also* sexual behavior
lorisiformes, 6, 128. *See also* prosimian
luteinizing hormone. *See* LH

Macaques, 53, 76, 78, 140, 170. *See also*
　Old World monkeys
male care, 180, 189, 190, 404, 457
male migration, 193, 228
male takeovers, 473
male tolerance, 263
mangabey, 80, 112, 145, 201
manners, 310
marmoset, 92. *See also* New World monkeys
marriage, 287
masculine, 271
masturbation, 150. *See also* sexual behavior
maternal care, 180. *See also* infant care
maternal instinct, 435
maternal restrictiveness, 180
mathematical ability, 438
mating season, 457. *See also* cyclicity
matriarchal unit, 206
matrifocal group, 226, 239
maturation, 434, 474
meme, 485
menopause, 116, 454. *See also* cyclicity
menstruation, 74, 99, 112, 117, 131, 323,
　324. *See also* cyclicity
migration, 228
mirror exposure, 481ff
MME, 144, 145. *See also* mounts

moan, 348. *See also* vocalizations
mobbing, 312
molting, 96, 455. *See also* cyclicity
monkey-human contact, 417
monogamy, 50, 83, 128, 129, 185, 219,
    221, 222, 283, 295, 310, 320, 327, 341
mood, 471
morning calls, 342. *See also* vocalizations
mortality, 423-428
mother-daughter associations, 285
mounts, 34, 127, 252. *See also* dominance
mouth groom, 151
mouth-to-mouth behavior, 308
multimale troops, 237
multiparous, 187, 192, 346
multiple births, 182
multiple-mount ejaculator (MME), 144, 145.
    See also mounts
mutual staring, 375
myopia, 368, 369, 425

neck-biting, 237
nuclear family, 188, 322, 407
neurology, 37, 60, 118, 441ff
neurons, 39, 443
neurotransmitters, 118, 119, 442
New World monkeys, 50, 74, 90, 92, 128,
    166, 183, 184, 221, 250, 295, 308, 320,
    359, 376, 407, 424, 443, 452, 468
noncyclical variability, 467-479, 491
non-primate data, 31

observational tonus, 366
Old World monkeys, 51, 75, 109, 140, 168,
    186ff, 199, 224, 236, 252, 262, 264,
    297, 309, 321, 322, 361, 389, 407, 425,
    444, 456, 471
one-male groups, 237, 238, 240
ontogeny, 157-218
oral sex, 142. *See also* sexual behavior
orangutan, 59, 151
orgasm, 115, 143, 150, 154, 174
oxytocin, 119

pair bonds, 437. *See also* monogamy
pant ho-ho, 342. *See also* vocalizations
pant hooting, 300
parity, 172. *See also* birth chapter, 165
parental behavior, 188. *See also* infant care
parental investment, 17

parents, 179. *See also* infant care
parking the infant, 181, 182
parturition, 43, 165ff, 320
patas, 146, 202
paternal care, 181, 189, 192, 201, 207
pathogens, 423
pathological conditions, 402-432
peak experience, 174
pelage color, 339, 473. *See also* coloration
penile erection, 446. *See also* genital display
people, 243, 301, 409, 483, 489
performance, 433-438
peripheralization, 205, 210, 220, 221, 222,
    223, 224, 227, 236, 238, 240, 241, 242,
    253, 255, 282, 284, 296, 299, 424, 435
personality, 473
personal space, 420
pets, 328
pheromone, 97ff, 117, 118, 153
photoperiod, 456
physical characteristics, 12ff
physical dimorphism in dentition, 19
physical sexual dimorphism, 12
physical visual communication, 358
pigtail monkey, 79. *See also* macaques
pituitary, 445, 446. *See also* LH, FSH,
    ICSH, ACTH
Platyrrhini, 6
placenta, 172, 173
play, 35, 49ff, 153, 185, 186, 188, 207,
    209, 211, 249, 324, 366, 490
play fighting, 391. *See also* play; aggression
play mothering, 52, 186, 187, 202
polygamous, 250
population density, 419
positive socialness, 347
postpartum estrus, 166, 175, 320
postpartum heat, 166, 175
postural communication, 356. *See also*
    visual communication
predation, 13, 98, 189, 280, 293, 300, 307,
    311-318, 335
prefrontal cortex, 441, 442
prenatal, 31ff, 35, 40, 42, 63, 72, 90, 374,
    490
presenting, 133, 252. *See also* dominance
prey, 313
primate, characteristics of, 3, 4, 5
primate order, 5-11, 489, 493ff

primate taxonomy, 493ff
primiparous, 187, 192, 346
principle of youngest ascendency, 190, 227, 255
priority of access, 307
proceptivity, 98, 125, 152, 459
process, 72
progesterone, 31, 34, 100. *See also* sex hormones
prolactin, 110
promiscuity, 149
prosimian, 49, 73, 91, 126, 165, 181, 220, 249, 294, 319, 359, 375, 443, 452, 467
protection, 293–306, 490
proximity, 220, 284
pseudohermaphrodite, 33, 36
psychosis, 428
psychosexual differentiation, 44
puberty, 72ff, 75, 78, 80, 82, 83, 253, 392, 408, 418, 428, 490
punishment, 405
purpose, 486
pygmy chimpanzee, 150. *See also* great apes

rally call,' 338. *See also* vocalizations
reaching back, 131
receptivity, 98, 125, 358, 369, 454, 459
recognition, 406
red light, 471
rejection from the family group, 376. *See also* peripheralization
reproductive success, 256, 267, 492
reproductive turnover, 452
respect for aging leaders, 264, 280
reversal learning, 434
rhesus, 54, 76, 90, 94, 97, 125, 132, 168, 186, 224, 252, 295, 309, 322, 336ff, 361, 377, 402, 415ff, 424, 434ff, 454, 469, 480
rocking, 403
rooming-in, 212

sand-digging task, 435
scapegoat role, 250
scent marking, 251, 295, 359
schizophrenia, 409
screams, 342. *See also* vocalizations
screech-bark, 336. *See also* vocalizations
seasonal cycles, 453. *See also* cyclicity
self-awareness, 1, 115, 479-489, 491, 492
self-biting, 362, 403

self-clasping, 404
self-concept, 482
self-consciousness, 483, 484
self-directed behavior, 480, 481
self-grooming, 322, 324
selfish gene, 485
self-object differentiation, 484
self-recognition, 481, 482, 486. *See also* self-awareness
self-threatening, 362. *See also* social deprivation
sentinel, 281, 297
separation, 323, 344, 346, 347, 349, 402, 405, 406
serotonin, 101
sex difference, 1, 2, 62
sex hormones, 31, 32, 37, 90-119, 154, 252, 254, 256, 263, 326, 374, 379, 382, 392, 394, 395, 415, 424, 441-446, 458, 460, 470, 490
sexism, 269
sex ratio, 219, 223, 227, 228, 238, 427
sex recognition, 341
sexual behavior, 61, 95ff, 125ff, 140ff, 328, 403, 407, 408, 428, 444, 470, 474, 490
sexual differentiation, 31ff
sexual dimorphism, 12ff, 118, 269, 360, 376, 394, 470, 490
sexual selection, 18
sexual training, 204. *See also* orangutan; sexual behavior
shape, 24
siamang, 148
sign language, 482
silver-back, 208
singing, 342. *See also* vocalizations
single-mount ejaculator. *See* SME
sleeping parties, 238
SME, 142, 144, 148. *See also* mounts
social, 286
social Darwinism, 485
social density, 419
social deprivation, isolation, 144, 345, 402-414, 424
social facilitation, 345, 347
social isolation, 481. *See also* social deprivation
social learning, 435
socialness, 328

social spacing, 219ff, 236ff, 490
social structure, 219ff
sociobiology, 336, 484, 485, 486, 492
sociosexual signals, 21, 23
soft social sounds, 337. *See also* vocalizations
solitary, 220, 221, 224, 225, 226, 228, 237, 238, 241, 242, 254, 392, 468
spacing, 219ff, 236ff
spatial density, 419
spatial harems, 220
species identification, 480
species identity, 480, 482
species names, 493ff
spectrographs, 344
spider monkey, 132. *See also* New World monkeys
squirrel monkey, 92. *See also* New World monkeys
S-R complementarity, 125
status, 249, 255
stereotyped movements, 402
St. Kitts Island, 391
storm and stress, 418
strangers, 222, 225, 346, 380, 396, 406
stress, 402, 415–520, 429
structure, 219ff, 236ff
stumptail macaque, 56. *See also* Old World monkeys
subfamilies, 8, 10
suborders, 5
suicide, 40
superfamilies, 6, 8
surrogate mothers, 405
sway walk, 393
swelling, 23, 267, 268
swinging gait, 366
synchrony, 459. *See also* cyclicity

tail coiling, 361
talapoin, 52, 392
tamarin, 129. *See also* New World monkeys; marmosets
Tarsiiformes, 6
taxonomy. *See* Chapter 2; *also* Appendix, p. 493ff
teeth chattering, 365
teeth grinding, 364, 381
terrestrial, 13, 15
territoriality, 293–306, 393

territorial calls, 338, 490
territorial marking, 301
territorial threats, 360, 364
testosterone, 32, 34, 37, 43, 93, 94ff, 100. *See also* sex hormones
threat, 357, 363, 367. *See also* visual communication
three-toned roar, 342. *See also* vocalizations
titi monkey, 130. *See also* New World monkeys
tolerance, 263
tongue movements, 128. *See also* 356-370
tool making, 436
tool using, 435, 436, 437
toque macaque, 79. *See also* macaque
tradition drift, 191, 365
traveling order, 241
triumph ceremony, 251
troop fission, 227, 265
troop mobilizing behavior, 281

unisexual subgroups, 283. *See also* isosexual subgroups
urine washing, 340
uterine activity, 456

variability, 451–488
ventro-ventral mount, 148, 150
verbal learning, 437
vertical tail carriage. *See* crook tail
vervets, 146, 201
vigilance, 293–306, 307, 313, 338, 366, 490, 491
visual access, 381
visual communication, 356-370, 468, 490
visual orientation, 358ff. *See also* looking; eye contact
visual-spatial skills, 282, 369, 438
vocalizations, 335–349, 369, 490
vulnerability, 368, 417, 423–428, 433, 474

warning barks, 337, 364
weaning, 206
whooping display, 297, 298, 337. *See also* vocalizations
women, 211, 382, 437
wounding, 377, 390, 427

xenophobia, 296

yawn, 357, 362, 364, 368

UE